Mass Spectrometry
Volume 6

A Specialist Periodical Report

Mass Spectrometry
Volume 6

A Review of the Recent Literature Published between July 1978 and June 1980

Senior Reporter
R. A. W. Johnstone, *Department of Organic Chemistry, University of Liverpool*

Reporters

T. Baer, *University of North Carolina, Chapel Hill, North Carolina, U.S.A.*
J. H. Bowie, *University of Adelaide, Australia*
R. H. Cragg, *University of Kent*
R. C. Dunbar, *Case Western Reserve University, Cleveland, Ohio, U.S.A.*
D. E. Games, *University College of Wales, Cardiff*
I. Howe, *University College of Wales, Swansea*
G. Klass, *University of Adelaide, Australia*
A. R. Krauss, *Argonne National Laboratory, Illinois, U.S.A.*
V. E. Krohn, *Argonne National Laboratory, Illinois, U.S.A.*
A. McCormick, *A.W.R.E., Aldermaston, Berkshire*
L. E. Martin, *Glaxo Group Research Ltd., Ware, Hertfordshire*
F. A. Mellon, *University of Sussex*
R. D. Sedgwick, *U.M.I.S.T., Manchester*
V. C. Trenerry, *University of Adelaide, Australia*
L. A. Viehland, *St. Louis University, Cahokia, Illinois, U.S.A.*

The Royal Society of Chemistry
Burlington House, London W1V 0BN

British Library Cataloguing in Publication Data

Mass spectrometry.—Vol. 6.—(A Specialist periodical report)
 1. Mass spectrometry—Periodicals
 I. Royal Society of Chemistry
 545'.33'05   QD96.M3

ISBN 0-85186-308-6
ISSN 0305-9987

Copyright © 1981
The Royal Society of Chemistry

All Rights Reserved
No part of this book may be reproduced or transmitted
in any form or by any means – graphic, electronic,
including photocopying, recording, taping or
information storage and retrieval systems – without
written permission from The Royal Society of Chemistry

Set in Times on Linotron and printed offset by
J. W. Arrowsmith Ltd., Bristol, England
Made in Great Britain

Foreword

Had J. J. Thomson been alive today he would have surely marvelled at the development in ion chemistry that has taken place since his initial observations on the behaviour of ions in magnetic and electric fields. Even those early pioneers Aston, Bainbridge, Dempster, Herzog, Mattauch, and Nier amongst others could scarcely have foreseen the developments in instrumentation which have occurred to such an extent that the mass spectrometer, usually with its attached computerized data system, can be operated routinely in analytical laboratories interested only in the results it gives and not in the underlying principles. Thus, some chapters of this Specialist Periodical Report are more concerned with the extraordinary ability of the mass spectrometer to yield molecular structural information and analytical identification and estimation of compounds available only in very small quantities. On the other hand, the successors of those early pioneers are making enormous strides in the understanding of ion chemistry, and the remainder of the chapters is given over to descriptions of these advances.

This dichotomy in outlook of the Reports together with the limited space available means that their usefulness will vary from reader to reader. However, it is hoped that the balance is reasonable for the majority. As introduced earlier, and continued in this volume, there are some short, authoritative accounts of developments in specialized areas. Ion mobilities and secondary ions are two such fields which have increasing practical importance in areas other than mass spectrometry alone. Somewhat more esoteric is the article on the effects of electromagnetic radiation on ions by which their structures can be probed. Along with the chapter on the theory and energetics of mass spectrometry these articles provide a strong American contribution to Volume 6 in the guise of Tomas Baer, Robert Dunbar, Alan Krauss, Victor Krohn, and Larry Viehland.

The international flavour of this series is enhanced by our regular Australian contribution from John Bowie and his colleagues. It is hoped to extend this international aspect of the Reporters to the series along with a continuation of the specialized topics. From this viewpoint the very helpful, objective criticisms and suggestions of reviewers are greatly appreciated.

Amongst our new contributors we welcome Ian Howe and Richard Cragg, who have taken over from stalwarts William Bentley and Trevor Spalding. Sadly, Volume 6 will see the last of contributions from Francis Mellon and Andrew McCormick, and it is not out of place to record how indebted we are to them for their efforts. Other new contributors, Donald Sedgwick and Leslie Martin, make a welcome appearance and have already given the old topics they took on a new, fresher aspect by the inclusion of microprocessor technology and pharmacokinetics, respectively.

As I seem so far to have mentioned every contributor bar an old acquaintance, David Games, it seems only right to recognize his continued exertions. David, your manuscripts may be always or nearly always last to arrive on my desk but they are never least – many thanks.

Finally, my whole-hearted thanks to the editorial office of The Royal Society of Chemistry for making my task much easier than it would be without them.

R. A. W. JOHNSTONE

Contents

Chapter 1 Theory and Energetics of Mass Spectrometry
By T. Baer

1 Introduction	1
2 Ion Thermochemistry	1
Molecular Orbital Calculations	1
Heats of Formation	7
Gas-phase Basicities and Acidities	12
Electron Affinities and Negative Ions	15
Dimers and Clusters	19
3 Ionization Processes	22
Charge Transfer	22
Photoionization	25
Autoionization and Predissociation	27
Multiphoton Ionization	31
4 Unimolecular Fragmentation	36
Advances in Unimolecular Decay Theory	36
Ion Lifetimes, Kinetic Energy Release, Breakdown Diagrams	41
Dissociation Pathways	47
5 Ion–Molecule Interactions	53
Collision Theory	53
Three-body Association Reactions	54
Energy Transfer and Internal Energy Effects	56

Chapter 2 Structures and Reactions of Gas-phase Organic Ions
By I. Howe

1 Introduction	59
2 Thermochemistry	59
3 Isomerization	64
Randomization	64
Rate-determining Isomerizations	66
Ion–Molecule Complexes	68
4 Decomposition	71
Specific Rearrangements	71
Fragmentation Rules	73
Translational-energy Release	74

Collisional Activation	77
Comparisons with Solution Chemistry	79
5 Summary	82

Chapter 3 Gas-phase Ion Mobilities, Ion–Molecule Reactions, and Interaction Potentials
By L. A. Viehland

1 Introduction	84
2 Drift-tube Experiments	84
Basic Ideas	84
Drift-tube Mass Spectrometers	85
Injected-ion Drift Tubes	86
Additional-residence-time Drift Tubes	86
Flow-drift Tubes	86
Selected-ion Flow-drift Tubes	87
3 Drift-tube Theories	87
Elementary Theories	88
One-temperature Kinetic Theories	89
Cold-gas Kinetic Theories	90
Monte Carlo Treatments	90
Two-temperature Kinetic Theories	91
Three-temperature Kinetic Theories	93
Kinetic Theories for Molecular Systems	94
4 Current Status	96
Measurement of Transport and Reaction Coefficients	96
Transferability of Transport and Reaction Coefficients	97
Determination of Cross-sections	97
Determination of Interaction Potentials	98
5 Prognosis	98

Chapter 4 Interaction of Electromagnetic Radiation with Gas-phase Ions
By R. C. Dunbar

1 Introduction	100
2 Ion Spectroscopy	100
Photodissociation in Ion-beam Instruments	101
Photodissociation in Ion-trap Instruments	105
Photodetachment and Photodissociation of Negative Ions	107
Spectroscopy as a Structural Tool	108
3 Photon Emission from Ions	110

4	Multiphoton Processes	112
5	Photofragmentation	115
6	Other Spectroscopic Methods	117

Chapter 5 Aspects of Secondary Ion Emission
By A. R. Krauss and V. E. Krohn

1	Introduction	118
2	The Development of the Equilibrium Treatment	119
3	Further Tests of the Validity of the LTE Model	125
4	Sensitivity Factors	132
5	Theory of Secondary Ion Emission	133
6	Secondary Ion Energies	138
7	Sputter Ion Sources	148
8	Fusion Reactor Applications	150
9	Secondary Ions in High-voltage Devices	152

Chapter 6 Development and Trends in Instrumentation in Mass Spectrometry
By A. McCormick

1	Introduction	153
2	'In-beam' Sample Introduction	155
3	Laser Applications	159
4	Field Ionization and Field Desorption	162
5	Chemical and Other High-pressure Ionization	165
6	Other Ionization Methods	167
7	Ion Detection and Measurement	168
8	G.C.–M.S. and L.C.–M.S.	170
9	Linked Scanning Methods	171
10	Miscellaneous	172

Chapter 7 Applications of Computers and Microprocessors in Mass Spectrometry
By R. D. Sedgwick

1	Introduction	174

2 Computer Control	176
Microprocessor Control	176
Minicomputer Control	179
Multiple Instrument Systems	180
Fourier Transform Ion Cyclotron Resonance	181
3 Data Acquisition and Processing	181
Pulse Counting	181
Focal Plane Detection	182
Reference Materials	183
Selected Ion Monitoring	183
Isotopic Analyses	184
G.C.–M.S.	184
Ionization Efficiency	185
Metastable Ion Spectra	185
Miscellaneous Applications	187
4 Interpretation of Mass Spectra	187
Library Search Methods	187
Pattern Recognition	191
Artificial Intelligence	193
Mixture Analysis	194

Chapter 8 Gas Chromatography–Mass Spectrometry and High-performance Liquid Chromatography–Mass Spectrometry
By F. A. Mellon

1 Introduction	196
2 Practical and Instrumental Considerations: G.C.–M.S.	196
Sensitivity, Interface Techniques, and Chromatographic Performance	196
Alternative Ionization Techniques	199
Data Processing	200
Quantification	201
3 Applications: G.C.–M.S.	201
Long-chain Compounds	201
Hydrocarbons	201
Fatty Acids, Aldehydes, Ketones, Alcohols, and Esters	202
Lipids and Sphingolipids	203
Prostaglandins and Related Compounds	204
Carbohydrates	206
Steroids and Terpenoids	207
Amino-acids and Peptides	210
Nucleotides, Nucleosides, and Related Compounds	212
Clinical and Metabolic Studies	213

Ecological and Environmental Studies	216
Air and Water Pollutants	217
Polycyclic Aromatic Hydrocarbons	218
Organochlorine Residues	219
Agrochemicals and Related Compounds	220
Organic Geochemistry	221
Food and Flavour Components	222
Allelochemicals and Other Secretions	223
Mammalian Allelochemicals	224
Invertebrates	224
Pyrolysis G.C.–M.S.	226
Miscellaneous	227

4 High-performance Liquid Chromatography–Mass Spectrometry   228

L.C.–M.S. Interfaces	228
The Atmospheric Pressure Ionization Source	229
Direct Introduction of L.C. Effluent	229
Jet-type L.C.–M.S. Interfaces	230
Membrane Interfaces	230
Mechanical Transfer	230
Applications	231
General Comments and Future Developments	231

Chapter 9 Reactions of Negative Ions in the Gas Phase
By J. H. Bowie, V. C. Trenerry, and G. Klass

1 Introduction	233
2 Fragmentation Mechanisms of Negative Ions Formed by Electron Capture (or Dissociative Electron Capture)	234
Organic Negative Ions	234
Organometallic Negative Ions	235
3 The Reactions of Negative Ions with Neutral Molecules	236
Negative Ion Chemical Ionization Mass Spectrometry	236
Ion Cyclotron Resonance and Flowing Afterglow Studies	237
Charge Inversion Spectra of Negative Ions	240
4 Concluding Remarks	240

Chapter 10 Natural Products
By D. E. Games

1 Introduction	241
2 Alkaloids	246
3 Aromatic Compounds and Oxygen Heterocycles	247

4	Isoprenoids	249
5	Steroids	251
6	Antibiotics	252
7	Nucleic Acid Components and Cytokinins	253
8	Pyrrole Pigments	254
9	Carbohydrates	255
10	Amino-acids and Peptides	257
	Amino-acids	257
	Peptides	257
11	Fatty Acids and Lipids	259

Chapter 11 The Use of Mass Spectrometry in Pharmacokinetic and Drug Metabolism Studies
By L. E. Martin

1 Introduction   262

2 Quantitative Mass Spectrometry   263
Internal Standards Used for Analysis by Mass
    Spectrometry   263
Direct Probe Methods   263
Selected Ion Recording *via* G.C.–M.S.—A Comparison
    Between G.C.–M.S. and Other Analytical Techniques   265
Internal Standards Yielding A Different Ion   266
High-resolution G.C.–M.S.   270
Homologous Internal Standard Yielding the Same Ion   271
Stable Isotope-labelled Analogues as Internal Standards   271
Selected Ion Recording Using Field Desorption   278
Quantitative Liquid Chromatography–Mass Spectrometry   278
The Use of Selected Metastable Peak Monitoring for
    Drug Analysis   278

3 Mass Spectrometry in Metabolite Identification   279
Metabolites Separated by Liquid Chromatography   279
Metabolites Isolated by Thin-layer Chromatography   282
Metabolites Characterized by Gas Chromatography–Mass
    Spectrometry   287

Chapter 12 Organometallic, Co-ordination, and Inorganic Compounds Investigated by Mass Spectrometry
By R. H. Cragg

1 Introduction   294

2 Main-group Organometallics 295
- Group II 295
- Group III 295
  - Compounds of Other Elements in Group III 297
- Group IV 298
  - Compounds of Other Elements in Group IV 302
- Group V 303
  - Compounds of Other Elements in Group V 306

3 Transition-metal Organometallics 306
- Transition-metal Cluster Compounds 306
- Metal Carbonyl and Related Compounds 307
- Complexes of Two-, Three-, and Four-electron Donors 308
- Cyclopentadiene Complexes 310
- Complexes Containing C_6 and Higher Rings 312

4 Co-ordination and Metal–Organic Compounds 314
- Compounds with Metal–Oxygen or Metal–Sulphur Bonds 314
  - β-Diketones and Related Complexes 314
  - Carboxylates and Related Compounds 316
  - Alkoxides and Related Compounds 316
- Compounds Containing Metal–Nitrogen Bonds 318

5 Inorganic Compounds 319
- Group I 319
- Group II 321
- Group III 322
  - Compounds Containing B—O Bonds 322
  - Boron–Nitrogen Compounds 322
  - Boron Hydrides and Related Compounds 322
  - Other Compounds 324
- Group IV 324
- Group V 325
- Group VI 326
- Group VII 327
- Group VIII 327
- Miscellaneous Compounds 327

1
Theory and Energetics of Mass Spectrometry

BY T. BAER

1 Introduction

A review of the theory and energetics in mass spectrometry is a formidable task because the field is so broad. Consistent with the theory and energetics chapters of previous volumes, I have tried to limit the review to those aspects of the literature during the past two years which are relevant to the fundamental understanding of ion dynamics. Particular emphasis has been given to those developments which will help us in our ultimate quest, the ability to predict qualitatively or quantitatively the behaviour of energized ions.

2 Ion Thermochemistry

As dynamical experiments and theories are becoming more sophisticated and precise, the need for accurate thermochemical data on molecules, ions, and fragments continues to grow. During the past few years the experimental effort has been supported by numerous calculations, most of which are of the *ab initio* type. With the advent of readily available high-level programs, numerous groups are now performing calculations. In combination with good experimental information, these results are of great value in extending our chemical knowledge because accompanying the calculated energy is an assumed structure. Although some of the theoretical work will be treated under a separate subheading, a large portion of it will be mixed in with the review of experimental results.

Molecular Orbital Calculations.—*Ab initio* calculations have decreased in cost to such an extent that few calculations are now being done with semiempirical programs. This is also partly as a result of the fact that the semiempirical programs are usually parametrized to do one job well, but at the expense of their predictive ability for other properties.

The most commonly used *ab initio* program is the STO-3G (Slater-type orbitals with 3 gaussian functions). This uses a minimal, split-valence set of basis functions. The split valence means that two basis functions are used for each valence atomic orbital. More sophisticated basis sets are ones belonging to the K-LMG family, in which K is the number of gaussians used to describe the inner-shell s-type functions, L is the number of gaussians for the s- and p-type valence functions, and M is the number of gaussians for the outer sp-type functions. A commonly used basis set has been the 4-31G[1] which is available through the

Quantum Chemistry Program Exchange (QCPE) of the University of Indiana. This and the other K-LMG programs have been developed by Pople and his co-workers.[1]

Some new K-LMG programs have been developed and are, or will soon be, available through the QCPE. Two of these are ones which use the 6-21G and 3-21G basis sets.[2] Either of these is claimed to be as good as the 4-31G or the PFPB 4-21G basis set.[3] The K-21G split-valence basis sets are definitely superior to the STO-3G minimal basis set. Equilibrium geometries are about as good as those of the 4-31G but superior with regard to the description of the bond angles involving heteroatoms. Vibrational frequencies are also equal to, or better than, those of the 4-31G. Similarly, electric dipole moments are better with either the 6- or the 3-21G than with the 4-31G. Happily, because the 3-21G has fewer primitive gaussian functions it is faster than the 4-31G set. It appears to be inferior to the 4-31G only in the calculation of reaction energies. Comparisons for over 20 molecules are given.[2]

Halgren et al.[4] have compared the speed and accuracy of a number of semiempirical and ab initio programs for calculations of various properties. The overall effectiveness *versus* speed curve is shown in Figure 1. This paper also

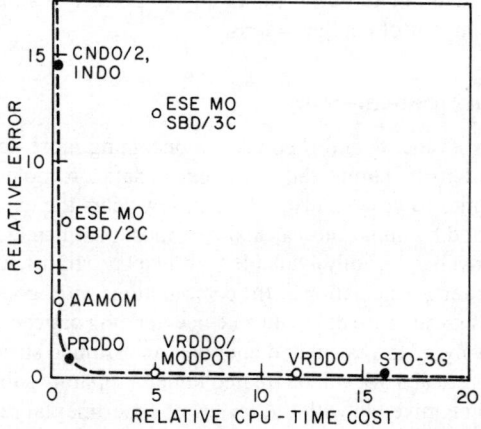

Figure 1 *Relative error versus relative CPU-time cost for the SCF methods discussed. The dashed line represents rel. error × rel. cost − 1, or unit cost efficiency*
(Reproduced with permission from *J. Am. Chem. Soc.*, 1978, **100**, 6595)

introduces a new semiempirical program, the PRDDO (partial retention of diatomic differential overlap). It is 16 times faster than one of the simplest *ab initio* programs, the STO-3G, yet it agrees very well with this program in relative energies, atomic charges, and dipole moments. In a follow-up to this paper, Dewar and Ford[5] have added their MNDO[6] program to this comparison. They

[1] R. Ditchfield, W. J. Hehre, and J. A. Pople, *J. Chem. Phys.*, 1971, **54**, 724.
[2] J. S. Binkley, J. A. Pople, and W. J. Hehre, *J. Am. Chem. Soc.*, 1980, **102**, 939.
[3] P. Pulay, G. Fogarasi, F. Pang, and J. E. Boggs, *J. Am. Chem. Soc.*, 1979, **101**, 2550.
[4] T. A. Halgren, D. A. Kleier, J. H. Hall, L. D. Brown, and W. N. Lipscomb, *J. Am. Chem. Soc.*, 1978, **100**, 6595.
[5] M. J. S. Dewar and G. P. Ford, *J. Am. Chem. Soc.*, 1979, **101**, 5558.
[6] M. J. S. Dewar and W. Thiel, *J. Am. Chem. Soc.*, 1979, **101**, 4989.

compared seven MO methods by listing the root mean square (r.m.s.) error of the energy, ionization energy (Koopmans' theory), and dipole moment with respect to experimental results. These are listed in Table 1 and should serve as a guide to experimentalists. Although impressive, the calculations must be used with care. Furthermore it is doubtful that single configuration calculations will reach experimental accuracies of say 1 kJ mol^{-1}. For such precision elaborate configuration interaction (CI) calculations must be carried out. These are still the domain of the theoreticians.

Table 1 Summary of r.m.s. error for molecular orbital methods relative to experiment*

Method	ΔE/kJ mol^{-1}	ΔIE/eV	$\Delta\mu$/D
CNDO/2	705	4.40	2.24
PRDDO	136	1.06	0.98
STO-3G	126	1.14	0.80
MBS	123	1.10	0.89
4-31G	59	0.57	0.41
Double Zeta	53	0.55	0.37
MNDO	45	1.01	0.83

* The r.m.s. errors are based on results from 23 molecules. Table taken from ref. 5.

The most studied molecular ion during the past two years has been CN^+. No less than four separate investigations were reported dealing primarily with the identity of the ground state. As with the isoelectronic C_2, the two states $^1\Sigma^+$ and $^3\Pi$ are very close in energy. Wu[7] did an SCF calculation and concluded that the $^3\Pi$ is lower in energy by 0.33 eV. The fact that this was a single configuration calculation makes it somewhat suspect. Yet Ha[8] using CI also found that the $^3\Pi$ is lower than the $^1\Sigma^+$ state by 0.41 eV. Murrell et al.[9] would not commit themselves, stating that the two states are extremely close. This caution is certainly justified because Hirst[10] using the ATMOL SCF calculation with CI found that either state could be made the ground state depending on the number of configurations used. Yet the bond distances are quite different (1.20 Å and 1.28 Å). Other diatomics studied[11] are the mixed alkali metals and alkali salts such as NaK^+, $NaRb^+$, $NaCs^+$, KBr^+, KCs^+, $RbCs^+$, Na_2^+, K_2^+, Rb_2^+, and Cs_2^+. A number of stable excited electronic states were found.

Often the most stable structure for an ion is not the same as the most stable neutral structure. These situations are sometimes difficult to establish experimentally, but they are quite amenable to calculations. In fact the calculation of energies of isomers is one of the most fruitful uses of MO calculations. Murrell and Derzi[12] have concluded that, although HCN is 0.5 eV more stable than HNC, in the ionic form HNC^+ is more stable than HCN^+ by 0.9 eV, and that the

---

[7] A. A. Wu, Chem. Phys. Lett., 1978, **59**, 457.
[8] T. K. Ha, Chem. Phys. Lett., 1979, **66**, 317.
[9] J. N. Murrell, A. Al-Derzi, and J. Tennyson, Mol. Phys., 1979, **38**, 1755.
[10] D. M. Hirst, Chem. Phys. Lett., 1979, **65**, 181.
[11] A. Valance, J. Chem. Phys., 1978, **69**, 355.
[12] J. N. Murrell and A. Al-Derzi, J. Chem. Soc., Faraday Trans. 2, 1980, **76**, 319.

activation for HCN^+ going to HNC^+ is about 0.45 eV. These results are based on an *ab initio* SCF calculation in which the basis functions consisted of contracted gaussians augmented by polarization functions. Another study[13] of HCN^+ addressed the problem of the $\tilde{A}$ and $\tilde{B}$ states and, in particular, the slow predissociation rate at 20.3 eV which results in sharp vibrational structure in the spectra from photoelectron spectroscopy (PES).

The isomers HNO^+ and NOH^+ have been investigated,[14,15] and in both cases HNO^+ was found to be more stable. Bruna and Marian[14] using the MRD-CI (multi-reference double-excitation) program developed by Buenker *et al.*[16] found the energy difference only about 12 kJ mol^{-1}. Both ions are bent; the angle in HNO^+ is 131° while in NOH^+ it is 124°. The other SCF study with CI found the energy difference to be 54 kJ mol^{-1} and the angles 126° and 116°, respectively.[15]

The MRD-CI program was also used[17] to calculate the relative stability of the HCS^+–CSH^+ system. HCS^+ was found more stable by 465 kJ mol^{-1}. In a follow-up to this study, CNDO/2 investigation of the same isomers found an energy difference of 520 kJ mol^{-1}.[18] Yet another study with the MRD-CI program was the calculation[19] of the low-lying states of NH_2^+. A rather strange result is that two states, the linear $^3\Sigma_g^-$ and the bent $^3B_1$ (150°), are candidates for the ground state. They differ in energy by only 330 cm^{-1} while the barrier is about 900 cm^{-1}. The reaction $N^+ + H^2 \rightarrow NH^+ + H$ was also investigated with CI by calculating triplet states of NH_2^+ which might correlate with the dissociation products.[20] In a study similarly relevant to dynamics, Hansoul *et al.*[21] investigated the higher-lying states of HCN^+ by CI. They found that the third state ($\tilde{B}$) at 19 eV cannot be assigned to the removal of a single 4σ electron. Instead at least two configurations, one of which is a two-electron excitation, are involved. On the basis of the experimentally observed H^+ onset the authors conclude that the $\tilde{B}$ state is strongly coupled with the $\tilde{A}$ state.

One of the most remarkable predictions comes from an INDO calculation on N_2O^+. The neutral N_2O is linear as is also the $^2\Pi$ state of the ion, which is thought to be the ground state. Yet Barber *et al.*[22] have found that for a bent state (N—N—O angle of 61.1°) the $^2A''$ is 306 kJ mol^{-1} more stable than the $^2\Pi$ state. The barrier is 220 kJ mol^{-1} which would explain why no experimental evidence for such a low-lying state of N_2O^+ has been reported. In view of these calculations it would be extremely interesting to carry out *ab initio* CI calculations and to determine onsets of N_2O^+ fragments from the dissociation of molecular ions containing N_2O. The $^4A''$ state of N_2O^+ has been carefully studied by Hopper[23] who found that it is stable but has an energy considerably above the linear $^2\Pi$ state.

[13] D. M. Hirst, *Mol. Phys.*, 1979, **38**, 2017.
[14] P. J. Bruna and C. M. Marian, *Chem. Phys.*, 1979, **37**, 425.
[15] A. D. McLean, G. H. Loew, and D. S. Berkowitz, *Mol. Phys.*, 1978, **36**, 1359.
[16] R. J. Buenker, S. D. Peyerimhoff, and W. Butscher, *Mol. Phys.*, 1978, **35**, 771.
[17] P. J. Bruna, S. D. Peyerimhoff, and R. J. Buenker, *Chem. Phys.*, 1978, **27**, 33.
[18] A. B. Sannigrahi, B. R. De, and R. Das, *Chem. Phys. Lett.*, 1980, **69**, 141.
[19] S. D. Peyerimhoff and R. J. Buenker, *Chem. Phys.*, 1979, **42**, 167.
[20] D. M. Hirst, *Mol. Phys.*, 1978, **35**, 1559.
[21] J. P. Hansoul, C. Galloy, and J. C. Lorquet, *J. Chem. Phys.*, 1978, **68**, 4105.
[22] M. Barber, D. G. Bounds, A. Hinchliffe, and R. D. Sedgwick, *J. Chem. Soc., Faraday Trans. 2*, 1978, **74**, 1042.
[23] D. G. Hopper, *J. Am. Chem. Soc.*, 1978, **100**, 1019.

Intermediate-size molecules whose structures and energies have been studied include four isomers of $C_2H_4O^+$. At the 4-31G level, vinylalcohol$^+$ was found most stable. Acetaldehyde$^+$, ethylene-oxide$^+$, and open ethylene-oxide$^+$ were found to be 46, 195, and 133 kJ mol^{-1} less stable, respectively.[24] The barrier for inversion in $N_2H_3^+$ has been calculated with an STO basis set to be 190 kJ mol^{-1}, which is much greater than the experimental value in solution of 55 kJ mol^{-1}.[25] This indicates that the transition state in solution is greatly stabilized by the solvent. LCAO MO SCF CI calculations have been carried out on the much studied cyclopropenyl cation $C_3H_3^+$.[26] The C—C bond length was found to be 1.389 Å. Substituent effects on the related propargyl ion have been carried out at the STO-3G level to determine the changes in the π electron charge distributions upon substitution of such groups as H, Me, F, and NH_2.[27] The formaldehyde and hydroxymethylene tautomeric ions [structures (1) and (2)] have been calculated.[28] Structure (1) was found to be more stable than structure (2) by 41 kJ mol^{-1}, and the inversion barrier *via* a 1,2 hydrogen shift was estimated to be 248 kJ mol^{-1}.

$$
\begin{array}{cc}
O^+ & O \\
\parallel & \parallel \\
C & C^+ \\
/ \ \backslash & / \ \backslash \\
H \quad H & H \quad H \\
(1) & (2)
\end{array}
$$

Other investigations carried out by MO calculations include such diverse studies as the origin of the molecular shapes of Me^+ and NH_3^+,[29] the potential energy surface for cyclopentadienyl cation,[30] the interactions of Li^+ ions with ethers, thioethers, and amide systems,[31] and the role of hyperconjugation in isotope effects.[32] *Ab initio* calculations on the structure and stability of SiH_n^+ ($n = 0$—4),[33] the structure of $C_4H_7^+$,[34] and the inductive effects of repelling groups such as Me, CCH, H, *etc.* on ions[35] have been performed.

Jefford *et al.*[36] calculated the various isomers of $C_8H_9^+$ by the MINDO/3 method, while a semiempirical X-alpha calculation on $C_2H_4^+$ and H_2O^+ was carried out by Barrow *et al.*[37] Semiempirical methods have been compared with *ab initio* calculations for $C_2H_7^+$.[38] The authors conclude that MINDO/3 is not

---

[24] W. J. Bauma, J. K. MacLeod, and L. Radom, *J. Am. Chem. Soc.*, 1979, **101**, 5540.
[25] V. Ya. Bespalov and M. A. Kuznetsov, *Theor. Exp. Chem. Engl. Transl.*, 1979, **15**, 434.
[26] T. Takada and K. Ohno, *Bull. Chem. Soc. Jpn.*, 1979, **52**, 334.
[27] M. Dorado, O. Mo, and M. Yanez, *J. Am. Chem. Soc.*, 1980, **102**, 947.
[28] W. J. Bouma, J. K. MacLeod, and L. Radom, *Int. J. Mass Spectrom. Ion Phys.*, 1980, **33**, 87.
[29] H. Nakatsuji, K. Matsuda, and T. Yonezawa, *Bull. Chem. Soc. Jpn.*, 1978, **51**, 1315.
[30] W. T. Borden and E. R. Davidson, *J. Am. Chem. Soc.*, 1979, **101**, 3771.
[31] G. Corongiu, E. Clementi, E. Pretch, and W. Simon, *J. Chem. Phys.*, 1980, **72**, 3096.
[32] D. J. DeFrees, M. Taagepera, B. A. Levi, S. K. Pollack, K. D. Summerhays, R. W. Taft, M. Wolfsberg, and W. J. Hehre, *J. Am. Chem. Soc.*, 1979, **101**, 5532.
[33] M. S. Gordon, *Chem. Phys. Lett.*, 1978, **59**, 410.
[34] B. A. Levi, E. S. Blurock, and W. J. Hehre, *J. Am. Chem. Soc.*, 1979, **101**, 5537.
[35] A. F. Marchington, S. C. R. Moore, and W. G. Richards, *J. Am. Chem. Soc.*, 1979, **101**, 5529.
[36] C. W. Jefford, J. Mareda, J. C. Perlberger, and U. Burger, *J. Am. Chem. Soc.*, 1979, **101**, 1370.
[37] W. L. Barrow, H. Sambe, and R. H. Felton, *Chem. Phys. Lett.*, 1979, **68**, 170.
[38] H. J. Köhler and H. Lischka, *Chem. Phys. Lett.*, 1978, **58**, 175.

useful for predicting the most stable structure of five-co-ordinate carbon atoms. The most stable structure predicted by the *ab initio* method is a protonated ethane in which the proton bridges the two carbon atoms. On the other hand, Lischka and Köhler[39] concluded that, in the case of Et^+ and $C_2H_3^+$ in which the nonclassical bridged structures are more stable by 16 kJ mol^{-1} as calculated by *ab initio* techniques, the MINDO/3 program significantly overestimates the bridged structure. Another comparison of *ab initio* with INDO was done by Mayr et al.[40] on 3,3-dimethylallyl cations. The *ab initio* results suggested a planar structure while INDO gave a 30° twisted ground state.[41]

MO calculations in support of PES have continued. Single configuration calculations of ionization energies may be adequate at low energies. However, at high energies the effect of doubly excited states and mixing of states becomes important so that either many configurations must be included or the many-body Green's function approach must be used. A comparison of the two methods has been carried out on the ionization energy of C_2H_4.[42] Both predict energies close to the experimental value but comparison for the higher-lying states has not been made. The Green's function method has been applied to the states between 20 and 70 eV in HCN^+ and $HCOOH^+$,[43] $C_nH_m^+$ $(n, m = 2—6)$,[44] and the isoelectronic CO_2^+ and N_2O^+ ions.[45] Many of the high-lying states are weak and diffuse and therefore cannot be described in terms of a 'quasi hole' or a one-electron excitation. This is especially true of the unsaturated hydrocarbons. Configuration interaction has been used in semiempirical calculations by Lauer et al.[46] to calculate the low-lying IE's of isobutene, *cis*-butene, ethylene, *etc.* Certainly the most esoteric study on IE's was that of Koller et al.[47] who determined that a quark in an H_2O or SO_2 molecule would significantly alter the IE. The problem lies in the experimental verification of this effect in view of the fact that the probability of finding a quark is only one in 10^{20}. An important word of caution has been raised by Heilbronner[48] concerning the fitting of a series of calculated IE's to the measured peaks in a spectrum. By the very fact of assuming the same ordering of states in the calculations and experiment, a significant statistical correlation is built into the comparison.

A rather useful application of MO calculations has been the determination of molecular polarizabilities[49] and electric dipole moments.[50] Both of these properties are difficult to measure experimentally, but are extremely important in calculating ion–molecule collision cross-sections and intensities of rotational and vibrational spectra.

[39] H. Lischka and H. J. Köhler, *J. Am. Chem. Soc.*, 1978, **100**, 5297.
[40] H. Mayr, W. Förner, and P. von R. Schleyer, *J. Am. Chem. Soc.*, 1979, **101**, 6032.
[41] N. L. Allinger and J. H. Siefert, *J. Am. Chem. Soc.*, 1975, **97**, 752.
[42] K. H. Thunemann, R. J. Buenker, S. D. Peyerimhoff, and S. K. Shih, *Chem. Phys.*, 1978, **35**, 35.
[43] J. Schirmer, L. S. Cederbaum, and W. Domcke, *Chem. Phys. Lett.*, 1978, **57**, 582.
[44] L. S. Cederbaum, W. Domcke, J. Schirmer, W. von Niessen, G. H. F. Diercksen, and W. P. Kraemer, *J. Chem. Phys.*, 1978, **69**, 1591.
[45] W. Domcke, L. S. Cederbaum, J. Schirmer, W. von Niessen, C. E. Brion, and K. H. Tan, *Chem. Phys.*, 1979, **40**, 171.
[46] G. Lauer, K. W. Schulte, and A. Schweig, *J. Am. Chem. Soc.*, 1978, **100**, 4925.
[47] J. Koller, B. Borstnik, and A. Azman, *Phys. Rev.*, 1978, **18A**, 1325.
[48] E. Heilbronner and A. Schmelzer, *Nouv. J. Chim.*, 1980, **4**, 23.
[49] P. J. Bounds, *Chem. Phys. Lett.*, 1980, **70**, 143.
[50] N. N. Haese and R. C. Woods, *Chem. Phys. Lett.*, 1979, **61**, 396.

Finally, properties of doubly charged ions have been calculated by *ab initio*[51] and semiempirical methods.[52]

Heats of Formation.—Probably the most important and generally useful thermodynamic property of a molecule is its heat of formation. A large body of ΔH_f° has been collected by mass spectrometric investigations. However, as is by now well known, results from electron impact (except for energy-selected electron impact) determinations have not been very accurate owing to a variety of factors. The advent of high-resolution photoionization and equilibrium measurements by ICR or high-pressure mass spectrometry (see following section) have changed this so that accuracies of less than 1 kJ mol^{-1} are now commonplace. This happy situation has also created problems because the ΔH_f° is a thermodynamic quantity while the ionization energy (IE) or fragment ion appearance energy (AE) are spectroscopic quantities. This problem has been treated by Lias[53] and Rosenstock.[54] At absolute zero the relationships between the energy and enthalpy for ionization and dissociative ionization can be written rigorously by equations (1) and (2) and the $\Delta H_{f0}^\circ(M^+)$ and $\Delta H_{f0}^\circ(A^+)$ will be given by equations (3) and (4).

$$\left.\begin{array}{l} M \to M^+ + e^- \\ IE = \Delta E_0 = \Delta H_0 = \Delta H_{f0}^\circ(M^+) + \Delta H_{f0}^\circ(e^-) - \Delta H_{f0}^\circ(M) \end{array}\right\} \quad (1)$$

$$\left.\begin{array}{l} AB \to A^+ + B + e^- \\ AE = \Delta E_0 = \Delta H_0 = \Delta H_{f0}^\circ(A^+) + \Delta H_{f0}^\circ(B) + \Delta H_{f0}^\circ(e^-) - \Delta H_{f0}^\circ(AB) \end{array}\right\} \quad (2)$$

$$\Delta H_{f0}^\circ(M^+) = IE + \Delta H_{f0}^\circ(M) - \Delta H_{f0}^\circ(e^-) \quad (3)$$

$$\Delta H_{f0}^\circ(A^+) = AE + \Delta H_{f0}^\circ(AB) - \Delta H_{f0}^\circ(B) - \Delta H_{f0}^\circ(e^-) \quad (4)$$

The assumption is, of course, that the measured IE or AE is a 0 K value, *i.e.* the energy from the ground state of M or AB to the ground state of M^+ or $A^+ + B$. In order to convert the 0 K value into a ΔH_{fT}°, one uses the usual thermodynamic relation (5).

$$\Delta H_{fT}^\circ = \Delta H_{f0}^\circ + \int_0^T [Cp(\text{products}) - Cp(\text{reactants})]\,dT \quad (5)$$

A problem arises in how the electron should be treated. In the JANAF Thermochemical Tables,[55] the electron is treated as a conventional element with its complement of $5/2\,RT$ of energy. On the other hand, in the Gaseous Ion Energetics[56] compilation, the electron is assumed to be at rest at all temperatures and thus its contribution to equation (5) is ignored. The difference is not at all negligible, being 6.2 kJ mol^{-1} at 300 K. The warning is clear: as values for ΔH_f° continue to improve in accuracy, these problems must be addressed if hopeless confusion and unnecessary disagreements are to be avoided.

---

[51] M. Cobb, T. F. Moran, R. F. Borkman, and R. Childs, *J. Chem. Phys.*, 1980, **72**, 4463.
[52] L. Asbrink, C. Fridh, and E. Lindholm, *Int. J. Mass Spectrom. Ion Phys.*, 1979, **32**, 93.
[53] S. G. Lias, in 'Kinetics of Ion–Molecule Reactions', ed. P. Ausloos, Plenum Press, New York, 1979, p. 223.
[54] H. M. Rosenstock in 'Kinetics of Ion–Molecule Reactions', ed. P. Ausloos, Plenum Press, New York, 1979, p. 246.
[55] JANAF Thermochemical Tables, Nat. Stand. Ref. Data Ser., Nat. Bur. Stand. (US), 1971, Vol. 37.
[56] H. M. Rosenstock, K. Draxl, B. W. Steiner, and J. T. Herron, *J. Phys. Chem.* (Ref. Data. Supplement 1), 1977, Vol. 6.

An example of such a disagreement involves the ΔH_f° of Pr^+. Traeger[57] and Baer[58] measured the photoionization (PI) onsets for Pr^+ from 2-iodopropane and obtained AE's (298 K) of 9.70 and 9.69 eV, respectively. In calculating a ΔH_f° (Pr^+), Traeger used ΔH_{f298}° values for I and PrI and obtained a ΔH_{f298}° (Pr^+) of 787 kJ mol^{-1}. Baer, on the other hand, corrected the room-temperature onset for the thermal energy of the PrI and obtained a AE(0 K) of 9.78 eV from which a ΔH_{f0}° (Pr^+) of 818 kJ mol^{-1} was calculated. Using the convention of an electron with zero energy, a ΔH_{f298}° (Pr^+) of 798 kJ mol^{-1} was obtained. This represents a difference of 11 kJ mol^{-1} resulting from data which differ by only 1 kJ mol^{-1}. To conclude this story it should be pointed out that in a re-analysis of their data Traeger and McLoughlin[59] obtained a room-temperature value of 802 kJ mol^{-1}, in reasonable agreement with the value of Baer. Using PES of the Pr radical Houle and Beauchamp[60] determined the ΔH_{f298}° (Pr^+) of 784 kJ mol^{-1}, a value which clearly depends on the knowledge of the ΔH_f° of the Pr radical and possibly its structure. The photochemistry and i.r. spectrum of the isopropyl radical indicate that two conformations of Pr are in equilibrium, at least in an Ar matrix,[61] thus the ΔH_f° of Houle and Beauchamp may well be too low.

Aside from equilibrium studies which yield relative heats of formation (after converting the ΔG into ΔH) at the experimental temperature, the bulk of the thermochemical data is derived from PI mass spectrometry, PES, or monoenergetic electron-impact mass spectrometry. In addition to the problem of converting a 0 K into 298 K ΔH_f°, such measurements are also subject to errors arising from kinetic shift and reverse activation energy, both of which give values for ΔH_f° which are too high.

The heats of formation of a number of hydrocarbon radical cations are uncertain. One of these is that of Et^+. Houle and Beauchamp[60] measured it to be 917 ± 4 kJ mol^{-1} by PES of Et. However, photoionization of EtI yields a ΔH_{f298}° of 901 ± 4 kJ mol^{-1}, a value corroborated by Traeger and McLoughlin.[59] Again an analysis of the neutral Et[62] may shed light on the difference between the PI and PES results. Similar uncertainty surrounds the enthalpy and identity of the $C_7H_7^+$ ion. The onset for this fragment by PI has been measured from a number of precursor molecules.[63] The ΔH_{f298}° ($C_7H_7^+$) obtained from the AE from toluene and cycloheptatriene was 866 ± 2 kJ mol^{-1}, which is lower than the ΔH_{f298}° of the benzyl structure of 882 kJ mol^{-1}.[64] From this it is concluded that the $C_7H_7^+$ has the tropylium structure.[63] In addition, six other alkyl benzene precursors yielded $C_7H_7^+$ with an average ΔH_{f298}° of 859 ± 8 kJ mol^{-1}, indicating that these also produce the tropylium structure.[65] Norborna-2,5-diene on the other hand produces a $C_7H_7^+$ ion with a ΔH_{f298}° of 917 kJ mol^{-1} and therefore does not appear to isomerize at threshold to toluene.[63] Similarly, tolyl rather than tropy-

---

[57] J. C. Traeger, *Int. J. Mass Spectrom. Ion Phys.*, 1980 **32**, 309.
[58] T. Baer, *J. Am. Chem. Soc.*, 1980, **102**, 2482.
[59] J. C. Traeger and R. G. McLoughlin, *J. Am. Chem. Soc.*, 1980, in press.
[60] F. A. Houle and J. L. Beauchamp, *J. Am. Chem. Soc.*, 1979, **101**, 4067.
[61] J. Pacansky and H. Coufal, *J. Chem. Phys.*, 1980, **72**, 3298.
[62] J. Pacansky and H. Coufal, *J. Chem. Phys.*, 1979, **71**, 2811.
[63] J. C. Traeger and R. G. McLoughlin, *Int. J. Mass Spectrom. Ion Phys.*, 1978, **27**, 319.
[64] F. P. Lossing, *Can. J. Chem.*, 1971, **49**, 357.
[65] R. G. McLoughlin, J. D. Morrison, and J. C. Traeger, *Org. Mass Spectrom.*, 1979, **14**, 104.

lium ions are produced by the ionization of fluorotoluenes as has been shown by collisional activation studies.[66] Their ΔH_f° is estimated to be 983 ± 12 kJ mol^{-1}. It should be pointed out that these heats of formation are uncertain because of the previously mentioned problem in conversion between 298 K and 0 K, and because of the possibility of a kinetic shift. The latter problem could be checked by the photoion–photoelectron coincidence (PIPECO) technique of Stockbauer and Rosenstock[67] in which ions are pulsed out of the ionization region at times delayed with respect to the electron signal.

Other alkyl ions investigated by PI mass spectrometry or by PIPECO include $C_7H_8^+$ from a variety of precursors,[68] from which the authors conclude that the toluene structure is the most stable one. The t-butyl cation was found to have a ΔH_{f298}° of 678 ± 3 kJ mol^{-1}.[69] $C_4H_4^+$ is one of the products in the fragmentation of the benzene ion. Neither its energy nor its structure has been established. However, Eland et al.[70] in a PIPECO study of the pyridine ion dissociation were able to fit the lifetime *versus* internal energy curve by assuming a ΔH_f° ($C_4H_4^+$) of 1172 ± 8 kJ mol^{-1}. In a similar study of the $C_6H_6^+$ dissociation, Baer et al.[71] derived a value of 1155 ± 8 kJ mol^{-1} which agrees to within the error. Both values are below the ΔH_f° of any known linear isomer of $C_4H_4^+$. this fact plus *ab initio* calculations[71] indicate that the $C_4H_4^+$ ion is cyclic.

Parr et al.[72] measured photoionization AE's for $C_3H_3^+$, $C_3H_2^+$, and C_3H^+ from cyclopropene, propyne, and allene, and determined average ΔH_f° values for these three fragment ions of 1083 ± 2, 1493 ± 8, and 1615 ± 15 kJ mol^{-1}, respectively. The ΔH_f° ($C_3H_3^+$) is about 8 kJ higher than is generally accepted[56] although the precise value is extremely difficult to establish. Another uncertain ΔH_f° is that of $C_6H_5^+$. Every dissociative ionization producing this fragment studied to date is associated with a large kinetic shift. That is, the parent ion lifetime at threshold is greater than the residence time of the ion in the mass spectrometer and therefore the true onset for $C_6H_5^+$ production has not been measured directly by either PI or monoenergetic electron impact. The only exception is the direct ionization of the C_6H_5 radical. However, this depends on a knowledge of the ΔH_f° ($C_6H_5^\cdot$). A re-evaluation of these results using proper correction for thermal energy leads to a value of ΔH_{f0}° ($C_6H_5^+$) = 1134 ± 12 kJ mol^{-1}, which corresponds to ΔH_{f298}° ($C_6H_5^+$) = 1116 ± 12 kJ mol^{-1}.[73] Two indirect determinations made involved the fitting of the dissociation rate *versus* energy curve for the reactions (6) and (7) studied by PIPECO with the RRKM theory.

$$C_6H_5Cl + h\nu \rightarrow C_6H_5^+ + Cl + e^- \qquad (6)$$

$$C_6H_6 + h\nu \rightarrow C_6H_5^+ + H + e^- \qquad (7)$$

---

[66] F. W. McLafferty and F. M. Bockhoff, *Org. Mass Spectrom.*, 1979, **14**, 181.
[67] R. Stockbauer and H. M. Rosenstock, *Int. J. Mass Spectrom. Ion Phys.*, 1978, **27**, 185.
[68] R. G. McLoughlin, J. D. Morrison, and J. C. Traeger, *Org. Mass Spectrom.*, 1978, **13**, 483.
[69] R. G. McLoughlin and J. C. Traeger, *J. Am. Chem. Soc.*, 1979, **101**, 5791.
[70] J. H. D. Eland, J. Berkowitz, H. Schulte, and R. Frey, *Int. J. Mass Spectrom. Ion Phys.*, 1978, **28**, 297.
[71] T. Baer, G. D. Willett, D. Smith, and J. S. Phillips, *J. Chem. Phys.*, 1979, **70**, 4076.
[72] (*a*) A. C. Parr, A. J. Jason, and R. Stockbauer, *Int. J. Mass Spectrom. Ion Phys.*, 1980, **33**, 243; (*b*) A. C. Parr, A. J. Jason, R. Stockbauer, and K. E. McCulloh, *ibid.*, 1979, **30**, 319; (*c*) A. C. Parr, A. J. Jason, and R. Stockbauer, *ibid.*, 1978, **26**, 23.
[73] H. M. Rosenstock, R. Stockbauer, and A. C. Parr, *J. Chem. Phys.*, 1979, **71**, 3708.

Rosenstock et al.[73] found a best fit when ΔH_{f0}° $(C_6H_5^+) = 1151 \pm 4$ kJ mol^{-1} while Baer et al.[71] in a similar study of the benzene ion dissociation used a value of 1122 ± 5 kJ mol^{-1}. These two values fall on either side of the direct determination by PES. One other value which is not subject to kinetic shift problems is the measurement of Beauchamp[74] which is based on an equilibrium study in an ICR instrument. That value is ΔH_{f0}° $(C_6H_5^+) = 1146$ kJ mol^{-1} which agrees rather well with that of Rosenstock.[73]

Other molecules which were studied by PI are the $C_7H_5O^+$ fragment from a number of benzoyl compounds,[75] Ge-, Sn-, and Pb-substituted benzenes,[76] COS,[77,78] CS_2,[78–80] KrF_2,[81] and t-butyl-lithium.[82] In addition, the IE's of some unusual or unstable compounds have been measured, many of them by PES. These include vinylketene,[83] SH,[84] SBr_2,[85] MeNHBr,[86] MeNBr$_2$,[86,87] Me$_2$NBr,[86] NH$_2$Br,[86,88] $(HCO)_2O$,[89] SBr_2,[90] SBr_2,[90] K_2,[91] LiH_2^+,[92] $N_2O_4^+$,[93] HCO^+,[94] SF_2,[95] NH_2^+,[96] AsX_n^+ (X = F, Cl, or Br),[97] Cs_2I^+,[98] HCS^+,[99] and SeX_2^+ (X = F, Cl, or Br).[100] Finally accurate IE's have been determined for allene,[101] NH,[102] and O_2^+[103] and NO$^+$.[103] The latter PES study lists the vibrational levels of these two ions up to $v = 34$ for the ground electronic state. Data such as these have allowed Albritton et al.[104] to generate accurate RKR potentials for eight potential energy curves of NO$^+$. A similar analysis of NO$^+$ has also been made by Rao and Reddy[105] using their own parametrized method. Tables of energy levels, inter-

---

[74] J. L. Beauchamp, *Adv. Mass Spectrom.*, 1974, **6**, 717.
[75] R. G. McLoughlin and J. C. Traeger, *Org. Mass Spectrom.*, 1979, **14**, 434.
[76] T. I. Evlasheva, V. K. Potapov, and A. N. Rodionov, *High Energy Chem.*, 1979, **13**, 146.
[77] M. J. Hubin-Franskin, D. Huard, and P. Marmet, *Int. J. Mass Spectrom. Ion Phys.*, 1978, **27**, 263.
[78] J. H. D. Eland and J. Berkowitz, *J. Chem. Phys.*, 1979, **70**, 5151.
[79] W. M. Trott, N. C. Blais, and E. A. Walters, *J. Chem. Phys.*, 1979, **71**, 1692.
[80] P. Coppens, J. C. Reynaert, and J. Drowart, *J. Chem. Soc., Faraday Trans.* 2, 1979, **75**, 292.
[81] J. Berkowitz and J. H. Holloway, *J. Chem. Soc., Faraday Trans.* 2, 1978, **74**, 2077.
[82] W. McLean, P. T. Murray, T. Baer, and R. C. Jarnagin, *J. Chem. Phys.*, 1978, **69**, 2715.
[83] J. K. Terlouw, P. C. Burgers, and J. L. Holmes, *J. Am. Chem. Soc.*, 1979, **101**, 225.
[84] S. J. Dunlavey, J. M. Dyke, N. K. Fayad, N. Jonathan, and A. Morris, *Mol. Phys.*, 1979, **38**, 729.
[85] D. M. DeLeeuw, R. Mooyman, and C. A. DeLange, *Chem. Phys. Lett.*, 1979, **61**, 191.
[86] D. Colbourne, D. C. Frost, C. A. McDowell, and N. P. C. Westwood, *Can. J. Chem.*, 1979, **57**, 1279.
[87] E. Nagy-Felsobuki, J. B. Peel, and G. D. Willett, *J. Electron Spectrosc. Relat. Phenom.*, 1978, **13**, 17.
[88] F. Carnovale, E. Nagy-Felsobuki, J. B. Peel, and G. D. Willett, *J. Electron Spectrosc. Relat. Phenom.*, 1978, **14**, 163.
[89] D. M. DeLeeuw, R. Mooyman, and C. A. DeLange, *Chem. Phys. Lett.*, 1979, **63**, 57.
[90] E. Nagy-Felsobuki and J. B. Peel, *Chem. Phys.*, 1980, **45**, 189.
[91] S. Leutwyler, A. Herrmann, L. Wöste, and E. Schumacher, *Chem. Phys.*, 1980, **48**, 253.
[92] C. H. Wu, *J. Chem. Phys.*, 1979, **71**, 783.
[93] K. Nomoto, Y. Achiba, and K. Kimura, *Bull. Chem. Soc. Jpn.*, 1979, **52**, 1614.
[94] J. M. Dyke, N. B. N. Jonathan, A. Morris, and M. J. Winter, *Mol. Phys.*, 1980, **39**, 629.
[95] D. M. DeLeeuw, R. Mooyman, and C. A. DeLange, *Chem. Phys.*, 1978, **34**, 287.
[96] S. J. Dunlavey, J. M. Dyke, N. Jonathan, and A. Morris, *Mol. Phys.*, 1980, **39**, 1121.
[97] R. E. Pabst, M. C. Sharpe, J. L. Margrave, and J. L. Franklin, *Int. J. Mass Spectrom. Ion Phys.*, 1980, **33**, 187.
[98] I. V. Sidorova, A. V. Gusarov, and L. N. Gorokhov, *Int. J. Mass Spectrom. Ion Phys.*, 1979, **31**, 367.
[99] A. G. Harrison, *J. Am. Chem. Soc.*, 1978, **100**, 4911.
[100] D. M. DeLeeuw, R. Mooyman, and C. A. DeLange, *Chem. Phys.*, 1979, **38**, 21.
[101] R. Stockbauer, K. E., McCulloh, and A. C. Parr, *Int. J. Mass Spectrom. Ion Phys.*, 1979, **31**, 187.
[102] L. G. Piper, *J. Chem. Phys.*, 1979, **70**, 3417.
[103] P. Natalis, J. E. Collin, J. Delwiche, G. Caprace, and M. J. Hubin, *J. Electron Spectrosc. Relat. Phenom.*, 1979, **17**, 205, 421.
[104] D. L. Albritton, A. L. Schmeltekopf, and R. N. Zare, *J. Chem. Phys.*, 1979, **71**, 3271.
[105] T. V. Ramakrishna Rao and R. Reddy, *Physika*, 1978, **95C**, 412.

nuclear distances, and rotational constants of the various electronic and vibrational states are displayed in a useful manner. These analyses now take the place of the much used, but now out of date, Gilmore curves for NO^+.[106]

In a most ingenious experiment, Bieri and Jonsson[107] determined the IE (HNC) = 12.5 ± 0.2 eV by charge exchange. They used $MeNH_2$ and MeNC as precursors for HCN^+ or HNC^+, which were then employed as primary ions in the investigation of charge transfer mass spectra of MeOH. They noted that the ions from these precursors gave mass spectra similar to H_2O^+, but very unlike HCN^+. From this they conclude that the HNC^+ ion is produced from both precursors and that it has a recombination energy of 12.5 eV as compared with 13.6 eV for HCN^+. The results also imply that HNC^+ is more stable than HCN^+ thereby experimentally verifying the previously mentioned theoretical results.[12]

An interesting method for obtaining a 'thermodynamic' IE from equilibrium studies by ICR has been proposed.[108] The values are, of course, relative, but if they can be related to a well known IE, such as that of NO, the results would give relative ΔG values. By using the calculated $\Delta S°$ for the reaction, the ΔH or 'thermodynamic' IE can be obtained. The advantage of this approach is that no correction must be made for thermal effects. Other thermal or equilibrium measurements of $\Delta H_f°$ have been made. Lias et al.[109] using proton transfer reactions in an ICR established the $\Delta H_{f298}°$ of a number of radical cations including Pr^+ for which they found a $\Delta H_f°$ of 772 ± 9 kJ mol^{-1}. These values are based on a $\Delta H_f°$ (Bu^{t+}) = 678 kJ mol^{-1}. However, the recently determined $\Delta H_f°$ (Pr^+) of 800 ± 3 kJ mol^{-1} by PI[57,58] suggests that the values of Lias et al.[41] will have to be increased by 28 kJ mol^{-1}. The $\Delta H_f°$ of some neutral free radicals have been determined by DeFrees et al.[110] Among these are cyclo-C_7H_7˙ (289 ± 13 kJ mol^{-1}), cyclo-C_3H_3˙ (440 ± 17 kJ mol^{-1}), C_3H_5˙ (183 ± 9 kJ mol^{-1}), cyclo-C_3H_5˙ (278 ± 11 kJ mol^{-1}), and C_2H_3˙ (299 ± 9 kJ mol^{-1}). By an independent method using the rate of reaction between Cl and cyclopropane, a $\Delta H_f°$ (cyclo-C_3H_5˙) = 280 ± 1 kJ mol^{-1} was found.[111] Thus the heat of formation of this radical seems to be well established now.

Leung and Harrison[112] using a high-pressure CI source estimated the $\Delta H_f°$ of a number of substituted phenyl cations including $MeC_6H_4^+$ (1058 kJ mol^{-1}), $H_2NC_6H_4$ (1058 kJ mol^{-1}), and $BrC_6H_4^+$ (1151 kJ mol^{-1}). The $MeC_6H_4^+$ energy is consistent with the tolyl structure as confirmed by a STO-3G calculation,[122] but significantly higher than the previously mentioned value of 983 kJ mol^{-1}.[66] These data indicate that the tolyl structure does not isomerize to the tropylium or benzyl ion, a fact confirmed by McLafferty and Bockhoff.[66] Holmes and Lossing[113] have found that rather consistently the enol form is more stable than the keto or acid form by 55 to 120 kJ mol^{-1}. The heats of formation of acids and ketones were determined from the IE and compared with the heats of formation of the same ion

---

[106] F. R. Gilmore, *J. Quant. Spectrosc. Radiat. Transfer*, 1965, **5**, 369.
[107] G. Bieri and B. O. Jonsson, *Chem. Phys. Lett.*, 1978, **56**, 446.
[108] S. G. Lias and P. Ausloos, *J. Am. Chem. Soc.*, 1978, **100**, 6027.
[109] S. G. Lias, D. M. Shold, and P. Ausloos, *J. Am. Chem. Soc.*, 1980, **102**, 2540.
[110] D. J. DeFrees, R. T. McIver, and W. J. Hehre, *J. Am. Chem. Soc.*, 1980, **102**, 3334.
[111] M. H. Baghal-Vayjooee and S. W. Benson, *J. Am. Chem. Soc.*, 1979, **101**, 2838.
[112] H. W. Leung and A. G. Harrison, *J. Am. Chem. Soc.*, 1979, **101**, 3168.
[113] J. L. Holmes and F. P. Lossing, *J. Am. Chem. Soc.*, 1980, **102**, 1591, 3732.

formed by dissociative ionization. Numerous ΔH_f° values for the keto and enol forms of such ions as $C_3H_6O^+$, $C_4H_8O^+$, $C_2H_4O_2^+$, $C_3H_6O_2^+$, and $C_4H_8O_2^+$ are tabulated. Finally, the heat of formation of CCl^+ has been determined to be 1515 ± 10 kJ mol^{-1} by measuring the onset of the endothermic reaction between Cl^+ and CO.[114]

An important update of the compilation of neutral molecules by Cox and Pilcher[115] has been published.[116] This compilation by Pedley and Rylance which is computer based is internally consistent. This means that future updates of individual compounds will automatically correct all compounds whose heats of formation depend on that one.

Gas-phase Basicities and Acidities.—The gas-phase basicity (GB) and acidity (GA) are defined by equations (8) and (9).

$$B + H^+ \rightleftharpoons BH^+ \qquad GB(B) = -\Delta G^\circ \qquad (8)$$

$$AH \rightleftharpoons H^+ + A^- \qquad GA(A) = -\Delta G^\circ \qquad (9)$$

The ΔG° for both reactions are large so that it is not convenient to perform equilibrium measurements on reactions (8) and (9) directly. Rather the relative ΔG° values are measured *via* reactions (10) and (11).

$$A + BH^+ \rightleftharpoons AH^+ + B \qquad GB(A) = GB(B) - \Delta G^\circ \qquad (10)$$

$$AH + B \rightleftharpoons BH^+ + A^- \qquad GA(A) = GA(B) - \Delta G^\circ \qquad (11)$$

Once one basicity or acidity is known, a whole series can be determined from relative measurements. A quantity related to the basicity is the proton affinity which is the negative of the enthalpy change in reaction (8). Most determinations of basicities and acidities are from equilibrium measurements in either high-pressure mass spectrometers or ICR spectrometers. In order to obtain the proton affinity, the ΔG° must be measured at several temperatures or the ΔS° must be calculated. Hartman and Lias[117] in an investigation on proton transfer equilibria in halobenzenes have discussed the calculation of ΔS°. They conclude that, although the rotational partition function is the major term in ΔS°, the internal partition functions (vibrations and internal rotations) must also be considered because they can contribute as much as 0.7 cal K^{-1} mol^{-1} ($\approx$ 1 kJ mol^{-1} at 300 K).

Direct measurements of proton affinities come from determination of heats of formation of ions. For instance, the ΔH_f° of Et^+ yields the proton affinity (PA) *via* reaction (12).

$$C_2H_4 + H^+ \rightarrow Et^+ \qquad PA = -\Delta H^\circ(\text{reaction}) \qquad (12)$$

The $\Delta H^\circ(\text{reaction}) = \Delta H_f^\circ(Et^+) - \Delta H_f^\circ(C_2H_4) - \Delta H_f^\circ(H^+)$. For this reason many of the articles referenced here could equally well have been reviewed in the previous section. Similarly, gas-phase acidities are related to the electron affinity.

---

[114] L. C. Frees, P. L. Pearl, and W. S. Koski, *Chem. Phys. Lett.*, 1979, **63**, 108.
[115] J. C. Cox and G. Pilcher, 'Thermochemistry of Organic and Organometallic Compounds', Academic Press, New York, 1970.
[116] J. B. Pedley and J. Rylance, 'Sussex-NPL Computer Analysed Thermochemical Data: Organic and Organometallic Compounds', Sussex University Press, Brighton, 1977.
[117] K. G. Hartman and S. G. Lias, *Int. J. Mass Spectrom. Ion Phys.*, 1978, **28**, 213.

Excellent reviews of gas-phase basicity[118,119] and acidity[120] measurements have been published with extensive tables on proton affinities[118] and acidities.[120]

Because equilibrium measurements give extremely accurate $\Delta G°$ values, and their determination is based on the relative schemes of equations (10) and (11), the absolute determination of proton or electron affinities of a few molecules to be used as standards is extremely important. Houle and Beauchamp[60] have measured the IE's of the free radicals Me, Et, Pr, and Bu by PES. A knowledge of the $\Delta H_f°$ of the radical allowed them to determine the $\Delta H_f°$ of the radical-cations. The PA(NH$_3$) was adjusted from the previously assumed 845 kJ mol^{-1} to 872 kJ mol^{-1} by combining the $\Delta H_f°$ (C$_3$H$_6$) with the relative PA's of C$_3$H$_6$ and NH$_3$. Because ammonia is one of the standards used in the PA scale, the whole scale is affected. However, the new $\Delta H_f°$(Pr$^+$) is not without possible sources of error. In particular, $\Delta H_f°$(Pr) is still under discussion and the determination of an adiabatic onset from a PES is not without its pitfalls. As mentioned in the previous section, $\Delta H_f°$(Et$^+$) determined by PES is as much as 8 kJ mol^{-1} higher than that determined by dissociative photoionization of EtI.[58] Nevertheless, the $\Delta H_f°$(t-butyl cation) determined by dissociative photoionization of C$_4$H$_{10}$ yields a value for the PA of NH$_3$ of 875 kJ mol^{-1} [69] which, within experimental error, is equal to Houle and Beauchamp's value.[60]

A novel measurement of the PA of HX (X = F, Cl, or Br) has recently been performed by measuring the dissociative PI onset of H$_2$X$^+$ from (HX)$_2$ dimers *via* reaction (13).[121]

$$(HX)_2 + h\nu \rightarrow H_2X^+ + X + e^- \quad (13)$$

The PA of HCl and HBr were measured as 564 and 589 kJ mol^{-1}, respectively. This method can be used if the binding energy of the dimer is known and if it can be established that the H$_2$X$^+$ comes from (HX)$_2$ and not from trimers, tetramers, *etc.* which invariably are formed in nozzle expansions. These conditions evidently are satisfied as the results of Teidemann *et al.*[121] agree well with other values for the PA of HX.

A number of empirical schemes relating PA's with other parameters have been suggested. Hodges *et al.*[122] noted that the ionization energies of cyclic phosphites are inversely related to their PA. DeKock and Barbachyn[123] also noted the inverse relationship between the PA and the IE. To improve the correlation they grouped molecules according to the nature of their frontier orbital electron density. Three categories created were (*a*) 2-electron lone-pair bases, (*b*) π or σ bond-pair bases, and (*c*) 1-electron bases. Yoder and Yoder[124] have correlated the gas-phase acidities and basicities with an electrostatic field model. They found

---

[118] D. H. Aue and M. T. Bowers in 'Gas Phase Ion Chemistry', ed. M. T. Bowers, Academic Press, New York, 1979, Vol. 2, Ch. 9.
[119] R. W. Taft in ref. 53, p. 271.
[120] J. E. Bartmess and R. T. McIver in ref. 118, Ch. 11.
[121] P. W. Tiedemann, S. L. Anderson, S. T. Ceyer, T. Hirooka, C. Y. Ng, B. H. Mahan, and Y. T. Lee, *J. Chem. Phys.*, 1979, **71**, 605.
[122] R. V. Hodges, F. A. Houle, J. L. Beauchamp, R. A. Montag, and J. G. Verkade, *J. Am. Chem. Soc.*, 1980, **102**, 932.
[123] R. L. DeKock and M. R. Barbachyn, *J. Am. Chem. Soc.*, 1979, **101**, 6516.
[124] C. S. Yoder and C. H. Yoder, *J. Am. Chem. Soc.*, 1980, **102**, 1245.

a high correlation between the measured enthalpy changes for gas-phase dissociations and the potential energies of the charge–dipole, charge–induced dipole, and dipole–dipole interactions. The dominant term is the charge–dipole. They used the charge–locked-dipole potential.

Some PA's determined by equilibrium studies include CCl_2,[125] CH_2CO,[126] and some $C_4H_6O^+$ isomers.[127] Aue et al.[128] have continued their extensive compilation of nitrogen-containing compounds with a study of amides and imidates. These PA's are all based on $PA(NH_3) = 845.6$ kJ mol^{-1}, and so will have to be increased by 27 kJ mol^{-1}.[60,69] Proton affinities can also be determined from a study of ion–molecule reactions in, for instance, a flowing afterglow (FA) system. The PA of CF_3NO was bracketed between 708 and 723 kJ mol^{-1} because it was found to react with H_3O^+ but not with H_2CN^+.[129] In another FA study, the PA's of MeOH, $MeNO_2$, and HCOOH were determined by measuring the forward and backward rate constants for proton transfer.[130] The equilibrium constant, which is equal to the ratio of the forward and reverse rate constants, was corrected for the differences in the collision rates using the method of Lias and Ausloos.[131] This correction factor has become rather controversial and is felt by some[132,133] to be unnecessary and to violate microscopic reversibility.[133] Other FA studies on H_2CO[134] and HCN[135] have also been reported.

Gas-phase acidities have been reported by McIver and co-workers.[136,137] An extensive scale of acidities from methanol to phenol is presented in a table which also includes $\Delta S°$, $\Delta H°$ for the process $AH \rightarrow A^- + H^+$, and $\Delta H_f°(A^-)$. From these data and $\Delta H_f°(A)$, the electron affinity (EA) of the radicals can be determined. Substituent and solvation effects on the gas-phase acidities are discussed, particularly for alkyl and phenyl groups.[137] It is shown that the phenyl group relative to the methyl substituent stabilizes the ion by between 20 and 150 kJ mol^{-1}. However no constant additive term is found. Similarly, only a qualitative correlation is found between the gas-phase acidity with the amount of delocalization as calculated by Hückel MO theory.

Numerous theoretical calculations of proton affinities and structures of protonated molecules have been performed. Formaldehyde[138,139] and thioformaldehyde[139,140] have been investigated. Protonation can occur in one of two

---

[125] P. Ausloos and S. G. Lias, *J. Am. Chem. Soc.*, 1978, **100**, 4594.
[126] W. R. Davidson, Y. K. Lau, and P. Kebarle, *Can. J. Chem.*, 1978, **56**, 1016.
[127] J. H. Vajda and A. G. Harrison, *Int. J. Mass Spectrom. Ion Phys.*, 1979, **30**, 293.
[128] D. H. Aue, L. D. Betowski, W. R. Davidson, M. T. Bowers, P. Beak, and J. Lee, *J. Am. Chem. Soc.*, 1979, **101**, 1361.
[129] C. G. Freeman, P. W. Harland, and M. J. McEwan, *Int. J. Mass Spectrom. Ion Phys.*, 1979, **30**, 285.
[130] C. G. Freeman, P. W. Harland, and M. J. McEwan, *Int. J. Mass Spectrom. Ion Phys.*, 1978, **28**, 377.
[131] S. G. Lias and P. Ausloos, *J. Am. Chem. Soc.*, 1977, **99**, 4831.
[132] T. Su in ref. 53, p. 250.
[133] J. M. Jasinski, R. N. Rosenfeld, D. M. Golden, and J. I. Brauman, *J. Am. Chem. Soc.*, 1979, **101**, 2559.
[134] C. G. Freeman, P. W. Harland, and M. J. McEwan, *Int. J. Mass Spectrom. Ion Phys.*, 1978, **28**, 19.
[135] C. G. Freeman, P. W. Harland, and M. J. McEwan, *Int. J. Mass Spectrom. Ion Phys.*, 1978, **27**, 77.
[136] J. E. Bartmess, J. A. Scott, and R. T. McIver, *J. Am. Chem. Soc.*, 1979, **101**, 6046.
[137] J. E. Bartmess, J. A. Scott, and R. T. McIver, *J. Am. Chem. Soc.*, 1979, **101**, 6556.
[138] J. D. Dill, C. L. Fischer, and F. W. McLafferty, *J. Am. Chem. Soc.*, 1979, **101**, 6531.
[139] T. Yamabe, K. Yamashita, K. Fukui, and K. Morokuma, *Chem. Phys. Lett.*, 1979, **63**, 433.
[140] J. D. Dill and F. W. McLafferty, *J. Am. Chem. Soc.*, 1979, **101**, 6526.

places yielding MeO^+ or H_2COH^+. It now appears established that the latter structure is more stable by as much as 170 kJ mol^{-1}.[139] Dill and McLafferty[140] also found the H_2CSH^+ ion to be more stable than MeS^+, but only by 42 kJ mol^{-1} and with a barrier in the endothermic direction of 167 kJ mol^{-1}. Yamabe et al.[139] found that the major difference between H_2COH^+ and H_2CSH^+ is that the carbon atom in the thio compound has a much larger electron density thereby shielding the C nucleus more effectively from interacting repulsively with the proton.

Two groups performed calculations on protonated H_2S.[141,142] Dixon and Marynick[142] found that the correlation effects on the inversion barrier in SH_3^+ are small, but that polarization functions are crucial for an accurate calculation of the barrier which they found to be 135 kJ mol^{-1}. The structures and energies of a number of protonated ions such as $C_3H_5^+$, $C_4H_9^+$, $C_5H_5^+$, $C_6H_7^+$, and $C_4H_{11}^+$ were determined by a STO-3G calculation.[143] Another STO-3G calculation was on protonated 3- and 4-substituted pyridines.[144] It was found that the protonation site is on the ring nitrogen and that no correlation exists between the net charge on the nitrogen and the PA as had been proposed by Davis and Shirley.[145]

(3)   (4)

Lischka and Köhler[146] found the protonated acetaldehyde structure (3) is more stable than the bridged oxiranium structure (4). This difference is less for the sulphur analogue EtS^+ where the energy difference is only 16—65 kJ mol^{-1}, depending on the basis set used in the calculation. Other calculations included investigation of intramolecular solvation effects,[147] the PA of R_2CO (R = Me, NH_2, OH, or F),[148] anisole,[149] and ethylidenimine and vinylamine,[150] and the use of the integral Hellman–Feynman theorem in calculating proton affinities.[151]

Electron Affinities and Negative Ions.—Only ten years ago the measurement of an accurate electron affinity for a molecule was considered a minor triumph. However, new experimental techniques as well as more sophisticated MO calculations now allow electron affinities to be determined in a routine fashion. The two major experimental approaches are photodetachment of electrons from negative ions, and collisional ionization in which neutral alkali-earth beams (e.g. K, Na, or Cs) are crossed with a molecular beam. Both methods are described in a

[141] S. P. So, Chem. Phys. Lett., 1979, **61**, 83.
[142] D. A. Dixon and D. S. Marynick, J. Chem. Phys., 1979, **71**, 2860.
[143] H. J. Köhler and H. Lischka, J. Am. Chem. Soc., 1979, **101**, 3479.
[144] J. Catalán, O. Mó, P. Pérez, and M. Yáñez, J. Am. Chem. Soc., 1979, **101**, 6520.
[145] D. W. Davis and D. A. Shirley, J. Am. Chem. Soc., 1976, **98**, 7898.
[146] H. Lischka and H. J. Köhler, Chem. Phys. Lett., 1979, **63**, 326.
[147] F. J. Winkler and D. Stahl, J. Am. Chem. Soc., 1979, **101**, 3685.
[148] J. E. Del Bene and S. Radovick, J. Am. Chem. Soc., 1978, **100**, 6936.
[149] J. Catalán and M. Yáñez, J. Am. Chem. Soc., 1979, **101**, 3490.
[150] M. R. Ellenberger, R. A. Eades, M. W. Thomsen, W. E. Farneth, and D. A. Dixon, J. Am. Chem. Soc., 1979, **101**, 7151.
[151] P. Politzer and K. C. Daiker, J. Chem. Phys., 1978, **68**, 5289.

recent review on electron affinities by Janousek and Brauman.[152] Calculations for electron affinities are difficult to perform because (a) the desired accuracy is greater than for ionization energies, (b) usually the adiabatic EA is desired requiring the geometry optimization of both negative and neutral species, and (c) the added electron increases the electron correlation. Although it has been claimed[153] that no CI is necessary and that a Koopman's approximation is adequate for calculating electron affinities, Garrett[154] asserts that results using the pseudopotential method show that the neglect of electron correlation in weakly bound anions can lead to serious errors. In addition, the magnitude of the long-range dipole moment must be properly represented in the case of dipolar molecules.

An interesting and valuable application of calculations is to molecular anions whose geometry is very different from the neutral counterpart. Examples are BeF_2 and MgF_2 both of which are linear. However, calculations indicate that a bent geometry in the case of MgF_2 leads to a positive EA.[155] The potential surface indicates that BeF_2^- which has an angle of 135° may be metastable as is the experimentally observed CO_2^-. This work was inspired by the fact that any molecule with a dipole moment in excess of 1.625 D can trap an electron by the electrostatic field of the dipole.[156] $LiOH^-$ is such a dipole-dominated anion whereas Li_2O being linear and symmetric is not.[157] When the bond angle is less than 135°, the dipole moment exceeds the critical value and the anion becomes more stable than the neutral. However, it is not expected to be a long-lived species because the energy is considerably above that of the linear neutral molecule.[157]

Comparisons of calculated and experimental EA's have been done on diatomic hydrides using a CI calculation.[158] Some calculated (and experimental) values are: BeH 0.48 eV (0.7); BH 0.03 eV (0.15); CH 1.04 eV (1.25); OH 1.51 eV (1.83); and SiH 1.13 eV (1.29). Taylor et al.[159] calculated the potential curves for CN and CN^- and obtained a EA(CN) = 3.85 eV which compares very well with the experimental value of 3.82 eV. Using eighth-order perturbation theory, the EA of HC_2 has been calculated to be 3.18 ± 0.25 eV.[160] The experimentally determined value is 2.94 ± 0.1 eV.[160] From these measurements the C—H bond energy of acetylene was calculated to be 554 ± 20 kJ mol^{-1}.

Although some EA's are being calculated, the bulk of the values obtained during the past two years has come from experiments. In addition to the photodetachment and collisional ionization methods, a new approach has been proposed by Nazarenko and Pokhodenko.[161] It is based on measuring the energy deficits of the molecular cation in field ionization (Δ_+) and the similar deficit for the anion in negative ion FI (Δ_-). Then, according to the theory the EA =

---

[152] B. K. Janousek and J. I. Brauman in ref. 118, Ch. 10.
[153] K. D. Jordan and J. J. Wendolski, *Chem. Phys.*, 1977, **21**, 145.
[154] W. R. Garrett, *J. Chem. Phys.*, 1979, **71**, 651.
[155] B. A. B. Seiders, W. L. Luken, R. P. Blickensderfer, and K. D. Jordan, *Chem. Phys.*, 1979, **39**, 285.
[156] O. H. Crawford, *Proc. Phys. Soc. (London)*, 1967, **91**, 279.
[157] R. P. Blickensderfer and K. D. Jordan, *Chem. Phys.*, 1979, **41**, 193.
[158] P. Rosmus and W. Meyer, *J. Chem. Phys.*, 1978, **69**, 2745.
[159] P. R. Taylor, G. B. Bacskay, N. S. Hush, and A. C. Hurley, *J. Chem. Phys.*, 1979, **70**, 4481.
[160] B. K. Janousek, J. I. Brauman, and J. Simons, *J. Chem. Phys.*, 1979, **71**, 2057.
[161] V. A. Nazarenko and V. D. Pokhodenko, *Int. J. Mass Spectrom. Ion Phys.*, 1979, **31**, 381.

IE $- (\Delta_+ + \Delta_-)$. This method gave an EA (tetracyanoethylene) = 2.8 eV which is within 0.1 eV of another determination.

Electron affinities measured by a variety of techniques are listed in Table 2. This table updates the compilation of Janousek and Brauman.[152] The values which are accurate to 0.02 eV or less are determined by laser photodetachment which is the most precise method available. In addition, it often gives vibrational frequencies of the anions as well.[163,170] When such structure is evident, the adiabatic EA can be determined with much greater confidence because the effects of anions not relaxed to the ground vibrational states can be accounted for.

Table 2 *Experimental electron affinities*/eV*

Molecule	EA	Ref.	Molecule	EA	Ref.
O_3	2.1028 ± 0.004	162	CF_2Cl_2	0.4 ± 0.3	165
MeS	1.861 ± 0.004	163	$CFCl_3$	1.1 ± 0.3	165
CD_3S	1.858 ± 0.006	163	CCl_4	2.0 ± 0.2	165
SH	2.31 ± 0.01	164	CF_3	1.9 ± 0.3	165
EtS	1.953 ± 0.006	164	CF_2Cl	1.6 ± 0.3	165
Pr^nS	2.00 ± 0.002	164	$CFCl_2$	1.1 ± 0.3	165
Pr^iS	2.02 ± 0.02	164	CCl_3	1.3 ± 0.3	165
Bu^nS	2.03 ± 0.02	164	$CHCl_2$	0.0 ± 0.3	165
Bu^tS	2.07 ± 0.02	164	CCl_2	1.8 ± 0.3	165
$n\text{-}C_5H_7S$	2.09 ± 0.02	164	FCl	1.5 ± 0.3	165
C_2H	2.94 ± 0.1	160	UF_6	5	166
$TiCl_4$	2.88 ± 0.15	169	WF_6	3.56 ± 0.1	167
$TiCl_3$	0.61 ± 0.2	169	FeCO	1.26 ± 0.02	168
$SnCl_4$	2.49 ± 0.15	169	$Fe(CO)_2$	1.22 ± 0.02	168
$SnCl_3$	3.69 ± 0.3	169	$Fe(CO)_3$	1.8 ± 0.2	168
$GeCl_3$	> 2.6	169	$Fe(CO)_4$	2.4 ± 0.3	168

* These are EA's reported between May 1978 and April 1980 which are not included in the compilation of Janousek and Brauman.[152]

Some EA's reported in 1978 which are included in the extensive compilation of Janousek and Brauman[152] are: MeO,[170,171] CD_3O,[170] MeS,[170] $MeNO_2$,[172] CF_3I,[172] CF_3Br,[172] Me_3CO,[171] Me_3CCH_2O,[171] $C_6O_2F_4$,[173] $C_6O_2Cl_4$,[173] $C_6O_2Br_4$,[173] C_5H_7,[174] and C_7H_9.[174]

In addition to electron affinities, Dispert and Lacmann[165] measured the absolute cross-section for negative ion formation in collisions between K

[162] S. E. Novick, P. C. Engelking, P. L. Jones, J. H. Futrell, and W. C. Lineberger, *J. Chem. Phys.*, 1979, **70**, 2652.
[163] B. K. Janousek and J. I. Brauman, *J. Chem. Phys.*, 1980, **72**, 694.
[164] B. K. Janousek, K. J. Reed, and J. I. Brauman, *J. Am. Chem. Soc.*, 1980, **102**, 3125.
[165] H. Dispert and K. Lacmann, *Int. J. Mass Spectrom. Ion Phys.*, 1978, **28**, 49.
[166] J. A. D. Stockdale, R. J. Warmack, and R. N. Compton, *Chem. Phys. Lett.*, 1979, **63**, 621.
[167] P. M. George and J. L. Beauchamp, *Chem. Phys.*, 1979, **36**, 345.
[168] P. C. Engelking and W. C. Lineberger, *J. Am. Chem. Soc.*, 1979, **101**, 5569.
[169] B. P. Mathur and E. W. Rothe, *Int. J. Mass Spectrom. Ion Phys.*, 1979, **31**, 77.
[170] P. C. Engelking, G. B. Ellison, and W. C. Lineberger, *J. Chem. Phys.*, 1978, **69**, 1826.
[171] B. K. Janousek, A. H. Zimmerman, K. J. Reed, and J. I. Brauman, *J. Am. Chem. Soc.*, 1978, **100**, 6142.
[172] R. N. Compton, P. W. Reinhart, and C. D. Cooper, *J. Chem. Phys.*, 1978, **68**, 4360.
[173] C. D. Cooper, W. F. Frey, and R. N. Compton, *J. Chem. Phys.*, 1978, **69**, 2367.
[174] A. H. Zimmerman, R. Gygax, and J. I. Brauman, *J. Am. Chem. Soc.*, 1978, **100**, 5595.

(1—30 eV) and fluorochlorocarbons. Some of the cross-sections are extremely large. Cl^- formation from $K + CCl_4$ collisions has a cross-section of 110 $Å^2$. On the other hand, CCl_4^- formation has a cross-section of only 0.2 $Å^2$. According to Riley and Hershbach,[175] in CCl_4 the electron enters an antibonding σ^* orbital with a node between the C and Cl^-. As a result, dissociation to Cl^- occurs readily. By contrast in $SnCl_4$, the extra electron goes into a low-lying vacant orbital giving rise to stable $SnCl_4^-$ as experimentally verified by Mathur and Rothe.[169] The cross-section and angular distributions of mass analysed product ions have been measured[176] for the collisional ionization reaction (14) and (15).

$$Cs + H_2O \rightarrow Cs^+ + H_2O^- \quad (14)$$

$$H_2O^- \rightarrow OH^- + H \quad (15)$$

Another study involving the use of crossed molecular beams was that between SbF_5 and RX (R = alkyl group; X = halide). The effect of vibrational energy was found to be significant, and isotropic angular distributions indicated that the reaction went via long-lived complexes.[177] In place of an alkali-earth atom, a beam of Ar metastables can be used to study collisional ionization. This has been done with I_2 in which angular distributions of the resulting Ar^+ ions were collected as well.[178]

Electron attachment to molecules cannot produce a stable anion without an accompanying dissociation. Illenberger et al.[179] investigated dissociative electron attachment in CF_2Cl_2, CF_3Cl, and $CFCl_3$ in order to ascertain the role of low-energy electrons and freons in the earth's atmosphere. Such small negative ions dissociate rapidly. On the other hand large molecules can form long-lived anions by electron attachment. They are long-lived for the same reason that large molecules have strong metastable peaks in mass spectra. It is because the energy gets distributed statistically among the many vibrational modes before dissociation or electron detachment takes place. The measurement of negative ion lifetimes is therefore of great interest in mass spectrometry. Cooper et al.[173] measured such lifetimes as a function of electron energy in fluoranil, chloranil, and bromanil ($C_6O_2X_4$). Owing to an unusual set of circumstances, the fragment negative ion from fluoranil ($C_4F_4CO^-$) is formed at an energy above that for neutral C_4F_4CO and therefore it can autodetach. That rate was measured also, but the structure of the $C_4F_4CO^-$ is not known. A quantity related to the lifetime is the electron attachment cross-section. This has been measured as a function of electron energy for SF_6, CCl_4, and $CFCl_3$ using PI of Xe as the source of low-energy electrons.[180]

Electron attachment to molecules can proceed via temporary negative ion states which produce strong resonances in the cross-section versus electron-energy curve. These states can be real states of the anion in which case they are called Feschbach resonances, or they may be the result of an electron trapping in a

[175] S. J. Riley and D. R. Herschbach, J. Chem. Phys., 1973, **58**, 27.
[176] R. J. Warmack, J. A. D. Stockdale, and R. N. Compton, Int. J. Mass Spectrom. Ion Phys., 1978, **27**, 239.
[177] A. Auerbach, R. J. Cross, and M. Saunders, J. Am. Chem. Soc., 1978, **100**, 4908.
[178] K. T. Gillen, T. D. Gaily, and D. C. Lorents, Chem. Phys. Lett., 1978, **57**, 192.
[179] E. Illenberger, H. V. Scheunemann, and H. Baumgärtel, Chem. Phys., 1979, **37**, 21.
[180] J. M. Ajello and A. Chutjian, J. Chem. Phys., 1979, **71**, 1079.

centrifugal barrier caused by the angular momentum, in which case they are referred to as shape resonances. The latter have been investigated theoretically for e-SF_6 scattering.[181] Feschbach resonances have been observed in cyclooctatetraenyl and perinaphthenyl anions[182] and in NaX^- (X = Cl, Br, or I).[183] In the latter study it was noted that the energy of the resonance is independent of the halide atom, indicating that the state is associated with the Na. Thus a picture of NaX^- emerges. The electron attaches almost exclusively on the Na in the neutral, ionically bonded, molecule while the resulting $NaCl^-$ appears to consist of a neutral Na attached to a negative halide ion.

Other studies on negative ions include photodetachment of a number of cyclic ketones,[184] the structure of the benzene anion distorted by Jahn–Teller interaction[185] and the thermochemistry of SO_2Cl^- and $(SO_2)_2Cl^-$ by high-pressure mass spectrometry.[186] The heats of formation of Cl_3^-, SO_2Cl^-, $SO_2Cl_3^-$, and $S_2O_4Cl^-$ have been determined by high-pressure mass spectrometry.[187] Schenk et al.[188] studied ion-pair formation by photoionization of CF_2Cl_2, CF_3Cl, and $CFCl_3$. Correlations with structure in the negative ion curves were associated with onsets for ion-pair processes. Finally ion-pair formation in O_2, NO, and CO between 17 and 30 eV has been investigated.[189]

Dimers and Clusters.—Clusters have been studied in high-pressure mass spectrometers for some time. The application of supersonic expansion nozzles to mass spectrometers has now expanded the field. Theoretical work has also multiplied. An example[190] of the co-operative nature of theory and experiment is the study of complexes $(N_2)_n H^+$, $(CO)_n H^+$, and $(O_2)_n H^+$ (n = 1—7). Equilibrium studies in a high-pressure mass spectrometer were carried out as a function of temperature for reactions (16)—(18).

$$(N_2)_{n-1}H^+ + N_2 \rightleftarrows (N_2)_n H^+ \quad (16)$$

$$(CO)_{n-1}H^+ + CO \rightleftarrows (CO)_n H^+ \quad (17)$$

$$(O_2)_{n-1}H^+ + O_2 \rightleftarrows (O_2)_n H^+ \quad (18)$$

From the ΔG and ΔS values obtained from van't Hoff plots, it was found that the stability of $B_n H^+$ toward B loss was in reverse order to the proton affinity of B. STO-3G calculations indicated that the O_2H^+ structure is non-linear with an angle of 111°. Hagler et al.[191] carried out an ICR study on formaldehyde dimers, and *ab initio* calculations established the structure and electron density maps. A PES study, supported by semiempirical calculations of the formic acid dimer,

---

[181] J. L. Dehmer, J. Siegel, and D. Dill, *J. Chem. Phys.*, 1978, **69**, 5205.
[182] R. Gygax, H. L. McPeters, and J. I. Brauman, *J. Am. Chem. Soc.*, 1979, **101**, 2657.
[183] S. E. Novick, P. L. Jones, T. J. Mulloney, and W. C. Lineberger, *J. Chem. Phys.*, 1979, **70**, 2210.
[184] A. H. Zimmerman, R. L. Jackson, B. K. Janousek, and J. I. Brauman, *J. Am. Chem. Soc.*, 1978, **100**, 4674.
[185] A. L. Hinde, D. Poppinger, and L. Radom, *J. Am. Chem. Soc.*, 1978, **100**, 4681.
[186] R. G. Keesee and A. W. Castleman, jun., *J. Am. Chem. Soc.*, 1980, **102**, 1446.
[187] R. Robbiani and J. L. Franklin, *J. Am. Chem. Soc.*, 1979, **101**, 764.
[188] H. Schenk, H. Oertel, and H. Baumgärtel, *Ber. Bunsenges. Phys. Chem.*, 1979, **83**, 683.
[189] H. Oertel, H. Schenk, and H. Baumgärtel, *Chem. Phys.*, 1980, **46**, 251.
[190] K. Hiraoka, P. P. S. Saluja, and P. Kebarle, *Can. J. Chem.*, 1979, **57**, 2159.
[191] A. T. Hagler, Z. Karpas, and F. S. Klein, *J. Am. Chem. Soc.*, 1979, **101**, 2191.

indicated that there is considerable interaction between the two monomer units via the σ orbitals.[192]

Hydrogen clusters in a supersonic expansion were studied by 100 eV EI.[193] Clusters up to H_{39}^+ were observed. The conclusions are that abundances generally decrease with n but that certain sizes are particularly abundant and therefore stable. These are H_9^+, H_{15}^+, and H_{27}^+. Only odd clusters were observed. These clusters were also studied theoretically.[194,195] The dissociation energies[195] of the H_7^+, H_9^+, and H_{11}^+ ions agreed reasonably well with experiment. By comparison with the odd H_n^+ clusters, the dimer H_2–H_2^+ interaction, investigated with the MOLE program, is weak.[196] The most stable structure (which is presumably the transition state for the $H_2^+ + H_2$ reaction) is the one in which the two diatomic species are perpendicular to each other.

Li_2^+ has been studied using single CI *ab initio* programs.[197] Potential energy curves were calculated taking into account, in particular, the long-range interaction between Li and Li$^+$. The isoelectronic $(NO_2)_2^+$ and $(CO_2)_2^-$ dimers were calculated to be most stable in the C_s geometry, it being 0.54 eV more stable than the isolated monomers.[198] An experimental PI study of the $(CS_2)_2$ dimer from a supersonic nozzle found that there is little interaction between the two CS_2 units in the neutral.[199] This is determined by the dimer autoionization structure which is very similar to that of the monomer, being only 0.1 eV lower in energy. Another dimer studied by PI was $(C_2H_4)_2$.[200] $C_4H_8^+$ is a stable ion as well as a transition state for reactions (19) and (20).

$$C_2H_4^+ + C_2H_4 \rightarrow C_4H_8^+ \rightarrow C_3H_5^+ + Me \quad (19)$$
$$\rightarrow C_4H_7^+ + H \quad (20)$$

The two fragment ions from C_4H_8 isomers had been investigated previously by PI[201,202] and the $C_3H_5^+$ fragment was found to be the dominant one at threshold. Yet, in the $(C_2H_4)_2^+$ dimer dissociation, the $C_3H_5^+$ ion was observed only 40 kJ mol^{-1} above the $C_4H_7^+$ onset. These results imply that the dimer does not isomerize to the most stable but-2-ene structure. It is also tempting to conclude that the transition state in the $C_2H_4^+ + C_2H_4$ reaction is not the butene ion. Yet the results of Meisels *et al.*[202] indicate that in the bimolecular reaction there is no 40 kJ mol^{-1} barrier for $C_3H_5^+$ formation. There is a real paradox here and more work clearly needs to be done to clear up this very interesting problem. The PI supersonic molecular beam combination was also used to investigate clusters of acetone and [$^2H_6$]-acetone by Trott *et al.*[203] The measured IE's of (MeCOMe)$_n$

[192] F. Carnovale, M. K. Livett, and J. B. Peel, *J. Chem. Phys.*, 1979, **71**, 255.
[193] A. Van Lumig and J. Reuss, *Int. J. Mass Spectrom. Ion Phys.*, 1978, **27**, 197.
[194] S. Yamabe, H. Hirao, K. Kitaura, *Chem. Phys. Lett.*, 1978, **56**, 546.
[195] H. Huber, *Chem. Phys. Lett.*, 1980, **70**, 353.
[196] M. Cobb, T. F. Moran, R. F. Borkman, and R. Childs, *Chem. Phys. Lett.*, 1978, **57**, 326.
[197] (*a*) D. D. Konowalow and M. E. Rosenkrantz, *Chem. Phys. Lett.*, 1979, **61**, 489; (*b*) D. D. Konowalow, W. J. Stevens, and M. E. Rosenkrantz, *ibid.*, p. 24.
[198] Y. Yoshioka and K. D. Jordan, *J. Am. Chem. Soc.*, 1980, **102**, 2621.
[199] M. E. Gress, S. H. Linn, Y. Ono, H. F. Prest, and C. Y. Ng, *J. Chem. Phys.*, 1980, **72**, 4242.
[200] S. T. Ceyer, P. W. Tiedemann, C. Y. Ng, B. H. Mahan, and Y. T. Lee, *J. Chem. Phys.*, 1979, **70**, 2138.
[201] T. Baer, D. Smith, B. P. Tsai, and A. S. Werner, *Adv. Mass Spectrom.*, 1978, **A7**, 56.
[202] G. G. Meisels, G. M. L. Verboom, M. J. Weiss, and T. Hsieh, *J. Am. Chem. Soc.*, 1979, **101**, 7189.
[203] W. M. Trott, N. C. Blais, and E. A. Walters, *J. Chem. Phys.*, 1978, **69**, 3150.

were 9.694 ± 0.006, 9.26 ± 0.03, 9.10 ± 0.03, and 9.02 ± 0.03 eV, for $n = 1—4$, respectively.

One of the most active areas in the cluster field involves rare-gas clusters. Dehmer and Dehmer[204] have obtained PES of Ar_2 and Kr_2. Using Morse potentials for both neutrals and ions, they fitted the Franck–Condon envelope of the PES assuming a known Morse function of the neutral. From this analysis, they obtained the Ar_2^+ and K_2^+ potential curves. These results were compared with curves calculated by *ab initio* MO programs with CI.[205] The derived dissociation energies agree well with those from PES. Another theoretical study of Xe_2^+ as well as Xe_2 and Xe_2^* includes relativistic effects.[206] Some mixed rare-gas cluster calculations have also been made on $Ne_n H^+$,[207] $Ar_n H^+$,[208] and $Ar_n K^+$.[209]

A MO study of the $CH_5^+(CH_4)_n$ cluster was performed at the 4-31G level[210] and the binding energy found to be small. An empirical relation between binding energy and the ionization energy difference between the two partners was found.[211] Arifov and Pozharov proposed a mechanism for the formation of $H^+(H_2O)_n OH$ involving a long-lived intermediate.[212] Although $(H_2O)_n H^+$ clusters are rather easily formed, the $(H_2O)_n^+$ cluster appears to be very unstable. However, keV ion impact on films of ice has produced $(H_2O)_n^+$ ($n = 1—30$),[213] but the mechanism is not clear because the process seems to be related to impurities in the ice.

Cook and Taylor[214] have investigated the stability of $(NH_3)_n$ clusters by PI of molecules in a supersonic beam. By varying the nozzle temperature, different cluster sizes were formed. From these results, they calculated ΔH_f° for reaction (21) but an unexplained pressure effect mars the easy interpretation of the results.

$$(NH_3)_n \rightleftharpoons (NH_3)_{n-1} + NH_3 \qquad (21)$$

Another problem would appear to be the temperature of the system. These clusters are formed, in part, in the expansion, during which the temperature is dropping. Whether a temperature under these conditions is known or even meaningful is not clear. The H—O bond in acetic acid was studied by dissociative ionization of acid dimers using PI with supersonically expanded gas.[215] The derived value of 4.48 ± 0.11 eV is considerably lower than the old value of 4.86 ± 0.2 eV.

A high-pressure mass spectrometric study of $M^+(NH_3)_n$ ($M = Na$ or Li) revealed that the most stable solvation shell consists of four ligands.[216] Clustering energies are obtained for $n = 0—5$. Finally a flowing afterglow study of solvated

---

[204] P. M. Dehmer and J. L. Dehmer, *J. Chem. Phys.*, 1978, **69**, 125.
[205] H. H. Michels, R. H. Hobbs, and L. A. Wright, *J. Chem. Phys.*, 1978, **69**, 5151.
[206] W. C. Ermler, Y. S. Lee, K. S. Pitzer, and N. W. Winter, *J. Chem. Phys.*, 1978, **69**, 976.
[207] R. L. Matcha, M. B. Milleur, and P. F. Meier, *J. Chem. Phys.*, 1978, **68**, 4748.
[208] R. L. Matcha and M. B. Milleur, *J. Chem. Phys.*, 1978, **69**, 3016.
[209] R. P. Pan and R. D. Etters, *J. Chem. Phys.*, 1980, **72**, 1741.
[210] S. Yamabe, Y. Osamura, and T. Minato, *J. Am. Chem. Soc.*, 1980, **102**, 2268.
[211] M. Meot-Ner, P. Hamlet, E. P. Hunter, and F. H. Field, *J. Am. Chem. Soc.*, 1978, **100**, 5466.
[212] V. A. Arifov and S. L. Pozharov, *High Energy Chem.*, 1978, **12**, 327, 331.
[213] E. N. Nikolaev and G. D. Tantsyrev, *High Energy Chem.*, 1978, **12**, 253.
[214] K. D. Cook and J. W. Taylor, *Int. J. Mass Spectrom. Ion Phys.*, 1979, **30**, 345.
[215] K. D. Cook and J. W. Taylor, *Int. J. Mass Spectrom. Ion Phys.*, 1979, **30**, 93.
[216] A. W. Castleman, P. M. Holland, D. M. Lindsay, and K. I. Peterson, *J. Am. Chem. Soc.*, 1978, **100**, 6039.

H_3S^+ and H_3O^+ indicated that H_3O^+ is stabilized more by solvation than is H_3S^+.[217]

3 Ionization Processes

Charge Transfer.—A long-standing hope has been that ionization by charge transfer [equation (22)] would produce an ion with internal energy equal to the reactant ion's (M^+) recombination energy (RE), which is usually associated with its IE.

$$M^+ + AB \rightarrow AB^+ + M \qquad (22)$$

This hope rests on the assumption that no, or little, momentum is transferred in the reaction. Although there are many examples to the contrary,[218—221] research in this area continues. Sunner and Szabo[222] studied the role of internal energy in the reaction $H_2O^+ + H_2O \rightarrow H_3O^+ + OH$. The H_2O ion was prepared in the $\tilde{X}$, $\tilde{A}$, and $\tilde{B}$ states by charge transfer. The cross-section did not vary significantly with the H_2O^+ electronic state. An equally small effect of the internal energy on the dissociative charge transfer spectrum of CH_4 was noted by Ardeleau and Mercea.[223] Jowko et al.[224] used ten primary ions including COS^+, $C_2H_2^+$, H_2O^+, Xe^+, and CO_2^+ to study the charge transfer mass spectra (CTMS) of H_2S and NH_3. The plot of the total cross-section with recombination energy followed the PES, indicating that adiabatic (no momentum transfer) transitions are dominant. Rather less satisfactory results were obtained for the CTMS of N_2O between 13 and 25 eV recombination energies.[225] For instance, C^+ and Hg^+ have RE = 16.58 and 16.7 eV, respectively, which is very close to the $v = 0$ level of the $\tilde{A}$ state of N_2O^+ at 16.38. Yet the cross-section for CT is quite small for these two primary ions. Consistent with the fact that autoionization can not play a role in CT, very little O^+ signal from N_2O^+ was observed. The O^+ comes mainly from autoionizing levels located between the $\tilde{X}$ and $\tilde{A}$ state[226] which are inaccessible by CT. Sunner[227] investigated the *cis* and *trans* isomers of but-2-ene by CT. Although their EI mass spectra are indistinguishable, the CTMS for some recombination energies were different. Oddly enough, high-energy species such as He^+, Ne^+, and Ar^+ gave identical spectra whereas low-energy species such as COS^+, Kr^+, N_2^+, and F^+ gave different spectra.

Other molecules studied by CTMS were benzene[228] and a number of chloromethanes.[229] The measured absolute cross-sections ranged from 0.8 Å^2 for NH_3^+

---

[217] D. K. Bohme, G. I. Mackay, and S. D. Tanner, *J. Am. Chem. Soc.*, 1979, **101**, 3724.
[218] G. Mauclaire, R. Derai, S. Fenistein, and R. Marx, *J. Chem. Phys.*, 1979, **70**, 4017.
[219] M. R. McMillan and M. A. Coplan, *J. Chem. Phys.*, 1979, **71**, 3063.
[220] T. F. Moran and J. B. Wilcox, *J. Chem. Phys.*, 1979, **70**, 1467.
[221] M. Gerard, T. R. Govers, and R. Marx, *Chem. Phys.*, 1979, **36**, 247.
[222] J. Sunner and I. Szabo, *Int. J. Mass Spectrom. Ion Phys.*, 1979, **31**, 213.
[223] P. Ardeleau and V. Mercea, *Int. J. Mass Spectrom. Ion Phys.*, 1978, **28**, 107.
[224] A. Jowko, M. Forys, and B. O. Jonsson, *Int. J. Mass Spectrom. Ion Phys.*, 1979, **29**, 249.
[225] J. Sunner and I. Szabo, *Int. J. Mass Spectrom. Ion Phys.*, 1979, **31**, 193.
[226] T. Baer, P. M. Guyon, I. Nenner, A. Tabché-Fouhaillé, R. Botter, L. F. A. Ferreira, and T. R. Govers, *J. Chem. Phys.*, 1979, **70**, 1585.
[227] J. Sunner, *Int. J. Mass Spectrom. Ion Phys.*, 1980, **32**, 285.
[228] J. M. Tedder and P. H. Vidaud, *Chem. Phys. Lett.*, 1979, **64**, 81.
[229] J. M. Tedder and P. H. Vidaud, *J. Chem. Soc., Faraday Trans. 2*, 1979, **75**, 1648.

on CCl_4 to 180 Å for NO^+ on benzene. Such large variations indicate that resonance is a major requirement for CT. When resonance exists, such as in the case of Ar^+ and Kr^+ on CH_4, the cross-section at thermal energies is that of the Langevin cross-section.[230]

Thomas et al.[231] have studied the kinetic energy dependence of reactions (23)—(26) in a drift tube, and have found that the cross-section varies by nearly two orders of magnitude between 0.04 and 1 eV.

$$Ar^+ + O_2 \rightarrow O_2^+ + Ar \qquad (23)$$

$$Ar^+ + N_2 \rightarrow N_2^+ + Ar \qquad (24)$$

$$Ar^+ + NO \rightarrow NO^+ + Ar \qquad (25)$$

$$Ar^+ + CO \rightarrow CO^+ + Ar \qquad (26)$$

At the other extreme are the results of Parker et al.[232–234] who studied CT cross-sections in the 1—5 keV translational energy range. Peaks in the cross-section *versus* translational energy were interpreted in terms of the adiabatic theory.[235] That is the energy defect ΔE is converted into a characteristic time (inverse frequency) by the relation $\Delta E = h\nu$. The adiabaticity factor is the ratio of this characteristic frequency, ν, to the collision time, a/v, where a is a characteristic distance of interaction and v is the velocity of the collision. When the adiabaticity factor $a\Delta E/vh \approx 1$, the cross-section is a maximum according to the adiabatic theory. Thus, one of the peaks at 2000 eV was associated with reaction (27) which has an energy defect $\Delta E = 2.36$ eV.[232]

$$H_2^+ (v = 3) + H_2O \rightarrow H_2 (v = 0) + H_2O^+(\tilde{B}) \qquad (27)$$

There are many product channels so that it is nearly always possible to find one whose energy defect can account for the resonance in the cross-section. Experiments which confirm the validity of the adiabatic theory would be desirable.

What appears to be necessary in the field of charge transfer are state-to-state experiments in which the final as well as the initial states are known and varied. Such results would clarify the role of Franck–Condon factors and the dependence of the cross-section on the energy mismatch.

Some progress toward this goal is evident. Mauclaire et al.[218] have studied the final states of the reactions $Ar^+ + O_2$ and $Ne^+ + O_2$ by observing fluorescence from the electronically excited O_2^+. Of course, only a fraction of the products can be observed in this manner because the products may end up in non-fluorescing states. Nevertheless, it was established that part of the reaction events produce O_2^+ accompanied by nearly 2 eV of translational energy. For $Ne^+ + O_2$, only O^+ was observed indicating that no reactions with an energy mismatch of more than 3 eV [IE(Ne) − AE(O^+)] occurred. For $He^+ + O_2$, the situation is more complicated because O_2^+ has no electronic state at the IE of He at 24.59 eV. Thus, all CT reactions must involve momentum transfer. However, very little O_2^+ is formed,

---

[230] Y. H. Li and A. G. Harrison, *Int. J. Mass Spectrom. Ion Phys.*, 1978, **28**, 289.
[231] R. Thomas, A. Barassin, and R. R. Burke, *Int. J. Mass Spectrom. Ion Phys.*, 1978, **28**, 275.
[232] J. E. Parker and R. G. Milner, *Int. J. Mass Spectrom. Ion Phys.*, 1979, **29**, 29.
[233] J. E. Parker and R. Y. Haddad, *Int. J. Mass Spectrom. Ion Phys.*, 1979, **31**, 103.
[234] J. E. Parker and R. Y. Haddad, *Int. J. Mass Spectrom. Ion Phys.*, 1978, **27**, 403.
[235] J. B. Hasted, *Adv. Electron. Phys.*, 1960, **13**, 1.

and 36% of the O^+ is electronically excited.[236] Another product fluorescence study was performed by Tsuji et al.[237] on reactions (28) and (29) in a He afterglow apparatus.

$$He(2^3s) + N_2O \rightarrow N_2O^+(\tilde{A}^2\Sigma^+) \rightsquigarrow h\nu \tag{28}$$

$$He^+ + N_2O \rightarrow N_2^+(\tilde{B}^2\Sigma_u^+) + O \rightsquigarrow h\nu \tag{29}$$

The analysis of the N_2^+ emission indicated that this ion was formed in vibrational levels up to $v = 4$. This level in the $\tilde{B}$ state lies at 21.6 eV which is 1.8 eV higher than the energy of the $He(2^3s)$ state. The $N_2^+(\tilde{B})$ was therefore presumed to come from the charge transfer reaction (29).

A crossed-molecular-beam study of $He^+ + CO_2$ dissociative CT revealed that at low relative energies the angular distributions are characteristic of complex formation with significant momentum transfer taking place.[219] However, at higher energies, the reactions went predominantly with no or very little momentum transfer. The study showed that 100 eV is probably an optimum energy for state selection by CT because it is high enough for little momentum transfer, but not so high (e.g. 1000 eV) where intimate collisions again lead to momentum transfer. An interesting result came from a CT study of $N^+(^3P)$ RE = 14.53 eV and $N^+(^1D)$ RE = 16.43 eV in their reactions with neutral molecules.[220] $N^+(^1D)$ cannot reach the ground state $N(^4S)$ by a simple one-electron attachment. Rather, the simplest neutralization would take the $N^+(^1D)$ to $N(^2D)$ with a resulting RE = 14.04 eV, nearly equal to the RE of $N^+(^3P)$. The preference for one-electron transitions (as opposed to an electron jump accompanied by an electronic transition in the nitrogen atom) results in nearly equal CT cross-sections for $N^+(^3P)$ and $N^+(^1D)$. This is further confirmed by a reaction with CO in which the reaction $N^+(^1D) \rightarrow N(^2D)$ is nearly isoenergetic [equations (30) and (31)]. For the reactions (30) and (31) at 1 keV relative energy, the cross-sections are 18 Å^2 and 30 Å^2, respectively. A similar study was carried out using $O^+(^4S)$ and $O^+(^2D)$.[238]

$$N^+(^3P) + CO \rightarrow N(^4S) + CO^+(\check{X}\Sigma^+); \quad \Delta E = 0.53 \text{ eV} \tag{30}$$

$$N^+(^1D) + CO \rightarrow N(^2D) + CO^+(\check{X}\Sigma^+); \quad \Delta E = 0.4 \text{ eV} \tag{31}$$

The role of ion vibrational energy in the symmetric CT reaction (32) was investigated between 1 and 100 eV and $v = 0-7$.[239]

$$O_2^+(\check{X}, v) + O_2 \rightarrow O_2 + O_2^+(\check{X}, v') \tag{32}$$

The comparison between the experimental results (obtained by photoion–photoelectron coincidence) and the multistate impact parameter theory[240] indicated that the theory underestimates the cross-section of the higher vibrational levels.

An intriguing use of CTMS has already been mentioned. This is the experimental establishment that HNC^+ has a lower ΔH_f° than HCN^+ and that the

[236] G. Manclaire, R. Rerai, S. Fenistein, R. Marx, and R. Johnson, *J. Chem. Phys.*, 1979, **70**, 4023.
[237] M. Tsuji, K. Tsuji, and Y. Nishimura, *Int. J. Mass Spectrom. Ion Phys.*, 1979, **30**, 175.
[238] T. F. Moran and J. B. Wilcox, *J. Chem. Phys.*, 1978, **69**, 1397.
[239] T. Baer, P. T. Murray, and L. Squires, *J. Chem. Phys.*, 1978, **68**, 4901.
[240] T. F. Moran, M. R. Flannery, and P. C. Cosby, *J. Chem. Phys.*, 1974, **61**, 1261.

IE(HNC$^+$) ≈ 12.5 eV.[107] Carrington et al.[241,242] have employed CT in measuring the absorption spectrum of CO$^+$. Absorption spectra of ions cannot be done by measuring the fraction of the light absorbed because the ion densities are far too small. These measurements always depend on being able to measure the concentration of the excited state or the depletion of the ground state. Photodissociation has become a classic method whereby ions whose excited states dissociate can be detected. This method does not apply to the stable CO$^+\tilde{A}$ state. However, the different cross-sections for CT of CO$^+$ in the $\tilde{X}$ and $\tilde{A}$ states were utilized to monitor the CO$^+\tilde{X}$ state absorption.[241]

Finally, Neuschäfer et al.[243] studied reactions of doubly charged Ar^{2+}(^3P) and Ar^{2+}(^1S) with N$_2$ to produce Ar$^+$ and N$_2^+$. Part of the N$_2^+$ was electronically excited. Singly charged ions reacting via two-electron transfer were investigated by Bursey et al.[244] Charge reversal reactions of negative ions such as reaction (33) were studied for possible use in structure determination similar to positive ion collision-induced dissociation (CID) studies.[245]

$$\text{EtO}^- + \text{M} \rightarrow \text{EtO}^+ + 2e^- \tag{33}$$

The results indicate that the structure of the negative ion is maintained in charge reversal so that structure determination is possible.[244]

Photoionization.—This section reviews the articles dealing with the fundamental aspects of photoionization such as absolute cross-sections. Papers utilizing photoionization (PI) to measure ΔH_f° or to study ion lifetimes are reviewed under their subject headings.

The PI cross-section is of fundamental interest to mass spectrometry because 70 eV electron impact is already in the optical limit which means that the molecule does not 'see' the electron as a particle, but as a wave much as it would in PI. This correspondence has been verified by a laser-induced fluorescence detection of the rotation distribution of N$_2^+\tilde{X}\ ^2\Sigma_g^+$ formed by EI of N$_2$.[246] The rotational temperature was found to be lower for 100 eV (323 K) than for 60 eV (342 K) EI ionization and the $\Delta J = 0, \pm 1, \pm 2$ rules held more rigorously at the higher electron energy. The effect of the selection rules is that $T(\text{ion}) \approx T(\text{neutral})$. However, at low electron energies, higher-order multipole contributions to the electron–molecule interaction increase and so increase the rotational temperature. This correspondence has been used by Brion et al.[247–250] to study the absolute total and partial PI cross-sections using an electron–electron coincidence technique. The total absorption cross-section is measured by impinging,

---

[241] A. Carrington, D. R. J. Milverton, and P. J. Sarre, *Mol. Phys.*, 1978, **35**, 1505.
[242] A. Carrington, D. R. J. Milverton, P. G. Roberts, and P. J. Sarre, *J. Chem. Phys.*, 1978, **68**, 5659.
[243] D. Neuschäfer, Ch. Ottinger, S. Zimmermann, W. Lindinger, F. Howorka, and H. Störi, *Int. J. Mass Spectrom. Ion Phys.*, 1979, **31**, 345.
[244] (a) M. M. Bursey, J. R. Hass, D. J. Harvan, and C. E. Parker, *J. Am. Chem. Soc.*, 1979, **101**, 5485; (b) M. M. Bursey, D. J. Harvan, C. E. Parker, L. G. Pedersen, and J. R. Hass, *J. Am. Chem. Soc.*, 1979, **101**, 5489.
[245] F. W. McLafferty, *Philos. Trans. R. Soc. London, Ser. A*, 1979, **293**, 93.
[246] J. Allison, T. Kondow, and R. N. Zare, *Chem. Phys. Lett.*, 1979, **64**, 202.
[247] C. E. Brion, K. H. Tan, M. J. van der Wiel, and Ph. E. van der Leeuw, *J. Electron. Spectrosc. Relat. Phenom.*, 1979, **17**, 101.
[248] C. E. Brion and K. H. Tan, *Chem. Phys.*, 1978, **34**, 141.
[249] K. H. Tan, C. E. Brion, Ph. E. van der Leeuw, and M. J. van der Wiel, *Chem. Phys.*, 1978, **29**, 299.
[250] A. P. Hitchcock, C. E. Brion, and M. J. van der Wiel, *Chem. Phys. Lett.*, 1979, **66**, 213.

for instance, 1 keV electrons on a sample gas and measuring the inelastically forward-scattered electrons. Measuring these electrons in coincidence with ions generated during impact produces the total ionization cross-section, while measuring them in coincidence with energy-analysed electrons gives the partial PI cross-sections to the various electronic states. Thus a complete description of electron impact can be obtained. It is a direct means for determining the elusive energy deposition function. In this manner O_2 between 5 and 300 eV,[247] N_2O and CO_2 between 20 and 60 eV,[248,251] and H_2O up to 60 eV[249] were investigated. Of interest is that the PI efficiency (ionization/absorption) is unity above about 20 eV for all four molecules studied. CO_2 was measured at a few energies as high as 291 eV and N_2O in the vicinity of 400 eV in order to determine if the nature of the fragments is related to the ionized inner-shell electron.[250]

Some theoretical progress in calculating PI cross-sections for molecules has been made. Hush et al.[252,253] utilized their ground-state inversion potential (GIP) method. In the approximation of the time-dependent perturbation theory, the PI probability is proportional to integrals of the type $\langle \Psi_i | \vec{r} | \Psi_f \rangle$ where the Ψ's are the total wavefunctions and the subscripts refer to the initial and final states. In the simplest theory, the electron wavefunction, which is a part of Ψ_f, is treated as a plane wave. A slightly better method is to orthogonalize this plane wave to the rest of the electron wavefunctions.[254] In the GIP method a high-quality Hartree–Fock one-electron bound-state wavefunction is used to construct a one-dimensional Schroedinger equation. This is inverted and the extracted potential is used to solve the one-particle Schroedinger equation which includes the continuum. These continuum wavefunctions are utilized in calculating the cross-section. The method can be generalized to molecules. Some examples of calculated and experimental cross-sections are for N_2 at 21.2 eV 12.6 Å^2 (exp) and 17.0 Å^2 (GIP), at 26.9 eV 10.6 Å^2 (exp) and 12.5 Å^2 (GIP). A variation of this approach has been developed by Nordholm[255] and called the absorbing boundary method. It makes use of the fact that the important part of the ionization process takes place when the electron is still quite close. Thus, pseudo-bound-state configurations with open boundaries are generated, a procedure which retains some of the ease of working with bound states while still allowing the wavefunction to be coupled to the continuum.

In a considerably more complicated set of calculations, McKoy and Langhoff[256-259] have used the static exchange and Stieltjes–Tchebycheff moment theory techniques. The overall cross-sections obtained agree reasonably well with the available experimental results; certainly they are better than the cross-sections from the GIP procedure. On the other hand, the added complexity is significant. The molecules investigated were O_2,[256] N_2,[257] F_2,[258] and H_2CO.[259]

[251] C. E. Brion and K. H. Tan, *J. Electron. Spectrosc. Relat. Phenom.*, 1979, **15**, 241.
[252] N. S. Hush, P. R. Hilton, and S. Nordholm, *J. Electron. Spectrosc. Relat. Phenom.*, 1978, **15**, 101.
[253] P. R. Hilton, S. Nordholm, and N. S. Hush, *J. Electron. Spectrosc. Relat. Phenom.*, 1980, **18**, 101.
[254] T. Fujikawa, T. Ohta, and H. Kuroda, *J. Electron. Spectrosc. Relat. Phenom.*, 1979, **16**, 285.
[255] S. Nordholm, *J. Electron. Spectrosc. Relat. Phenom.*, 1979, **15**, 109.
[256] A. Gerwer, C. Asaro, B. V. McKoy, and P. W. Langhoff, *J. Chem. Phys.*, 1980, **72**, 713.
[257] T. N. Rescigno, A. Gerwer, B. V. McKoy, and P. W. Langhoff, *Chem. Phys. Lett.*, 1979, **66**, 116.
[258] A. E. Orel, J. N. Rescigno, B. V. McKoy, and P. W. Langhoff, *J. Chem. Phys.*, 1980, **72**, 1265.
[259] P. W. Langhoff, A. E. Orel, T. N. Rescigno, and B. V. McKoy, *J. Chem. Phys.*, 1978, **69**, 4689.

O'Neal and Reinhardt[260] have used CI wavefunctions to calculate the PI cross-sections and distribution of vibrational levels in H_2. The results are good at high energies but deteriorate significantly near threshold where autoionization dominates the ionization process.

Considerable interest has been generated by the phenomena of shape resonances.[261,262] A shape resonance arises when the outgoing electron is temporarily bound or trapped by the centrifugal barrier generated by its angular momentum. It is therefore a *one*-electron phenomenon as opposed to autoionization which involves two electrons. The autoionization resonance is often called a Feshbach resonance. The barrier in the shape resonance is a strong function of the nuclear positions. As a result, shape resonances in ionization processes lead to significant departures from Franck–Condon distributions of vibrational levels. This effect has been verified experimentally for CO[261] and N_2.[262] Between 15 and 40 eV, the approximate range of the shape resonance in N_2, the branching ratios between $v = 1$ and $v = 0$ vary between 4 and 20%. It appears that these shape resonances, which are generally very broad, extending over several eV's, are ubiquitous and may account for a number of experimental results in PI which to date have not been satisfactorily explained. The calculations of the shape resonance can be done with relatively simple models which separate the molecule into three regions, the core of each atom, an intermediate region which is very anisotropic and sensitive to the position of the nuclei, and the rest of space from which the molecule or ion is essentially a point. An alternative calculation of shape resonances is contained in the previously mentioned calculations of Langhoff and McKoy. These authors claim that for polyatomics, such as H_2CO, the Stieltjes–Tchebycheff approach has greater predictive power.[259]

Rescigno et al.[263] have calculated the PI cross-section of the $^1\Sigma_n^+$ excimer state of Ar_2. Normal closed-shell calculations of PI cross-sections work well with a frozen core approximation. However, for an open-shell excited state, the cross-section may be dominated by electronically induced autoionizing states. The calculated IE is 3.15 eV but the cross-section peaked at 3.9 eV.

Autoionization and Predissociation.—A great deal of progress has been made in the past two years in our understanding of the relationship between autoionization, preionization, shape resonances, and direct ionization.[264,265] The excellent monograph by Berkowitz on these subjects reviews much of this field.[266] This progress has come mainly from the study of di- and tri-atomics in which there is some hope of understanding these phenomena in terms of electronic states.

---

[260] S. V. O'Neal and W. P. Reinhardt, *J. Chem. Phys.*, 1978, **69**, 2126.
[261] R. Stockbauer, B. E. Cole, D. L. Ederer, J. B. West, A. C. Parr, and J. L. Dehmer, *Phys. Rev. Lett.*, 1979, **43**, 757.
[262] J. B. West, A. C. Parr, B. E. Cole, D. L. Ederer, R. Stockbauer, and J. L. Dehmer, *J. Phys. B*, 1980, **13**, L105.
[263] T. N. Rescigno, A. U. Hazi, and A. E. Orel, *J. Chem. Phys.*, 1978, **68**, 5283.
[264] C. Jungen, *J. Chim. Phys.*, 1980, **77**, 27.
[265] P. M. Guyon in 'Trends in Physics Conf. EPS IV York 1978', ed. M. M. Woolfson, Hilget, Bristol, 1979.
[266] J. Berkowitz, 'Photoabsorption, Photoionization and Photoelectron Spectroscopy', Academic Press, 1979.

Mass spectrometrists have noted for many years that onsets for numerous fragment ions are observed at their thermochemical onsets in regions of Franck–Condon gaps in their photoelectron spectra. By direct ionization, there is essentially zero probability for forming these ions but, with the aid of autoionization, there is some probability for producing ions at all internal energies. A particularly dramatic example of this effect is in N_2O. Between the ground $\tilde{X}\,^2\Pi$ state at 12.89 eV and the first excited $\tilde{A}\,^2\Sigma^+$ state at 16.39 eV, there is more than a 3 eV gap. Yet the NO^+ and O^+ onsets are observed at 15 and 15.3 eV respectively by either PI or EI. Baer et al.[226] measured PES at various photon energies in this Franck–Condon gap region. These are shown along with the total PI spectrum and the threshold PES (TPES) in Figure 2. The PES, obtained by electron

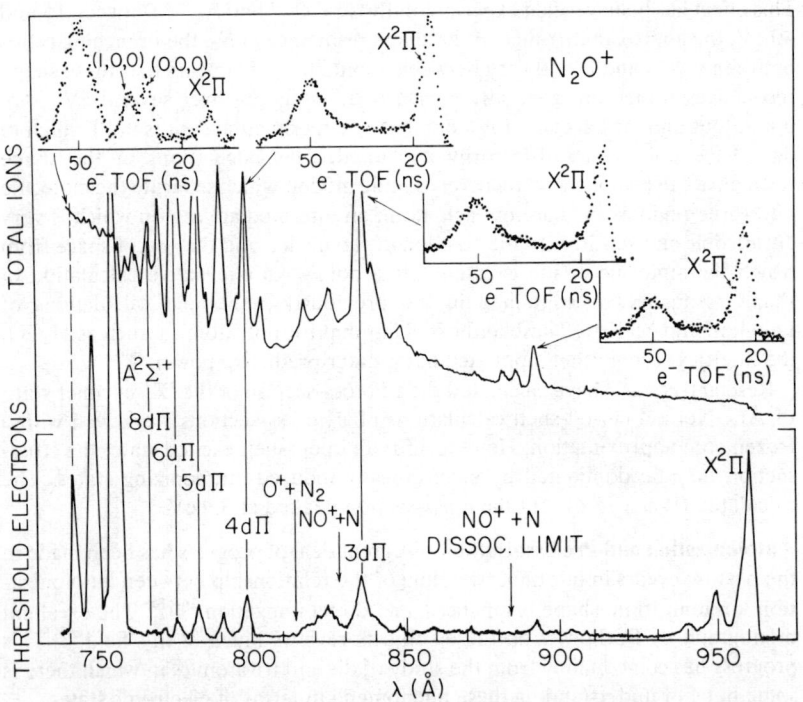

Figure 2 *Total ionization and threshold electron spectra of N_2O. Insets are electron TOF distributions at various photon energies which illustrate the two paths available for autoionization below the $\tilde{A}$ state*
(Reproduced with permission from *J. Chem. Phys.*, 1979, **70**, 1585)

time-of-flight, indicate that at all photon energies between the $\tilde{A}$ and $\tilde{X}$ states, two groups of electrons are formed: quasi-zero energy electrons consisting of approximately 10% of all the electrons, and the energetic electrons associated with formation of the $N_2O^+\tilde{X}$ state. Evidently a mechanism exists which forms ions in highly vibrationally excited states accompanied by the ejection of electrons of zero or near-zero energy. A mechanism was proposed which involves predissociation to neutral products interrupted by autoionization before the predissociation has progressed beyond the point of no return. The phenomenon of low-energy electron production and the proposed mechanism are quite general as low-energy electrons have been observed in a number of molecules.[267-269] The details of the mechanism in terms of perturbation theory and specific states must still be worked out, but the connection between predissociation and autoionization in producing excited ions appears now to be established.

Predissociation to neutral products has also been investigated, primarily by fluorescence of one of the excited products. H_2 predissociation both below and above the ionization limit was monitored by L_α fluorescence.[270-273] By photolysing the gas with dispersed light and comparing the emission intensities to the total absorption, the fraction of absorbed molecules which decay *via* predissociation [equation (37)] can be determined. Some of the possible decay paths are given in equations (34)—(38).

$$H_2 + h\nu \begin{array}{l} \nearrow H_2^+ + e^- \to H^+ + H + e^- \quad (34) \\ \\ \searrow H_2^* \left\{ \begin{array}{l} \to H_2^+ + e^- \quad (35) \\ \to H^+ + H + e^- \quad (36) \\ \to H^* + H \quad (37) \\ \to H^+ + H^- \quad (38) \end{array} \right. \end{array}$$

The decay path was found to depend strongly on the type of level excited. Some rotational lines above the IE appear to decay exclusively by predissociation. The results have been analysed in terms of perturbation theory.[272,274]

Böse and Linder[275] examined some of the same H_2 states by EI using an electron–photon coincidence technique to look at H^* production. A comparison of the photon[273] and electron[275] impact results suggests that triplet states are produced with far greater probability in EI. Other EI studies observed fluorescence from $H(2s)$ by electric field quenching, a necessary perturbation to induce the metastable $2s$ level to fluoresce.[276] Certain states produced by PI have been found to fluoresce from the molecular state.[277] A Birge–Sponer plot of this $3p\Pi\ ^1\Pi_u^-$ state shows that molecular fluorescence is observed up to the last bound level ($v = 17$). The complexity of these results for a molecule as simple as H_2 dramatically demonstrates how complicated these processes can be at intermediate energies. It may be that at higher energy the problem simplifies because relatively slow reactions such as predissociation to neutrals cease to play a role.

---

[267] P. T. Murray and T. Baer, *Int. J. Mass Spectrom. Ion Phys.*, 1979, **30**, 165.
[268] W. B. Peatman, B. Gotchev, P. Gürtler, E. E. Koch, and V. Saile, *J. Chem. Phys.*, 1978, **69**, 2089.
[269] R. Stockbauer, *J. Chem. Phys.*, 1979, **70**, 2108.
[270] M. Glass-Maujean, J. Breton, and P. M. Guyon, *Chem. Phys. Lett.*, 1979, **63**, 591.
[271] P. M. Dehmer and W. A. Chupka, *Chem. Phys. Lett.*, 1980, **70**, 127.
[272] P. M. Guyon, J. Breton, and M. Glass-Maujean, *Chem. Phys. Lett.*, 1979, **68**, 314.
[273] M. Glass-Maujean, K. Köllmann, and K. Ito, *J. Phys. B*, 1979, **12**, L453.
[274] M. Glass-Maujean, *Chem. Phys. Lett.*, 1979, **68**, 320.
[275] N. Böse and F. Linder, *J. Phys. B*, 1979, **12**, 3805.
[276] S. R. Ryan, J. J. Spezeski, O. F. Kalman, W. E. Lamb, jun., L. C. McIntyre, and W. H. Wing, *Phys. Rev.*, 1979, **19A**, 2192.
[277] J. Breton, P. M. Guyon, and M. Glass-Maujean, *Phys. Rev.*, 1980, **A21**, 1909.

Kirby et al.[278] have calculated the H_2 resonant dissociative PI to $H^+ + H$ in the vicinity of 26 eV. The predicted kinetic energy release agrees with the experimental results.

Predissociation in H_2O has been studied by monitoring the L_α and Balmer series fluorescence from the excited H atoms.[279,280] The peak in the fluorescence intensity is at about 700 Å (17.7 eV), well above the IE of 12.6 eV. The cross-section for predissociation to H* at 700 Å is 1.7×10^{-18} cm^2.[279] Wu et al.[281] monitored the production of $O_2^+(\tilde{A}\,^2\Pi_u)$ and $O_2^+(\tilde{b}\,^4\Sigma_g^-)$ during PI by measuring the fluorescence from these states to the ground $\tilde{X}\,^2\Pi_g$ and $\tilde{a}\,^4\Pi_u$ states, respectively. The full range of autoionization structure is evident in these excitation spectra. Autoionization structure is also evident in the total absorption cross-section. Relative total cross-sections also have been measured in the range from 175 to 760 Å for a number of fluorochlorocarbons.[282]

Autoionization is less obvious in electron impact yet it surely exists there as well. Locht et al.[283] have investigated proton production from CH_4^+ and CD_4^+ where H^+ onsets are observed at 21.3 ± 0.3 eV and 22.17 ± 0.1 eV with 0 kinetic energy release. Because the adiabatic IE of the $\tilde{A}\,^2A_1$ state of CH_4^+ is at 22.39 eV, the H^+ onsets must be formed via autoionization. Locht et al.[283] propose the mechanism of equation (39) which implies dissociative autoionization rather than initial formation of an excited CH_4^+ followed by dissociation.

$$CH_4 + e^- \rightarrow CH_4^* + e^- \\ \hookrightarrow Me + H^+ + e^- \qquad (39)$$

Wilden et al.[284] have developed an e–e coincidence experiment to investigate autoionization in EI more directly. Incident electrons of say 100 eV are used to ionize the gas. Forward-scattered electrons are energy analysed with a 180° dispersive analyser while ejected electrons scattered at 90° and measured in coincidence with the forward-scattered electrons are energy analysed by their TOF. This TOF spectrum is a PES for ions with an energy equal to 100 eV minus the energy of the 180° dispersive analyser. Autoionization states in CO were investigated and analysed in terms of the Bardsley theory of molecular autoionization.[285]

By use of angular distributions in PES for various photoionizing energies, Mintz and Kuppermann[286] investigated autoionization in N_2. They found that the β values for the $\tilde{X}\,\Sigma_g^+$ state exhibit anomalous behaviour. A model based on autoionizing states near 22 eV is used to explain the results and works reasonably well for the $v = 0$ state but fails for the $v = 1$ and 2 states. This anomaly may well be associated with the aforementioned shape resonance in N_2.[262]

[278] K. Kirby, S. Guberman, and A. Dalgarno, J. Chem. Phys., 1979, 70, 4635.
[279] C. Y. R. Wu, E. Phillips, L. C. Lee, and D. L. Judge, J. Chem. Phys., 1979, 70, 601.
[280] J. E. Mental, G. R. Möhlmann, and P. M. Guyon, J. Chem. Phys., 1978, 69, 3735.
[281] C. Y. R. Wu, E. Phillips, L. C. Lee, and D. L. Judge, J. Chem. Phys., 1979, 71, 769.
[282] C. Y. R. Wu, L. C. Lee, and D. L. Judge, J. Chem. Phys., 1979, 71, 5221.
[283] R. Locht, J. L. Olivier, and J. Momigny, Chem. Phys., 1979, 43, 425.
[284] D. G. Wilden, P. J. Hicks, J. Comer, and A. Weingartshofer, J. Phys. B, 1978, 11, 3693.
[285] J. N. Bardsley, Chem. Phys. Lett., 1967, 1, 229; ibid., 1968, 2, 329.
[286] D. M. Mintz and A. Kuppermann, J. Chem. Phys., 1978, 69, 3953.

Autoionizing peak shapes are generally understood in terms of the configuration interaction theory of Fano[287] which is most easily applied to atoms. For molecules in which there are often a number of overlapping resonance states and numerous final channels, the problem can become very complicated.[285] Eland et al.[288] have collected a number of high-resolution PI spectra of resonances in N_2O, COS, and neopentane. They found that often the resonances for parent and daughter ions are very different; a peak in one corresponds to a valley in the other, etc. A relatively simple parametrization is proposed to derive consistent resonance energies and widths independent of which channel is monitored.

Other large molecules investigated are ethylene and *trans*-but-2-ene.[289] A Rydberg series converging to the fourth electronic state is observed in ethylene but the structure is very diffuse.

Multiphoton Ionization.—During the past two years multiphoton ionization (MPI) has expanded into an extraordinarily active field. It is not yet clear what role MPI will have in mass spectrometry but certain aspects such as its selectivity for particular molecules at specific wavelengths may well attract the interests of mass spectroscopists in the next few years.

The experimental approach used by workers in MPI has been primitive by current standards in mass spectrometry. Most workers have used the visible and tunable dye lasers pumped either by the pulsed nitrogen or the more powerful Nd:YAG lasers (10—20 Hz, 5—10 ns pulse width). The simplest detection system for ionization is an ionization region consisting of a parallel plate capacitor from which the ionizing current is measured by an electrometer. Gas pressures are between 1 and 50 Torr. Pulse counting has been used with gas cells at low pressure (less than 10^{-3} Torr) or with molecular jets. Finally, mass analysis with either quadrupoles or by ion time-of-flight has been used in order to understand the MPI process better.

Although the mechanism for MPI in molecules is still uncertain, some aspects are becoming clear. One of the most studied molecules has been NO and Cremaschi et al.[290] have discussed the theory as applied to it. A review of it here will serve as an introduction to this topic. Suppose that four visible photons are needed to excite NO above its ionization energy of 9.27 eV.

$$\sigma = \left| \sum_m \frac{\langle f|\bar{r}|m\rangle}{\Delta E_{m0} - 3h\nu + i\Gamma} \sum_l \frac{\langle m|\bar{r}|l\rangle}{\Delta E_{l0} - 2h\nu + i\Gamma} \sum_k \frac{\langle l|\bar{r}|k\rangle\langle k|\bar{r}|o\rangle}{\Delta E_{k0} - h\nu + i\Gamma} \right|^2 \quad (40)$$

Fourth-order perturbation theory gives the cross-section for four-photon MPI as given in equation (40), where m, l, and k are indices which are summed over all the states, $h\nu$ is the photon energy, Γ is a parameter related to the lifetime or width of an intermediate state, ΔE_{i0} are energy differences between the ground and real intermediate states, and $\langle i|\bar{r}|j\rangle$ are matrix elements between the bound or continuum states. If $nh\nu = \Delta E_{i0}$ for any value of $nh\nu$, the denominator becomes very small and the cross-section becomes large. This situation occurs when $nh\nu$ is in resonance with one of the excited states of the molecule. The difference

---

[287] U. Fano, *Phys. Rev.*, 1961, **124**, 1866.
[288] J. H. D. Eland, J. Berkowitz, and J. E. Monahan, *J. Chem. Phys.*, 1980, **72**, 253.
[289] K. V. Wood and J. W. Taylor, *Int. J. Mass Spectrom. Ion Phys.*, 1979, **30**, 307.
[290] P. Cremaschi, P. M. Johnson, and J. L. Whitten, *J. Chem. Phys.*, 1978, **69**, 4341.

between resonant and non-resonant MPI can be as large as 1000 so that the non-resonant signal often is not measurable above the noise. It is precisely for this reason that MPI can be selective in ionizing particular molecules at a given wavelength. Crance[291] and Geltman[292] have treated MPI of atoms in the vicinity of resonances. Selection rules in MPI are very different from those of the equal-energy one-photon transitions. Grynberg[293] has discussed selection rules and line intensities in three-photon absorption.

Equation (40) is a starting point for understanding molecular MPI when the laser pulse is extremely short. However, most pulsed laser systems used for MPI have pulses of the order of 5—10 ns. We have then the fundamental question as to whether each MPI process takes place in 10^{-15} s, or whether long-lived excited intermediate neutral states are formed which absorb additional photons sometime during the 10^{-8} s laser pulse. This question is not easily answered because lasers of variable pulse width in this time domain do not generally exist. However, in one experiment, Avouris et al.[294] showed that the MPI process in nitromethane is a function of peak power rather than total laser fluence. This was shown by noting that the ion signal is ten times greater with the 200 ns laser pulse than with the equal number of photons but 1 μs long laser pulse. Other relevant experiments using two different lasers,[295–299] some with the second laser pulse delayed from the first,[300,301] have been done.

If the intermediate state has a sufficiently long lifetime, then an appropriate dynamical description of the MPI process is one based on rate equations. For a four-photon process resonant at the three-photon level (3 + 1 resonantly enhanced MPI or REMPI) the simplest set of rate equations is given by equations (41)—(43), where X, R, and C are the ground, resonant, and continuum or ionized states, I is the photon intensity, and σ_I and σ_{II} are cross-sections connecting the ground and resonant, and resonant and continuum states.

$$\frac{dX}{dt} = -\sigma_I I^3 X + \sigma_I I^3 R \qquad (41)$$

$$\frac{dR}{dt} = \sigma_I I^3 X - \sigma_I I^3 R - \sigma_{II} I R \qquad (42)$$

$$\frac{dC}{dt} = \sigma_{II} I R \qquad (43)$$

This set of equations predicts that overall power dependence at intermediate laser intensities is third order, i.e. the rate-determining step is formation of the resonant

---

[291] M. Crance, J. Phys. B, 1980, 13, 101.
[292] S. Geltman, J. Phys. B, 1980, 13, 115.
[293] G. Grynberg, J. Phys., 1979, 40, 965.
[294] P. Avouris, I. Y. Chan, and M. M. T. Loy, J. Chem. Phys., 1979, 70, 5315.
[295] A. Herrmann, S. Leutwyler, E. Schumacher, and L. Wöste, Helv. Chim. Acta, 1978, 61, 453.
[296] A. Herrmann, M. Hofmann, S. Leutwyler, E. Schumacher, and L. Wöste, Chem. Phys. Lett., 1979, 62, 216.
[297] P. Esherick and R. J. M. Anderson, Chem. Phys. Lett., 1980, 70, 621.
[298] D. H. Parker and M. A. El-Sayed, Chem. Phys., 1979, 42, 379.
[299] A. D. Williamson and R. N. Compton, Chem. Phys. Lett., 1979, 62, 295.
[300] D. M. Lubman, R. Naaman, and R. N. Zare, J. Chem. Phys., 1980, 72, 3034.
[301] U. Boesl, H. J. Neusser, and E. W. Schlag, J. Chem. Phys., 1980, 72, 4327.

level. As Zakheim and Johnson[302] point out, the intermediate state may fluoresce or predissociate with a rate, k. To account for this, equation (42) should be written as shown in (44).

$$\frac{dR}{dt} = \sigma_I I^3 X - \sigma_I I^3 R - \sigma_{II} IR - kR \qquad (44)$$

In this case, the rate equations become more complicated and the power dependences will not be an integral number. In fact at low laser intensity, a $2 + 2$ REMPI process will have a fourth-order power dependence. At intermediate laser intensities the power dependence reduces to 2 while at high laser intensity it drops still further due to saturation. The spacial and temporal structure of the laser beam can also affect the measured power dependence.[302,303] Eberly and O'Neil[304] have treated the problem of coherence and incoherence in two-photon ionization $(1 + 1)$ and have described conditions for the validity of the rate equation approach. In a 'simple model for MPI', Swain[305] found that if the laser has sufficiently broad line width, rate equations are appropriate.

Experimental evidence for fluorescence was obtained by Parker and Avouris[306] who studied the $2 + 2$ REMPI process in 1,4-diazabicyclo[2,2,2]octane (DABCO). The fluorescence excitation spectrum (FES), obtained by monitoring the fluorescence *versus* exciting laser wavelength, and the MPI spectrum are shown in Figure 3. By use of interference filters, it was determined that the fluoresence was between 280 and 350 nm, just twice the exciting energy. From this analysis and the similarity between the FES and MPI spectra, the authors

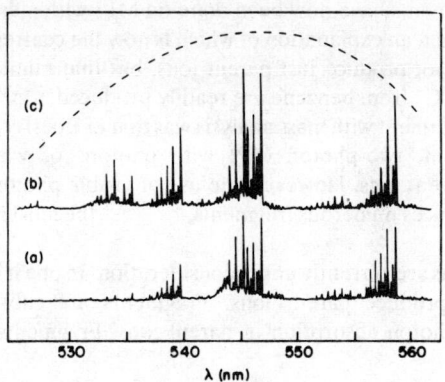

Figure 3 (a) *The two-photon fluorescence excitation spectrum of the* $\tilde{A} \leftarrow \tilde{X}$ *transition of DABCO.* (b) *The corresponding multiphoton ionization spectrum.* (c) *The dye-laser intensity profile*
(Reproduced with permission from *J. Chem. Phys.*, 1979, **71**, 1241)

---

[302] D. S. Zakheim and P. M. Johnson, *Chem. Phys.*, 1980, **46**, 263.
[303] P. Zoller and P. Lambropoulous, *J. Phys. B*, 1980, **13**, 69.
[304] J. H. Eberly and S. V. O'Neil, *Phys. Rev.*, 1979, **19A**, 1161.
[305] S. Swain, *J. Phys. B*, 1979, **12**, 3201.
[306] D. H. Parker and P. Avouris, *J. Chem. Phys.*, 1979, **71**, 1241.

concluded that the DABCO REMPI process is 2 + 2. Similar results on aniline lead Brophy and Rettner[307] to conclude that the rate equation approach is an appropriate one.

The depletion of the intermediate state by predissociation has a marked effect on the MPI spectrum (ion signal *versus* photon energy). Heath et al.[308] have compared the one-photon u.v. spectrum of a number of polycyclic hydrocarbons such as naphthalene, cyclohexane, and *trans*-decalin with the 2 + 1 REMPI spectra. The peaks in the two spectra are at the same energies but the intensities are often different and the widths in the MPI spectra are generally narrower. This sharper structure is due not only to the inherently high resolution of the lasers used in MPI, but also to the fact that, in MPI, long-lived states are preferentially ionized, *i.e.* in a rate equation model, short-lived valence states are depleted relative to the longer-lived Rydberg states, and thus the ion signal from the latter is enhanced. This property makes REMPI a useful tool for investigating Rydberg states, or for differentiating a Rydberg from a valence state. One such investigation involved the previously unobserved Rydberg states in fluorobenzene.[309]

Much of the REMPI work on molecules has been done by spectroscopists interested in using the ionization process simply as a tool for monitoring absorption to the resonant level. Thus, tortional modes in Rydberg states of tetrachloroethylene,[310] a new state in NH_3,[311] butadiene,[312] alkyl iodides,[313] I_2,[314] hexatriene,[315] and pyridine and pyrazine[316] have been investigated primarily from the spectroscopic point of view. Different selection rules for two- and one-photon absorption have allowed new states to be discovered.

Of interest to mass spectrometrists are the fragment ions observed in MPI. A number of experiments have now been done on MPI with mass analysis of ions. The remarkable fact, an explanation of which is now the centre of heated debate, is that MPI does not produce just parent ions, but that numerous high-energy fragments such as C^+ from benzene are readily produced. One of the first, if not the first, MPI experiment with mass analysis was that of Boesl et al.[317] on benzene. In that experiment, two-photon MPI with photons of wavelength 2500 Å produced only parent ions. However, the use of visible photons with three and four photons produces numerous fragments,[301,318,319] the most intense of which is sometimes C^+.

Two mechanisms are currently under consideration. In one it is assumed that all ionization steps produce parent ions. Fragments are subsequently formed through further photon absorption of parent ions. Fragments, especially open

---

[307] J. H. Brophy and C. T. Rettner, *Chem. Phys. Lett.*, 1979, **67**, 351.
[308] B. A. Heath, N. A. Kuebler, and M. B. Robin, *J. Chem. Phys.*, 1979, **70**, 3362.
[309] K. Krogh-Jespersen, R. P. Rava, and L. Goodman, *Chem. Phys.*, 1980, **47**, 321.
[310] B. A. Heath and M. B. Robin, *J. Am. Chem. Soc.*, 1980, **102**, 1796.
[311] G. C. Nieman and S. D. Colson, *J. Chem. Phys.*, 1979, **71**, 571.
[312] J. O. Berg, D. H. Parker, and M. A. El-Sayed, *J. Chem. Phys.*, 1978, **68**, 5661.
[313] D. H. Parker, R. Pandolfi, P. R. Stannard, and M. A. El-Sayed, *Chem. Phys.*, 1980, **45**, 27.
[314] K. K. Lehmann, J. Smolarek, and L. Goodman, *J. Chem. Phys.*, 1978, **69**, 1569.
[315] D. H. Parker, J. O. Berg, and M. A. El-Sayed, *Chem. Phys. Lett.*, 1978, **56**, 197.
[316] J. O. Berg, D. H. Parker, and M. A. El-Sayed, *Chem. Phys. Lett.*, 1978, **56**, 411.
[317] U. Boesl, H. J. Neusser, and E. W. Schlag, *Z. Naturforsch.*, *Teil A*, 1978, **33**, 1546.
[318] L. Zandee and R. B. Bernstein, *J. Chem. Phys.*, 1979, **70**, 2574.
[319] L. Zandee and R. B. Bernstein, *J. Chem. Phys.*, 1979, **71**, 1359.

shell ions,[320] can in turn absorb more photons and dissociate. In support of this, Fisanick et al.[320] state that most open-shell ions have low-lying excited states which can absorb the visible light of the laser. More direct evidence has been claimed by Boesl et al.[301] A 2 + 2 REMPI process in benzene with 5043.8 Å photons produces numerous fragments, while the two-photon ionization with 2416.1 Å u.v. photons gives only parent ions. However, photons with just one half of the latter energy (4832.2 Å) not being in resonance with a benzene state produce no measurable ionization. Yet when the u.v. and the 4832.2 Å lasers were both used, a large number of fragments was produced. This occurred even when the visible laser pulse was delayed by 17 ns and certainly shows that $C_6H_6^+$ ions were photodissociated. From this result the authors conclude that the total REMPI process in benzene proceeds in this fashion but the results are not conclusive because it was not determined whether photodissociation of the $C_6H_6^+$ ions accounts for 100% of the REMPI process taking place at 5043.8 Å. In fact, the mass spectra from the 5043.8 Å and 2416.1 + 4832.2 Å processes are quite different. The main peak in the former is the C^+ while mass 12 in the latter mass spectrum is only a few percent of the major peak.

In the other mechanism under consideration, it is assumed that the molecule absorbs photons faster than it can ionize and that highly excited autoionization states are produced. At each level there is a competition between absorption of another photon and ionization. Data of Lubman et al.,[300] in which a first laser pulse was followed by a second one, delayed by between 10 and 1000 ns, support this mechanism, because no MPI signal was observed unless the laser pulses overlapped. The results of other workers[318,319] also tend to support this mechanism, but the evidence is less compelling.

The most direct test of the two mechanisms will undoubtedly come from photoelectron spectra (PES) in MPI. If the first mechanism is correct, only low-energy electrons should be produced, while if a climbing of the autoionization tree takes place, some high-energy electrons will be observed. Such experiments are beginning to be done[321] and the preliminary results support the first mechanism but it may well be that all REMPI processes are not alike and that some molecules react primarily via the first mechanism while others react via the second. The fact that cross-sections for REMPI differ by as much as they do [e.g. $\sigma(NO) \approx 100\ \sigma(H_2S)$] suggests that a single mechanism will not apply to all processes.

In an effort to understand better the identity of the intermediate state and the REMPI process, a number of workers have utilized supersonic expansion nozzles.[295,296,302,322–325] These cooled beams decrease the number of hot bands and greatly simplify the analysis. The MPI of clusters of Li_2[325] and Na_n[295,296] (n = 1—14) have been studied by this method. In the latter experiment, the ionization energies of the metal clusters were determined as a function of cluster size. In the limit of infinite size, the IE should approach the work function of the metal. However, even for n = 14 the cluster IE was 3·5 eV, considerably above

---

[320] G. J. Fisanick, T. S. Eichelberger, B. A. Heath, and M. B. Robin, *J. Chem. Phys.*, 1980, **72**, 5571.
[321] R. N. Compton, J. C. Miller, A. E. Carter, and P. Kruit, *Chem. Phys. Lett.*, 1980, **71**, 87.
[322] L. Zandee, R. B. Bernstein, and D. A. Lichtin, *J. Chem. Phys.*, 1978, **69**, 3427.
[323] T. G. Dietz, M. A. Duncan, M. G. Liverman, and R. E. Smalley, *Chem. Phys. Lett.*, 1980, **70**, 246.
[324] J. Murakami, K. Kaya, and M. Ito, *J. Chem. Phys.*, 1980, **72**, 3263.
[325] B. P. Mathur, E. W. Rothe, G. P. Reck, and A. J. Lightman, *Chem. Phys. Lett.*, 1978, **56**, 336.

the work function of 2.3 eV. The even-numbered clusters were found to be slightly more stable than the odd ones. Flad *et al.*[326] carried out calculations for Na_n up to $n = 8$ using a pseudopotential scheme and a density functional approximation for the correlation energy. They calculated the IE's and found good agreement with the experimental results. An interesting alternative to spectrum simplification by gas cooling is to use a two-laser experiment in which the first laser excites the molecule to a well defined state from which MPI can take place.[297,299]

Other, more esoteric, MPI experiments have involved i.r. laser MPI of nitromethane, methanol, *etc.* in which a CO_2 laser was used.[327] However, it seems that this ionization is aided by collisional processes. Collisions surely also play a role in the MPI of liquid benzene studied by Scott *et al.*[328] Finally, the influence of nuclear spin on the angular distribution and polarization of photoelectrons in two-photon ionization of Na has been reported.[329]

4 Unimolecular Fragmentation

Advances in Unimolecular Decay Theory.—The statistical theory (RRKM or QET) has formed the conceptual framework for our understanding of unimolecular reactions since its inception in the late 1920's. Its basic assumption is that energy flow among all the vibrational degrees of freedom is random and much faster than dissociation, so that the lifetime of an activated ion or molecule depends only on the total energy and angular momentum but not on the manner in which it was excited. These concepts have been reviewed recently for neutral systems by Oref and Rabinovitch.[330] As pointed out by Kay,[331] some of the original restrictions of the theory are very severe. In particular, it is generally assumed that the interaction matrix elements involving the exit state are statistically uncorrelated. This means that the exit states (those states leading to dissociation) are equally accessible from all regions in the quasi-bound phase space. This runs counter to our intuition as well as to Franck–Condon factors, because 'the interaction matrix elements are likely to be large only when the classical turning points for relative motion of the fragments in the initial and final states coincide'.[331] Kay relaxes this assumption in this paper, and finds that an RRKM type expression emerges which in fact converges to the usual RRKM theory in the classical limit. The intramolecular vibrational relaxation rates are difficult to determine. However, Beck *et al.*[332] have succeeded in doing so by measuring the linewidths of laser-induced fluorescence (LIF) signals in supersonically cooled naphthalene. In the low-lying vibrational states (≈ 500 cm^{-1}), the relaxation rates are slow. But at higher energies (*e.g.* 4300 cm^{-1}) peaks broaden, indicating that the lifetimes are $\approx 2 \times 10^{-12}$ s. In virtually all unimolecular reactions, the energy is far greater than 5000 cm^{-1} so that one can expect the relaxation times to be considerably shorter than 2×10^{-12} s.

[326] J. Flad, H. Stoll, and H. Preuss, *J. Chem. Phys.*, 1979, **71**, 3042.
[327] P. Avouris, I. Y. Chan, and M. M. T. Loy, *J. Chem. Phys.*, 1980, **72**, 3522.
[328] T. W. Scott, A. J. Twarowski, and A. C. Albrect, *Chem. Phys. Lett.*, 1979, **66**, 1.
[329] M. P. Strand, J. Husen, R. L. Chien, and R. S. Berry, *Chem. Phys. Lett.*, 1978, **59**, 205.
[330] I. Oref and B. S. Rabinovitch, *Acc. Chem. Res.*, 1979, **12**, 166.
[331] K. G. Kay, *J. Chem. Phys.*, 1978, **69**, 434.
[332] S. M. Beck, D. L. Monts, M. G. Liverman, and R. E. Smalley, *J. Chem. Phys.*, 1979, **70**, 1062.

The transition state has long been an invaluable, yet embarrassing, species. It is embarrassing because it cannot be experimentally observed, nor its frequencies or structure determined. Yau and Pritchard[333] have reformulated the unimolecular reaction theory (for thermalized systems) so that the transition state no longer enters as a real species. However, the degrees of freedom or adjustable parameters are associated with the molecule (or ion) in 'reactive' and 'nonreactive' regions. Alternatives to the statistical theory are extremely difficult to formulate, except when applied to triatomics which are the simplest molecules of interest in unimolecular reactions. Some effort has been devoted to investigating full quantum mechanical theories.[334,335] An R-matrix method[334] has been proposed which is reasonably efficient. Non-exponential decay for molecules selected in a pure state has been shown to be the rule rather than the exception. Kulander and Heller[335] have investigated the dissociation of H_3^+, a problem involving three potential surfaces, but the problem was greatly simplified by ignoring the coupling of these surfaces by nuclear motion and the breakdown of the Born–Oppenheimer approximation. Another simplification is the assumption of an isosceles triangle configuration (C_{2v}) for H_3^+ during the dissociation. Thus, H_3^+ was assumed to absorb light into one of the two excited states and to dissociate on that surface. Final vibrational states of H_2^+ as a function of photon energy (12—30 eV) were calculated.

Unimolecular rate theory has received an enormous impetus during the past few years from experiments on the lifetimes of van der Waals molecules such as HeI_2[336,337] and $Cl_2\cdots Cl_2$.[338] These complexes are particularly significant for the progress of the theory because they represent an extreme situation in which, for instance, the I—I stretching frequency is very large compared to that of the I_2—He bond, and the coupling between the I_2 and He groups is extremely weak. Non-statistical effects would therefore be likely. Lifetimes as long as 10^{-4} s have been measured for activated $Cl_2\cdots Cl_2$ which is far longer than can be accounted for by RRKM theory. Beswick and Jortner[339–343] have modelled the dimer with van der Waals type potentials and have solved the corresponding Schrödinger equation which consists of a set of close coupling equations. Various models including linear and T-shaped complexes have been considered. An energy gap law [equation (45)] was proposed which relates the lifetime, τ, to the energy mismatch, ε, i.e. the energy released as kinetic energy.

$$\tau = A \exp(-\xi \varepsilon) \tag{45}$$

This is actually not unlike QET in which the low kinetic energy releases are more probable.

---

[333] A. W. Yau and H. O. Pritchard, Can. J. Chem., 1978, **56**, 1389.
[334] R. W. Numrich and K. G. Kay, J. Chem. Phys., 1979, **70**, 4343.
[335] K. C. Kulander and E. J. Heller, J. Chem. Phys., 1978, **69**, 2439.
[336] K. E. Johnson, L. Wharton, and D. H. Levy, J. Chem. Phys., 1978, **69**, 2719.
[337] J. E. Kenny, D. V. Brumbaugh, and D. H. Levy, J. Chem. Phys., 1979, **71**, 4757.
[338] D. A. Dixon and D. R. Hershbach, Ber. Bunsenges. Phys. Chem., 1977, **81**, 145.
[339] J. A. Beswick and J. Jortner, J. Chem. Phys., 1978, **68**, 2277; ibid., 1978, **69**, 512.
[340] J. A. Beswick and J. Jortner, Chem. Phys. Lett., 1979, **65**, 240.
[341] J. A. Beswick and J. Jortner, J. Chem. Phys., 1979, **71**, 4737.
[342] J. A. Beswick and J. Jortner, Mol. Phys., 1980, **39**, 1137.
[343] J. A. Beswick, G. Delgado-Barrio, and J. Jortner, J. Chem. Phys., 1979, **70**, 3895.

A simple physical model in terms of the reduced mass and the energy difference, ε, has also been developed.[344] Woodruff and Thompson[345] used the classical analogue of the Hamiltonian of Beswick and Jortner in carrying out classical trajectories on the I_2He dissociation and, in all cases, exponential decay was found. The results agree extremely well with quantum results which suggests that this is the appropriate approach to extending the calculations to three dimensions as well as more complex molecules. The experimental and classically and quantum mechanically calculated lifetimes of I_2He for I_2 in $v = 25$ are: 5×10^{-10} s, 3×10^{-10} s, and 2.5×10^{-10} s, respectively. The Beswick–Jortner model appears to be less satisfactory for describing the dynamics of $(N_2O)_2$ dimers.[346] The additional degrees of freedom in $(N_2O)_2$ may require a modification of equation (45). Finally, scaling theory has been used to characterize the vibration-to-translation de-excitation rate as a function of the transition frequency for I_2He.[347]

A viable alternative to RRKM/QET for intermediate-sized molecules is that of classical trajectory calculations. Polanyi and Sathyamurthy[348,349] have continued their investigation of the effect of barrier position on kinetic energy release. Hase[350] has used this approach to study the relationship between lifetime and relative translational energy distributions in HCCCl. This relationship is not simple because in energy regions where non-RRKM lifetime distributions are predicted the distribution of kinetic energy remains statistical. The implication is that kinetic energy release distributions cannot be used as a test of the fundamental hypothesis of random energy flow. Classical trajectors were also used to study H loss in the Et radical. Intrinsically non-RRKM behaviour was found.[351] The role of the potential surface in HCC dissociation was investigated.[352]

The virtue of the RRKM theory is its ease of use because of its insensitivity to details of the reaction path. Its drawback is that very little information about the dissociation dynamics can be extracted. On the other hand, trajectory calculations and quantum mechanical theories provide a great deal of dynamical information. However, its relevance to real systems containing a number of atoms (e.g. more than 5) is often questionable unless the dissociation path has been thoroughly explored by good calculations. Extensive calculations have been done by Lorquet et al.[353–355] using CI on the dissociation of H_2CO^+ and CH_2^+. The dissociation of excited H_2CO^+ from the $\tilde{A}$ state involves a predissociation yet the experimental onset for loss of H_2 is observed at the thermochemical dissociation limit. The calculations show that, by the use of conical intersections, a low-energy path via a C_s symmetry is possible.[353] The conical intersections arise from two surfaces

[344] G. E. Ewing, J. Chem. Phys., 1979, **71**, 3143.
[345] S. B. Woodruff and D. L. Thompson, J. Chem. Phys., 1979, **71**, 376.
[346] T. E. Gough, R. E. Miller, and G. Scoles, J. Chem. Phys., 1978, **69**, 1588.
[347] R. Ramaswamy and A. E. DePristo, J. Chem. Phys., 1980, **72**, 770.
[348] J. C. Polanyi and N. Sathyamurthy, Chem. Phys., 1978, **33**, 287.
[349] J. Polanyi and N. Sathyamurthy, Chem. Phys., 1979, **37**, 259.
[350] W. L. Hase, Chem. Phys. Lett., 1979, **67**, 263.
[351] W. L. Hase, R. J. Wolf, and C. S. Sloane, J. Chem. Phys., 1979, **71**, 2911.
[352] R. J. Wolf and W. L. Hase, J. Chem. Phys., 1980, **72**, 316.
[353] M. Vaz Pirez, C. Galloy, and J. C. Lorquet, J. Chem. Phys., 1978, **69**, 3242.
[354] C. Galloy and J. C. Lorquet, Chem. Phys., 1978, **30**, 169.
[355] M. Desouter-Lecomte, C. Galloy, J. C. Lorquet, and M. Vaz Pirez, J. Chem. Phys., 1979, **71**, 3661.

which repel each other to form adiabatic surfaces. The origin of the interaction is similar to the Jahn–Teller effect except that the latter arises from an exact degeneracy while the former arises from an accidental degeneracy.[355] The interactions between adiabatic surfaces have been treated in terms of the Massey function, $T = \Delta E/\hbar v g$, where g (in units of cm^{-1}) is a coupling parameter and v is the relative velocity of the two fragments.[356] The model is similar to the Landau–Zener approximation.

Angular momentum constraints in unimolecular dissociations were investigated experimentally by Meisels et al.[202] The product ion distribution from the dissociation of state-selected $C_4H_8^+$ produced by photoion–photoelectron coincidence (PIPECO) was compared with the products of the $C_2H_4^+ + C_2H_4$ ion–molecule reaction at the same total energy. In the latter, the angular momentum was considerably larger. As a result, the branching ratios $C_4H_7^+/C_3H_5^+$ at a given energy were 0.39 and 0.11, respectively, for the unimolecular and bimolecular case. These ratios were calculated to be 0.42 and 0.11 using the RRKM theory with angular momentum conservation. The rates of production of $C_3H_5^+$ and $C_4H_7^+$ have been calculated explicitly as a function of J, the angular momentum, by Klots[357] and are shown to vary in opposite directions. The $C_3H_5^+$ path has a higher rate at $J = 0$ and increases with increasing J. Angular momentum was also investigated for radiationless transitions by Novak and Rice.[358] The symmetric top molecule was assumed to have the same symmetry in the two electronic states, which has the effect of keeping the vibrations and rotations separated during the relaxation. They conclude that the initial rotational level influences the radiationless transition rate, in agreement with experimental results of Howard and Schlag.[359]

An important paper has been published describing corrections to unimolecular rate constants due to tunnelling.[360] The application is to neutral formaldehyde dissociating to $H_2 + CO$, a channel associated with a significant reverse activation energy. Using an accurate SCF-CI calculated surface,[361] and an Eckart potential tunnelling probability, Miller obtained the rate expression (46):

$$k(E) = \frac{(s-1)!\prod_{i=1}^{s}\hbar\omega_i}{2\prod \hbar E^{s-1}} \sum_n P[E - V_0 - \hbar\omega^\ddagger(n + \tfrac{1}{2})] \qquad (46)$$

in which

$$\omega^\ddagger(n + \tfrac{1}{2}) = \sum_{i=1}^{s-1} \omega_i^\ddagger(n_i + \tfrac{1}{2}) \qquad (47)$$

and P is the tunnelling probability. The functional form of P for various tunnelling models is discussed. Near threshold, the rate without tunnelling is $10^9\,s^{-1}$ and drops to zero just at the onset but, with tunnelling, the rate is still $10^6\,s^{-1}$ at an energy 33 kJ below the onset. Above the onset, the rates with and without tunnelling are about the same. It is evident from these results that it is extremely

---

[356] M. Desouter-Lecomte and J. C. Lorquet, *J. Chem. Phys.*, 1979, **71**, 4391.
[357] C. E. Klots in ref. 53, p. 69.
[358] F. A. Novak and S. A. Rice, *J. Chem. Phys.*, 1979, **71**, 4680.
[359] W. E. Howard and E. W. Schlag, *Chem. Phys.*, 1976, **18**, 123; *J. Chem. Phys.*, 1978, **68**, 2679.
[360] W. H. Miller, *J. Am. Chem. Soc.*, 1979, **101**, 6810.
[361] J. D. Goddard and H. F. Schaefer, *J. Chem. Phys.*, 1979, **70**, 5117.

difficult to determine the energy barrier experimentally in the case of dissociations with large reverse activation barriers.

The relationship between the energy and structure of the transition state was treated in a qualitative manner by Hammond[362] in 1955. The basic idea is very simple. If the transition state has the same energy as say the products of a reaction, then the structure of the transition state is close to that of the products. This idea has been made quantitative by two independent workers, Agmon[363] and Miller.[364] According to the Hammond–Agmon–Miller (HAM) theory, the structure of the reactants during the course of the reaction is described in terms of a single parameter, the bond order, n. In terms of the energies shown in Figure 4,

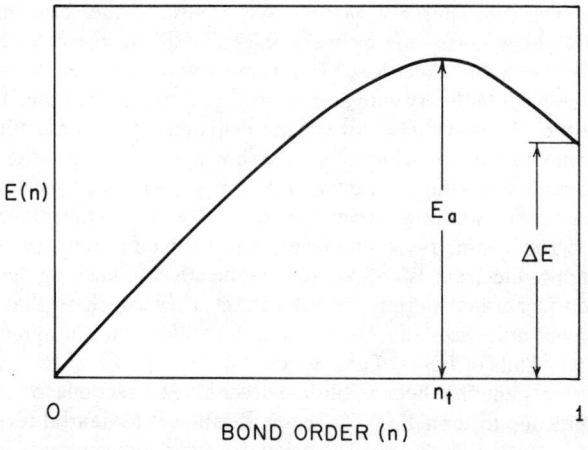

Figure 4 *The relationship between the bond order at the transition state, n_t, and the energies E_a and ΔE according to the HAM theory*[362–364]

the bond order at the transition state, n_t, is given by equation (48). The derivation of this equation is based on the minimization of the reaction path between reactants and products.

$$n_t = \frac{1}{2 - (\Delta E/E_a)} \qquad (48)$$

Another aspect to carrying out RRKM/QET calculations is the problem of the statistical factor or the reaction multiplicity. Coulson[365] has discussed this in some detail. Finally, Alexandra[366] has analysed the mass spectral fragmentation processes of normal hydrocarbons using quantum field theory. The results suggest that the mass spectra of the larger hydrocarbons should be similar.

---

[362] G. S. Hammond, *J. Am. Chem. Soc.*, 1955, **77**, 334.
[363] N. Agmon, *J. Chem. Soc., Faraday Trans. 2*, 1978, 388.
[364] A. R. Miller, *J. Am. Chem. Soc.*, 1978, **100**, 1984.
[365] D. R. Coulson, *J. Am. Chem. Soc.*, 1978, **100**, 2992.
[366] G. Alexandru, *Int. J. Mass Spectrom. Ion Phys.*, 1980, **32**, 369.

Ion Lifetimes, Kinetic Energy Release, Breakdown Diagrams.—Photoion-photoelectron coincidence (PIPECO) studies have continued to play a major role in advancing our understanding of ion dynamics. Powis et al.[367] have described their new PIPECO spectrometer which incorporates an effusive jet that cools the translational energy perpendicular to the gas flow to 12 K. This allows for a greatly improved resolution in determining the kinetic energy released upon dissociation by the ion TOF technique. Kinetic-energy release distributions (KERD's) were determined for acetone ion dissociation[368] and CF_3^+ production from the ground states of CF_3Cl^+ and CF_3Br^+.[369] In both cases, good agreement between experiment and the statistical theory (Klots version)[370] were noted. The acetone ion dissociation has now become a standard as it has been measured in four different laboratories,[368,371] and good agreement among the three earlier experimental results[371] has been noted.[372] However, the average release energy is considerably higher than the theoretically predicted one. No definitive resolution of this discrepancy has been advanced.

The excited states of CF_3X^+ (X = F, Cl, or Br) dissociate directly as shown by very non-statistical KERD's.[373] A most interesting observation is that the production of CF_3^+ from the $\tilde{A}$ state of CF_3Cl^+ at 15.4 eV produces a KERD peak at about 1 eV with hardly a trace of low-energy dissociation products when a Ne(I) light source is used. With the He(I) source, the same ion internal energy results in a bimodal KERD, indicating perhaps two dissociation paths. Evidently in one or the other case, autoionization forms a different state (but of the same energy). This is *prima facie* evidence for non-statistical behaviour because the KERD depends on how the energy is initially deposited in the ion and is a particularly intriguing problem because only one dissociation limit exists at 15.4 eV and whether autoionization contributes to the Ne(I) or the He(I) case is not clear.

Another unresolved problem was encountered in $COCl_2^+$ and COF_2^+ dissociations.[374] The dissociation limit to form $COCl^+$ almost coincides with the ionization energy yet the resulting KERD's are very non-statistical *and* the ion is metastable. The COF^+ onset is about 1.4 eV above the IE, again too small an activation energy to make the COF_2^+ ion metastable on the basis of RRKM theory, yet the ion does exhibit metastable characteristics and the KERD's are non-statistical. In a rather difficult experiment because of the unfavourable daughter ion to parent ion mass ratio, Powis investigated the KERD's for the dissociation of CH_4^+ and CD_4^+.[375] Previous work reported only the average release energy.[376] The KERD's were also analysed using a surprisal by plotting

---

[367] I. Powis, P. I. Mansell, and C. J. Danby, *Int. J. Mass Spectrom. Ion Phys.*, 1979, **32**, 15.
[368] I. Powis and C. J. Danby, *Int. J. Mass Spectrom. Ion Phys.*, 1979, **32**, 27.
[369] I. Powis and C. J. Danby, *Chem. Phys. Lett.*, 1979, **65**, 390.
[370] C. E. Klots, *J. Chem. Phys.*, 1976, **64**, 4269.
[371] C. S. T. Cant, J. C. Danby, and J. H. D. Eland, *J. Chem. Soc., Faraday Trans.* 3, 1975, **71**, 1015; R. Stockbauer, *Int. J. Mass Spectrom. Ion Phys.*, 1977, **25**, 89; D. M. Mintz and T. Baer, *Int. J. Mass Spectrom. Ion Phys.*, 1977, **25**, 39.
[372] T. Baer in ref. 118, Ch. 5, p. 153.
[373] I. Powis, *Mol. Phys.*, 1980, **39**, 311.
[374] K. M. Johnson, I. Powis, and C. J. Danby, *Int. J. Mass Spectrom. Ion Phys.*, 1979, **32**, 1.
[375] I. Powis, *J. Chem. Soc., Faraday Trans.* 2, 1979, **75**, 1294.
[376] R. Stockbauer, *Int. J. Mass Spectrom. Ion Phys.*, 1977, **25**, 401.

$I(\varepsilon_t)$ according to equation (49) in which $P(\varepsilon_t)$ is the KERD, $P°(\varepsilon_t)$ is the statistically expected KERD, and ε_t is the energy released in translation.

$$I(\varepsilon_t) = \ln \frac{P(\varepsilon_t)}{P°(\varepsilon_t)} \quad (49)$$

In order to place KERD's at all ion internal energies on the same surprisal plot, a reduced energy was defined as $f_t = \varepsilon_t/\varepsilon_{ex}$ where ε_{ex} is the total available energy, the excess of energy. The surprisal plot with this reduced energy is shown in Figure 5. A surprisal of 0.0 means a statistical value. The departure from statistical behaviour is evidently the same for all ion internal energies. However, it should be pointed out that the departure from statistical behaviour is slight in comparison with ions such as CF_3Cl^+.[373]

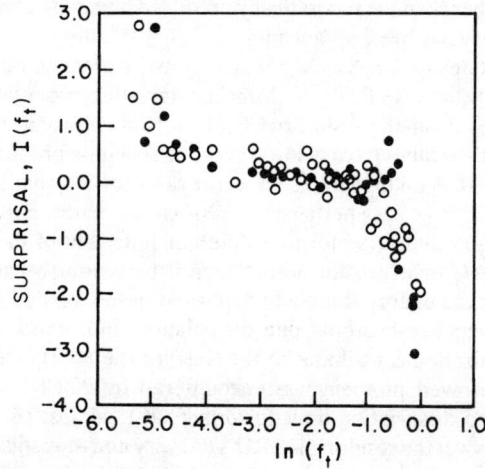

Figure 5 *Kinetic energy distribution surprisals. Open circles*: Me^+ *formation*; *filled circles*: CD_3^+ *formation*
(Reproduced from *J. Chem. Soc., Faraday Trans.* 2, 1979, **75**, 1294)

An extreme departure from the statistically expected results was found for $C_2F_6^+$ dissociation.[377] CF_3^+ production, as shown in the breakdown curve, goes from 1.0 at 15.5 eV to nearly 0.0 and then increases at 17.5 eV to nearly 1.0. The authors considered the possibility that the high-energy source of CF_3 was the consecutive reaction (50).

$$C_2F_6^+ + h\nu \rightarrow C_2F_5^+ + F + e^- \\ \hookrightarrow CF_3^+ + CF_2 \quad (50)$$

However, this explanation was rejected because the average KER of CF_3^+ was found to be only 0.2 eV whereas the $C_2F_5^+$ precursor has a measured KER of 1.4 eV and this is not compatible with only 0.2 eV of release energy. Inghram *et al.*[377] conclude that this is a case of an isolated state, the $\tilde{D}$ state at 18 eV. These

---

[377] M. G. Inghram, G. R. Hanson, and R. Stockbauer, *Int. J. Mass Spectrom. Ion Phys.*, 1980, **33**, 253.

data and their interpretation are consistent with earlier work of Lifshitz[378] and Eland.[379] Another fluorocarbon with an isolated state has been found by Dannacher et al.[380] The $\tilde{C}(^2A'')$ state of vinylfluoride at 16 eV decays almost exclusively to $C_2H_3^+ + F$. At low energies, the rate as a function of ion internal energy was also obtained.

Another halocarbon of considerable interest is 1-chloropropyne. This ion is an isomer of 3-chloropropyne (propargyl chloride). However, these two ions behave very differently. The 1-chloropropyne ion competitively fragments and fluoresces from the $\tilde{A}$ state at 13.2 eV.[381] The measured lifetime is 18 ns.[382] From the branching ratios of fragmentation to $C_3H_3^+$ and fluorescent stabilization, a dissociation rate of $4 \times 10^{-7}\,s^{-1}$ was determined which is consistent with a radiationless decay to the ground electronic state. Two other small ions of interest, $C_2H_2^+$ and H_2S^+, have been studied by Eland.[383] The onset for H loss from $C_2H_2^+$ is still uncertain. The data suggest that at the PI onset of 17.36 eV, which is 0.5 eV higher than the presumed thermochemical onset, the KER is 0.01 eV and is remarkably small, suggesting that the dissociation proceeds directly from the excited state. No other evidence for this inference exists. In H_2S^+, the spin-forbidden formation of $S^+(^4S_u) + H_2(^1\Sigma_g^+)$ is shown to occur almost exclusively from the $\tilde{A}^2A_1$ state. Another small ion, H_2^+, has been studied by PIPECO[384] and a KERD peaking at 0 eV was found at the AE for H^+.

Lifetimes of ions with selected internal energies have been measured for $C_5H_5N^+$ (pyridine),[70] $C_8H_8^+$ isomers,[385] and $C_6H_6^+$ isomers.[71] The last study has laid to rest many of the problems associated with the question of isolated states in benzene. The story began with Andlauer and Ottinger's observation that benzene ions dissociated via two different reactions at two different rates.[386] One path produced $C_3H_3^+$ and $C_4H_4^{+\cdot}$ ions, the other produced $C_6H_4^{+\cdot}$ and $C_6H_5^+$. Although Baer et al.[71] did not actually measure the rate of production of $C_6H_5^+$, indirect measurements and RRKM calculations show that (a) hexa-1,5-diyne and hexa-2,4-diyne rearrange to benzene prior to dissociation and (b) $C_6H_6^+$ ions decay to all channels with a single decay rate. That is, there is no isolated state in benzene. It was also found that, at higher energies, hexa-1,5-diyne does not rearrange to benzene but rather dissociates directly.

Weiss et al.[387] have investigated the dissociation of SO_2^+ in the vicinity of the dissociation limit. The SO^+ onset of 15.965 eV agrees reasonably well with the 16.02 eV based on the IE of SO and the heats of formation of SO_2, SO, and O and we may safely discard the lower onset of 15.81 eV of Dibler and Liston.[388] The

---

[378] C. Lifshitz and F. A. Long, *J. Phys. Chem.*, 1965, **63**, 3746.
[379] I. G. Simm, C. J. Danby, and J. H. D. Eland, *J. Chem. Soc., Chem. Commun.*, 1973, 838; *Int. J. Mass Spectrom. Ion Phys.*, 1974, **14**, 285.
[380] J. Dannacher, A. Schmelzer, J. P. Stadelmann, and J. Vogt, *Int. J. Mass Spectrom. Ion Phys.*, 1979, **31**, 175.
[381] J. Dannacher and J. P. Stadelmann, *Chem. Phys.*, 1980, **48**, 79.
[382] J. P. Maier, O. Marthaler, and E. Kloster-Jensen, *J. Electron Spectrosc. Relat. Phenom.*, 1980, **18**, 251.
[383] J. H. D. Eland, *Int. J. Mass Spectrom. Ion Phys.*, 1979, **31**, 161.
[384] S. Strathdee and R. Browning, *J. Phys. B*, 1979, **12**, 1789.
[385] D. Smith, T. Baer, G. D. Willett, and R. C. Ormerod, *Int. J. Mass Spectrom. Ion Phys.*, 1979, **30**, 155.
[386] B. Andlauer and C. Ottinger, *J. Chem. Phys.*, 1971, **55**, 1471; *Z. Naturforsch., Teil A*, 1972, **27**, 293.
[387] M. J. Weiss, T. C. Hsieh, and G. G. Meisels, *J. Chem. Phys.*, 1979, **71**, 567.
[388] V. H. Dibeler and S. K. Liston, *J. Chem. Phys.*, 1968, **49**, 482.

coincidence results of Weiss et al.[387] indicate that SO_2^+ is observed up to 16.44 eV, consistent with fluorescence data of both Wu and Yencha[389] and more recent results of Tsuji et al.[390] in which SO_2^+ fluorescence was observed to at least 16.26 eV. Fluorescence-stabilized SO_2^+ was not observed by Brehm et al.[391] using the He(I) PIPECO technique. Quite recently, Tsuji et al.[392] repeated the fluorescence measurements and concluded that the previously assumed fluorescence of SO_2^+ is due to the process $SO^+ \tilde{A}\ ^2\Pi \rightarrow \tilde{X}\ ^2\Pi$. More work is clearly called for.

One of the exciting new developments during the past two years has been the measurement of angular distributions of fragment ions in a PIPECO experiment by Eland.[393] In this experiment, a He(I) or He(II) lamp is used. Because the 180° hemispherical energy analyser accepts only electrons ejected into a small solid angle, ions of a particular orientation are selected by the electron ion coincidence condition. It is well known that a delta function energy release in the absence of discrimination by apertures results in an approximately rectangular ion TOF distribution. A departure from such a distribution can be attributed to non-isotropic ejection of fragment ions. Among the ions studied was O_2^+. Guyon et al.[394] had earlier investigated O_2^+ dissociation by threshold PIPECO. Because there is no discrimination against zero-energy electrons, ions measured in coincidence with such electrons are not oriented, except by the circularly polarized photon beam. Hence, far less angularly dependent effects can be observed. These two PIPECO techniques are complementary in that the threshold approach gives correct energy releases while the He(I) or He(II) approach can determine angular distributions once the release energy is known. This symbiosis was utilized by Eland[393] to determine that the $^2\Pi_u$ state of O_2^+ at 24.2 eV, a dissociative state leading to $O^+(^2P) + O(^3P)$ at 23.75 eV,[394] dissociates with a strongly anisotropic angular distribution peaking broadly at $\theta = 90°$. By contrast, the $\tilde{B}\ ^2\Sigma_g^-$ state at 20.5 eV dissociates with an isotropic distribution. Another anisotropic distribution is found in formation of N^+ from $NO^+ \tilde{C}\ ^3\Pi$ at 21.7 eV. The dashed line in Figure 6 is the expected isotropic distribution. At 23 eV the $\tilde{B}\ ^1\Pi$ state evidently dissociates with an approximately isotropic angular distribution.

The aforementioned study by Guyon et al.[394] on O_2^+ investigated the dissociative paths of O_2^+ from 18 eV up to 25 eV. The $\tilde{C}\ ^4\Sigma_u^-$ state was shown to be 99% predissociated and to produce principally $O(^1D) + O^+(^4S°)$. Tanaka and Yoshimine[395] have confirmed this with extensive CI calculations. The upper bound on the $v = 0$ lifetime of 2×10^{-8} s also agrees with the experimental results. The predissociation of N_2^+ has been investigated theoretically by Roche and Tellinghuisen.[396] Although the $\tilde{B}\ ^2\Sigma_u^+$ state at 18.75 eV and the $\tilde{C}\ ^2\Sigma_u^+$ state at 25.3 eV have a large energy separation, they are strongly coupled by the nuclear

[389] K. T. Wu and A. J. Yencha, Can. J. Phys., 1977, **55**, 767.
[390] M. Tsuji, H. Fukutome, K. Tsuji, and Y. Nishimura, Int. J. Mass Spectrom. Ion Phys., 1978, **28**, 257.
[391] B. Brehm, J. H. D. Eland, R. Frey, and A. Küstler, Int. J. Mass Spectrom. Ion Phys., 1973, **12**, 197.
[392] M. Tsuji, C. Yamagiwa, M. Endoh, and Y. Nishimura, Chem. Phys. Lett., 1980, **73**, 407.
[393] J. H. D. Eland, J. Chem. Phys., 1979, **70**, 2926.
[394] P. M. Guyon, T. Baer, L. F. A. Ferreira, I. Nenner, A. Tabché-Fouhaillé, R. Botter, and T. R. Govers, J. Phys. B, 1978, **11**, L141.
[395] K. Tanaka and M. Yoshimine, J. Chem. Phys., 1979, **70**, 1626.
[396] A. L. Roche and J. Tellinghuisen, Mol. Phys., 1979, **38**, 129.

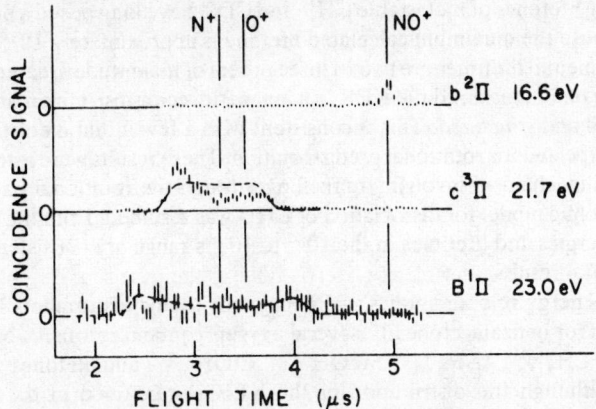

Figure 6 *Photoelectron–photoion coincidence spectra for NO dissociative photoionization using the He(II)$_\alpha$ light. The curves through the data are calculated for formation of $N^+ + O$ ground states with isotropic distributions*
(Reproduced with permission from *J. Chem. Phys.*, 1979, **70**, 2926)

kinetic energy. The calculations indicate that the $\tilde{B}$ state homogeneously predissociates the $\tilde{C}$ state, resulting in a rather unusual isotope effect on the ratio of predissociation to fluorescence. The latter has been observed experimentally.[397]

Breakdown curves of a number of ions have been published. The breakdown curve of MeOH was determined directly by PIPECO[398,399] and by differentiating PI curves.[400] Both studies agree that Me$^+$ is formed *via* an isolated state. Berkowitz[400] also concluded that HCOH$^+$, and not CH$_2$O$^+$, is formed preferentially. The breakdown curve of allene in the vicinity of the first dissociation limit, using time-delay ion extraction, was measured for two different extraction delays.[67] The resulting shift, due to the slow dissociation rate, was found to be 0.030 eV. This is an interesting and useful technique for determining the kinetic shift, as long as it is not too great. Other breakdown diagrams were obtained for BF$_3$,[401] formic acid,[402] methyl ethyl ketone,[403] and isomers of C$_3$H$_4$.[72] Propyne, allene, and cyclopropene have identical breakdown curves, indicating that they isomerize to a common structure prior to dissociation,[72] a result that appears to hold up to 20 eV. However, collisional activation studies have shown that the barrier to isomerization is significant.[404] McKeen and Taylor[405] have calculated the breakdown diagram for ethylene and propane ions and have compared them with experimental results.[406]

[397] T. R. Govers, C. A. van de Runstraat, and F. J. de Heer, *Chem. Phys.*, 1975, **9**, 285.
[398] Y. Niwa, T. Nishimura, H. Nozoye, and T. Tsuchiya, *Int. J. Mass Spectrom. Ion Phys.*, 1979, **30**, 63.
[399] T. Nishimura, Y. Niwa, T. Tsuchiya, and H. Nozoye, *J. Chem. Phys.*, 1980, **72**, 222.
[400] J. Berkowitz, *J. Chem. Phys.*, 1978, **69**, 3044.
[401] C. F. Batten, J. A. Taylor, B. P. Tsai, and G. G. Meisels, *J. Chem. Phys.*, 1978, **69**, 2547.
[402] A. V. Golovin, M. E. Akopyan, F. I. Vilesov, and Yu. L. Sergeev, *High Energy Chem.*, 1979, **13**, 171.
[403] A. V. Golovin, M. E. Akopyan, F. I. Vilesov, and Yu. L. Sergeev, *Theor. Exp. Chem.*, 1979, **15**, 163.
[404] W. Wagner, K. Levsen, and C. Lifshitz, *Org. Mass Spectrom.*, 1980, **15**, 271.
[405] L. W. McKeen and J. W. Taylor, *Int. J. Mass Spectrom. Ion Phys.*, 1980, **33**, 167.
[406] R. Stockbauer and M. G. Inghram, *J. Chem. Phys.*, 1976, **65**, 4081.

The long lifetimes of metastable CH_4^+ and CD_4^+ have long posed a problem for QET because the maximum calculated lifetime is approximately 10^{-8} s whereas the experimental lifetimes are two to three orders of magnitude greater than that. Flamme et al.[407] measured the KER in a magnetic sector instrument and found it to be small and *structured*. This is consistent with a few initial metastable CH_4^+ states as expected for rotational predissociation. These results were interpreted in terms of a mechanism involving tunnelling through the rotational barrier.[408] A quasi-diatomic model for dissociation of CH_4^+ was assumed. Both the calculated release energies and lifetimes in the 10^{-5} to 10^{-8} s range are consistent with the experimental results.

Kinetic energy release studies in EI magnetic sector instruments have been carried out for benzalacetones,[409] several oxygen-containing ions,[410] $MeOH^+$,[411] $C_6H_6^+$,[411] CH_4^+,[412] $C_2H_6^+$,[412] $MeOH^+$,[412] $EtOH^+$,[412] and ketones and alcohols.[413] Although the distribution of the KER is contained in the data, few workers report the KERD which is unfortunate because the KERD is a very sensitive probe into the dissociation path or paths. Holmes and Osborne[414] have reported a new technique, based on a Monte Carlo computation, for extracting the KERD from the shape of the metastable peak. The KERD for reaction (51) has been obtained.[415]

$$C_3H_7 \rightarrow C_3H_5^+ + H_2 \qquad (51)$$

It was found that 90% of the reverse activation energy is channelled to KER and it was concluded that the same non-statistical partitioning applies to the so-called non-fixed energy, the energy above the activation barrier. Very little experimental information is available relating to this interesting problem. The KERD for production of $C_3H_3^+$ from propargyl chloride and bromide has also been determined.[416] An earlier PIPECO study had been reported on the KER of these dissociations.[417] Although the two techniques are not totally comparable because the former does not select states, there appears to be a real discrepancy between the two investigations for dissociation of propargyl bromide. Holmes and Lossing concluded that, at the threshold for $C_3H_3^+$ ion formation, only the cyclopropenium ion is energetically possible whereas Tsai et al.[417] concluded that both cyclopropenium and propargyl ions are energetically possible. Both propargyl halide ions release a large fraction of their energy as kinetic energy. Why this should be so while other ions release little energy continues to be an important and interesting question.

Progress toward understanding the partitioning of the reverse activation barrier among the final products continues to be made.[418,419] In this calculation,

[407] J. P. Flamme, H. W. Wankenne, R. Locht, and J. Momigny, *Chem. Phys.*, 1978, **27**, 45.
[408] J. P. Flamme and J. Momigny, *Chem. Phys.*, 1978, **34**, 303.
[409] B. Schaldach and H. Fr. Grützmacher, *Int. J. Mass Spectrom. Ion Phys.*, 1979, **31**, 271.
[410] M. Kumakura and T. Sugiura, *Ber. Bunsenges. Phys. Chem.*, 1978, **82**, 1343.
[411] M. D. Migahed and F. H. Abd El-Kader, *Int. J. Mass Spectrom. Ion Phys.*, 1978, **28**, 225.
[412] M. D. Burrows, S. R. Ryan, W. E. Lamb, and L. C. McIntyre, *J. Chem. Phys.*, 1979, **71**, 4931.
[413] M. D. Migahed and F. H. Abd El-Kader, *Int. J. Mass Spectrom. Ion Phys.*, 1979, **31**, 373.
[414] J. H. Holmes and A. D. Osborne, *Int. J. Mass Spectrom. Ion Phys.*, 1978, **27**, 271.
[415] J. L. Holmes and A. D. Osborne, *Org. Mass Spectrom.*, 1978, **13**, 133.
[416] J. L. Holmes and F. P. Lossing, *Can. J. Chem.*, 1979, **57**, 249.
[417] B. P. Tsai, A. S. Werner, and T. Baer, *J. Chem. Phys.*, 1975, **63**, 4384.
[418] J. R. Christie, P. J. Derrick, and G. J. Richard, *J. Chem. Soc., Faraday Trans.* 2, 1978, **74**, 304.
[419] N. W. Cole, G. J. Richard, J. R. Christie, and P. J. Derrick, *Org. Mass Spectrom.*, 1979, **14**, 337.

the atomic motions at the transition state are used to estimate the fraction of the available energy which will go into KER. The transition state is characterized by MO calculations from which one can obtain not only the energy but the momentum of the various atoms. Semiempirical MINDO/3 calculations have also been used by Day et al.[420] in order to correlate the experimental KER for dissociation of protonated methanol with a calculated potential energy surface.

Kinetic energy release and ion lifetimes following high-energy collisions have been measured. Morgan et al.[421] found that 99% of the ions fragment with rate constants in excess of $2 \times 10^7 \text{ s}^{-1}$, consistent with the ability of collision-induced dissociation (CID) to differentiate isomers. The KER and the effects of internal energy in the ion on the CID spectrum have been measured.[422] Ions such as NO^+, H_2S^+, and NH_3^+ were obtained by photoionization at various energies and the 2.5 keV collision products were analysed for the kinetic energy released. It was found that the KER in the $N^+ + O$ products from $NO^+(v)$ decreased with increasing internal energy in the ion. This is clearly related to the excited-state potential surface and may aid in identifying the latter.

Negative ions have been investigated in the 10^{-5}—10^{-11} s time-scale by field ionization (FI)[423] but it was found that FI is much more limited in negative ion studies. For instance, only molecules with a positive EA give a signal and, even when this is satisfied, the signals are weak. Usually only the parent negative ion is observed. Alexandru[424] has investigated theoretically the probability that more than one fragment is ejected in a fragmentation event.

Finally, a new source of breakdown diagrams has been developed by Laramée et al.[425–427] Following CID, the mass spectrum is studied as a function of the scattering angle. The larger the scattering angle, the greater is the internal energy imparted to the ion. The correlation is good enough so that the branching ratios plotted as a function of scattering angle look similar to a breakdown diagram. It is not clear at this time whether this experiment will be of most interest for the dynamical information concerning CID or for the breakdown diagrams themselves. Most likely it is the former because the energy resolution for the internal energy in the ion is very modest by present standards.

Dissociation Pathways.—Papers dealing with dissociation pathways are more fully reviewed in a separate chapter on reaction mechanisms. Two recent review papers on even-electron ion decompositions have been published,[428,429] as well as a monograph on organic mass spectrometry.[430] In this section, some key features are discussed, particularly as they relate to the application of statistical theory to

---

[420] R. J. Day, D. A. Krause, W. L. Jorgensen, and R. G. Cooks, *Int. J. Mass Spectrom. Ion Phys.*, 1979, **30**, 83.
[421] R. P. Morgan, A. G. Brenton, and J. H. Beynon, *Int. J. Mass Spectrom. Ion Phys.*, 1979, **29**, 195.
[422] R. V. Manvelyan, A. A. Perov, S. E. Kupriyanov, and V. K. Potapov, *Theor. Exp. Chem.*, 1978, **14**, 645.
[423] J. van der Greef and N. M. M. Nibbering, *Int. J. Mass Spectrom. Ion Phys.*, 1979, **31**, 71.
[424] Gr. Alexandru, *Int. J. Mass Spectrom. Ion Phys.*, 1980, **32**, 375.
[425] J. A. Laramée, P. H. Hemberger, and R. G. Cooks, *J. Am. Chem. Soc.*, 1979, **101**, 6462.
[426] J. A. Laramée, J. J. Carmody, and R. G. Cooks, *Int. J. Mass Spectrom. Ion Phys.*, 1979, **31**, 333.
[427] J. A. Laramée, P. H. Hemberger, and R. G. Cooks, *Int. J. Mass Spectrom. Ion Phys.*, 1980, **33**, 231.
[428] F. W. McLafferty, *Org. Mass Spectrom.*, 1980, **15**, 114.
[429] M. Karni and A. Mandelbaum, *Org. Mass Spectrom.*, 1980, **15**, 53.
[430] K. Levsen, 'Fundamental Aspects of Organic Mass Spectrometry', Verlag Chemie, Weinheim, 1978.

unimolecular decay. In applying the theory, it is essential that the energies and structures of the ion and its transition state be known. Thus, if an ion isomerizes to a more stable structure prior to dissociation, neither the energy nor the structure is generally known. A great deal of ingenuity and guess work is then often necessary to apply the theory.

Isomerization in metastable ions is commonplace. It is often identified by noting the similarity in the metastable peak intensities of a series of isomers. This can sometimes be misleading in EI work because the internal energies are not very well controllable. This problem is removed in photoion–photoelectron coincidence studies.[431] Smith et al.[385] noted that cyclo-octatetraene$^+$ (COT) ions isomerize to styrene$^+$ ions prior to undergoing the rate-determining dissociation to $C_6H_6^+$ and C_2H_2. This was established by measuring the decay rate of COT$^+$ and styrene$^+$ at the same total internal energy. They were identical over the whole metastable energy range, indicating that isomerization had taken place. On the other hand, Fu and Dunbar[432] on the basis of the dissimilar photodissociation spectra of COT and styrene cations concluded that they had not isomerized (a discussion and review of photodissociation appears in a different chapter of this volume). A follow-up to this study added barrelene to the list of $C_8H_8^+$ ions.[433] This isomer also had a unique photodissociation spectrum which was similar to the PES of barrelene, thereby indicating no isomerization. The discrepancy is resolved if one postulates stable isomeric ions up to some activation barrier for isomerization. As long as that barrier is high enough to encompass the first band of the PES, thereby assuring a substantial number of stable ions, but still lower than the first dissociation limit, both sets of conclusions are correct. Metastable ions usually have activation barriers for dissociation of the order of 2 eV, which allows the activation barriers for isomerization to be anywhere between 1 and 2 eV. It is interesting that Dunbar et al.[433] found that the protonated isomers, $C_8H_9^+$, which photodissociate by losing H_2, have identical photodissociation spectra. The most likely common structure is that of the styryl cation. The different behaviour of the $C_8H_8^+$ isomers and their protonated analogues is clearly a function of the activation energy for isomerization. Another $C_8H_8^+$ isomer, from tetralin, has been investigated by field ionization kinetics (FIK).[434,435] Most likely it also isomerizes to styrene prior to dissociation at low internal energies. However, the FIK studies indicate that loss of C_2H_2 proceeds by more than one mechanism.

Another apparent discrepancy which can be resolved by an understanding of the energetics involves loss of CH_4 from saturated alkanes. Wolkoff and Holmes[436] investigated loss of both Me and CH_4 from propane, butane, methylpropane, pentane, methylbutane, and hexane using monoenergetic EI. An earlier study by Lifshitz and Shapiro[437] had indicated that metastable MeCD$_2$Me loses 100% CH_4 but a PIPECO study of the same reaction

---

[431] T. Baer, J. Electron. Spectrosc. Relat. Phenom., 1979, **15**, 225.
[432] E. W. Fu and R. C. Dunbar, J. Am. Chem. Soc., 1978, **100**, 2283.
[433] R. C. Dunbar, M. S. Kim, and G. A. Olah, J. Am. Chem. Soc., 1979, **101**, 1368.
[434] R. Stolze and H. Budzikiewicz, Org. Mass Spectrom., 1978, **13**, 75.
[435] K. Levsen, F. Borchers, R. Stolze, and H. Budzikiewicz, Org. Mass Spectrom., 1978, **13**, 510.
[436] P. Wolkoff and J. L. Holmes, J. Am. Chem. Soc., 1978, **100**, 7346.
[437] C. Lifshitz and M. Shapiro, J. Chem. Phys., 1967, **46**, 4912.

indicated that large amounts of MeD and CH_2D_2 are lost as well.[438] Wolkoff and Holmes[436] repeated the metastable work and agree with the conclusion of Lifshitz and Shapiro. They further conclude that within 0.15 eV of the onset a 1,3-elimination of methane from propane occurs. Above this energy, the middle hydrogens can begin to participate. Because the metastable propane ions are at an energy very close to the dissociation limit, while the PIPECO experiment sampled higher energies, the discrepancy between the two types of studies is removed. For loss of CH_4 from butane and methylpropane, Wolkoff and Holmes[436] conclude that higher-energy cyclopropane$^+$ rather than propene$^+$ fragments are formed because the KER is only 0.037 eV, a value too small to be reasonable for propene production. However, Mead et al.[439] in a PI study of loss of CH_4 from methylpropane found an onset of 10.89 eV, which is lower than a previous PI study and is 31 kJ mol^{-1} above the thermochemical dissociation limit for propene$^+$ but below the onset for cyclopropane$^+$ formation. The problem is further complicated by the fact that H loss occurs at 10.68 eV, implying a kinetic shift for all higher dissociation channels. Mead et al.[439] find that the tertiary hydrogen does not leave with the CH_4 and propose a reaction sequence (52) to account for this. An RRKM calculation of the isotope effect supported this mechanism. By analogy with propane, a 1,3-hydrogen shift would also appear to account for the non-participation of the tertiary hydrogen.

$$\text{Me}-\underset{CH_2}{\overset{Me}{CH}}{\diagdown}H \quad]^{\ddagger} \longrightarrow \text{Me}-\underset{\underset{CH_2}{|}}{CH}\cdots H \quad \overset{Me}{} \longrightarrow \text{MeCH}=CH_2 \quad]^{\ddagger} + CH_4 \quad (52)$$

Hydrocarbon rearrangements have been studied by FIK on the ps timescale.[440] It was shown that H/D scrambling in butadiene ions takes place in $10^{-10.7}$ s. Carbon atoms scramble more slowly. At 10^{-11} s 80% of the carbons in Me are from the terminal position. At long lifetimes, this number nearly reaches 50%. Methyl cyclopropene, cyclobutene, and buta-1,3-diene are shown to interconvert readily.[441] The IE of methylcyclopropene was measured to be 9.15 ± 0.5 eV[440] and the onset for loss of methyl 9.73 ± 0.05 eV. Hept-1-ene has been investigated by FI.[442,443] A comparison of its fragmentation with that following EI ionization showed that the high electric field 'dresses' the ion, causing it to dissociate in a manner different from that in low-energy EI. On the other hand, metastable spectra for FI and EI were rather similar, indicating that, at long reaction times, energy considerations dominate. Simple C—C bond ruptures were dominant at short times.[443]

One approach to learning about a reaction mechanism is to calculate the potential energy surface. The major advantages are that (a) details not available

---

[438] R. Stockbauer and M. G. Inghram, J. Chem. Phys., 1976, **65**, 4081.
[439] P. J. Mead, K. F. Donchi, J. C. Traeger, J. R. Christie, and P. J. Derrick, J. Am. Chem. Soc., 1980, **102**, 3364.
[440] D. H. Russell, M. L. Gross, J. van der Greff, and N. M. M. Nibbering, J. Am. Chem. Soc., 1979, **101**, 2086.
[441] M. L. Gross and D. H. Russell, J. Am. Chem. Soc., 1979, **101**, 2082.
[442] P. Tecon, D. Stahl, and T. Gäumann, Int. J. Mass Spectrom. Ion Phys., 1978, **27**, 83.
[443] P. Tecon, D. Stahl, and T. Gäumann, Int. J. Mass Spectrom. Ion Phys., 1979, **29**, 363.

from experiment are accessible and (b) the calculations always give a definite answer. The disadvantages are that, in the absence of good experimental data or guide posts, the calculations may be very inaccurate without informing us of this. In this spirit Krause et al.[444] have undertaken MINDO/3 calculations on the allyl cation dissociation to $C_3H_3^+ + H_2$. Two mechanisms were considered. These are ones involving 1,2- and 1,3-hydrogen shifts previously discussed by Holmes et al.[445] and Hvistendahl and Williams[446] and shown here as reactions (53) and (54).

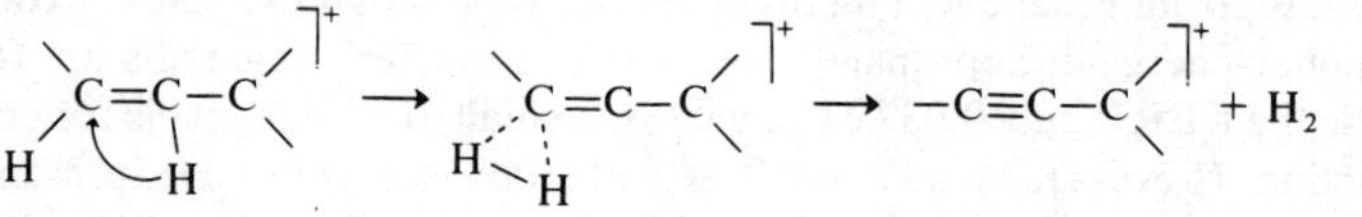

$$\left[\begin{array}{c}\diagdown\!\!\diagup\\C=C-C\\H\!\!\diagup\,\diagdown\!H\end{array}\right]^+ \rightarrow \left[\begin{array}{c}\diagdown\!\!\diagup\\C=C-C\\H\!\!\diagdown\,H\end{array}\right]^+ \rightarrow \left[-C\equiv C-C\diagup\right]^+ + H_2 \quad (53)$$

$$\left[\begin{array}{c}\diagdown\!\!\diagup\\C=C-C\\H\quad H\end{array}\right]^+ \rightarrow \left[\begin{array}{c}\diagdown\!\!\diagup\\C\cdots C=C\\H\cdots C\diagdown\end{array}\right]^+ \rightarrow \left[\begin{array}{c}\diagdown\!\!\diagup\\C\cdots C\\C\end{array}\right]^+ + H_2 \quad (54)$$

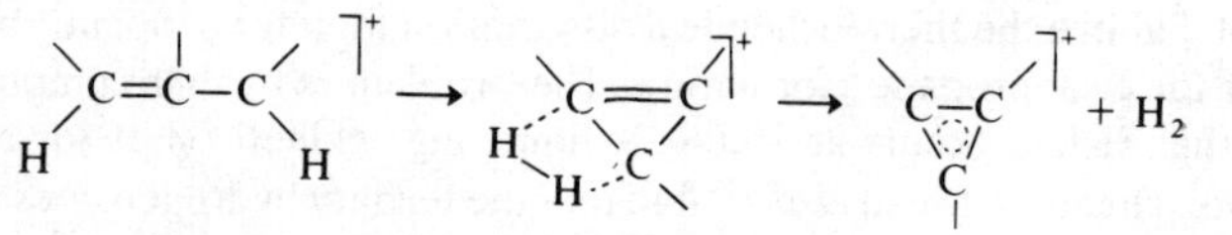

There are no isotopic labelling studies to support these mechanisms. The MINDO/3 calculations indicate that the activation energies to the two products are about the same. Both reactions are predicted to release 100% of their reverse activation energy as kinetic energy. Qualitative and semiquantitative relationships between tightness of activated complex, Woodward–Hoffman rules, and KER are discussed.

Bowen et al.[447] have investigated the energetics of $C_3H_5^+$ isomers via CID of $C_3H_5^+$ produced from cyclopropyl bromide, 1-propylbromide, and 2-propylbromide. The allyl ion was found to be most stable and the activation energies for isomerization to the various isomers were high. Both the cyclopropyl bromide and 1-propenylbromide produce the allyl $C_3H_5^+$ structure, but with considerable energy release. On the other hand, ionization of 2-propenylbromide produces a $C_3H_5^+$ ion with little energy release. The authors concluded that isomerization is the rate-determining step. In a similar study on butyl halides and alkanes, Shold and Ausloos[448] concluded that all ions formed from precursor molecules having a Bu^i or Bu^t group exist primarily as Bu^{t+}. The method for determining this was via ion–molecule reactions rather than CID.

A more facile isomerization occurs in $C_6H_{10}^+$ isomers where the metastable peak shapes for $C_5H_7^+$ fragment ions from a variety of precursors such as structures (5)—(9) are all similar.[449] The conclusion is that they all go through a common transition state. The most probable structures for isomerized $C_6H_{10}^+$ and product $C_5H_7^+$ ions are methylcyclopentene and cyclopentenium ions respectively.[449]

---

[444] D. A. Krause, R. J. Day, W. L. Jorgensen, and R. G. Cooks, *Int. J. Mass Spectrom. Ion Phys.*, 1978, **27**, 227.
[445] J. L. Holmes, A. D. Osborne, and G. M. Weese, *Org. Mass Spectrom.*, 1975, **10**, 867.
[446] G. Hvistendahl and D. H. Williams, *J. Chem. Soc., Perkin Trans.* 2, 1975, 881.
[447] R. D. Bowen, D. H. Williams, H. Schwartz, and C. Wesdemiotis, *J. Am. Chem. Soc.*, 1979, **101**, 4681.
[448] D. M. Shold and P. Ausloos, *J. Am. Chem. Soc.*, 1978, **100**, 7915.
[449] P. Wolkoff and J. L. Holmes, *Can. J. Chem.*, 1979, **57**, 348.

(5)   (6)   (7)   (8)   (9)

Keto–enol tautomerism has been investigated by a number of workers. Weber and Levsen[450] found that, of the four structures (10)—(13) of $C_2H_3O^+$, the most

MeCO$^+$ CH$_2$——CH   CH$_2$=C=OH$^+$ $^+$CH$_2$CHO
                    \ /
                     O$^+$

(10)   (11)   (12)   (13)

stable one is probably the acetyl ion (10). This conclusion was based on the identical collisional activation (CA) spectra of $C_2H_3O^+$ ions from a variety of precursors such as acetone, acetyl chloride, acetaldehyde, ethyl vinyl ether, and so on. However, the enol from (12) with a heat of formation of 648 kJ mol^{-1} (there is a misprint in the paper) may be produced from glycerol. Schwartz and Wesdemiotis[451] have determined the activation energy for keto–enol isomerization in methyl acetate. The mechanism, given by reaction (55), involves a 1,3-hydrogen shift with an estimated energy barrier from left to right of 172 kJ mol^{-1} but the heats of formation of the enol and keto forms have been determined to be 477 and 582 kJ mol^{-1} respectively,[113] a difference far greater than was assumed by Schwartz and Wesdemiotis.[451]

$$H_2C=C\begin{matrix}O-H\\OMe\end{matrix}\Big]^+ \longrightarrow H_2C\begin{matrix}H\\C\\OMe\end{matrix}O \longrightarrow Me-C\begin{matrix}O\\OMe\end{matrix} \quad (55)$$

A similar discrepancy exists in the assumed energetics of acetic acid and its enol. The mechanism and barrier to 1,3-hydrogen migration was discussed by Schwartz et al.[452] However, it was assumed that the acid form was more stable whereas Holmes and Lossing[113] found the enol form to be more stable by 92 kJ mol^{-1}. Other studies on keto–enol isomerizations have been done on esters and unsaturated acids,[453] and $C_6H_6O^+$ radical ions.[454] Splitter and Calvin[455] claim that the tautomerism may be a result of a Woodward–Hoffman allowed 1,2-hydrogen shift rather than a 1,3-shift.

Although isomerization is often fast compared with dissociation, there are examples in which the isomerization step is actually rate determining. There are therefore three situations: (a) isomerization is faster than dissociation, (b) isomerization is slower than dissociation, but takes place nevertheless as the initial step to dissociation, and (c) no isomerization.

---

[450] R. Weber and K. Levsen, *Org. Mass Spectrom.*, 1980, **15**, 138.
[451] H. Schwartz and C. Wesdemiotis, *Org. Mass Spectrom.*, 1979, **14**, 25.
[452] H. Schwarz, D. H. Williams, and C. Wesdemiotis, *J. Am. Chem. Soc.*, 1978, **100**, 7052.
[453] P. C. Burgers, J. K. Terlouw, P. C. Vijfhuizen, and J. L. Holmes, *Org. Mass Spectrom.*, 1978, **13**, 470.
[454] D. H. Russell, M. L. Gross, and N. M. M. Nibbering, *J. Am. Chem. Soc.*, 1978, **100**, 6133.
[455] J. S. Splitter and M. Calvin, *J. Am. Chem. Soc.*, 1979, **100**, 7329.

It has been found that $C_{10}H_9O^+$ isomers do not isomerize prior to the production of 2-methylbenzopyrylium ions.[456] On the other hand, the isomers of BuO^+ isomerize *via* a rate-determining step,[457] while $C_3H_8N^+$ and PrO^+ ions isomerize and dissociate with equal facility.[458] This latter condition gives rise to rather complicated kinetics. Although it is difficult to say how a given ion will react, some progress towards a predictive model has been made.[459]

Isolated states continue to be discussed in the literature[460] despite the rather weak evidence for their existence. Certainly small ions do dissociate from excited states,[373,377,380,383] but there is no firm evidence for isolated states in large organic ions. The case of benzene, which now appears to have been resolved, has been discussed in the previous section. McAdoo *et al.*[461] have proposed an isolated state in the loss of Me from butanoic acid$^+$ ions. This conclusion was based on the study of 16 different isotopically labelled butanoic acid species. It was found that Me loss occurs with no H/D scrambling, while C_2H_4 loss occurred with a statistical H/D ratio. In addition, although the AE for loss of Me is slightly higher than for loss of C_2H_4, the former fragmentation path dominates in the metastable mass spectrum. This is most likely the result of the low entropy of activation associated with the simple C—C bond break. The isotopic scrambling results are difficult to account for without postulating an isolated state yet an isolated electronic state for a rather loose ion such as butanoic acid is difficult to reconcile with the behaviour of other organic ions. It would appear to be an excellent candidate for further study. Another possible isolated state was found with butyrophenone.[462] However, Kim *et al.*[463] have pointed out that this dissociation is complicated by a very rapid protonation reaction in the ICR cell and the 'isolated state' may have been the protonated butyrophenone ion.

Ion–dipole interactions in carbonium ion isomerizations have been discussed.[464–466] Using the ion–dipole interaction energy, E, given by expression (56),

$$E = \frac{q\mu \cos \theta}{r^2} \tag{56}$$

Bowen and Williams[464] point out that a polarizable group such as CH_2O can reduce the transition-state energy by as much as 75 kJ mol^{-1} when r, θ, and μ are assumed to be 0.3 mm, 0, and 2.3 D, respectively.

[456] B. Schaldach and H. Fr. Grützmacher, *Int. J. Mass Spectrom. Ion Phys.*, 1979, **31**, 257.
[457] R. D. Bowen and D. H. Williams, *J. Am. Chem. Soc.*, 1978, **100**, 7454.
[458] R. D. Bowen, D. H. Williams, G. Hvistendahl, and J. R. Kalman, *Org. Mass Spectrom.*, 1978, **13**, 721.
[459] R. D. Bowen, B. J. Stapleton, and D. H. Williams, *Org. Mass Spectrom.*, 1978, **13**, 330.
[460] L. J. Hendricks and R. H. Shapiro, *Int. J. Mass Spectrom. Ion Phys.*, 1980, **33**, 81.
[461] D. J. McAdoo, D. N. Witiak, F. W. McLafferty, and J. D. Dill, *J. Am. Chem. Soc.*, 1978, **100**, 6639.
[462] R. Gooden and J. J. Brauman, *J. Am. Chem. Soc.*, 1977, **99**, 1977.
[463] M. S. Kim, R. C. Dunbar, and F. W. McLafferty, *J. Am. Chem. Soc.*, 1978, **100**, 4600.
[464] R. D. Bowen and D. H. Williams, *Int. J. Mass Spectrom. Ion Phys.*, 1979, **29**, 47.
[465] R. D. Bowen and D. H. Williams, *J. Am. Chem. Soc.*, 1980, **102**, 2752.
[466] T. H. Morton, *J. Am. Chem. Soc.*, 1980, **102**, 1596.

5 Ion–Molecule Interactions

Ion–molecule reactions are reviewed in another chapter. Here, reactions are discussed only so far as they relate to unimolecular reactions or to the statistical theory of unimolecular decay.

Collision Theory.—An excellent review of ion–molecule collision theory has been published.[467] The collision cross-section can be calculated by use of simple classical mechanics once the interaction potential is known. Some of the terms contributing to the potential energy are shown in equation (57) in which q is the electronic charge, α the neutral polarizability, μ_D the neutral dipole moment, Q the neutral quadrupole moment, θ the angle the dipole or quadrupole makes with the ion, and c a constant which depends on a number of parameters.[467,468]

$$V(r) = V_L(r) + V_D(r, \theta) + V_Q(r, \theta) + V_I(r)$$
$$= -\frac{q^2\alpha}{2r^4} - \frac{9\mu_D}{r^2}\cos\theta + \frac{3Qq}{2r^3}(\cos^2\theta - \tfrac{1}{3}) - \frac{c}{r^6} \quad (57)$$

The most important term is the $V_L(r)$ (Langevin) or ion-induced dipole interaction. This is the only term for which the cross-section can be solved exactly. For ion–dipolar collisions, the second term $V_D(r, \theta)$ must be included.[469,470] Recent work has been directed at higher-order terms and at conserving angular momentum. In the average ion dipole (ADO) theory of Su and Bowers,[469] the average orientation (θ) of the dipole in the ion field is calculated at a given temperature. This average orientation specifies an angle θ between the dipole and a vector connecting the dipole and the point charge. When this is combined with the potential between a point charge and a polarizable neutral, and the centrifugal potential, the effective potential is given by equation (58) where $L = \mu bv$ for an impact parameter, b.

$$V_{\text{eff}} = \frac{L^2}{2\mu r^2} - \frac{\alpha q^2}{2r^4} - \frac{q\mu_D}{r^2}\cos\bar{\theta}(r) \quad (58)$$

The cross-section for a collision is usually defined as the maximum πb^2 which results in a true collision, *i.e.* the orbiting collision. It is that collision which just surmounts the centrifugal barrier.

Now, in a collision between a point charge and a dipole, the rotational energy of the dipole rotor increases as a result of the torque exerted by the ion. In order to conserve angular momentum, the orbital angular momentum (which is responsible for the centrifugal barrier) must decrease. The net result is that the collision cross-section increases. Su and Bowers[471] have developed an equation in parametrized form similar to the ADO equation, but which includes conservation of angular momentum. The resulting AADO cross-sections are greater than the ADO cross-sections by about 20% and represent a significant improvement in the theory. It now agrees with the experimental cross-sections for proton transfer

---

[467] T. Su and M. T. Bowers in ref. 118, Ch. 3.
[468] T. Su, E. C. F. Su, and M. T. Bowers, *Int. J. Mass Spectrom. Ion Phys.*, 1978, **28**, 285.
[469] T. Su and M. T. Bowers, *J. Chem. Phys.*, 1973, **58**, 3027.
[470] R. A. Barker and D. P. Ridge, *J. Chem. Phys.*, 1976, **64**, 4411.
[471] T. Su, E. C. F. Su, and M. T. Bowers, *J. Chem. Phys.*, 1978, **69**, 2243.

reactions (59) and (60), both of which had been underestimated by the ADO theory.

$$XH^+ + NH_3 \rightarrow NH_4^+ + X \tag{59}$$

$$XH^+ + HCN \rightarrow H_2CN^+ + X \tag{60}$$

The anisotropic nature of the polarizability has also been considered.[472,473] The difference between $\alpha_\parallel$ and $\alpha_\perp$ can be quite large for some molecules, giving rise to as much as an 8% difference in the calculated cross-section.[472] Bass et al.[473] claim that the average polarizability should be used, not the maximum, and they find that the cross-section increases by only about 2%.

Finally, the fact that molecules and ions are not point particles has been considered.[474] Such collisions in non-central field will be most important at short internuclear distances between a small ion and a large neutral. The results have been compared with trajectory calculations. Short-range forces have been treated by Bearman and Gentry.[475] Because the cross-sections of ion–molecule encounters are usually determined from fast reactive collisions which give only a lower limit to the collision cross-section, the factors governing reaction are obviously important. Bearman and Gentry[475] concluded that, even at collision energies much less than the potential well depth, information on the dynamics of the short-range part of the interaction may persist in the angular distribution despite the influence of the strong long-range attractive forces. It is unfortunate that inelastic scattering cross-sections obtainable from crossed molecular beams are not of sufficient accuracy to allow comparison of these results with theory.

Three-body Association Reactions.—The relationship between association and unimolecular reactions is an extremely intimate one. The general mechanism is given by (61) and the rate of formation of AM^+ by (62).

$$A^+ + M \underset{k_b}{\overset{k_a}{\rightleftharpoons}} (AM^+)^* \xrightarrow[M]{k_s} AM^+ \tag{61}$$

$$\frac{d(AM^+)}{dt} = k_s(AM^+)^*M \tag{62}$$

If the associated complex, $(AM^+)^*$, is treated as a steady-state species, the resulting rate of product formation is given by equation (63).

$$\frac{d(AM^+)}{dt} = \frac{k_a k_s}{k_b + k_s[M]}[A^+][M]^2 \tag{63}$$

At low pressure, the rate is third order while at sufficiently high pressure it becomes second order. The second-order rate constant expression, $k_a k_s[M]/(k_b + k_s[M])$, can be calculated from first principles using the collision cross-sections discussed in the previous section and the RRKM theory. The association rate constant, k_a, is just the collision rate, while the collisional stabilization rate, k_s, can

---

[472] R. Clair and T. B. McMahon, Int. J. Mass Spectrom. Ion Phys., 1978, **28**, 365.
[473] L. Bass, T. Su, and M. T. Bowers, Int. J. Mass Spectrom. Ion Phys., 1978, **28**, 389.
[474] W. J. Chesnavich, T. Su, and M. T. Bowers, J. Chem. Phys., 1980, **72**, 2641.
[475] G. H. Bearman and W. R. Gentry, J. Chem. Phys., 1979, **71**, 1128.

also be assumed to be the collision rate. This assumption is often referred to as the strong collision assumption in which every collision deactivates or stabilizes the complex $(AM^+)^*$. Therefore, k_a and k_s can be calculated at the various levels of sophistication outlined in the previous section.

The evaluation of k_b presents a more formidable obstacle. The first approach was with the classical RRK theory. Bates[476] and Herbst[477] have justifiably criticized this practice and suggest the use of RRKM theory but these workers have ignored a large body of work in which the RRKM theory has in fact been used. In addition, Herbst employs the Langevin rather than the more accurate ADO theory for the collision cross-sections. Although the RRKM theory certainly is a big improvement in calculating k_b, it has some serious shortcomings unless angular momentum is conserved.

Bass et al.[478] have discussed the problem of angular momentum in the context of both RRKM and phase space theory. The latter is based on the separated products while the former is based on the transition state. A difference in the two approaches arises because separated fragments have angular momenta which cannot be readily accommodated in the transition-state treatment. The large distribution of energies and angular mometa $P(E, J)$ in a collision complex leads to a wide range of collision complex lifetimes. Bass et al.[478] obtained general expressions for $P(E, J)$ as well as for the high- and low-pressure limits. Comparisons between the RRKM and the phase space results for a number of reactions between AH^+ and A [A = NH_3, $MeNH_2$, Me_2NH, or Me_3N] were made. In the pressure and temperature regime investigated, the two theories are about equally good in reproducing the experimental second-order rate constant versus pressure data. The agreement between 10^{-4} Torr and 1 Torr was generally within 20%.

Use of RRKM theory in determining k_b was also made by Jasinski et al.[133] in their analysis of the kinetics and thermodynamics of ion–molecule association reactions of H_2O, H_2S, and C_6H_6. In a similar study, the reaction complex dissociation of protonated pyridine bases was modelled using RRKM theory.[479] Neilson et al.[480] investigated reactions such as $NH_3^+ + NH_3$, $NH_4^+ + NH_3$, and $(NH_3)_2H^+ + NH_3$. For the analysis, k_b was determined for $(NH_3)_2H^+$ to be 1×10^8 at 210 K and 6×10^8 at 302 K. The rates for decomposition of $[Me_2NH]_2H^+$ were found to be a factor of three lower than those of $(NH_3)_2H^+$, consistent with statistical theory in that larger ions dissociate more slowly.

Woodin and Beauchamp[481] have measured three-body attachment rates of Li^+ to carbonyl compounds. Pressure-dependent studies at low pressures in the ICR cell indicate that a deactivation path other than collisional was effective in stabilizing the LiM^+ adduct. They concluded that i.r. emission (rate $\approx 10^2 \text{ s}^{-1}$) could account for this deactivation but they did not directly observe this emission.

---

[476] D. R. Bates, J. Phys. B, 1979, **12**, 4135.
[477] E. Herbst, J. Chem. Phys., 1979, **70**, 2201.
[478] L. Bass, W. J. Chesnavich, and M. T. Bowers, J. Am. Chem. Soc., 1979, **101**, 5493.
[479] J. M. Jasinski and J. I. Brauman, J. Am. Chem. Soc., 1980, **102**, 2906.
[480] P. V. Neilson, M. T. Bowers, M. Chau, W. R. Davidson, and D. H. Aue, J. Am. Chem. Soc., 1978, **100**, 3649.
[481] R. L. Woodin and J. L. Beauchamp, Chem. Phys., 1979, **41**, 1.

Meot-Ner[482] investigated reactions such as (64) in the presence of CH_4 as a third body.

$$Pr^+ + HCN \rightleftharpoons (PrHCN^+)^* \rightarrow PrNCH^+ \tag{64}$$

These experiments were carried out over a range of pressures, much of it in the high-pressure second-order rate constant limit. An unexplained feature in this study is that reaction (64), at some given pressure, is in the third-order limit at low temperature but is second-order at high temperatures.

Energy Transfer and Internal Energy Effects.—The statistical theory of energy transfer has been the basis of our understanding of energy transfer processes in polyatomic molecule collisions. In its simplest form it is assumed that the collision complex is long-lived compared with the time for energy randomization so that the products separate with all energies statistically partitioned. It has been recognized that this theory overestimates the energy transfer cross-section. Part of the problem is in the need to conserve angular momentum. Another part is related to the collision lifetime. Bhattacharjee and Forst[483] have pointed out that, for dynamical reasons, the collision complex duration may be orders of magnitude smaller than the collision complex lifetime as estimated by RRKM theory. They claim that the short collision duration results in fewer vibrational modes participating in the energy flow process. The reduced density of states results in less energy flow per collision. Their modified theory has been applied to the dynamics of chemically activated $(C_5H_9)^*$.[484] This ion, once formed in a super excited state, can decay via the reactions (65).

$$(C_5H_9^+)^* \rightarrow \begin{array}{l} C_3H_5^+ + C_2H_4 \\ C_5H_7^+ + H_2 \\ C_5H_9^+ \end{array} \tag{65}$$

The predicted ratios of stabilized $C_5H_9^+$ ions to dissociated products is in reasonable agreement with the experimental results.[485]

A similar modification of the statistical theory has been carried out by Micha et al.[486] and Estes and Toennies.[487] It was pointed out that long-lived collision complexes are not a sufficient criterion for energy equilibration. In any reaction system, a range of impact parameters will result in a range of collision types from grazing to head-on. A Massey criterion or adiabaticity relation is introduced according to which the collision time should always be smaller than $h/\Delta E$, where ΔE is the energy transferred.[487] The theory has been applied to collisions of Li^+ and CH_4 and good agreement noted with observations.[488]

---

[482] M. Meot-Ner, *J. Am. Chem. Soc.*, 1978, **100**, 4694.
[483] R. C. Bhattacharjee and W. Forst, *Chem. Phys.*, 1978, **30**, 217.
[484] W. Forst and R. C. Bhattacharjee, *Chem. Phys.*, 1979, **37**, 343.
[485] P. G. Miasek and A. G. Harrison, *J. Am. Chem. Soc.*, 1975, **97**, 714; R. Houriet and J. H. Futrell, *Adv. Mass Spectrom.*, 1978, **7A**, 335.
[486] D. Micha, E. Vilallonga, and J. P. Toennies, *Chem. Phys. Lett.*, 1979, **62**, 238.
[487] W. Eastes and J. P. Toennies, *J. Chem. Phys.*, 1979, **70**, 1644.
[488] W. Eastes, U. Ross, and J. P. Toennies, *J. Chem. Phys.*, 1979, **70**, 1652.

The statistical theory for molecular collisions has been rederived based on the maximization of the entropy, but subject to various constraints such as conservation of total flux.[489] Bottcher[490] has considered energy transfer in ion–polar molecule collisions using an impact parameter and straight-line path approximation. In order to solve for the number of eigenstates, a new technique called 'multiple time method' was developed. It was applied to the rotational excitation of HCN in collisions with Na^+ ions in the relative energy range 0.1—10.0 eV. This theory also predicts a collision cross-section for ion–dipole interactions. In particular, it predicts a temperature dependence of the collision rate which is proportional to $TT_R^{-1/2}$ where T and T_R are the temperature and rotational temperature respectively. This appears to be contrary to ADO theory in which this rate has a $T^{-1/2}$ dependence.

Not a great deal of experimental information on energy transfer in ion–molecule encounters exists. The voluminous data on neutral systems cannot be applied to ionic systems because ions and neutrals behave very differently. A large part of the difference may be a result of the open-shell nature of most ions. For this reason, electronically excited molecules may have properties more akin to ions. One of these is the increased vibrational relaxation rates. Therefore, the results on the vibrational relaxation rates of electronically excited aniline reported by Chernoff and Rice[491] are of relevance to ion–molecule interactions. The $^1B_2$ state of aniline was studied by laser-induced fluorescence under single collision conditions. Fluorescence from a state other than the one initially excited is evidence for energy transfer to that state. According to the statistical hypothesis, each vibrational state has equal probability of being formed, subject of course to energy conservation. However, the data clearly show that the vibrational modes of $^1B_2$ aniline fall into two non-communicating groups. Within each group the rate of energy transfer proceeds at approximately the hard-sphere collision rate but, between the two groups, the transfer probability is low.

Vibrational energy transfer has also been measured in ionic systems. Kim and Dunbar[492] determined relative vibrational relaxation efficiencies for collisions between $(PhBr^+)^*$ and various bath molecules. As expected, PhBr was about 50 times more efficient at relaxing the excited ions than CH_4, presumably because of its greater vibrational density of states. However, why $PhNO_2$ was about as effective as CH_4, and considerably less efficient than C_6H_6, is not clear.

An interesting study of vibrational energy transfer between $\tilde{A}\ ^2\Pi CO^+(v)$ and He has been reported.[493] While the cross-section for collisional relaxation from $CO^+(v = 0)$ is less than 0.04 Å^2, the $v = 1, 2,$ and 3 states have cross-sections of 0.11, 0.27, and > 0.6 Å^2. The results are interpreted in terms of a collisional perturbation which mixes the ground and excited electronic states.

The role of vibrational energy in bimolecular reactions continues to be a topic which generates more interest than data. The reaction between $H_2^+(v) + He$ has been investigated by quasi-classical trajectory calculations.[494] In a similar study,

---

[489] E. Pollak and R. D. Levine, *J. Chem. Phys.*, 1980, **72**, 2990.
[490] C. Bottcher, *Chem. Phys. Lett.*, 1979, **66**, 126.
[491] D. A. Chernoff and S. A. Rice, *J. Chem. Phys.*, 1980, **70**, 2521.
[492] M. S. Kim and R. C. Dunbar, *Chem. Phys. Lett.*, 1979, **60**, 247.
[493] V. E. Bondybey and T. A. Miller, *J. Chem. Phys.*, 1978, **69**, 3597.
[494] Ch. Zuhrt, F. Schneider, V. Havemann, L. Zülicke, and Z. Hermann, *Chem. Phys.*, 1979, **38**, 205.

Stroud and Raff[495] considered the effect of the potential energy surface on the cross-section of this reaction.

Larger ions have been studied experimentally. The influence of vibrational energy on reaction (66) has been investigated by varying the electron impact energy in the preparation of Bu^+ by dissociative ionization of neopentane (C_5H_{12}).[496]

$$(Bu^+)^* + NH_3 \to NH_4^+ + C_4H_8 \qquad (66)$$

Tanaka et al.[497,498] have studied the reactions of $C_2H_2^+$ (v) and $C_3H_4^+$ (v) by photoionization. This method of ionization leads to considerably more control of internal energy in the ion. The most effective general method for investigating the role of internal energy is by PIPECO. Charge transfer results on O_2^+ (v) + O_2 have already been mentioned previously in the section on charge transfer. Tanaka and Koyano[499] have reported a PIPECO investigation of the $H_2^+(v)$ + H_2 proton transfer reaction and have found that the cross-section goes down with increasing v. These results agree with earlier published work. Finally, reaction (67) has been investigated by PIPECO.[500]

$$CO^+(v) + D_2 \to COD^+ + D \qquad (67)$$

It was found that the cross-sections for $v = 0$ and $v = 1$ are about the same.

---

[495] C. Stroud and L. M. Raff, *Chem. Phys.*, 1980, **46**, 313.
[496] W. J. Chesnavich, T. Su, and M. T. Bowers, *J. Am. Chem. Soc.*, 1978, **100**, 4362.
[497] K. Honma, I. Koyano, and I. Tanaka, *Bull. Chem. Soc. Jpn.*, 1978, **51**, 1923.
[498] K. Honma and I. Tanaka, *J. Chem. Phys.*, 1979, **70**, 1893.
[499] K. Tanaka and I. Koyano, *J. Chem. Phys.*, 1978, **69**, 3422.
[500] I. Koyano and K. Tanaka in 'Electronic and Atomic Collisions', ed. N. Oda and K. Takayanagi, North-Holland, Amsterdam, 1980, p. 547.

2
Structures and Reactions of Gas-phase Organic Ions

BY I. HOWE

1 Introduction

The purpose of this chapter is to review some of the publications in the recent literature that have attempted, by a variety of means, to discover information about the structures, rearrangements, and decompositions of organic ions produced in a mass spectrometer. The scope of the review is wide and considerable selection has been necessary. The references have been cited both from a point of view of interest and to avoid excessive overlap with other chapters.

Following the first attempts over twenty years ago to identify ion structures in the mass spectrometer, many new experimental methods have evolved. A resumé of techniques employed for ion structure and reaction mechanism determination is presented in Table 1. As general introductory texts, refs. 1—5 are particularly relevant. A wide coverage of recent references on topics in gas-phase ion chemistry is found in ref. 6.

The reviewed topics are broadly divided into three sections. The thermochemistry of ion formation is considered, with reference mainly to mass spectrometric measurements but with brief recognition of molecular orbital calculations. The subsequent section covers isomerization reactions and incorporates some novel advances in the understanding of gas-phase organic ion chemistry (e.g. recognition of the importance of ion–dipole interactions in ionic intermediates). The final main section reviews decomposition reactions. Collisional activation spectra and measurements of translational-energy release are widely employed and form a substantial part of the research reported in this section. The review is concluded by a summary.

2 Thermochemistry

A strong clue to the structure of an ion may often be obtained from its heat of formation, measured by a mass spectrometric method. The experimental techniques are fraught with difficulties and measured ionization and appearance energies (hence heats of formation) are frequently not the optimum values obtainable. Problems arise for several reasons.[1—3] (i) Low ion currents at the

---

[1] I. Howe, in this series, 1971, Vol. 1, Chapter 2; 1973, Vol. 2, Chapter 2.
[2] I. Howe, D. H. Williams, and R. D. Bowen, 'Mass Spectrometry: Principles and Applications', McGraw-Hill, 1981.
[3] K. Levsen, 'Fundamental Aspects of Organic Mass Spectrometry', Verlag Chemie, 1978.

Table 1 *Techniques for investigating gas-phase ion structures and reaction mechanisms*

Technique/Measurement	Features/Applications	Basic Refs.
Thermochemical measurements	Ion structures at threshold	1, 2, 7, 8
Electron-impact mass spectra	Higher-energy ions: isotopic labelling useful	2
Variable-electron-energy studies	Identifies rearrangements	1, 3
Field-ionization kinetics	Wide range of reaction rate constants	3, 9
Unimolecular metastables	Narrow, low-energy range of decompositions	1—4, 10
Translational-energy release	Structure of metastable ions	2—4, 10
Isotopic labelling	Rearrangement reactions	1—3
Kinetic isotope effects	Mechanisms: particularly of hydrogen transfers	1—3
Collisional activation	Structures of stable and metastable ions	1—4, 11
High-pressure mass spectrometry	Equilibrium constants; acidities, basicities, and proton affinities	12, 13
Ion cyclotron resonance	As above	12—14
Rate determinations	Reaction mechanisms	3, 15
Photodissociation	Energy states and reaction mechanisms	16
Molecular orbital calculations	Ion thermochemistry	8

threshold for formation of an ion are difficult to measure. (ii) Ionic decompositions in the source occur from ions of lifetime less than 10 μs which therefore possess some excess of energy above the minimum decomposition threshold; critical energy measurements for these decompositions may be too high (*q.v.* the kinetic shift). (iii) Even if the correct threshold for ion formation is identified, the products may be formed with excess of internal energy. (iv) Heats of formation of neutral compounds employed in thermochemical cycles are sometimes uncertain. All these problems may compromise the results.

---

[4] R. G. Cooks, J. H. Beynon, R. M. Caprioli, and G. R. Lester, 'Metastable Ions', Elsevier, 1973.
[5] 'Gas-Phase Ion Chemistry', ed. M. T. Bowers, Academic Press, 1979, Vols. 1 and 2.
[6] A. L. Burlingame, T. A. Baillie, P. J. Derrick, and O. S. Chizov, *Anal. Chem.*, 1980, **52**, 214R.
[7] H. M. Rosenstock, *Int. J. Mass. Spectrom. Ion Phys.*, 1976, **20**, 139.
[8] T. W. Bentley, in this series, 1979, Vol. 5, Chapter 2, p. 71—87.
[9] P. J. Derrick, in 'Mass Spectrometry', International Review of Science, Phys. Chem., Series II, ed. A. Maccoll, Butterworths, London, 1976, Vol. 5.
[10] J. L. Holmes and J. K. Terlouw, *Org. Mass Spectrom.*, 1980, **15**, 383.
[11] F. W. McLafferty, in 'Chemical Applications of High-Performance Mass Spectrometry', ed. M. L. Gross, Am. Chem. Soc., Washington, D.C., 1977.
[12] D. H. Aue and M. T. Bowers, in ref. 5, Vol. 2, Chapter 9.
[13] J. E. Bartmess and R. T. McIver, in ref. 5, Vol. 2, Chapter 11.
[14] T. A. Lehman and M. M. Bursey, 'ICR Spectrometry', Wiley-Interscience, New York, 1976.
[15] T. Baer, Chapter 1 of this volume.
[16] R. C. Dunbar, in ref. 5, Vol. 2, Chapter 14.

The quality of the data obtained for critical energies varies with the type of instrumentation employed. In one of the continual studies on the $C_7H_7^+$ ion, appearance energies have been measured from photoionization efficiency curves using signal averaging to enhance sensitivity near threshold.[17] The heats of formation determined for $C_7H_7^+$ from the parent C_7H_8 molecules (1) and (2) are consistent with tropylium formation rather than benzyl formation [a possibility from structure (3)]. Similarly, for alkylbenzenes, a mean heat of formation for

		$\Delta H_f / \text{kJ mol}^{-1}$
Ph—CH₃ (1)		865
cycloheptatriene (2)	→ $C_7H_7^+$	868
norbornadiene (3)		917

$C_7H_7^+$, determined by the above method, of 859 ± 8 kJ mol^{-1} is consistent with tropylium ion formation.[18] These conclusions refer to generation of $C_7H_7^+$ at threshold. Collisional activation studies indicate that, in addition, benzyl ions are formed from structures (1) and (2) at higher energies.[19]

Rearrangements to lower-energy product ion isomers during decomposition can frequently be identified by measurements of heat of formation. Formation of the 2-propyl ion from 1-propyl halides at the minimum energy for fragmentation has been confirmed by such methods.[20] Similarly, heats of formation indicate that 2-alkyl ions are formed from alkanes *via* 1,2-hydrogen migrations.[21] Rate-determining reactions have also been identified from critical energies; the rate-determining step for loss of H_2O from protonated cyclic ethers, *e.g.* ion (4), involves the C—O ring-cleavage reaction (1).[22]

$$\text{(4)} \longrightarrow {}^+(CH_2)_4OH \quad \text{(5)} \tag{1}$$

The gas-phase heats of formation of 13 enol positive ions from aliphatic aldehydes, ketones, carboxylic acids, and esters, determined by mass spectrometric methods, have been compared with those of the corresponding keto

---

[17] J. C. Traeger and R. G. McLoughlin, *Int. J. Mass Spectrom. Ion Phys.*, 1978, **27**, 319.
[18] R. G. McLoughlin, J. D. Morrison, and J. C. Traeger, *Org. Mass Spectrom.*, 1979, **14**, 104.
[19] F. W. McLafferty and F. M. Bockhoff, *J. Am. Chem. Soc.*, 1979, **101**, 1783.
[20] J. C. Traeger, *Int. J. Mass Spectrom. Ion Phys.*, 1980, **32**, 309.
[21] P. Wolkoff and J. L. Holmes, *J. Am. Chem. Soc.*, 1978, **100**, 7346.
[22] H. E. Audier, A. Milliet, C. Perret, and P. Varenne, *Org. Mass Spectrom.*, 1979, **14**, 129.

ions.[23] The enolic ions all have lower energy, by amounts ranging from 59 to 130 kJ mol^{-1}. In neutral chemistry, the keto forms are usually more stable.

Studies of enol–keto tautomerism have been extensively reported in the recent literature for small positive ions in the gas phase and there have been several discrepancies between the various results obtained. For example, two significantly different values were determined[24,25] for the heat of formation of the enol tautomer of methyl acetate. The lower value indicates that this tautomer (produced by C_2H_4 loss from ionized methyl butyrate) is 100 kJ mol^{-1} more stable than the keto tautomer.[25] Discrepancies between published heats of formation arise also for the tautomers of ionized acetic acid.[26–28] The lowest value[26] for the appearance energy of the enol isomer (6) was obtained by a method employing an energy-resolved electron beam together with an appropriate data system. This apparatus yields values for appearance energies of fragment ions that are consistently lower and probably more reliable than measurements obtained by many other means.

MO calculations have contributed to the understanding of enol–keto tautomerism in the mass spectrometer.[28,29] This arrangement in radical cations is usually expressed in terms of a 1,3-hydrogen shift. The tautomerism is symmetry-forbidden but an alternative mechanism has been proposed on theoretical grounds which involves two consecutive 1,2-hydrogen shifts [see, for example, isomers of ionized acetic acid (6), (7), and (8) in reaction (2)].[28] MO theories have

$$\begin{array}{ccc} \text{HO}\diagdown\diagup\text{OH} \rceil^{+\cdot} & \text{HO}\diagdown\diagup\text{O}^{\cdot} & \\ \text{C} & \text{CH} & \\ \parallel & \mid & \rightarrow \text{MeCO}_2\text{H}\rceil^{+\cdot} \quad (2) \\ \text{CH}_2 & ^+\text{CH}_2 & \\ (6) & (7) & (8) \end{array}$$

also been used to suggest possible isomeric forms of the acetone radical ion.[29] Again the enol form was calculated to be more stable than its keto tautomer (by 20.3 kJ mol^{-1}). A total of 17 isomeric $C_3H_6O^{+\cdot}$ species were suggested by theory and some of them identified by ion cyclotron resonance (ICR) spectrometry.

Distinction between ionized vinyl ketene and other $C_4H_4O^{+\cdot}$ isomers is possible *via* measurement of heats of formation.[30,31] In these, as in other cases,[32,33] it is sometimes possible to employ appropriate thermochemical cycles to estimate bond energies or heats of formation of the neutral and ionic species involved.

The complicating factor of the kinetic shift in appearance energy measurements has been re-emphasized.[34–36] In theory, any kinetic shift should be less for

[23] J. L. Holmes and F. P. Lossing, *J. Am. Chem. Soc.*, 1980, **102**, 1591.
[24] H. Schwarz and C. Wesdemiotis, *Org. Mass Spectrom.*, 1979, **14**, 25.
[25] J. L. Holmes and F. P. Lossing, *Org. Mass. Spectrom.*, 1979, **14**, 512.
[26] J. L. Holmes and F. P. Lossing, *J. Am. Chem. Soc.*, 1980, **102**, 3732.
[27] H. Schwarz, D. H. Williams, and C. Wesdemiotis, *J. Am. Chem. Soc.*, 1978, **100**, 7052.
[28] J. S. Splitter and M. Calvin, *J. Am. Chem. Soc.*, 1979, **101**, 7329.
[29] W. J. Bouma, J. K. MacLeod, and L. Radom, *J. Am. Chem. Soc.*, 1980, **102**, 2246.
[30] J. L. Holmes and J. K. Terlouw, *J. Am. Chem. Soc.*, 1979, **101**, 4973.
[31] J. K. Terlouw, P. C. Burgers, and J. L. Holmes, *J. Am. Chem. Soc.*, 1979, **101**, 225.
[32] U. Büchler and J. Vogt, *Org. Mass Spectrom.*, 1979, **14**, 503.
[33] J. C. Traeger and R. G. McLoughlin, *J. Chem. Thermodyn.*, 1978, **10**, 505.
[34] H. M. Rosenstock, R. Stockbauer, and A. C. Parr, *J. Chem. Phys.*, 1979, **71**, 3708.
[35] M. A. Baldwin, *Org. Mass Spectrom.*, 1979, **14**, 601.
[36] R. G. McLoughlin and J. C. Traeger, *Org. Mass Spectrom.*, 1979, **14**, 434.

decompositions of metastable ions compared with those for higher-energy reactions occurring in the source. It has been contended that the semi-log plot for metastable ions (formed in the second field-free region of a reverse geometry instrument) can give reliable results for appearance energies, free of a significant kinetic shift.[35]

Another source of thermochemical measurements in recent years has evolved from the establishment of chemical equilibria in either ICR or high-pressure mass spectrometers. This has led to the determination of gas-phase acidities[13,37,38] and proton affinities.[12,39–41] The heat of formation of the t-butyl cation, an important reference energy for gas-phase proton affinity measurements,[42] has been measured by photoionization of methylpropane to be $678 \pm 3 \text{ kJ mol}^{-1}$,[43] considerably lower than previously accepted values. The adiabatic ionization energy of the t-butyl radical, measured from its photoelectron spectrum, has provided confirmatory evidence, giving a heat of formation for the t-butyl cation of $681 \pm 5 \text{ kJ mol}^{-1}$.[44]

Known heats of formation for hypothetical product ions and neutrals from mass spectral reactions can be employed simply to predict which decompositions are likely to be preferred. Transition-state energies and metastable peak shapes may also be suggested by such studies, which have been performed for series of C_nH_{2n-3} ions.[45] Conversely, heats of formation can sometimes be estimated by observing preferred reaction products. For example, approximate heats of formation have been assigned to substituted phenyl cations (10) by a simple iterative procedure.[46] Protonated disubstituted benzenes (9), generated by H_2 chemical ionization, were found in general to form structure (10) and/or (11), reaction scheme (3). The substituent X was varied and the propensities for the two reaction channels (a) or (b) were determined. Where the two reactions occurred with approximately equal tendency it was concluded that

$$\Delta H_f(10) + \Delta H_f(HX) \simeq \Delta H_f(11) + \Delta H_f(X)$$

[37] J. E. Bartmess, J. A. Scott, and R. T. McIver, *J. Am. Chem. Soc.*, 1979, **101**, 6056.
[38] F. J. Winkler and D. Stahl, *J. Am. Chem. Soc.*, 1979, **101**, 3685.
[39] P. Longevialle, J.-P. Girard, J.-C. Rossi, and M. Tichý, *Org. Mass Spectrom.*, 1979, **14**, 414.
[40] P. P. S. Saluja and P. Kebarle, *J. Am. Chem. Soc.*, 1979, **101**, 1084.
[41] J. H. Vajda and A. G. Harrison, *Int. J. Mass Spectrom. Ion Phys.*, 1979, **30**, 293.
[42] S. G. Lias, D. M. Shold, and P. Ausloos, *J. Am. Chem. Soc.*, 1980, **102**, 2540.
[43] R. G. McLoughlin and J. C. Traeger, *J. Am. Chem. Soc.*, 1979, **101**, 5791.
[44] F. A. Houle and J. L. Beauchamp, *J. Am. Chem. Soc.*, 1979, **101**, 4067.
[45] R. G. Bowen, B. J. Stapleton, and D. H. Williams, *Org. Mass Spectrom.*, 1978, **13**, 330.
[46] H.-W. Leung and A. G. Harrison, *J. Am. Chem. Soc.*, 1979, **101**, 3168.

An estimate of $\Delta H_f(10)$ from the other three terms is therefore possible and the use of further substituents (X) establishes limits for this parameter. For example, the two reaction channels occur with almost equal tendency for fragmentation of m-bromotoluene, giving an estimate for the heat of formation of the tolyl cation of 1058 kJ mol^{-1}, in good agreement with MO calculations. For m-fluoro- and m-chloro-toluene product (10) is favoured and for m-iodotoluene ion (11) is preferred. These data establish that the above heat of formation lies between 983 and 1112 kJ mol^{-1}.

Recent improvements in the reliability of MO calculations as applied to organic ions have meant that this technique is a powerful adjunct to measured heat of formation data in mass spectrometry.[8] Such calculations can give information on the energies and charge distributions for isomeric ions. However, the features of individual types of calculation and the results obtained from them will not be elaborated here. Incidental references to MO calculations on specific ions are to be found elsewhere in this chapter.

The publications discussed in this section form only a fraction of those in the recent literature that have employed thermochemical measurements. Such data have also been acquired to support arguments in many of the papers reported below.

3 Isomerization

It is an established fact that organic ions frequently undergo isomerization reactions before reaching the collector of the mass spectrometer. Evidence has been accumulated over the past fifteen years from a variety of techniques (see Table 1, p. 60). An isolated organic ion may have several reaction channels open but the availability and relative importance of these various channels depend, of course, on the internal energy. At low internal energy most organic ions in the mass spectrometer exist in a potential well, unable to react in any way, whereas at higher energies isomerization and/or decomposition reactions may occur. A general qualitative discussion of the competition between isomerization and decomposition in terms of potential-energy diagrams can be found in ref. 3, p. 212—215.

Isomerization reactions in the mass spectrometer are identified either from energy measurements (translational and internal), or from the product ion abundances, of the decomposition reactions undergone by the isomerizing ion. Techniques which involve mass selection of ionic species prior to decomposition are particularly valuable for the study of isomerization reactions (see, for example, *Collisional Activation*, p. 77, which is employed in many of the publications reported below).

Randomization.—This type of reaction (sometimes known as 'scrambling') is defined for the present purposes as the loss of positional identity of an atom or group of atoms prior to decomposition. It is frequently observed that randomization is more prevalent in lower-energy metastable ions compared with those observed from higher-energy reactions in the ion source. This is a simple consequence of the Quasi-Equilibrium theory of mass spectra[1—3] and arises where the randomization reactions have lower critical energies (and usually tighter transition states) than those which result in decomposition.

Some significant observations of reversible hydrogen transfers between remote positions of organic ions have been noted in the recent literature. For example, intramolecular ring-to-ring proton transfer reactions have been observed in ions (12)[47,48] generated by chemical ionization (CI). Equilibration takes place between all of the aromatic hydrogens, without involving those from the aliphatic chain. These processes occur in long-chain homologues (*e.g.* $n = 20$) of (12). The explanation for this phenomenon may be related to the stability conferred on ion–molecule complexes (in this case probably an intramolecular protonated dimer) as discussed below (see pp. 68—69).

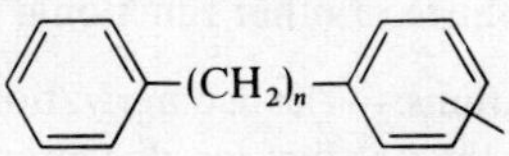

(12)

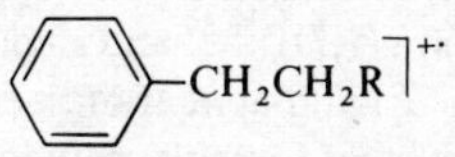

(13) R = CH_2Ph, $o,m,p\text{-}C_6H_4'CH_3$, or 1-cycloheptatrienyl

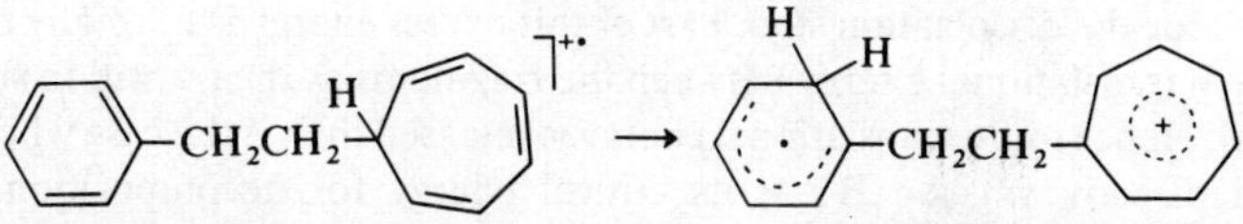

(14) → (15) (4)

In the radical ions (13), however, ring-to-ring reversible hydrogen transfer is prominent only for the cycloheptatrienyl isomer, occurring even in short-lived ions.[49] The reason for this isomerization is not entirely clear but it is argued that the 1-cycloheptatrienyl hydrogen is favourably disposed spatially and energetically for transfer to the phenyl ring [reaction (4)]. Together with intra-ring hydrogen transfers, this mechanism accounts for the extensive randomization.

Hydrogen–deuterium exchange reactions have been investigated for a variety of other labelled aromatic systems. The fullest information is usually obtained by comparing results from ions having 'short' (10 ps to 1 μs) and 'long' (>1 μs) lifetimes. In particular, the technique of field ionization kinetics[9] determines the fragmentation behaviour of ions over a wide range of lifetimes from 10 ps to 10 μs. By such means, different C- and/or H-randomization reactions have been distinguished.[50—54] For example, it was established that, as ion lifetime increases,

---

[47] D. Kuck, W. Bather, and H.-F. Grützmacher, *J. Am. Chem. Soc.*, 1979, **101**, 7154.
[48] D. Kuck and H.-F. Grützmacher, unpublished work.
[49] D. Kuck and H.-F. Grützmacher, *Org. Mass Spectrom.*, 1979, **14**, 86.
[50] P. J. Derrick, A. M. Falick, S. Lewis, and A. L. Burlingame, *J. Phys. Chem.*, 1979, **83**, 1567.
[51] J. Van der Greef and N. M. M. Nibbering, *Org. Mass Spectrom.*, 1979, **14**, 537.
[52] F. Borchers, K. Levsen, C. Wesdemiotis, and H. Schwarz, *Int. J. Mass Spectrom. Ion Phys.*, 1978, **28**, 203.
[53] P. Tecon, D. Stahl, and T. Gäumann, *Int. J. Mass Spectrom. Ion Phys.*, 1979, **29**, 363.
[54] D. H. Russell, M. L. Gross, J. Van der Greef, and N. M. M. Nibbering, *J. Am. Chem. Soc.*, 1979, **101**, 2086.

two distinct hydrogen interchange processes compete more successfully with loss of H_2O from the molecular ion of 3-phenylpropanol (16).[51] One randomization mechanism involves the benzylic and hydroxylic hydrogens, while the ortho-hydrogens participate in the second interchange process. This example illustrates

$$\text{C}_6\text{H}_5\text{-CH}_2\text{CH}_2\text{CH}_2\text{OH}$$

(16)

that randomization processes in bifunctional aromatic ions are rarely simple. The mass spectra and metastable ion decompositions of labelled nitrogen heterocycles have also revealed intricate hydrogen randomizations[55—57] which are often strongly modified by the presence of other functional groups.[55,57]

Rate-determining Isomerizations.—The isomerization of one ion structure to another may occur such that the product ion decomposes at a rate which is fast compared with the rate of isomerization. This type of reaction is known as a rate-determining isomerization (RDI)[58,59] and is most conveniently identified *via* translational-energy release (T) and appearance-energy measurements.

The potential-energy diagram for this category of reaction occurring in metastable ions is illustrated in Figure 1. When isomerization occurs from A^+ to B^+ *via* transition state (X), excess (non-fixed) energy is present in the transition state (Y) for the dissociation step. Part of this excess energy is normally released as relative translational energy between the fragments as the transition state (Y) is passed. Hence more translational energy is released than is the case when (Y) is reached directly *via* ion B^+ at its critical energy for decomposition. These differences in T values can be monitored *via* measurements of metastable peak profiles. Moreover, in the identification of RDI's by these methods, the determination of accurate critical energies [*e.g.* those of (X) and (Y)] strongly substantiates the evidence. Isotopic labelling data are also useful in removing ambiguities.

Where losses of identical species from isomeric ions lead to substantially different T values, it is sometimes a valid suggestion that an RDI has occurred. Tentative conclusions may be reinforced by critical-energy data. $C_3H_5^+$ ion formation from some $C_3H_5X^{+\cdot}$ parent ions (where X = Br or CO) appears to be accompanied by such an isomerization.[60,61] The occurrence of a rate-determining hydrogen migration in a reaction, which formally might be expected to be a simple cleavage reaction, has been identified in the molecular ion of methyl isobutyrate (17).[62,63] Loss of Me occurs with a T value which is at least an order of

---

[55] J.-L. Aubagnac and P. Campion, *Org. Mass Spectrom.*, 1979, **14**, 425.
[56] M. Corval, *Org. Mass Spectrom.*, 1979, **14**, 213.
[57] T. A. Molenaar-Langeveld, N. P. E. Vermeulen, N. M. M. Nibbering, R. P. Morgan, A. G. Brenton, J. H. Beynon, D. K. Sen Sharma, and K. R. Jennings, *Org. Mass Spectrom.*, 1979, **14**, 524.
[58] D. H. Williams, *Acc. Chem. Research*, 1977, **10**, 280.
[59] R. D. Bowen, D. H. Williams, and H. Schwarz, *Angew. Chem., Int. Ed. Engl.*, 1979, **18**, 451.
[60] R. D. Bowen, D. H. Williams, H. Schwarz, and C. Wesdemiotis, *J. Am. Chem. Soc.*, 1979, **101**, 4681.
[61] H. Schwarz, C. Wesdemiotis, K. Levsen, R. D. Bowen, and D. H. Williams, *Z. Naturforsch., Teil B*, 1979, **34**, 488.
[62] C. Wesdemiotis and H. Schwarz, *Angew. Chem., Int. Ed. Engl.*, 1978, **17**, 678.
[63] P. H. Hemberger, J. C. Kleingeld, K. Levsen, N. Mainzer, A. Mandelbaum, N. M. M. Nibbering, H. Schwarz, R. Weber, A. Weisz, and C. Wesdemiotis, *J. Am. Chem. Soc.*, 1980, **102**, 3736.

magnitude too large for a direct C—C cleavage. A hydrogen migration from the β-Me to the carbonyl group is invoked and is neatly supported by a primary kinetic deuterium isotope effect.

$$\underset{\text{Me}}{\overset{\text{Me}}{>}}\text{CH}-\text{CO}_2\text{Me}$$

(17)

Another consequence of an RDI is the discrimination, in the subsequent decomposition, against reactions with a stringent geometrical requirement. This is because of the higher excess internal energy present in the product isomeric ion (B^+ in Figure 1).

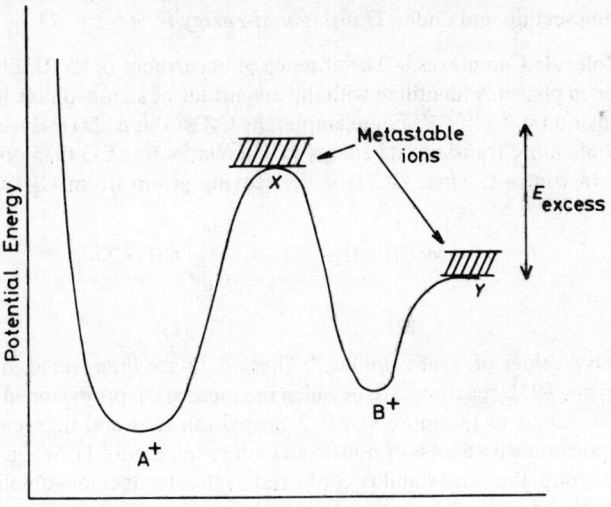

Figure 1 *A potential-energy diagram illustrating a rate-determining isomerization*

The difficulties in identifying an RDI from T measurements alone have been emphasized for some $C_3H_3O^+$ isomers.[64] The much larger value of T for CO loss from (18) compared with the same reaction from (19) might have been taken as an indication that an RDI (18) → (19) had occurred. Variable-electron-energy measurements established that the metastable peak from (18) is composite, having a narrow component of the same width as the peak for CO loss from (19). It was concluded that fragmentation of (18) occurs from ions of two different structures, one of which is ion (19). Isotopic labelling results similarly suggested a

$$\text{CH}\equiv\text{C}-\overset{+}{\text{CHOH}} \quad \text{CH}_2=\text{CH}-\text{CO}^+ \quad \text{MeCH}_2\overset{+}{\text{CHOH}} \quad \overset{+}{\text{CH}_2}\text{OCH}_2\text{Me}$$
(18) (19) (20) (21)

[64] J. L. Holmes, J. K. Terlouw, and P. C. Burgers, *Org. Mass Spectrom.*, 1980, **15**, 140.

composite metastable peak for C_2H_4 loss from (20)[65] and were accommodated in terms of isomerization reactions (20) → (21), not involving an RDI.

Measurements of T for decompositions of isomeric saturated hydrocarbon ions are sometimes consistent with the occurrence of threshold isomerizations to form decomposing ions with excess of energy. For example, critical-energy measurements suggest that the butane metastable molecular ion can isomerize to form ionized methylpropane with excess of non-fixed energy.[21] This mechanism requires that CH_4 loss from the butane molecular ion occurs with a greater T value than from methylpropane. This is borne out in practice. However, the decomposition reactions of ionized methylpropane itself are complex and non-classical structures are believed to be involved.[66]

The research reported in this section has demonstrated one use of the measurement of T in investigating gas-phase organic structures and reaction mechanisms. Further reference to the use of this parameter can be found in the following section and under *Translational-energy Release*, p. 74.

Ion–Molecule Complexes.—The absence or occurrence of an RDI has in certain cases been plausibly identified with the magnitude of an ion–dipole interaction in the transition state.[59,67–69] For example, the $C_4H_7O^+$ ion (22) releases an order of magnitude more translational energy (T) in eliminating CO than does its isomer (23).[67] In contrast, when OCH_2 is the leaving group (from $C_4H_9O^+$ ions) the

$$MeCH_2\overset{+}{C}H_2\cdots CO \rightarrow \overset{Me}{\underset{Me}{\diagdown}}\overset{+}{CH}\cdots CO \qquad (5)$$

(22)   (23)

respective values of T are similar.[68] These data are interpreted in terms of an exothermic RDI [reaction (5)], in which the inchoate 1-propyl ion in the complex (22) rearranges to the more stable 2-propyl ion (23) and then undergoes fast decomposition with excess of non-fixed energy (*cf.* Figure 1). When OCH_2 is the leaving group, the extra stability conferred by the stronger ion–dipole interaction renders the analogous reaction [*cf.* reaction (5)] reversible at energies corresponding to metastable transitions.

Ion–dipole stabilization has been invoked to account for the products of the slow, unimolecular dissociations of 'onium' ions of the general formula (24).[68–71]

$$R^1CH=\overset{+}{X}CR^2R^3CHR^4R^5$$

(24) R = H, alkyl

X = O, S, NH, or NMe

---

[65] J. L. Holmes, R. T. B. Rye, and J. K. Terlouw, *Org. Mass Spectrom.*, 1979, **14**, 606.
[66] P. T. Mead, K. F. Donchi, J. C. Traeger, J. R. Christie, and P. J. Derrick, *J. Am. Chem. Soc.*, 1980, **102**, 3364.
[67] D. H. Williams, B. J. Stapleton, and R. D. Bowen, *Tetrahedron Lett.*, 1978, 2919.
[68] R. D. Bowen and D. H. Williams, *Int. J. Mass Spectrom. Ion Phys.*, 1979, **29**, 47.
[69] R. D. Bowen and D. H. Williams, *J. Chem. Soc., Perkin Trans. 2*, 1980, 1411.
[70] R. D. Bowen and D. H. Williams, *J. Am. Chem. Soc.*, 1980, **102**, 2752.
[71] R. D. Bowen, *J. Chem. Soc., Perkin Trans. 2*, 1980, 1219.

A generalized mechanism is postulated in terms of a loose intermediate ion–molecule complex. Consider a specific example, the fragmentation of ion (25).[70] A complex is first formed in which acetaldehyde is loosely bound to the 1-butyl ion and stabilized by ion–dipole interaction. Isomerization then occurs,

$$\text{MeCH}=\overset{+}{\text{O}}-\text{CH}_2\text{CH}_2\text{CH}_2\text{Me} \longrightarrow \text{MeCH}=\text{O}\cdots\overset{+}{\text{H}}\begin{pmatrix}\text{Me}\\\text{Me}\end{pmatrix}$$

(25)   (26)

leading to complexes (e.g. 26) in which acetaldehyde is bound, via a common proton, to isomers of C_4H_8. In the subsequent decomposition, the proton is retained by the component of greater proton affinity (see Figure 2 and p. 71).[75]

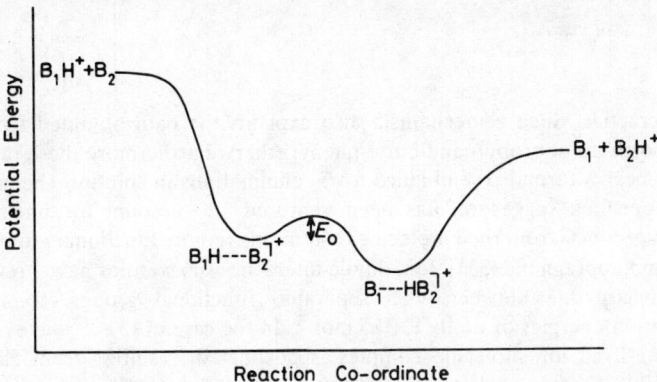

Figure 2 *A potential-energy diagram for proton transfer*

Although the above general explanation can involve a complex series of rearrangements it nevertheless elegantly explains a series of experimental facts, including the preferences between reaction products, results from deuterium-labelling studies, and measurements of T values. The approach is certainly worthy of general consideration for a wide variety of reactions in organic mass spectrometry and there are, of course, many precedents for hydrogen-bridged ions (e.g. in high-pressure ion sources) and neutrals (e.g. in boron chemistry).

Further support for the intermediacy of ion–molecule complexes in fragmentations has been presented for the molecular ion radicals of alkyl phenyl ethers.[72] The experimental evidence relies on the ability to detect and identify structurally (in an electron-bombardment flow reactor) the neutral olefins released in the transition. Consider, for example, the decomposition of the alkyl phenyl ether (27). Traditionally, the elimination of C_5H_{10} from (27) would be rationalized via a site-specific hydrogen transfer from one position of the alkyl chain to the oxygen atom. Concomitant with, or subsequent to, this rearrangment O—C bond fission occurs to form the products. An alternative mechanism

[72] T. H. Morton, *J. Am. Chem. Soc.*, 1980, **102**, 1596.

incorporates the intermediate ion–molecule complex (28; detailed structure not specified) in which the hydrocarbon ionic component has rearranged from 2,2-dimethylpropyl to the more stable 2-methyl-2-butyl structure. Proton transfer then occurs from the C_5H_{10} component to the phenoxide radical, liberating neutral C_5H_{10}. The isomeric distributions of the C_5H_{10} deprotonation products from the 2-methyl-2-butyl cation are known as a result of mass spectrometric identification of the neutrals from the authentic precursor ion. Hence this distribution can be compared with that obtained in the mass spectrometric decomposition of (27). The predicted ratio of about 1 : 1 for [(29)]:[(30)] is borne

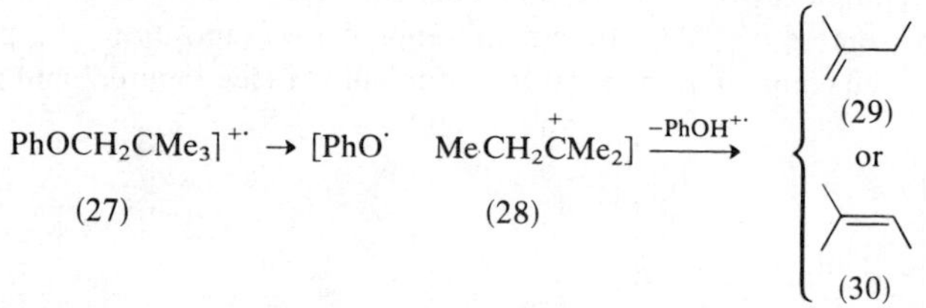

$$PhOCH_2CMe_3]^{+\cdot} \rightarrow [PhO^{\cdot} \quad Me\overset{+}{C}H_2CMe_2] \xrightarrow{-PhOH^{+\cdot}} \begin{cases} (29) \\ \text{or} \\ (30) \end{cases}$$

(27)   (28)

out in practice. Such a mechanism also explains the data obtained from the decompositions of propyl- and butyl-phenyl ethers. Furthermore these rationalizations bear a formal resemblance to E_1 eliminations in solution chemistry.[72]

An appealing suggestion has been ventured[73] to account for unexpected reaction products from rigid molecules containing remote functional groups [e.g. 3,20-diaminopregnane (31)]. Ion–dipole interactions in steroids have previously been suggested,[74] between well separated functional groups, to involve stabilization energies of up to 130 kJ mol^{-1}. In the case of (31),[73] the existence of a long-lived ion–molecule complex is postulated, resulting from cleavage of the 17—20 bond as shown. If the two resultant fragments move apart with sufficiently high relative velocity, then formation of (32) results. However, if the relative velocity is low, then an ion–molecule complex results between the two fragments, such that rotational interaction occurs. This postulate explains: (i) formation of (33) *via* a hydrogen transfer, (ii) hydrogen exchange between the 3- and 20-positions (~1 nm apart) prior to formation of (33), requiring rotation

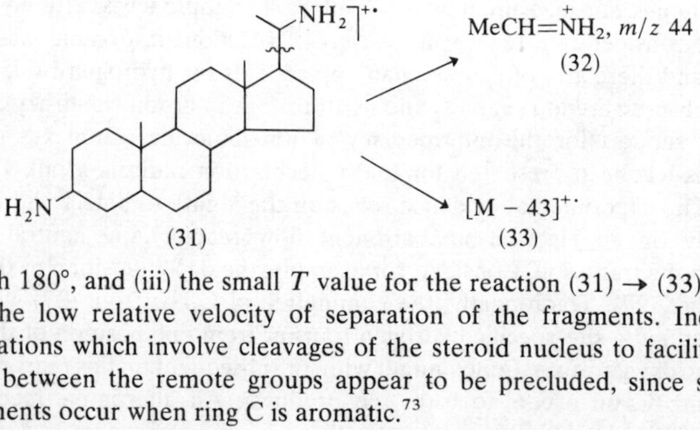

$MeCH=\overset{+}{N}H_2$, m/z 44   (32)

$[M-43]^{+\cdot}$ (33)

(31)

through 180°, and (iii) the small T value for the reaction (31) → (33), resulting from the low relative velocity of separation of the fragments. Incidentally, explanations which involve cleavages of the steroid nucleus to facilitate interaction between the remote groups appear to be precluded, since such rearrangements occur when ring C is aromatic.[73]

[73] P. Longevialle and R. Botter, *J. Chem. Soc., Chem. Commun.*, 1980, 823.

The ion-molecule complexes reported in this section are rationalized as stable species, existing in potential wells. They also correspond, in the cases discussed, to the intermediates in proton-transfer equilibria between two bases. One model which has been found to explain some proton-transfer data (obtained from an ICR instrument) between pyridine bases is expressed as a potential-energy diagram in Figure 2.[75] Two intermediate complexes exist in separate potential wells for this system and even the less stable of these complexes is thought to require a critical energy E_0 of ~20 kJ mol^{-1} for isomerization to the more stable complex.

4 Decomposition

Specific Rearrangements.—Perhaps the most powerful means for probing the mechanisms of ionic decompositions involving rearrangement remains isotopic labelling.[2] Determination of the fate of a labelled isotope, located at a specific position in the parent molecule, yields information on: (i) the bonds made and broken during the decomposition and/or (ii) the occurrence of atom randomization reactions. In appropriate situations, kinetic isotope effects, usually from investigations employing deuterium labelling, provide additional details about the fragmentation.

The stereochemical requirements of the McLafferty rearrangement occurring in series of 4-cyclohexylacetates (34) have been investigated using deuterium-labelled compounds.[76] A good correlation was found between the conformer distribution from which fragmentation occurs and that of the precursor neutrals before ionization. The results show that: (i) fragmentation occurs from the chair conformers and (ii) γ-hydrogen abstraction appears to be faster than chair–chair interconversion. Conformer populations in some cyclic and bicyclic β-amino alcohols have been evaluated *via* a CI method in an independent study.[77] A series of 3- and 4-arylcyclohexyl acetates has also been investigated for stereochemical variations.[78] In contrast to the t-butyl derivatives (34; R = CMe$_3$), the aryl derivatives showed substantial stereospecific differences in their mass spectra between the *cis* and *trans* unlabelled isomers.

OCOMe

R

(34) R = CMe$_3$, CHMe$_2$

Attention is being directed increasingly to the elucidation, by isotopic labelling, of some of the more complex fragmentation reactions in mass spectrometry. Extensive labelling in different positions of the molecule often accentuates the intricacy of the reactions. For example, various mechanisms have been ventured

---

[74] J. R. Dias, R. Ramachandra, and B. Nassim, *Org. Mass Spectrom.*, 1978, **13**, 307.
[75] J. M. Jasinski and J. I. Brauman, *J. Am. Chem. Soc.*, 1980, **102**, 2906.
[76] R. N. Rej, E. Bacon, and G. Eadon, *J. Am. Chem. Soc.*, 1979, **101**, 1668.
[77] P. Longevialle, J.-P. Girard, J.-C. Rossi, and M. Tichý, *Org. Mass. Spectrom.*, 1980, **15**, 268.
[78] J. Sharvit and A. Mandelbaum, *Org. Mass Spectrom.*, 1978, **13**, 303.

to explain the loss of Me from the molecular ion of 2-cyclohexenol (35).[79,80] Eleven labelled compounds were necessary to reveal that most of the Me fragments expelled in fast reactions contain the C-2 atom, together with the hydrogen atoms from C-1, C-2, and C-3.[79]

OH

(35)

The reactions of isotopically labelled alkanes, both short-[21] and long-chain,[81] have been investigated. High-energy direct cleavages of the molecular ions in the source occur largely without rearrangement of the carbon skeleton.[81] Many specific hydrogen transfers have been identified recently by deuterium labelling, including those in $C_3H_7O^+$,[65] $C_4H_5O^+$,[82] $C_2H_5S^+$,[83] $C_3H_7S^+$,[84] 1- and 2-acetoxytetralins (36; R = Ac),[85] alkylnitrobenzenes,[86] bicyclic diols,[87] and protonated

OR

(36) R = H or Ac

long-chain aldehydes.[88] From the molecular ion of benzyl cyanide, 1,1-elimination of HCN specifically occurs at short lifetimes (<1 ns) but the reaction is less specific for decompositions of long-lived ions.[89]

In aromatic systems *ortho* effects continue to supply instances of unusual rearrangement reactions which are diagnostically useful for structural identification.[86,90—97] The mechanisms are evaluated with the aid of isotopic labelling and metastable peak measurements, including determination of T values (see also *Translational-energy Release*, p.74).

[79] D. Braem, F. O. Gülacar, U. Burger, and A. Buchs, *Org. Mass Spectrom.*, 1979, **14**, 609.
[80] T. J. Adley, G. Matteo, and R. T. B. Rye, *Org. Mass Spectrom.*, 1979, **14**, 562.
[81] A. Lavanchy, R. Houriet, and T. Gäumann, *Org. Mass Spectrom.*, 1979, **14**, 79.
[82] P. C. Burgers, H. Hommes, and J. K. Terlouw, *Org. Mass Spectrom.*, 1979, **14**, 571.
[83] W. J. Broer, W. D. Weringa, and W. C. Nieuwpoort, *Org. Mass Spectrom.*, 1979, **14**, 543.
[84] W. J. Broer and W. D. Weringa, *Org. Mass Spectrom.*, 1979, **14**, 36.
[85] S. F. Wojinski and M. L. Gross, *Org. Mass Spectrom.*, 1979, **14**, 135.
[86] A. R. Butcher and C. B. Thomas, *Org. Mass Spectrom.*, 1979, **14**, 448.
[87] J. Deutsch and M. Chriki, *Org. Mass Spectrom.*, 1979, **14**, 482.
[88] H. E. Audier, A. Milliet, C. Perret, and P. Varenne, *Org. Mass Spectrom.*, 1979, **14**, 132.
[89] J. Van der Greef, T. A. Molenaar-Langeveld, and N. M. M. Nibbering, *Int. J. Mass Spectrom. Ion Phys.*, 1979, **29**, 11.
[90] H. Schwarz, *Top. Curr. Chem.*, 1978, **73**, 232.
[91] H. Florêncio and W. Heerma, *Org. Mass Spectrom.*, 1978, **13**, 368.
[92] S. Tajima, T. Azami, H. Shizuka, and T. Tsuchiya, *Org. Mass Spectrom.*, 1979, **14**, 499.
[93] M. A. Baldwin, *Org. Mass Spectrom.*, 1980, **15**, 109.
[94] R. Schubert and H.-F. Grützmacher, *Org. Mass Spectrom.*, 1980, **15**, 122.
[95] M. A Rabbih and E. T. M. Selim, *Org. Mass Spectrom.*, 1980, **15**, 195.
[96] D. V. Ramana and N. Sundaram, *Org. Mass Spectrom.*, 1980, **15**, 220.
[97] H. Schwarz and C. Köppel, in 'The Chemistry of Ketenes, Allenes and Related Compounds,' ed. S. Patai, Wiley-Interscience, 1980, Chapter 6, p. 203.

Fragmentation Rules.—This section contains a brief survey of some rules in mass spectrometry which govern fragmentation reactions of positive organic ions. The discussion is confined to those which have been covered in the recent literature.

Although even-electron ions in most cases decompose to products in which all electrons are paired (the 'even-electron' rule),[2] anomalies are constantly being uncovered. Many known exceptions have been enumerated in a recent review article.[98] For example, the dimethylstilbene molecular ion (37) eliminates successively two methyl radicals to generate ion (39) [reaction (6)], relative abundance 74%, which formally corresponds to the phenanthrene radical ion. The

$$\text{(37)} \xrightarrow{} [M-\text{Me}]^+ \xrightarrow{} \text{(39)} \quad (6)$$

(37)   (38)   (39)

second step in this overall process is confirmed by a metastable peak and violates the even-electron rule. In this and many other examples quoted, the contravening reaction [e.g. (38) → (39)] apparently results in formation of an ionized aromatic species. The precise explanations for these effects of course lie in the energetic and kinetic requirements of the reactions from the individual even-electron ions [e.g. (38)].

One convenient means of examining the reactions of even-electron ions is *via* collisional activation (CA) mass spectrometry. Ions are first mass-selected and the products of their collision-induced dissociations are then identified (see also *Collisional Activation*, p. 77). Such reactions for even-electron ions have been characterized following either ionization by electron impact[99] or chemical ionization[99-101] (usually protonation). Several general features emerge. The fragmentations of protonated molecules involve loss of alkanes, alkenes, and such structure-dependent species as ammonia and amines (from protonated amines) and alcohols and acids (from protonated esters). The reactions are often best rationalized in terms of 4-centre transition states. Alkyl radicals are also eliminated, contravening the even-electron rule. In bifunctional protonated molecules there is evidence for the formation of cyclic ions prior to decomposition.[102] Finally, it is becoming apparent that energetic considerations (*e.g.* proton affinities, see also *Ion–Molecule Complexes*, p. 68, and Figure 2) are often useful in predicting preferred fragmentations of even-electron ions.

In the retro Diels–Alder reaction, the charge should remain on the fragment of lower ionization energy, assuming that any reverse activation energies are negligible. This is one statement of the Stevenson–Audier rule.[2] Several such reactions have been studied recently with the aid of deuterium labelling, for compounds where formally the same products result whichever portion retains

---

[98] M. Karni and A. Mandelbaum, *Org. Mass Spectrom.*, 1980, **15**, 53.
[99] F. W. McLafferty, *Org. Mass Spectrom.*, 1980, **15**, 114.
[100] M. L. Sigsby, R. J. Day, and R. G. Cooks, *Org. Mass Spectrom.*, 1979, **14**, 273.
[101] M. L. Sigsby, R. J. Day, and R. G. Cooks, *Org. Mass Spectrom.*, 1979, **14**, 556.
[102] C. C. Van der Sande, F. Van der Gaever, L. Dhaenens, and R. Myngheer, *Org. Mass Spectrom.*, 1979, **14**, 191.

the charge.[103] Compound (40) fits this criterion. Where the ratio of the two possible fragment ions {*e.g.* [(41)]:[(42)]} differs substantially from unity it is contended that rotationally excited (rather than planar) dienes are formed in one of the products.[103] For example, in the reactions of (40), the discrepancy from unity is interpreted by formation of the rotationally excited diene (42) and planar (41). The ionization energy of neutral (42) would therefore be lower than that of neutral (41): and (42) would tend to bear the charge. In a related study, formally

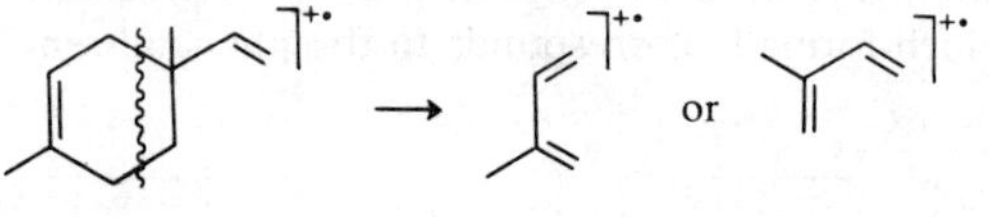

(40)   (41)   (42)

identical products of the McLafferty rearrangement from a hydrocarbon molecular ion were formed with equal probability.[104] This confirms that the structures and energy states of the respective ions are identical. The relative propensities of formation of the two possible McLafferty rearrangement products (not formally identical) have been investigated from a series of aliphatic aldehydes.[105] The reaction was suggested to be stepwise and both field ionization kinetics (FIK) and measurements of heats of formation were used to substantiate the arguments.

Translational-energy Release.—One of the most extensively measured parameters for the extraction of information on ion structures and fragmentation mechanisms is the translational-energy release (T) accompanying an ionic decomposition. Provided that standard means are employed to extract the experimental value, this parameter is reproducible from different laboratory systems. T is readily measured from metastable peak widths on magnetic sector instruments,[4] even of Mattauch–Herzog geometry.[106]

The simplest application of this measurement lies in the direct comparison of isomeric structures as a criterion for identity or non-identity. Ions of the same structure but generated by different means will have very similar internal energies and lifetimes when they decompose unimolecularly in a field-free region. (Ions formed *via* an RDI are exceptions—see pp. 66—68.) Therefore, in general the equality of T values from isomeric ions indicates identical structures. For example, loss of HCN from the isomeric molecular ions (43)—(45) occurs with the

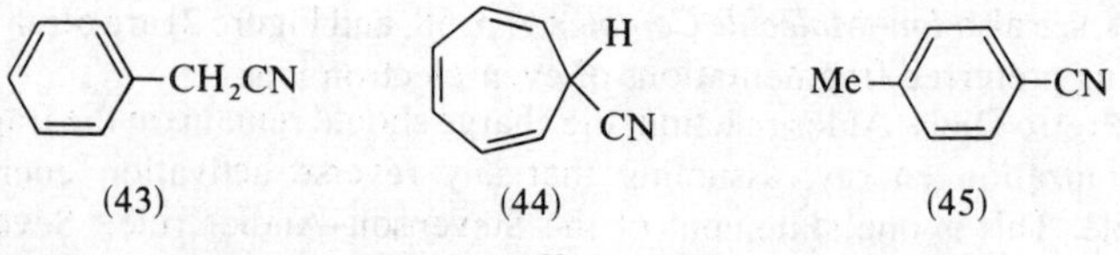

(43)   (44)   (45)

same release of translational energy.[89] Additional evidence, from FIK and isotopic labelling studies, suggests that the common ion has the benzyl cyanide structure (43).

---

[103] F. Tureček and V. Hanuš, *Org. Mass Spectrom.*, 1980, **15**, 4.
[104] F. Tureček and V. Hanuš, *Org. Mass Spectrom.*, 1980, **15**, 8.
[105] R. P. Morgan, P. J. Derrick, and A. G. Loudon, *J. Chem. Soc., Perkin Trans. 2*, 1980, 306.
[106] U. Herzig and P. Krenmayr, *Org. Mass Spectrom.*, 1979, **14**, 75.

Isomerization of the molecular ions of polychlorinated biphenyls (PCB's) has been suggested as a reason for the lack of specificity in their spectra but T measurements for Cl and Cl_2 loss provide some correlations with the structure before ionization.[107] For example, a large value of T (~100 meV) accompanying Cl loss is diagnostic of *meta*-chloro without *ortho*-chloro substituents. There appears to be a correlation between T and the fundamental C—Cl vibrational stretching mode, which is larger for the above-mentioned PCB's.

Although T values usually show a good degree of reproducibility from one instrument to another, there remain contradictions in the literature. For example, some controversy has surrounded the measurement of T for loss of CO from ionized phenol[108,109] and these data have a bearing on whether or not the decomposition occurs from the cyclohexa-2,4-dienone structure.

It is commonly observed that T for a given transition increases when the reaction is collision-induced as opposed to unimolecular. In simple terms, this is merely a consequence of greater non-fixed energy in the transition state following collisional excitation (see in addition the following section). Reverse effects have been observed, however, and require explanation in terms of energy partitioning (between translational and internal) when the activated complex breaks to form products. In an unusual example, the loss of H_2 from protonated methanol $MeOH_2^+$ releases 930 meV in the unimolecular dissociation but only 330 meV following collision.[110] It is suggested from MINDO/3 calculations that a smaller proportion of available energy passes into translation following collisional activation, owing to a loosening of the activated complex at higher energy.

Attempts have been made[94,111] to relate the 'energy partitioning quotient' (the fraction of reverse critical energy released as translation in the products) to the position of the transition state on the reaction co-ordinate.[112] For example, this quotient for reaction (7) from the molecular ions of 2-benzoylpyridines is large

(46)   (47)   (7)

(~0.4) for H and Me substituents but considerably smaller (~0.1) when X = Cl, Br, I, or NO_2.[94] A bond is formed as well as broken in the reaction as depicted and it is contended that: (i) a larger fraction of energy is channelled into translation when the transition state is 'late' and most of the energy has been used to break the C—X bond and (ii) the energy quotient is lower when the transition state is

---

[107] J. R. Hass, M. M. Bursey, L. A. Levy, and D. J. Harvan, *Org. Mass Spectrom.*, 1979, **14**, 319.
[108] D. H. Russell, M. L. Gross, J. Van der Greef, and N. M. M. Nibbering, *Org. Mass Spectrom.*, 1979, **14**, 474.
[109] A. Maquestiau, R. Flammang, G. L. Glish, J. A. Laramée, and R. G. Cooks, *Org. Mass Spectrom.*, 1980, **15**, 131.
[110] R. J. Day, D. A. Krause, W. L. Jorgensen, and R. G. Cooks, *Int. J. Mass Spectrom. Ion Phys.*, 1979, **30**, 83.
[111] B. Schaldach and H.-F. Grützmacher, *Int. J. Mass Spectrom. Ion Phys.*, 1979, **31**, 271.
[112] A. R. Miller, *J. Am. Chem. Soc.*, 1978, **100**, 1984.

'early' on the reaction co-ordinate and the C—X bond is weak, directing most of the energy into internal energy of the products. If this explanation is correct, then the energy quotient depends largely on the C—X bond energy in (46).

A large difference was observed between the translational-energy releases for the losses of H_2O (R = H) and $MeCO_2H$ (R = Ac) via a 1,4-elimination from the molecular ion of the 1-isomer of (36).[85] In the case of the alcohol, the large T value (270 meV) is postulated to indicate a 'late' barrier for H_2O loss.

Examples given above show that the effect of substituents on the T value for a common reaction is not straightforward. Nevertheless, attempts have been made to relate T to an appropriate substituent constant. For example, T for loss of Me from the molecular ions of p-substituted anisoles (48) correlates reasonably well

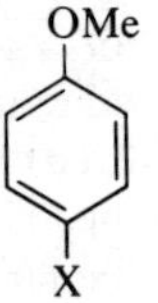

OMe

X

(48)

with a dual-parameter substituent constant.[113] However, an assumption was made that the energy-partitioning quotient is constant. In this and related cases there is a requirement for accurate heat of formation data to clarify the validity of simplifying assumptions.

The partition of reverse activation energy into translation has been approached using a 'dynamical theory'.[114] The approach considers the atomic motions comprising the transition-state reaction co-ordinate. As applied to hydrogen rearrangements,[115] the theory predicts that: (i) elimination of H_2 is likely to proceed with a significant energy-partitioning quotient but (ii) hydrogen rearrangements in large ions leading to fragments of comparable mass are not likely to partition much reverse critical energy into translation.

During loss of ketene from the acetanilide molecular ion (49), motion of the oxygen atom occurs towards the incipient aniline ion, in the opposite direction to

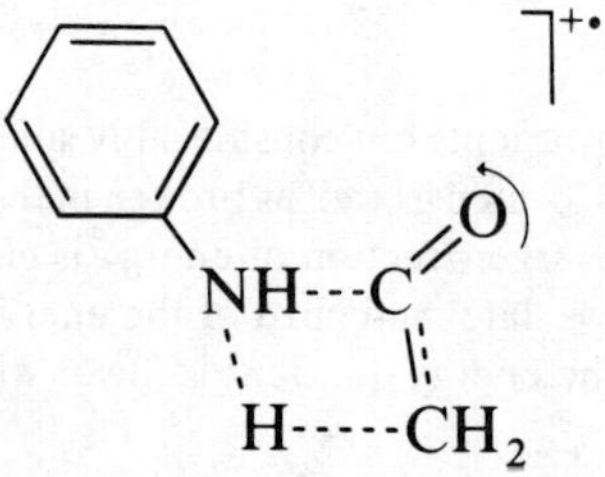

NH---C=O
  ⋮   ‖
  H----CH$_2$

(49)

that required for fragmentation. The translation energy released in this motion does not therefore contribute to fragment separation and probably appears as

---

[113] H. Florencio, P. C. Vijfhuizen, W. Heerma, and G. Dijkstra, Org. Mass Spectrom., 1979, **14**, 198.
[114] J. R. Christie, P. J. Derrick, and G. J. Rickard, J. Chem. Soc., Faraday Trans. 2, 1978, **74**, 304.

vibration. Consequently, the energy-partitioning quotient into translation is less than 10%.[115] In contrast, the 1,2-elimination of H_2 from $C_2H_6^{+\cdot}$ leads to a large energy-partitioning quotient and conservation of momentum demands that H_2 carries away 28/30 of T. Furthermore, the departing H atoms are both moving in the correct direction for fragment separation.[115]

It is evident from the coverage in this section and in the section *Rate-determining Isomerizations*, pp. 66—68, that there is a variety of factors influencing the T value for a reaction and the overall situation is now well understood (see also ref. 116). Evidently a better understanding of the factors affecting T will not be forthcoming without a careful determination of the geometry of the transition state for individual decomposing ions.

Collisional Activation.—Ions reaching the field-free region(s) of the mass spectrometer normally have lifetimes of 5—10 μs. They consist largely of stable ions, having insufficient energy to decompose, together with metastable ions, which decompose unimolecularly between source and collector. Ions passing through the field-free regions can be given a boost of internal energy *via* collisional activation[11,117,118] (or *via* laser radiation,[119] which adds a controlled amount of energy). This extra excitation energy causes additional decomposition, and information on ion structures and fragmentation mechanisms is then forthcoming from the ion products and their relative abundances. Such collisional activation (CA) spectra of ions of a particular m/z value are obtained by (i) selection in a magnetic field,[4,118,120,121] (ii) appropriate 'linked' scanning,[122] and (iii) the use of tandem quadrupole mass spectrometers.[123,124] The type of information obtainable from the last technique requires further evaluation but appears to be promising.

The internal excitation energy acquired by collision originates from a fraction of the translational energy of the incident ion. The average energy added increases with increasing incident translational energy and decreases with increasing ion mass.[125] The most probable internal energy added is between 1 and 5 eV[125] although there is ample evidence that some collisions add 10—50 eV and cause high-energy decompositions, including charge-inversion[126—130]

---

[115] N. W. Cole, G. J. Rickard, J. R. Christie, and P. J. Derrick, *Org. Mass Spectrom.*, 1979, **14**, 337.
[116] D. Stahl and H. Schwarz, *Int. J. Mass Spectrom. Ion Phys.*, 1980, **34**, 387.
[117] F. W. McLafferty, *Philos. Trans. R. Soc. London, Ser. A*, 1979, **293**, 93.
[118] R. W. Kondrat and R. G. Cooks, *Anal. Chem.*, 1978, **50**, 81A.
[119] E. S. Mukhtar, I. W. Griffiths, F. M. Harris, and J. H. Beynon, *Org. Mass Spectrom.*, 1980, **15**, 51.
[120] F. W. McLafferty, *Acc. Chem. Res.*, 1980, **13**, 35.
[121] F. W. McLafferty, P. J. Todd, D. C. McGilvery, and M. A. Baldwin, *J. Am. Chem. Soc.*, 1980, **102**, 3360.
[122] U. P. Schlunegger, 'Advanced Mass Spectrometry', Pergamon Press, 1980.
[123] R. A. Yost, C. G. Enke, D. C. McGilvery, D. Smith, and J. D. Morrison, *Int. J. Mass Spectrom. Ion Phys.*, 1979, **30**, 127.
[124] D. F. Hunt, J. Shabanowitz, and A. B. Giordani, *Anal. Chem.*, 1980, **52**, 386.
[125] M. S. Kim and F. W. McLafferty, *J. Am. Chem. Soc.*, 1978, **100**, 3279.
[126] M. M. Bursey, J. R. Hass, D. J. Harvan, and C. E. Parker, *J. Am. Chem. Soc.*, 1979, **101**, 5485.
[127] M. M. Bursey, D. J. Harvan, C. E. Parker, L. G. Pedersen, and J. R. Hass, *J. Am. Chem. Soc.*, 1979, **101**, 5489.
[128] I. Howe, J. H. Bowie, J. E. Szulejko, and J. H. Beynon, *J. Chem. Soc., Chem. Commun.*, 1979, 983.
[129] J. E. Szulejko, J. H. Bowie, I. Howe, and J. H. Beynon, *Int. J. Mass Spectrom. Ion Phys.*, 1980, **34**, 99.
[130] G. A. McClusky, R. W. Kondrat, and R. G. Cooks, *J. Am. Chem. Soc.*, 1978, **100**, 6045.

and charge-stripping[131–135] reactions. Because of the wide range of internal energies added to the ion by CA, the hypothesis[117] that the relative abundance of ions in the CA spectrum shows a negligible dependence on the internal-energy distribution of the incident ion is generally accepted. There has, however, been some disagreement in this respect over the CA spectra of benzoyl ions obtained from different sources.[136,137]

CA studies on ions containing C, H, and O have revealed a remarkable number of stable isomeric species. For example, the $C_4H_5O^+$ ion has been found from CA spectra to exist in more than ten isomeric forms,[82,138–140] including four ions of the type $C_3H_5CO^+$, four ether-type ions, and several alcohol-type ions. $C_8H_6O^{+\cdot}$ ions have been identified in several structural forms.[141] In contrast, $C_2H_3O^+$ ions generated from various sources yield identical CA spectra, with the exception of that generated from glycerol, which appears to yield the structurally distinct isomer $CH_2=C=OH^+$.[142] Other isomers interconvert below the critical energy for decomposition, the acetyl ion being the most stable. In the case of COH_3^+ [143] and CSH_3^+ [144] ions, support from MO calculations was given to conclusions from CA spectra. In the former case, $H_2C=OH^+$ is the most stable isomer. A variety of $C_6H_5N^{+\cdot}$ isomers has also been distinguished by means of their CA spectra from 26 different precursors.[145]

Sometimes CA constitutes the only feasible technique for investigating the decompositions of stable, even-electron ions formed by chemical ionization (see in addition the following section). In one joint application of CI and CA the most stable structure of $C_3H_5^+$ formed in the methane CI plasma was identified as the allyl cation.[146] Other small hydrocarbon ions investigated (by CA/EI) include $C_3H_4^{+\cdot}$ [147] and $C_3H_6^{+\cdot}$.[134]

A procedure employing CA has been devised to obtain a rough estimate of the isomeric composition of $C_7H_7^+$ ion mixtures derived from different sources.[19,148] The method requires that reference CA spectra of 'pure' ions from appropriate molecules be available. Characteristic ions and their relative abundances are

[131] T. Ast and J. H. Beynon, *Int. J. Mass Spectrom. Ion Phys.*, 1980, **32**, 385.
[132] A. Maquestiau, Y. Van Haverbeke, R. Flammang, and A. Menu, *Bull. Soc. Chim. Belg.*, 1979, **88**, 53.
[133] J. L. Holmes, J. K. Terlouw, P. C. Burgers, and R. T. B. Rye, *Org. Mass Spectrom.*, 1980, **15**, 149.
[134] R. D. Bowen, M. P. Barbalas, F. P. Pagano, P. J. Todd, and F. W. McLafferty, *Org. Mass Spectrom.*, 1980, **15**, 51.
[135] Ref. 3, p. 265.
[136] C. J. Porter, R. P. Morgan, and J. H. Beynon, *Int. J. Mass Spectrom. Ion Phys.*, 1978, **28**, 321.
[137] F. W. McLafferty, A. Hirota, M. P. Barbalas, and R. F. Pegues, *Int. J. Mass Spectrom. Ion Phys.* 1980, **35**, 299.
[138] H. Hommes and J. K. Terlouw, *Org. Mass Spectrom.* 1979, **14**, 51.
[139] J. K. Terlouw, P. C. Burgers, and H. Hommes, *Org. Mass Spectrom.*, 1979, **14**, 387.
[140] J. K. Terlouw, P. C. Burgers, and H. Hommes, *Org. Mass Spectrom.*, 1979, **14**, 574.
[141] C. G. Van den Heuvel, N. M. M. Nibbering, H. Heimbach, and K. Levsen, *Org. Mass Spectrom.*, 1979, **14**, 550.
[142] R. Weber and K. Levsen, *Org. Mass Spectrom.*, 1980, **15**, 138.
[143] J. D. Dill, C. L. Fischer, and F. W. McLafferty, *J. Am. Chem. Soc.* 1979, **101**, 6531.
[144] J. D. Dill and F. W. McLafferty, *J. Am. Chem. Soc.*, 1979, **101**, 6526.
[145] A. Maquestiau, Y. Van Haverbeke, R. Flammang, A. Menu, and C. Wentrup, *Org. Mass Spectrom.*, 1978, **13**, 518.
[146] C. Wesdemiotis, R. Wolfschütz, G. Höhne, and H. Schwarz, *Org. Mass Spectrom.*, 1979, **14**, 231.
[147] W. Wagner, K. Levsen, and C. Lifshitz, *Org. Mass Spectrom.*, 1980, **15**, 271.
[148] F. W. McLafferty and F. M. Bockhoff, *Org. Mass Spectrom.*, 1979, **14**, 181.

selected from each of the pure spectra and the composition of unknown mixtures is determined by interpolation between these extreme abundances. For example, the ratio m/z 77:74 is selected as a parameter to identify the structural components of $C_7H_7^+$ mixtures, being 3.1 for benzyl and 0.16 for *ortho*-tolyl. Intermediate values of this parameter for the CA spectra from (50) are used to

Me

X

(50)

determine the percentage of tolyl ions by simple interpolation. For reliability the method requires that (i) pure CA spectra can be identified and (ii) CA spectra are independent of the internal energy of the precursor ion.

An alternative means of identifying product ions having different internal energy requirements following CA is afforded by selection of different scattering angles through which the ions are directed.[149-152] This method in effect selects decomposing ions of different energy.

It has been emphasized that caution should be exercised when interpreting sharp peaks in CA spectra; they may be artefact peaks due to decompositions along the flight path before reaching the collision region.[153,154]

Comparisons with Solution Chemistry.—One fundamental contribution that mass spectrometry has made, and will continue to make, to mechanistic organic chemistry is to the understanding of the role of solvent in liquid-phase ionic organic reactions. This is effected by comparison of reactions, or parameters measured from those reactions, between the gas phase and solution. Success in this field has been achieved mainly since the advent of ICR and CI mass spectrometry, because equilibrium and protonation reactions are conveniently observed *via* these techniques.[12-14]

Proton-catalysed rearrangements, a class of reactions widely occurring in solution, can be investigated conveniently in a CI source, employing an ion such as CH_5^+ or $C_4H_9^+$ as the protonating reagent. The method relies on being able to identify a rearrangement product *via* its decomposition reactions, which for chemically ionized species in the ion source are often few. Collisional activation (see above) has sometimes been employed to characterize the rearrangement ions. For example, the acid-catalysed Fischer indole synthesis, shown for the conversion of acetonephenylhydrazone (51) to 2-methylindole (52), has been demonstrated to occur in the gas phase.[155] The CA spectrum of m/z 132

---

[149] J. A. Laramée, P. H. Hemberger, and R. G. Cooks, *Int. J. Mass Spectrom. Ion Phys.*, 1980, **33**, 231.
[150] D. M. Fedor and R. G. Cooks, *Anal. Chem.*, 1980, **52**, 679.
[151] J. A. Laramée, P. H. Hemberger, and R. G. Cooks, *J. Am. Chem. Soc.*, 1979, **101**, 6460.
[152] J. A. Laramée, J. J. Carmody, and R. G. Cooks, *Int. J. Mass Spectrom. Ion Phys.*, 1979, **31**, 333.
[153] T. Ast, M. H. Bozorgzadeh, J. L. Wiebers, J. H. Beynon, and A. G. Brenton, *Org. Mass Spectrom.*, 1979, **14**, 313.
[154] B. Shaldach and H.-F. Grützmacher, *Org. Mass. Spectrom.*, 1980, **15**, 166.
[155] G. L. Glish and R. G. Cooks, *J. Am. Chem. Soc.*, 1978, **100**, 6720.

[M + H−NH₃]⁺ from (51) was found to be identical with the CA spectrum of m/z 132 [M + H]⁺ from (52). The Beckmann rearrangement of ketoximes has

(51) → (52)

been neatly demonstrated to occur in the gas phase.[156,157] The CA spectrum of (54) produced by protonation of oxime (53) in the gas phase is identical to that produced by gas-phase ethylation of PhNC. The [M + Et]⁺ ion produced from PhCN, however, yields a different spectrum. These results illustrate the stereospecific nature of the transposition in the gas phase.

$$\underset{Et}{\overset{Ph}{>}}C=N\underset{OH}{\diagdown} \xrightarrow{H^+} Et-C\equiv \overset{+}{N}-Ph + H_2O$$

(53)   (54)

Proton-catalysed rearrangements of 1,2-diols (e.g. the pinacol rearrangement) occur readily in solution and have been shown to occur in the gas phase.[155,158,159] Migration of groups appears to occur as in the solution reactions, for the cases studied. Intramolecular gas-phase migrations have also been demonstrated in the isomerization of protonated aldehydes to protonated ketones, known to occur in superacid media.[158] For example, the CA spectra of m/z 197 [M + H]⁺, from diphenylacetaldehyde (55) and deoxybenzoin (58), are almost identical. Prominent peaks occur in the CA spectra at m/z values of 105 (PhCO⁺) and 91

(C₇H₇⁺), indicating that decomposition occurs from structure (58). Furthermore, protonated stilbene oxide (56) generates the same ion, as shown by its CA

---

[156] A. Maquestiau, Y. Van Haverbeke, R. Flammang, and P. Meyrant, *Org. Mass Spectrom.*, 1980, **15**, 80.
[157] A. Maquestiau, Y. Van Haverbeke, C. de Meyer, C. Duthoit, P. Meyrant, and R. Flammang, *Nouv. J. Chim.*, 1979, **3**, 517.
[158] A. Maquestiau, R. Flammang, M. Flammang-Barbieux, H. Mispreuve, I. Howe, and J. H. Beynon, *Tetrahedron*, 1980, **36**, 1993.
[159] R. Wolfschutz, H. Schwarz, W. Blum, and W. J. Richter, *Org. Mass Spectrom.*, 1979, **14**, 462.

spectrum. The gas-phase proton-catalysed dehydration of the 1,4-diketones (59; R = CH_3 or Ph) to protonated furans (60) has also been demonstrated to occur by combined application of CI and CA.[160] The gas-phase ring-opening reaction (60) → (59) was not achieved and the reaction of (59) with NH_4^+ did not result in formation of a pyrrole.

$$RCOCH_2CH_2COR \xrightarrow{H^+} R-\underset{O}{\text{furan}}-R + H_2O$$

(59)   (60)

The mechanisms of aromatic substitution reactions can be conveniently studied in the gas phase since the reagent ions can often be generated in a CI source. For example, NH_4^+ reacts as an electrophile with certain monosubstituted benzenes, presumably forming a sigma complex.[161] The mechanism of the gas-phase nitrosation of benzene has been studied in detail, utilizing photodissociation, ICR, and CA spectroscopy.[162] Expected intermediates in the reaction pathway, $C_6H_6NO^+$ (π-complex) and $C_6H_5NOH^+$, were identified as distinct species and their roles in the reaction co-ordinate discussed. The isomeric $C_{12}H_{18}NO^+$ ions produced from the nitrosation of (61) and (62) in a CI source were distinct, since their CA spectra are different.[163]

(61)   (62)

Collisional activation has also been employed to determine the positions of protonation and ethylation on a series of aromatic amines.[164] For example, protonation or ethylation of pyridine in the gas phase appears to occur on the nitrogen atom, since the isomeric ions generated by ethylation of pyridine (63) and protonation of the ethylpyridines (64) yield quite different CA spectra.

(63)   (64)

The above discussion has been confined to examples where 'solvent effects' are completely absent. However, the reactions of partially solvated ions have also been monitored in the mass spectrometer.[37,38,165] For example, the kinetics of the

---

[160] H. Schwarz, C. Wesdemiotis, and R. Wolfschütz, *Tetrahedron*, 1980, **36**, 929.
[161] J. Van Thuijl, W. C. M. M. Luijten, and W. Onkenhout, *J. Chem. Soc., Chem. Commun.*, 1980, 106.
[162] W. D. Reents and B. S. Freiser, *J. Am. Chem. Soc.*, 1980, **102**, 271.
[163] R. Wolfschütz and H. Schwarz, *Int. J. Mass Spectrom. Ion Phys.*, 1980, **33**, 291.
[164] A. Maquestiau, Y. Van Haverbeke, H. Mispreuve, R. Flammang, J. A. Harris, I. Howe, and J. H. Beynon, *Org. Mass Spectrom.*, 1980, **15**, 144.
[165] D. K. Bohme, G. I. MacKay, and S. D. Tanner, *J. Am. Chem. Soc.* 1979, **101**, 3724.

reactions of $H_3O^+ \cdot nH_2O$ with various organic substrates have been investigated in a flowing afterglow apparatus.[165]

5 Summary

Mass spectrometry has progressed enormously since the early 1960's when virtually the only fundamental information obtainable concerning gas-phase ion chemistry was *via* the electron-impact mass spectrum, with perhaps limited use of metastable peaks and measurements of heats of formation to augment the studies. Nowadays, a large amount of supplementary information can be acquired. Positively or negatively charged ions can be generated by several different means and they and their fragment ions analysed and interconnected by appropriate combinations of magnetic, electric, and radiofrequency fields. High- or low-energy reactions can be studied by selection of an appropriate ionization method and/or 'time window' for analysis of the ions. Molecular or fragment ions of individual m/z value can be selected for study. Excitation of ions along the flight path is achieved, if required, by employing a collision gas or laser radiation. Energy parameters (both internal and translational) can be measured for a reaction.

A wealth of fundamental information is therefore obtainable for any particular ion or ionic reaction mechanism by appropriate selection from the techniques discussed in this chapter. In many cases, a satisfactory answer to a particular problem in gas-phase ion chemistry is attainable only from a combination of methods. This fact is amply illustrated in ref. 63, where ICR, CA, FIK, ^{2}H- and ^{13}C-labelling, and deuterium isotope effect studies were all utilized to study the reactions of methyl isobutyrate (17). In many cases, application of more than one technique is desirable owing to the shortcomings accompanying some of the measurements. The most glaring of these in the recent literature are measurements of heats of formation, which can be seriously in error unless obtained *via* methods employing photoionization or energy-resolved electron beams.

Significant progress has been made in the period covered by this report in identifying intermediates in the gas-phase decomposition of ions. In particular, the importance of ion–molecule complexes as reaction intermediates has been recognized (see pp. 68—71). There has been a continuing tendency to tackle problems involving intricate ionic decompositions requiring a variety of measurements. Isotopic labelling usually forms an essential part of such studies. Notable advances have been made in quantifying the mixture composition of isomeric ions (see pp. 78—79). Favourite ions such as $C_7H_8^{+\cdot}$, $C_7H_7^+$, and $C_3H_6O^{+\cdot}$ have been re-examined, as have the mechanisms of the McLafferty and retro Diels–Alder reactions and perennial reactions such as the loss of CO from ionized phenol and CH_2CO from the acetanilide molecular ion.

Mass spectrometry remains primarily a tool for structural analysis. However, fundamental studies, such as those reported in this chapter, continue to offer new fields for the analytical chemist or structural investigator to exploit. The high sensitivity of the mass spectrometric method adds extra impetus to any new technique developed. One example of a technique that has passed from being largely of esoteric interest to being a potentially important tool for mixture analysis is collisional activation (pp. 77—79). Some of the other techniques listed in Table 1 can be applied to individual problems in structural elucidation and are frequently overlooked. For example, measurements of translational-energy release in ionic reactions can yield structural information (*q.v.* polychlorinated biphenyls).[107] In general terms, there is probably always another mass spectral technique or measurement, maybe an obscure one, that will give additional structural information where the normal electron-impact mass spectrum fails.

3
Gas-phase Ion Mobilities, Ion–Molecule Reactions, and Interaction Potentials

BY L. A. VIEHLAND

1 Introduction

Increased interest during the past twenty years in the upper atmosphere, lasers, and gaseous electronics has led to considerable interest in the behaviour of ions in neutral gases. For many of these applications it is important to know how the macroscopic properties of a particular ion–neutral system change with temperature. Below about 900 K, the experimental methods of mass spectrometry, ion cyclotron resonance, and stationary and flowing afterglow can provide the needed data. At very high temperature, usually no lower than 10^4 K, these data can be calculated from the microscopic information provided by ion–neutral beam experiments. In general, however, the gap between 900 and 10^4 K can be covered only by means of drift tubes.

This chapter will review the experimental and theoretical advances which have allowed drift tubes to become, in the past few years, accurate and versatile probes of ion–neutral systems. Topics covered will include gas-phase ion mobility and diffusion coefficients, ion–molecule reaction rate coefficients, and the extraction of information about ion–neutral interaction potentials and reactive and non-reactive cross-sections from measurements of these coefficients. Although many of the same experimental and theoretical ideas can be applied to charge-exchange situations and to electron–neutral gas systems, these topics will not be covered here.

2 Drift-tube Experiments

Basic Ideas.—In simplest form, a drift tube consists of the following: an enclosure containing a dilute gas at constant temperature T, an ion source which can produce, either continuously or in pulses, trace amounts of ions, a set of electrodes that establish a uniform, axial, electrostatic field E along which the ions drift, and an ion detector which electronically measures the spectrum of ion arrival times.

In a drift tube the ions quickly acquire an average velocity, the drift velocity v_d, as a result of a balance between the momentum gained from the electric field and that lost by collisions with the neutral gas molecules of number density N. It is customary to report the reduced ion mobility $K_0 = NK/N_0$, where the standard number density N_0 is 1.687×10^{25} molecules m^{-3} and where the ion mobility is defined by $K = v_d/E$. The importance of K_0 is that it can be rigorously shown[1] to be a function only of T and of E/N, for a particular ion–neutral system.

During its transit of the drift tube, an ion signal changes as a result of diffusion and ion–neutral reactions. Because the ions diffuse as a result of both random thermal motion and of motion induced by the electric field, the diffusion must be characterized by two coefficients: $D_\|$ for diffusion parallel to the electric field, and $D_\perp$ for perpendicular diffusion. It can be shown[1] that the quantities $ND_\|$, $ND_\perp$, and the reaction rate coefficients k in a drift tube are functions only of T and of E/N.

In the limit of low E/N, the parallel and perpendicular diffusion coefficients become identical and may be represented by the common symbol D. Moreover, K, D, and the various k become independent of E/N in this limit. Finally, the ion mobility and diffusion coefficients are related in this limit by the well known Einstein (or Nernst–Townsend) relation (1).[1]

$$qD/K = k_B T \qquad (1)$$

In the limit of both low E/N and low T, all of the transport coefficients approach constant values. For single-component neutral gases, these values depend only on the polarizability of the neutral molecules and the ion–neutral reduced mass. For mixtures of neutral gases, these coefficients can be obtained from the values appropriate to each of the individual gases by Blanc's law.[1]

If all ion–molecule collisions lead to reaction, the rate coefficient for ion reactions with non-polar molecules approaches a 'Langevin' limit[2] at low E/N and low T. The value of k obtained from this expression serves as an upper bound on the true rate coefficient.

Although the use of drift tubes began before the turn of the century, few of the data obtained before 1930 are of more than historical interest. This is due to the difficulty in basic drift tubes of separating the effects of ion drift, ion diffusion, and especially ion–neutral reactions. Use of the Bradbury–Nielsen time-of-flight method has allowed these difficulties to be overcome in some cases,[3] but modification of the basic drift-tube technique seems preferable in more general circumstances.

Drift-tube Mass Spectrometers.—The effects of ion drift, ion diffusion, and ion–neutral reactions can be separated[4] if the arrival-time spectrum for each ion can be mapped independently. Drift-tube mass spectrometry, DTMS, was developed in 1962[5] in order to accomplish this. In DTMS the ion-current detector of a basic drift tube is replaced by a sampling orifice leading to a mass spectrometer, which is followed by a current or particle detector. The usefulness of DTMS can be seen in the accurate and extensive data[6,7] that have been obtained for ion mobility and diffusion coefficients using this technique. However, DTMS

---

[1] E. W. McDaniel and E. A. Mason, 'The Mobility and Diffusion of Ions in Gases', Wiley, New York, 1973.
[2] G. Gioumousis and D. P. Stevenson, *J. Chem. Phys.*, 1958, **29**, 294.
[3] M. T. Elford, in 'Case Studies of Atomic Collision Physics', ed. E. W. McDaniel and M. R. C. McDowell, North-Holland, Amsterdam, 1972, Vol. 1, Ch. 2.
[4] I. R. Gatland, *Case Stud. At. Phys.*, 1974, **4**, 369.
[5] E. W. McDaniel, D. W. Martin, and W. S. Barnes, *Rev. Sci. Instrum.*, 1962, **33**, 2.
[6] H. W. Ellis, R. Y. Pai, E. W. McDaniel, E. A. Mason, and L. A. Viehland, *At. Data Nucl. Data Tables*, 1976, **17**, 177.
[7] H. W. Ellis, E. W. McDaniel, D. L. Albritton, L. A. Viehland, S. L. Lin, and E. A. Mason, *At. Data Nucl. Data Tables*, 1978, **22**, 179.

still suffers from three drawbacks that limit its chemical versatility and its usefulness for determining ion–neutral reaction rate coefficients. These are:

(i) The ion source is inside the drift tube.

(ii) In order to obtain a meaningful ion signal at the detector, almost all of the ion–neutral collisions in the drift tube must be non-reactive.

(iii) Since an ion signal is not produced by the source in the steady-state condition that it eventually attains in the drift tube, effects that are due to the necessary 'thermalization' of the ion signal must be 'subtracted out' by comparing results obtained with drift tubes of different lengths.

The modifications of DTMS discussed below were developed in order to overcome these limitations.

Injected-ion Drift Tubes.—Injected-ion drift tubes were developed in 1966[8] by placing a mass analyser in front of a DTMS apparatus. This separation of the ion source and the drift tube is important for two reasons. First, a much greater variety of ion sources (and hence ions) can be used, especially since the ions of interest need not even be the predominant ions produced by the source. Second, a non-reactive (buffer) gas can be added to the drift tube without fear that other, interfering ions will be produced by the ion source. This is crucial for studying ion–neutral systems in which reactive collisions predominate, since a large excess of buffer gas can be used to moderate the loss of ions.

Although any combination of ions and neutral systems can, in principle, be studied in injected-ion drift tubes, their chemical versatility is often limited by problems with thermalization of the ions. Various modifications of the apparatus have been introduced[9] to minimize this problem.

Additional-residence-time Drift Tubes.—Separation of the ion source and the drift tube and the use of a buffer gas are also features of the additional-resonance-time drift tube developed in 1969.[10] In this type of drift tube the length of time during which the ions are in the drift region and can react with the neutral gas molecules is varied by proper switching of the electric field direction. This eliminates any problems with thermalization of the ions as they leave the source, but only by introducing similar transient effects at each reversal of the field direction. Progress has been made[11] in understanding the velocity and energy relaxations that occur in these 'ping-pong' experiments, and most of the data obtained with this technique agree well with those obtained using other drift-tube techniques.

Flow-drift Tubes.—Flow-drift tubes were developed in 1973[12] by combining a flowing afterglow apparatus with DTMS. The flowing afterglow apparatus can be used as the source of a very wide variety of ions which are already thermalized before they enter the drift tube. Much of the accurate data now available[13] for

[8] Y. Kaneko, L. R. Megill, and J. B. Hasted, *J. Chem. Phys.*, 1966, **45**, 3741.
[9] Y. Kaneko, in 'Electronic and Atomic Collisions', ed. N. Oda and K. Takayanagi, North-Holland, Amsterdam, 1980, p. 109.
[10] J. Heimerl, R. Johnsen, and M. A. Biondi, *J. Chem. Phys.*, 1969, **51**, 5041.
[11] S. L. Lin, L. A. Viehland, E. A. Mason, J. H. Whealton, and J. N. Bardsley, *J. Phys. B*, 1977, **10**, 3567.
[12] M. McFarland, D. L. Albritton, F. C. Fehsenfeld, E. E. Ferguson, and A. L. Schmeltekopf, *J. Chem. Phys.*, 1973, **59**, 6610, 6620, 6629.
[13] D. L. Albritton, *At. Data Nucl. Data Tables*, 1978, **22**, 1.

ion–neutral reaction rate coefficients as a function of energy between 0.04 and a few eV has been obtained with flow-drift tubes. The major limitations of this technique are that the ions are not mass selected before entering the DTMS apparatus and that gases from the flowing afterglow are carried into the drift tube and can sometimes cause side reactions which confuse the analysis.

Selected-ion Flow-drift Tubes.—Recently an even more versatile and accurate type of drift tube has been developed.[14,15] A selected-ion flow-drift tube, shown schematically in Figure 1, is basically a combination of a flowing afterglow

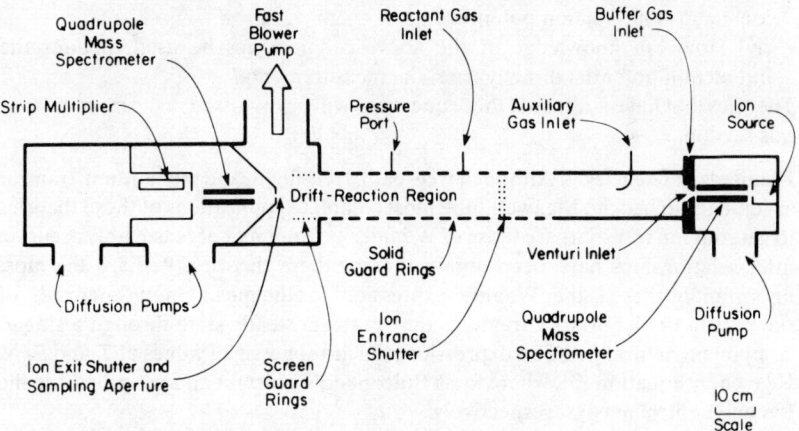

Figure 1 *Schematic diagram of a selected-ion flow-drift tube*
(Reproduced with permission from *J. Phys. B*, 1979, **12**, 4189)

apparatus, DTMS, and a device[16] that allows for mass-selected venturi-inlet ion injection. This apparatus seems capable of overcoming all of the difficulties encountered previously with drift tubes, and a dramatic increase in the range of ion–neutral systems for which accurate drift-tube data are available seems imminent.

3 Drift-tube Theories

Because of the experimental advances discussed above, excellent drift-tube data are now available for a large number of ion–neutral systems over wide ranges of E/N at fixed (usually low) T. Such data are important because variation of E/N has roughly the same effect as variation of T (both serve to change the mean energy of the ion swarm) and is more easily performed over the wide ranges that must be covered in order to probe fully ion–neutral interactions at the molecular

---

[14] F. Howorka, F. C. Fehsenfeld, and D. L. Albritton, *J. Phys. B*, 1979, **12**, 4189.
[15] F. Howorka, I. Dotan, F. C. Fehsenfeld, and D. L. Albritton, *J. Chem. Phys.*, 1980, **73**, 758.
[16] N. G. Adams and D. Smith, *Int. J. Mass Spectrom. Ion Phys.*, 1976, **21**, 349.

level. The fundamental difficulties in exploiting these data have been theoretical, with the following questions needing to be answered:

(i) What is the relationship between the T and E/N dependences of the measured transport and reaction coefficients? In particular, how can information about the T dependence at constant, low E/N be inferred from measurements of the E/N dependence at constant, low T?

(ii) What is the relationship between the E/N dependence of the measured transport and reaction coefficients and the reactive and non-reactive cross sections that govern ion–neutral collisions?

(iii) What is the relationship between the non-reactive cross-sections and the ion–neutral interaction potential?

(iv) How can knowledge of the above relationships be used to infer the interaction potential directly from the measurements?

Theories that have addressed these questions will be discussed in the remainder of this section.

Elementary Theories.—Although free-path, free-flight, and momentum-transfer arguments go back to Maxwell,[1] the most complete applications of these theories to gaseous ion transport are those of Wannier.[17] A number of reasonably accurate inter-relationships have been obtained using these theories. Perhaps the most important of these is the 'Wannier expression' for the mean kinetic energy E_k of the ions in the laboratory frame as they move in steady state through a single-component neutral gas. This expression, which applies at all values of T and E/N, is given by equation (2), where k_B is Boltzmann's constant and m and M are the ion and neutral masses, respectively.

$$E_k = \tfrac{3}{2}k_B T + \tfrac{1}{2}mv_d^2 + \tfrac{1}{2}Mv_d^2 \tag{2}$$

The average ion energy is therefore represented as the sum of the thermal energy, energy that the ions have acquired from the electric field and that is exhibited as motion along the direction of the field, and energy that has been similarly acquired but that is exhibited as random motion due to ion–neutral collisions in the drift tube.

These elementary theories also lead to generalizations to high E/N of the Einstein relation in the forms (3) and (4).

$$\frac{qD_\parallel}{K} = k_B T_\parallel \left[1 + \frac{d \ln (K)}{d \ln (E/N)} \right] \tag{3}$$

$$qD_\perp/K = k_B T_\perp \tag{4}$$

The constant mean-free-time model has been used to obtain expressions[1] for the parallel and perpendicular 'temperatures' $T_\parallel$ and $T_\perp$.

In principle, the Wannier expression provides a solution to the first question posed at the beginning of this section: the transport and reaction coefficients should be the same at any combination of T and E/N that leads to the same

---

[17] G. H. Wannier, *Phys. Rev.*, 1951, **83**, 281; *Phys. Rev.*, 1952, **87**, 795; *Bell System Tech. J.*, 1953, **32**, 170; *Aust. J. Phys.*, 1973, **16**, 897.

relative kinetic energy in the centre-of-mass frame. For a single-component neutral gas, this relative kinetic energy is given by equation (5).

$$E_{rel} = \tfrac{3}{2}k_B T + \tfrac{1}{2}Mv_d^2 \tag{5}$$

Table 1, which contains results calculated to an accuracy of ± 3% from the three-temperature kinetic theory discussed below, shows that this prediction is accurate within about 10%. However, little use of this prediction was made until recently because other predictions of these elementary theories are only semi-quantitative and because there does not seem to be any systematic way of improving the accuracy of results calculated using these theories.

Table 1 *Calculated values for the reduced mobility K_0 of Cs^+ ions in Ar at different combinations of the gas temperature T and the field strength parameter E/N, as compared with the relative kinetic energy E_{rel} from equation (5) of the text*

T/K	$E/N/10^{-21}$ V m^2	$K_0/10^{-4}$ m^2 V^{-1} s^{-1}	$E_{rel}/$kJ mol^{-1}
298	132.9	2.449	19.00
298	168.8	2.438	28.14
2980	18.29	2.267	37.41
2980	73.23	2.253	41.09
298	222.1	2.344	42.82
2980	93.05	2.233	43.39
2980	153.0	2.159	52.91

One-temperature Kinetic Theories.—The relationship between the T and E/N dependences of gaseous ion transport and reaction coefficients and the cross-sections governing the ion–neutral collisions can be obtained by a kinetic-theory solution of the appropriate[1] Boltzmann equation. Since the ions are present only in trace amounts in a drift tube, the Boltzmann equation must be linear in the ion velocity distribution function. Nevertheless, an exact solution of this integro-differential equation for the distribution function is not possible.

The general procedure used to solve the Boltzmann equation for gaseous ion transport is to convert it into a coupled set of moment equations, which are also known as transport-relaxation equations or Maxwell's equations of change. These equations are then solved for the few moments of interest (the transport and reaction coefficients) by a successive approximation scheme. The crucial step for obtaining rapid convergence to accurate results is the choice of basis functions used to form the moment equations.

One-temperature kinetic theories[18-21] of gaseous ion transport use the so-called Burnett functions[1] to form moment equations from the Boltzmann equation. These functions are eigenfunctions of the collision operator in the Boltzmann equation for the special case (the Maxwell model) where the ion–neutral potential varies as the inverse fourth power of the separation, r. Moreover, these

---

[18] T. Kihara, *Rev. Mod. Phys.*, 1953, **25**, 844.
[19] E. A. Mason and H. W. Schamp, *Ann. Phys. (N.Y.)*, 1958, **4**, 233.
[20] K. Kumar and R. E. Robson, *Aust. J. Phys.*, 1973, **26**, 157, 187.
[21] J. H. Whealton and E. A. Mason, *Ann. Phys. (N.Y.)*, 1974, **84**, 8.

functions are orthogonal in velocity space with respect to the weighting function (6), where v is the ion velocity in the laboratory frame.

$$g_1(v) = (m/2\pi k_B T)^{3/2} \exp(-mv^2/2k_B T) \qquad (6)$$

This weighting function must represent,[22] in the one-temperature theories, a zeroth-order approximation to the true ion distribution function (normalized to unit number density). It should be noted that the description of these theories as 'one-temperature' arises from their use of the neutral gas temperature T to characterize $g_1(v)$ at all E/N values.

The use of Burnett functions means that the one-temperature theories are exact for the Maxwell model. This model is particularly important in gaseous ion transport theory, since the long-range interaction potential between any ion and any neutral molecule must vary as r^{-4}. Consequently, results calculated from the one-temperature theories must be exact in the limit of low T and low E/N, where such low energies are involved that the long-range term in the potential dominates.

In the past, one-temperature kinetic theories of gaseous ion transport have always yielded expressions for the transport coefficients that are power series or ratios of polynomials in E/N. Although these expressions are very accurate in the limit of low E/N, they diverge at higher E/N. The ability of one-temperature theories to answer the questions posed at the beginning of this section is therefore limited.[23]

Cold-gas Kinetic Theories.—At high E/N, the thermal energy contribution to the mean energy of the ions in a drift tube is negligible. This idea led Wannier to formulate[17] a cold-gas kinetic theory in which thermal motions of the neutral gas molecules are totally neglected. This assumption leads to a drastic simplification of the collision operator in the Boltzmann equation, and allows the transport coefficients in the limit of high E/N to be calculated accurately and efficiently.

In recent work,[24] Wannier's cold-gas kinetic theory has been extended to finite gas temperatures by the inclusion of thermal motions in a perturbation approach. For light ions in a heavy gas, this treatment yields rapidly converging results for all but the very lowest values of E/N of experimental interest, as expected on physical grounds. For heavier ions and lighter neutral molecules, this theory complements the one-temperature kinetic theories. However, even the combination of one-temperature and cold-gas theories does not provide completely satisfactory answers to the theoretical questions given above.

Monte Carlo Treatments.—A Monte Carlo study of gaseous ion transport is a computer simulation of the way a single ion moves under the influence of an electric field while experiencing a very large number of collisions with neutral molecules. Afterwards, the time-averaged behaviour of the probe ion is equated to the ensemble-averaged behaviour of the many ions of a real system. The 'Monte Carlo' label is applied to this technique because the time intervals between collisions and the magnitudes of the velocity changes that occur during

---

[22] D. Mintzer, *Phys. Fluids*, 1965, **8**, 1076.
[23] H. B. Milloy, R. O. Watts, R. E. Robson, and M. T. Elford, *Aust. J. Phys.*, 1974, **27**, 787.
[24] H. R. Skullerud and L. R. Forsth, *J. Phys. B*, 1979, **12**, 1881.

collisions are determined by sets of random numbers drawn from an appropriate distribution of possible values.

Monte Carlo treatments of gaseous ion transport have become practical in the past ten years owing to advances in computer hardware and software, especially with the development of the method of null collisions.[25,26] This method has been used[27-29] to verify the accuracy of equations (2)—(4), to test and improve the accuracy of proposed ion–neutral interaction potentials, and to reconcile reaction rate coefficients measured in drift tubes filled with different buffer gases.

Although Monte Carlo treatments are very powerful, they do have drawbacks. Since the procedure in effect duplicates the experiment on a computer, much more computer time is required than would be needed for any of the kinetic theories discussed here. In addition, each ion–neutral system is a special case in a Monte Carlo treatment, and each new interaction potential requires a full calculation from the beginning. This makes it difficult to develop physical insight and detect trends using this procedure; although it can be used to verify proposed answers to the questions posed at the beginning of this section, it cannot readily be used to obtain new predictions. Nevertheless, Monte Carlo methods are valuable complements to kinetic-theory methods.

Two-temperature Kinetic Theories.—The essential feature of a two-temperature kinetic theory is the explicit recognition that at high E/N the trace ions in a drift tube can have a temperature appreciably greater than the neutral gas temperature. Accordingly, the basis functions used by Viehland and Mason[30] to solve the Boltzmann equation by a moment technique were Burnett-like functions orthogonal with respect to a weighting function $g_2(v)$ that is identical with $g_1(v)$ except that the gas temperature T is replaced by an ion temperature T_k. It seems plausible that this parameter for solving the moment equations should be set equal to the kinetic temperature of the ions in the laboratory frame, at least in first approximation.[31] When this is done, the final expressions for the transport coefficients become well behaved at all E/N values.

According to the two-temperature theory of Viehland and Mason,[30] the mobility of a swarm of atomic ions moving through a single-component atomic gas is given at all T and all E/N values by equations (7) and (8).

$$K = \frac{3q}{8N}\left(\frac{\pi}{2\mu k_B T_{\text{eff}}}\right)^{1/2} \frac{1+\alpha}{\Omega(T_{\text{eff}})} \tag{7}$$

$$\tfrac{3}{2}k_B T_{\text{eff}} = E_{\text{rel}} = \tfrac{3}{2}k_B T + \tfrac{1}{2}Mv_d^2(1+\beta) \tag{8}$$

Here μ is the ion–neutral reduced mass and $\Omega(T_{\text{eff}})$ is the temperature-dependent momentum-transfer collision integral discussed below. The correction terms α and β depend in a complicated way on T, E/N, m, M, and the ion–neutral

---

[25] H. R. Skullerud, *J. Phys. D*, 1968, **1**, 1567; *J. Phys. B*, 1973, **6**, 728.
[26] S. L. Lin and J. N. Bardsley, *Comput. Phys. Commun.*, 1978, **15**, 161.
[27] H. R. Skullerud, *J. Phys. B*, 1978, **6**, 918.
[28] S. L. Lin and J. N. Bardsley, *J. Chem. Phys.*, 1977, **66**, 435.
[29] D. L. Albritton, I. Dotan, W. Lindinger, M. McFarland, J. Tellinghuisen, and F. C. Fehsenfeld, *J. Chem. Phys.*, 1977, **66**, 410.
[30] L. A. Viehland and E. A. Mason, *Ann. Phys. (N.Y.)*, 1975, **91**, 499; *Ann. Phys. (N.Y.)*, 1978, **110**, 287.
[31] S. L. Lin and E. A. Mason, *J. Phys. B*, 1979, **12**, 783.

interaction potential. Numerical calculations[30] have shown that in most practical cases α and β are substantially less than 0.1, so the accuracy obtained by setting them equal to zero in the first of a series of successive approximations is generally good. Thus, equation (8) provides a more rigorous foundation for the Wannier expression (5), which had already been shown to be accurate within 10% by Monte Carlo studies. However, these calculations have also shown that it is usually necessary to go to third approximation in order to ensure a converged result that is as accurate as measured mobilities.

In the limit of low E/N, both T_k and T_{eff} become equal to T, and the two-temperature theory reduces to a one-temperature theory. In the limit of high E/N, both T_k and T_{eff} become independent of T, and the two-temperature theory reduces to a cold-gas theory. The value of the two-temperature theory is that, without appreciably more effort, it gives not only the correct limiting values but also accurate results at intermediate E/N.

Equations (7) and (8) show that in first approximation all of the influence of the ion–atom interaction potential upon gaseous ion mobility arises through the momentum-transfer collision integral.[1] This is an integration over energy of the well known[1] momentum-transfer cross-section. There is therefore a direct link between the ion mobility and the ion–neutral interaction potential. Moreover, the value of $\Omega(T_{\text{eff}})$ at a given T_{eff} value is essentially determined by values of the potential over the small range of separation distance that leads to both a large scattering angle and a large impact parameter. This idea was first incorporated into a simple computational scheme for the direct determination of interaction potentials from transport data by Smith and co-workers.[32] Recently it has been extended to gaseous ion mobility data[33] and placed on a more secure theoretical foundation.[34] In this regard, it should be noted that the results in Table 1 were calculated using a Cs^+–Ar interaction potential obtained[35] by direct inversion of mobility data taken at room temperature.

In first approximation, the two-temperature kinetic theory of Viehland and Mason leads[36] to expressions (9) and (10) for the reaction rate coefficient between a swarm of atomic ions in a drift tube filled with buffer atoms of mass M_b and a small proportion of reactive atoms of mass M_R.

$$k = \frac{1}{k_B T_R}\left(\frac{8}{\pi \mu_R k_B T_R}\right)^{1/2} \int_0^\infty \exp\left(-\frac{\varepsilon}{k_B T_R}\right) Q^*(\varepsilon)\varepsilon\, d\varepsilon \qquad (9)$$

$$\tfrac{3}{2}k_B T_R = \tfrac{3}{2}k_B T + (\tfrac{1}{2}Mv_d^2)\left(\frac{m + M_b}{m + M_R}\right) \qquad (10)$$

Here $Q^*(\varepsilon)$ is the energy-dependent cross-section for the reaction, μ_R is the ion–reactive atom reduced mass, and T_R is the effective temperature that characterizes the relative kinetic energy between the ions and reactive neutrals. These equations, and their higher-order analogues, provide a method[36] for

[32] D. W. Gough, G. C. Maitland, and E. B. Smith, *Mol. Phys.*, 1972, **24**, 151.
[33] L. A. Viehland, M. M. Harrington, and E. A. Mason, *Chem. Phys.*, 1976, **17**, 433.
[34] G. C. Maitland, E. A. Mason, L. A. Viehland, and W. A. Wakeham, *Mol. Phys.*, 1978, **36**, 797.
[35] I. R. Gatland, M. G. Thackston, W. M. Pope, F. L. Eisele, H. W. Ellis, and E. W. McDaniel, *J. Chem. Phys.*, 1978, **68**, 1775.
[36] L. A. Viehland and E. A. Mason, *J. Chem. Phys.*, 1977, **66**, 422.

extracting thermal rate coefficients at elevated temperatures from swarm measurements made at low T and high E/N; the numerical value of k is the same if T_R is the same. The difficult problems of finding the ion velocity distribution function and of unfolding the reaction cross-sections from the rate data can thus be by-passed, if desired.

From the comments above it might appear that, at least for atomic ions and neutrals, two-temperature kinetic theories can answer the theoretical questions about gaseous ion transport. There is, however, one major difficulty. Two-temperature theories are not nearly as good for gaseous ion diffusion as for mobility and reaction rate coefficients. When the ion/atom mass ratio is less than one, the convergence of successive approximations to the coefficients $D_\parallel$ and $D_\perp$ is nearly as good as for the mobility. When the mass ratio is between about one and four, convergence is good only at low and intermediate E/N, and low approximations at high E/N must be viewed with suspicion. Finally, when the mass ratio is greater than about four, the two-temperature theories are essentially useless for diffusion.[30]

Three-temperature Kinetic Theories.—It seems likely that the inability of a two-temperature theory to describe adequately gaseous ion diffusion stems from two facts. First, compared with the mobility, the diffusion coefficients are more intimately connected with the anisotropic nature of the ion motion that becomes increasingly manifest as E/N and the ion/neutral mass ratio increase. Second, the two-temperature theories make use from the start of a single ion temperature to characterize the energy distribution, thereby assuming essentially isotropic conditions.

The arguments just presented led to the development[37] of a three-temperature kinetic theory of gaseous ion transport. This theory is based on the use of basis functions for solving the Boltzmann equation which are orthogonal to the weighting function (11).

$$g_3(v) = (m/2\pi k_B T_\perp)(m/2\pi k_B T_\parallel)^{1/2} \exp[-m(v_x^2 + v_y^2)/2k_B T_\perp - m(v_z - v_{dis})^2/2k_B T_\parallel]$$

(11)

The parameters $T_\perp$, $T_\parallel$, and v_{dis} may be chosen so as to facilitate the solution of the moment equations for the transport coefficients. Rapid convergence is obtained when these parameters are defined, in first approximation,[38] to be the temperatures characterizing the ion energies perpendicular and parallel to the electric field and the ion drift velocity, respectively.

One major advantage of the three-temperature kinetic theory is that it leads to the generalized Einstein relations (3) and (4) if the mobility K in these equations is interpreted as the differential mobility.[39] In general, however, this theory does not lead to equations with the physical transparency of those obtained from other theories.

The three-temperature theory has been used[38] to test the Cs^+–Ar interaction potential directly determined[35] from mobility data and the converged third

---

[37] S. L. Lin, L. A. Viehland, and E. A. Mason, *Chem. Phys.*, 1979, **37**, 411.
[38] L. A. Viehland and S. L. Lin, *Chem. Phys.*, 1979, **43**, 135.
[39] M. Waldman and E. A. Mason, to be published.

approximation of the two-temperature theory. Naturally, there is excellent agreement between the mobility calculated (see Table 1) from the converged fourth approximation of the three-temperature theory and the experimental data. Of more importance is the excellent agreement between the calculated and measured parallel diffusion coefficients. This was a theoretical 'first', and it also provides added support for the claimed accuracy ($\pm 10\%$) of the Cs^+–Ar interaction potential.

Comparisons[37,38] of ion mobility and diffusion coefficients calculated from the three-temperature theory with exact or very accurate results obtained by other means (such as Monte Carlo treatments) have established that, by fourth approximation, the three-temperature theory has converged to results that have a higher accuracy than is found in experimental data. Consequently, the three-temperature theory is capable of providing answers to all of the questions posed at the beginning of this section, provided that the ions and neutral molecules are atomic.

Kinetic Theories for Molecular Systems.—The relationship between the E/N dependence of the transport coefficient of molecular ion–neutral systems and the cross-sections governing the ion–neutral collisions may be obtained by solving the Wang Chang–Uhlenbeck–de Boer (WUB) equation.[40] This equation is a generalization of the Boltzmann kinetic equation that takes into account the presence of anisotropic intermolecular forces and internal degrees of freedom that can result in inelastic collisions. It does not take into account the quantum-mechanical interference effects that can occur when the internal states are degenerate,[41] but such effects on the types of transport coefficients of interest here are small.[42]

Solving the WUB equation for gaseous ion transport is not very different from solving the Boltzmann equation. Indeed, Lin et al.[43] have presented a two-temperature theory for electrons in molecular gases. However, for more general molecular systems one additional ion temperature is needed in treatments analogous to the preceding kinetic theories: an ion temperature T_i that characterizes the energy held in the internal degrees of freedom.

A kinetic theory of gaseous ion transport for molecular systems that is analogous to the two-temperature theory for atomic systems can be based on the use of spherical basis functions that are orthogonal with respect to the weighting function (12).

$$h_2^{(\alpha)}(v) = g_2(v) \exp(-\varepsilon^{(\alpha)}/k_B T_i)/\sum_\alpha \exp(-\varepsilon^{(\alpha)}/k_B T_i) \qquad (12)$$

Here the superscripts label the internal states and $\varepsilon^{(\alpha)}$ is the internal energy of an ion in state α. Similarly, a kinetic theory analogous to the three-temperature theory can be based on the use of Cartesian basis functions that are orthogonal with respect to the weighting function $h_3^{(\alpha)}(v)$, which is defined as in equation (12)

---

[40] C. S. Wang Chang, G. E. Uhlenbeck, and J. de Boer, in 'Studies in Statistical Mechanics', ed. J. de Boer and G. E. Uhlenbeck, North-Holland, Amsterdam, 1964, Vol. II.
[41] L. Waldmann, in 'Statistical Mechanics of Equilibrium and Non-Equilibrium', ed. J. Meixner, North-Holland, Amsterdam, 1965.
[42] J. J. Beenakker and F. R. McCourt, *Rev. Phys. Chem.*, 1970, **21**, 47.
[43] S. L. Lin, R. E. Robson, and E. A. Mason, *J. Chem. Phys.*, 1979, **71**, 3483.

except that $g_2(v)$ is replaced by $g_3(v)$. Although comprehensive tests of these approaches must still be performed, preliminary indications[44] are that no special difficulties are likely to arise just because inelastic collisions and internal degrees of freedom have been introduced. The spherical basis functions should give good results for those quantities that are not sensitive to the anisotropy of the ion distribution function at high E/N: ion mobility, total ion energy, and ion–neutral reaction rate coefficients. For greater accuracy, and in order to obtain good results for ion diffusion coefficients and for the partitioning of the ion energy in directions parallel and perpendicular to the electric field, the Cartesian basis functions will need to be used.

Using spherical basis functions, the kinetic theory for molecular systems leads[44] to first approximation results, (13) and (14), for the ion mobility in a single-component neutral gas that are very similar to equations (7) and (8).

$$K = \frac{3q}{8N}\left(\frac{\pi}{2\mu k_B T'_{\text{eff}}}\right)^{1/2} \frac{1}{\Omega'(T'_{\text{eff}})} \tag{13}$$

$$\tfrac{3}{2}k_B T'_{\text{eff}}(1 + M\xi/m) = E_{\text{rel}} = \tfrac{3}{2}k_B T + \tfrac{1}{2}M v_d^2 \tag{14}$$

The collision integral Ω' differs only slightly from the corresponding integral Ω for elastic collisions, in the same manner that the collision integrals for interdiffusion of atomic and molecular gases differ.[45] This minor difference means that there is nothing remarkable about the self-diffusion coefficients of molecular gases as compared with atomic gases, and that these coefficients contain little information about the anisotropic interaction potentials. However, this may not be true of gaseous ion mobility. In neutral gases all species are at the same temperature regardless of whether or not inelastic collisions occur, while in drift tubes the value of T'_{eff} can depend strongly on inelastic collisions.

Equation (14) represents a generalization to molecular species of the expression (5) for the relative kinetic energy in the centre-of-mass frame between the ions and the single-component neutral molecules. The quantity ξ is a dimensionless ratio[43,44] of the collision integral for inelastic energy loss to that for momentum transfer. In combination with the factor of M/m in equation (14), we can expect the presence of anisotropic potentials and inelastic collisions in molecular systems to have the largest impact on ion mobility in the case of light ions in heavy neutral gases.

The internal temperature T_i is the temperature that results when the difference between the pre- and post-collision ion internal energies averages to zero. In general, this average energy balance will depend on the cross-sections for inelastic collisions, as compared with the elastic collisions. A special case that is of great importance is that of molecular ions in an atomic neutral gas: the ions acquire an internal temperature T_i that is equal to their effective kinetic temperature T'_{eff}. A simple physical explanation can be given for this. Energy is fed into the internal degrees of freedom of the ions by collisions with the structureless neutrals; the source of the ion internal energy is thus its translational motion. Energy leaks out of both the internal and translational degrees of freedom of the ions only through

---

[44] L. A. Viehland, S. L. Lin, and E. A. Mason, *Chem. Phys.*, 1981, **54**, 341.
[45] L. Monchick, K. S. Yun, and E. A. Mason, *J. Chem. Phys.*, 1963, **39**, 654.

the translational motion of the neutrals. Since the leak is the same for both forms of energy, and since the internal energy is fed by the translational, it is not surprising that T_i must equal T'_{eff} in steady state.

The spherical basis functions lead to an expression[44] for the reaction rate coefficient between a swarm of molecular ions in a drift tube filled with a buffer gas and a small proportion of reactive molecules that is, in first approximation, a generalization of equation (9) and is simply related to the usual theoretical expression[46] for a near-equilibrium thermal rate coefficient. The importance of this expression is that it equates the experimental rate coefficient, measured in a drift tube as a function of E/N and T, to an equivalent thermal rate coefficient that depends on T, T_i, and a reactive translational temperature T'_R. Although only two of these temperatures can be independent, the relationship between the three is simple only in the special case where the buffer gas is atomic. In this situation we have

$$(m + M_b)M_R(T_i - T) = (m + M_R)M_b(T'_R - T) \tag{15}$$

This result suggests that the effects of internal and translational energies on reaction rates between molecular ions and neutrals can be at least partially separated by making measurements in two or more buffer gases. This conclusion is strikingly consistent with some recent experimental work summarized by Albritton.[47]

With the exception of a brief mention of Blanc's law, little attention has been devoted in this section to gaseous ion transport in mixtures of neutral gases. This is because a theoretical treatment of mixtures is, in general, a simple (but messy) extension of that for pure gases. One important difference arises when molecular ions move through gas mixtures: no simple result is obtained for the ion internal temperature T_i, even when all of the neutrals are atomic. The reason is that each neutral species has a different effective kinetic temperature T'_{eff} (relative to the ions), and there is no *a priori* way to tell how the energy loss of the ion upon collision will be partitioned between internal and translational modes unless relative cross-sections are known for each neutral species. This inability to predict T_i in a mixture suggests that it may be unprofitable to study ion–neutral reactions in mixtures of buffer gases if the ion is molecular. However, there should be no difficulty involved with the use of a mixture of reactant species, since these should act independently in the presence of a large excess of buffer gas.

4 Current Status

Measurement of Transport and Reaction Coefficients.—As discussed in Section 2, recent experimental advances have led to a great increase in the variety of ion–neutral systems that can be studied in drift tubes, and in the ranges of T and E/N that can be used. It is quite feasible to obtain results that are accurate within 2% for the ion mobility,[6] 5% for the parallel diffusion coefficient,[7] 30% for the ion–molecule reaction rate coefficient,[13] and 4% for the branching ratios of the

---

[46] J. C. Light, J. Ross, and K. E. Shuler, in 'Kinetic Processes in Gases and Plasmas', ed. A. R. Hochstim, Academic, New York, 1969.
[47] D. L. Albritton, in 'Kinetics of Ion–Molecule Reactions', ed. P. Ausloos, Plenum Press, New York, 1979.

reaction.[15] The major source of inaccuracy in measuring the transport coefficients is usually the determination of the gas pressure or density. The major source of inaccuracy in measuring the reaction coefficients arises from the fact that the decrease in amount of a reactant (or the increase in amount of a product) is very small compared to the large amount of buffer gas present.

Transferability of Transport and Reaction Coefficients.—It has been clearly established from experimental results, Monte Carlo studies, and kinetic theory calculations that the Wannier expression (2) holds within 10% for atomic ion–neutral systems. Consequently, it is possible to transfer transport and reaction coefficients from one combination of E/N and T to another with reasonable accuracy. It is only necessary that the relative kinetic energy in the centre-of-mass frame be the same; this energy is approximated by equation (5) when there is only a single neutral species present, and by equation (10) when there is a trace amount of one reactant species immersed in a large amount of non-reactive buffer gas.

It is customary when analysing drift tube data for molecular ion–neutral systems to assume that the Wannier expression applies. Equation (14) shows that this assumption is correct only when the quantity $M\xi/m$ is negligible, for example, when heavy ions move through light neutral gases or when molecular ions move through atomic gases. In more general situations, equation (14) must be used in place of the Wannier expression. It should be noted that most of the special effects of inelastic collisions and anisotropic interaction potentials upon the relative kinetic energy of an ion–neutral system arise through the quantity ξ. The hope for the near future is that theoretical studies will provide a means for estimating this quantity without a detailed analysis of the ion–neutral interactions.

There is a second factor that complicates transfer of transport and reaction coefficients when molecular species are involved – the internal energy of the ions. As discussed in Section 3, the temperature characterizing this internal energy is approximately equal to the relative kinetic temperature given by equation (5) if the single neutral species present is atomic. If a small amount of a reactant neutral species is present in a large amount of atomic buffer gas, then equation (15) connects these two temperatures with the gas temperature. Similarly, simple relations do not appear to exist when the predominant neutral species is molecular or when a mixture of neutral gases is used. In order to transfer transport and reaction coefficients in general situations, it appears that it will be necessary to study first the behaviour of the ion–neutral systems when immersed in a number of separate atomic buffer gases, so that equation (15) can be used to disentangle the effects of internal and kinetic energy.

Determination of Cross-sections.—The kinetic theories discussed in Section 3 allow energy averages of the momentum-transfer and total reaction cross-sections to be extracted, with an accuracy of about 10%, directly from measurements of the transport and reaction coefficients. Equations (7) and (9) show this explicitly for the case of atomic ions and neutrals. By making use of data obtained over a wide range of E/N, it is possible to invert these integral relationships to obtain reasonably accurate results for the cross-sections over a wide range of relative kinetic energies.

In practice, cross-sections have seldom been unfolded from drift-tube measurements involving atomic or molecular species. In the case of the reaction cross-section, this is due to the limited accuracy of the rate data and to the fact that the only purpose for determining Q^* would be to calculate from it the rate coefficient at combinations of T and E/N different from those used in the original measurements, a task which can be done more simply and with equal accuracy using the transferability of reaction coefficients just discussed. Momentum-transfer cross-sections are not unfolded because the accuracy of the transport data warrants a more accurate analysis in terms of the ion–neutral interaction potential.

Determination of Interaction Potentials.—The Monte Carlo technique allows a proposed ion–neutral interaction potential to be tested by comparison of calculated and measured transport coefficients, although the technique is sufficiently complex that few applications to molecular systems have been made. Trial-and-error modification of the potential has been used[28] to produce acceptable agreement of the calculated and measured values, but in general this does not seem to be a practical way to determine the potential.

The two- and three-temperature kinetic theories discussed in Section 3 can also be used to test proposed potentials for atomic ion–neutral systems.[48-50] In addition, the collision integrals that arise in these theories provide a means[32-34] for iteratively determining the potential directly from the transport data; this method does not involve assuming some functional form for the potential and adjusting its parameters to make the calculated and measured results agree. It has been shown for Cs^+-Ar[35] and for Li^+-He[49] that a potential accurate within 10% over wide ranges of ion–neutral separation distance (ranging from the long-range tail to well up on the repulsive wall) can be obtained from gaseous ion transport data accurate within 2%. A word of caution is in order here: some of the potentials obtained recently in this manner suffer from inaccuracies introduced by the computer codes used, a situation which will hopefully be remedied in the near future.

Using a kinetic theory of gaseous ion transport to test or determine potentials for molecular ion–neutral systems has not yet been attempted. The main reason for this is that the relationship between the molecular cross-sections and the ion–neutral potential is very complicated because a large number of internal states can be involved, resulting in very large sets of coupled, quantum-mechanical equations to solve. The hope for the future is that approximate schemes for accurately calculating molecular cross-sections can be developed and applied to ion–neutral systems.

5 Prognosis

In the near future we can expect a continuation of the rapid growth in the variety of ion–neutral systems and the ranges of T and E/N for which accurate drift-tube

---

[48] I. R. Gatland, L. A. Viehland, and E. A. Mason, *J. Chem. Phys.*, 1977, **66**, 537.
[49] I. R. Gatland, W. F. Morrison, H. W. Ellis, M. G. Thackston, E. W. McDaniel, M. H. Alexander, L. A. Viehland, and E. A. Mason, *J. Chem. Phys.*, 1977, **66**, 5121.
[50] S. L. Lin, I. R. Gatland, and E. A. Mason, *J. Phys. B*, 1979, **12**, 4179.

results are available. We are beginning to understand how the effects of internal and kinetic energies in molecular systems can be disentangled, so this growth should be especially rapid in the area of using drift tubes to study ion–neutral reaction rate coefficients over a wide range of relative kinetic energy.

On the theoretical side, the near future should bring a dramatic increase in the number of atomic ion–neutral systems for which previously proposed potentials have been tested and/or for which accurate potentials have been directly derived from transport data. Considerable effort will undoubtedly be devoted to understanding molecular systems, but progress in this area will probably be slow.

In short, the future looks very promising for gas-phase studies of ion mobilities and ion–molecule reactions in drift tubes, and for using such data to probe ion–neutral interaction potentials.

4
Interaction of Electromagnetic Radiation with Gas-phase Ions

BY R. C. DUNBAR

1 Introduction

The past two years have seen great activity and interest in the photon interactions of ions, and the literature has perhaps doubled in volume. In part this is due to increasing realization that techniques are now well developed to obtain very detailed and satisfactory spectroscopic information about a wide variety of gas-phase ions; in part to the exceptional advantages that gas-phase ions offer for studying photophysical and photochemical events in isolated molecules; in part to the rapid emergence of ion-beam experiments offering extreme and almost unprecedentedly high optical resolution; and in part to the increasing inclusion of ion-spectroscopic tools among those available to the mass spectroscopist.

Described first is the large volume of recent work concerned with the optical-spectroscopic properties of gas-phase ions. Several recent reviews have covered aspects of this.[1-3] Next is a description of the complementary and rapidly emerging area of optical emission observed from ions, both following ionization into excited states (recently reviewed[4]) and also using techniques of laser-induced fluorescence. Third to be surveyed is the emerging area of ion multiphoton photochemistry. Next is described progress in photofragmentation studies, *i.e.* those experiments where emphasis is on measuring the kinetic energy and/or the angular distribution of the products of ion photodissociation, a substantially reviewed area.[3] The final section covers a few experiments of exceptional interest which fall outside these classifications.

2 Ion Spectroscopy

With a few exceptions, nearly all the information we have about the absorption spectroscopy of gas-phase ions has come from photodissociation experiments. Only in rare and heroic experiments has it been possible to observe optical absorption directly since ion concentrations greater than about $10^8 \, \text{cm}^{-3}$ are difficult to achieve, so that the indirect experiments in which photon absorption is signalled by subsequent dissociation of the ion have been the most fruitful. It is

---

[1] R. C. Dunbar, in 'Gas Phase Ion Chemistry', ed. M. T. Bowers, Academic Press, 1979, Vol. II, p. 182.
[2] R. C. Dunbar, in 'Kinetics of Ion–Molecule Reactions', ed. P. Ausloos, Plenum Press, New York, 1979, p. 463.
[3] J. T. Moseley and J. Durup, *J. Chim. Phys. Phys.-Chim. Biol.*, 1980, **77**, 673.
[4] J. P. Maier, in ref. 2, p. 437.

useful to divide photodissociation experiments into those performed in ion traps, where the time scale (milliseconds or seconds) is long and the emphasis has tended to be on larger ions, and those performed in ion beams, where the time scale is milliseconds or less and the emphasis has been on small ions. Two traps in use currently for work involving photons are the ion cyclotron resonance spectrometer and the r.f. quadrupole trap. Ion beams primarily are instruments in which a free-flying beam of mass-selected ions interacts with a laser in either crossed or coaxial geometry; also included here can be the tandem quadrupole spectrometer and the drift-tube laser photodissociation spectrometer. Most beam instruments have been specifically constructed but recently a laser has been added to a more or less conventional, high-quality double-focusing mass spectrometer.[5]

Photodissociation in Ion-beam Instruments.—Some of the most dramatic recent developments in all of spectroscopy have come from the combination of fast ion beams (keV or greater kinetic energies) coaxial with single-mode laser light sources of very high stability and monochromaticity.[3] Two aspects of this technique lead to its extraordinary power. One is the phenomenon of kinetic compression: when the ions are accelerated to keV energies (or higher[6]), their random thermal velocities become an insignificant fraction of their total velocity in the laboratory frame, so that Doppler-broadening is reduced by a large factor, and the spectra are essentially Doppler-free. The other is the ability to vary the beam velocity so as to Doppler-tune the ion through a spectral region. Used in combination with a single-mode laser at a fixed frequency with a spread of less than 0.001 cm^{-1}, Doppler-tuning allows spectra to be scanned at a realized resolution better than 50 MHz. Photo-absorption by the beam is usually detected by photodissociation of the ion, but some elegant recent experiments have looked at perturbations of charge-transfer or ion–molecule reaction rates as the signal of light absorption.

Most ion-beam spectroscopy has been performed with less spectacular, but often more convenient, lower-energy instruments. In addition to conventional ion-beam instrumentation, two other types of instruments have proven useful. One is the tandem quadrupole, in which three quadrupole mass filters are strung together in tandem. The first acts to mass select ions emerging from the ion source, the second runs without mass selectivity and provides the ion-laser interaction region, and the third mass analyses the photodissociation product ions. The other type of instrument is the drift-tube spectrometer, in which the ion source is a high-pressure drift tube yielding thermalized ions. Emerging from the drift tube, the ions are crossed by a laser, and then mass analysed by (typically) a quadrupole mass spectrometer.

An imposing amount of effort has been spent on the spectroscopy of O_2^+ quartet states, which has been adopted by a number of investigators as a proving ground for experimental techniques and theoretical understanding. It is these experiments which have dramatized the possibilities of interacting an ion beam with a laser, and for the configuration in which a fast ion beam (10^3—10^4 eV) is

---

[5] E. S. Mukhtar, I. W. Griffiths, F. M. Harris, and J. H. Beynon, *Org. Mass. Spectrom.* 1980, **15**, 51.
[6] M. Horani, H. H. Bukow, M. Carre, M. Druetta, and M. L. Gaillard, *J. Phys. (Paris), Colloq.*, Supplement No. 2, 1979, **40**, C1-57.

coaxial with a monochromatic laser, the spectral resolution obtained is phenomenal, less than 50 MHz. Only the O_2^+ quartet manifold originating in the metastable $\tilde{a}^4\Pi_u$ state has been explored, since photodissociation of the ground $\tilde{X}^2\Pi_g$ state requires an inconveniently energetic photon. Figure 1 shows schematically the relevant known quartet states. The first type of transition observed[7,8]

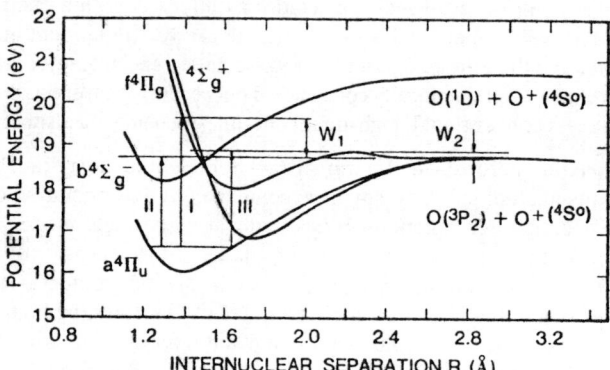

Figure 1 *Relevant quartet states of* O_2^+
(Reproduced with permission from *J. Chem. Phys.*, 1978, **68**, 2406)

was I, which is a direct dissociative excitation to the repulsive wall of the newly confirmed $\tilde{f}\,^4\Pi_g$ state. Soon after, predissociative transitions to the $\tilde{b}\,^4\Sigma_g$, type II in the figure, were reported[9] at extreme resolution. These coaxial-beam high-resolution results have been expanded[6,10—15] to include a large number of lines of the $\tilde{a}(v'', J'') \rightarrow b(v', N')$ system. From these data the molecular constants for the $\tilde{a}$ and $\tilde{b}$ states have been calculated with great precision,[12] and the agreement of theory and experiment for this system now seems satisfactory. The $\tilde{a} \rightarrow \tilde{f}$ direct dissociation I appears as an unstructured background beneath the sharp $\tilde{a} \rightarrow \tilde{b}$ predissociation peaks. A tandem quadrupole spectrometer with a coaxial laser has been used to scan the photodissociation spectrum over the entire visible region at somewhat lower resolution than the fast-beam technique permits, and a large number of $\tilde{a} \rightarrow \tilde{b}$ vibrational bands have been seen.[16,17] Discrete structured

---

[7] A. Tabche-Fouhaille, J. Durup, J. T. Moseley, J.-B. Ozenne, C. Pernot, and M. Tadjeddine, *Chem. Phys.*, 1976, **17**, 81.
[8] M. L. Vestal and G. H. Mauclaire, *J. Chem. Phys.*, 1977, **67**, 3767.
[9] A. Carrington, P. G. Roberts, and P. J. Sarre, *Mol. Phys.*, 1977, **34**, 291.
[10] C. Pernot, J. Durup, J.-B. Ozenne, J. A. Beswick, P. C. Cosby, and J. T. Moseley, *J. Chem. Phys.*, 1979, **71**, 2387.
[11] J. T. Moseley, P. C. Cosby, J.-B. Ozenne, and J. Durup, *J. Chem. Phys.*, 1979, **70**, 1474.
[12] P. C. Cosby, J.-B. Ozenne, J. T. Moseley, and D. L. Albritton, *J. Mol. Spectrosc.*, 1980, **79**, 203.
[13] R. Abouaf, B. A. Huber, P. C. Cosby, R. P. Saxon, and J. T. Moseley, *J. Chem. Phys.*, 1978, **68**, 2406.
[14] J. T. Moseley, P. C. Cosby, H. Helm, and D. L. Albritton, *Bull. Am. Phys. Soc.*, 1979, **24**, 1202.
[15] A. Carrington, P. G. Roberts, and P. G. Sarre, *Mol. Phys.*, 1978, **35**, 1523.
[16] D. C. McGilvery, J. D. Morrison, and D. L. Smith, *J. Chem. Phys.*, 1979, **70**, 4761.
[17] D. C. McGilvery, J. D. Morrison, and D. L. Smith, *J. Chem. Phys.*, 1978, **68**, 4759.

peaks, apparently of type III for predissociative excitation into two vibrational levels supported by the shallow well in $\tilde{f}^4\Pi_g$, have been reported and molecular constants for this state derived.[18]

The spectrum of CO^+ has been studied by Doppler-tuned coaxial-beams methods. Two different means of detecting photon absorption were demonstrated, which, since CO^+ is not photodissociated, is not easy to accomplish. In the first work,[19] an earlier idea[20] for HD^+ was adopted, in which the rate of charge transfer from the ion to a suitable reagent molecule is monitored and perturbation of the charge-transfer rate following photoexcitation of the ions is observed. Using H_2, air, or CO as collision gases, it was possible to observe several CO^+ lines near 4579 Å for $\tilde{X}^2\Sigma^+ \to \tilde{A}\,^2\Pi$ at better than 100 MHz resolution. Later, it was shown[21] that the same experiment could be done using the perturbation of ion–molecule reaction rates to monitor photo-absorption. Reaction rates of ground-state CO^+ with H_2 to yield products including HCO^+ and H_3^+ were found to differ from excited-state CO^+ reactants to a sufficient extent to be usable as spectroscopic indicators. A line of the H_2O^+, $\tilde{X}^2B_1 \to \tilde{A}^2A_1$, transition was also observed by the same method. These techniques have great promise for the high-resolution spectroscopy of ions, since they are not limited to dissociating ion transitions. Ion infrared spectroscopy is an area likely to benefit from these possibilities; an observation of infrared transitions in HeH^+ by perturbed charge transfer is a first example of this.[22]

A large number of sharp predissociation peaks have been observed in CH^+ using Doppler-tuning at several laser wavelengths around 350 nm.[23] The transition is thought to be $\tilde{X}\,^1\Sigma^+ \to \tilde{A}\,^1\Pi$, but the predissociation mechanism is not definitely understood. Understanding of the interstellar formation of CH^+ by radiative association of C^+ with H may be affected by the unexpectedly large number of quasi-bound CH^+ states found in this work.

Laser photodissociation spectroscopy of the N_2O^+, $\tilde{X}^2\Pi \to \tilde{A}^2\Sigma$, system in the region around 300 nm was studied and carefully analysed using a low-energy beam; many vibrational lines were identified and assigned.[24] A recent study by Doppler-tuned coaxial-beams spectroscopy at 337.5 nm refined part of the spectrum to high resolution.[25] High-resolution data on NO^+ dissociation spectra are just beginning to appear and to be studied.[26]

MeI^+ has a richly detailed photo-predissociation spectrum in the visible region. Long vibrational progressions appear which, by use of a tandem quadrupole,[27,28] have been assigned to the transitions $\tilde{X}^2E_{1/2} \to \tilde{A}^2E_{1/2}$, centred near

---

[18] H. Helm, P. C. Cosby, and D. L. Huestis, *J. Chem. Phys.*, in press.
[19] A. Carrington, D. R. J. Milverton, and P. R. Sarre, *Mol. Phys.*, 1978, **35**, 1505.
[20] W. H. Wing, G. A. Ruff, W. E. Lamb, jun., and J. J. Spezeski, *Phys. Rev. Lett.*, 1976, **36**, 1488.
[21] A. Carrington, P. G. Roberts, P. J. Sarre, and R. J. Milverton, *J. Chem. Phys.*, 1978, **68**, 5659.
[22] D. E. Tolliver, G. A. Kyrala, and W. H. Wing, *Phys. Rev. Lett.*, 1979, **43**, 1719.
[23] P. C. Cosby, H. Helm, and J. T. Moseley, *Astrophys. J.*, 1980, **235**, 52.
[24] T. F. Thomas, F. Dale, and J. F. Paulson, *J. Chem. Phys.*, 1977, **67**, 793.
[25] M. Larzilliere, M. Carre, M. Gaillard, J. Rostas, M. Horani, and M. Velghe, *J. Chim. Phys. Phys.-Chim. Biol.*, 1980, **77**, 689.
[26] H. Helm and P. C. Cosby, *Bull. Am. Phys. Soc.*, 1979, **24**, 1201.
[27] D. C. McGilvery and J. D. Morrison, *J. Chem. Phys.*, 1977, **67**, 368.
[28] D. C. McGilvery and J. D. Morrison, *Int. J. Mass Spectrom. Ion Phys.*, 1978, **28**, 81.

550 nm, and $\tilde{X}^2E_{3/2} \rightarrow \tilde{A}^2E_{1/2}$, centred below 450 nm. An abundance of rotational structure in the vibrational peaks is resolved in high-resolution coaxial-beams Doppler-tuned spectra.[29] EtI$^+$ and PrI$^+$ have also been studied.[30] The $\tilde{X}^2E_{3/2} \rightarrow \tilde{A}^2E_{1/2}$ transition in MeI$^+$ is reflected in the EtI$^+$ spectrum by a broad peak between 550 and 600 nm but, in contrast to the MeI$^+$ case, this peak showed no indication of resolved vibrational structure, perhaps because of a short $\tilde{A}$ state lifetime. The absence of $\tilde{X}^2E_{1/2} \rightarrow \tilde{A}$ absorption was attributed to rapid internal conversion of $\tilde{X}^2E_{1/2}$ to the $\tilde{X}^2E_{3/2}$ ground state. In PrI$^+$, a photodissociation peak near 600 nm was assigned as $\tilde{X}^2E_{3/2} \rightarrow \tilde{X}^2E_{1/2}$, and in this case also no vibrational structure was evident.

Methane ion has been examined several times, recently in the tandem quadrupole,[31] both for its intrinsic interest as a simple Jahn–Teller distorted molecule, and also spurred by suggestions of Herzberg that it could be observable in astrophysical spectroscopy. The spectrum appears to be a featureless one, rising gradually and smoothly from a value below 10^{-20} cm^2 in the red to a value around 3×10^{-19} cm^2 near 330 nm, where it probably has a broad maximum. The electronic transition responsible is a transition (or transitions) between the three Jahn–Teller split components of the $\tilde{X}^2T_2$ ground state.

The rare-gas dimer ions, especially Ar$_2^+$, have been well studied.[32–38] The photodissociation spectrum reflects two bound-to-repulsive state transitions, the $1(1/2)u \rightarrow 1(1/2)g$ transitions in the visible and the $1(1/2)u \rightarrow 2(1/2)g$ transition in the u.v. The structureless spectral features observed appear to be in good agreement with theoretical calculations for these electronic transitions. Studies of the photofragmentation spectra[34,35] and the pressure dependence of the photodissociation spectra[36] have allowed a detailed understanding of these processes, and have given information about the role that vibrational excitation plays.

An overwhelming abundance of photodissociation and photodetachment data has been obtained in drift-tube spectrometers, mostly for ions of atmospheric interest.[39–47] Spectra and cross-section data for photodestruction of dozens of positive and negative ions have been reported, including a large number of cluster ions, in which an ionic core like O_3^-, CO_3^-, N_2O^+, etc. is clustered by electrostatic

---

[29] P. C. Cosby, personal communication.
[30] S. P. Goss, D. C. McGilvery, J. D. Morrison, and D. L. Smith, to be published.
[31] D. C. McGilvery, J. D. Morrison, and D. L. Smith, *J. Chem. Phys.*, 1978, **68**, 3949.
[32] L. C. Lee and G. P. Smith, *Phys. Rev. A*, 1979, **19**, 2329.
[33] L. C. Lee, G. P. Smith, T. M. Miller, and P. C. Cosby, *Phys. Rev. A*, 1978, **17**, 2005.
[34] R. Abouaf, B. A. Huber, P. C. Cosby, R. P. Saxon, and J. T. Moseley, *J. Chem. Phys.*, 1978, **68**, 2406.
[35] J. T. Moseley, R. P. Saxon, B. A. Huber, P. C. Cosby, R. Abouaf, and M. Tadjeddine, *J. Chem. Phys.*, 1979, **67**, 1659.
[36] T. L. Rose, D. H. Katayama, J. A. Welsh, and J. F. Paulson, *J. Chem. Phys.*, 1979, **70**, 4542.
[37] R. O. Hunter, J. Oldenettel, C. Howton, and M. V. McCusker, *J. Chem. Phys.*, 1979, **70**, 1492.
[38] J. A. Vanderhoff, *J. Chem. Phys.*, 1978, **68**, 3311.
[39] G. P. Smith, L. C. Lee, and P. C. Cosby, *J. Chem. Phys.*, 1979, **71**, 4464.
[40] G. P. Smith, L. C. Lee, and J. T. Moseley, *J. Chem. Phys.*, 1979, **71**, 4034.
[41] G. P. Smith and L. C. Lee, *J. Chem. Phys.*, 1979, **71**, 2323.
[42] L. C. Lee, G. P. Smith, J. T. Moseley, P. C. Cosby, and J. A. Guest, *J. Chem. Phys.*, 1979, **70**, 3237.
[43] L. C. Lee and G. P. Smith, *J. Chem. Phys.*, 1979, **70**, 1727.
[44] G. P. Smith, L. C. Lee, P. C. Cosby, J. R. Peterson, and J. T. Moseley, *J. Chem. Phys.*, 1978, **68**, 3818.
[45] G. P. Smith and L. C. Lee, *J. Chem. Phys.*, 1978, **69**, 5393.
[46] R. V. Hodges and J. A. Vanderhoff, *J. Chem. Phys.*, 1980, **72**, 3517.
[47] T. F. Thomas, T. L. Rose, and J. F. Paulson, *J. Chem. Phys.*, 1979, **71**, 552.

forces with one or more molecules of O_2, N_2, H_2O, CO_2, etc. The photodissociation behaviour of the cluster ion is often similar to that of the core ion, with perturbations and perhaps some loss of vibrational structure. However, some ions like $O_2^{+} \cdot O_2$ show distinctive visible-wavelength photodissociation spectra. These have been accounted for as transitions to dissociative dimer excited states associated with the electrostatic dimer binding potential function, considering the cluster ion as a pseudo-diatomic.[46] The theory of these transitions seems to be successful, and to give good fits to the observed band shapes and positions for a number of dimer ions.

CO_3^{-} was among the first ions to be studied by the drift-tube instrument, and is notable in showing very prominent vibrational structure in its photodissociation spectrum. It has recently been the subject of a tandem quadrupole study[48] which seems to have clarified a confused situation regarding the photodissociation threshold; it was shown that, in instruments like the tandem quadrupole and the drift tube where ion-laser interaction time is relatively long, two-photon photodissociation takes place below the one-photon threshold at 2.2 eV. Unless this is taken into account, energetic arguments from apparent thresholds cannot be valid.

Photodissociation in Ion-trap Instruments.—Essentially all the ion spectroscopy recently reported with ion-trap techniques has used the ion cyclotron resonance (ICR) instrument. Ions are formed by electron impact and trapped for times typically between 100 ms and 100 s. Ions are detected, and ion–molecule reactions and photochemical processes are followed, by the *in situ* resonance detection methods of the ICR electronics. As compared with ion-beam techniques, ion-trap experiments (i) work on a much slower time scale, with processes occurring on a time scale of many seconds being observable, and (ii) have a wide range of possible manipulations of the collisional history and ion–molecule reactions of the ion, before, during, and after optical irradiation. Between the relaxing effects of electronic fluorescence, collisions, and i.r. radiation on the ICR time scale, the ion in a typical trapped-ion experiment can be expected to be more or less well thermalized, in contrast to the frequent observation in ion-beam studies of large fractions of highly excited and electronically metastable ions (drift-tube spectrometers, however, are the method of choice for observing well thermalized ions).

Using a pulsed dye laser in conjunction with a pulsed ICR spectrometer, the photodissociation of $EtCl^{+}$ in the visible was studied.[49] A band with poorly resolved vibrational structure was observed in the red, which can apparently be identified with the ion $\tilde{A}$ state from photoelectron spectra; and a band in the blue apparently corresponds to the $\tilde{B}$ state observed by PES. Both $C_2H_4^{+}$ and Et^{+} products were observed in a ratio which varied strongly with wavelength, and was entirely inconsistent with RRKM statistics. The product branching ratios agreed with those from charge-transfer fragmentation, and seem to be explained by direct dissociation from one or more excited electronic states. This ion is particularly interesting, in being one of very few cases known of a polyatomic ion

[48] J. F. Hiller and J. L. Vestal, *J. Chem. Phys.*, 1980, **72**, 4713.
[49] L. N. Morgenthaler and J. R. Eyler, *J. Chem. Phys.*, 1979, **71**, 1486.

showing very different photodissociation behaviour in different electronic transitions. A report suggesting analogous behaviour in different electronic bands of butyrophenone ion has apparently been ruled out by later work.[50]

The spectroscopy of radical ions of substituted benzenes has received substantial attention. Spectra were reported[51] of the entire series of methylbenzene ions, with attention to determining the small substituent perturbations of the positions and intensities of the $\pi-\pi$ transition near 400 nm and the $\pi-\pi^*$ transition near 270 nm. First-order perturbation theory was found to account well for the $\pi-\pi$ substituent effects, but within a perturbation-theory framework the $\pi-\pi^*$ band showed more complex substituent shifts.[52] Spectra were also reported for several monoalkyl benzene ions, where it was suggested that a new side-chain-to-ring charge-transfer transition becomes important in the visible for larger alkyl groups.[53] A study of the spectroscopy of halogen-substituted benzenes, using one- and two-photon photodissociation techniques, was reported.[54] The lone-pair orbitals of all the halogens except fluorine play a prominent spectroscopic role, with the visible-region spectroscopy being quite different from that of benzene ions with weakly interacting substituents. By contrast, adding a methyl group to a halogen-substituted benzene ion gives no significant change in the spectrum. Finally, spectra of several styrene ion derivatives were reported.[55] The spectroscopy of these ions seems well accounted for by a scheme involving strong interaction between the ring and the ethylene side-chain.

Spectra of alkane and cycloalkane ions from C_5 to C_{10} show strong photodissociation in the visible region,[56,57] attributed to the overlapping of numerous electron-promotion transitions carrying electrons from the dense set of high-lying occupied C_{2p} orbitals into the half-filled highest orbital. MINDO/3 molecular orbital calculations were quite successful in accounting for the overall features of the spectra, and for the differences between photodissociation and photoelectron spectroscopic envelopes observed for the C_{2p} band.[58] The strongly coloured nature of the alkane radical ions is one of the most striking illustrations of the difference between the spectroscopy of neutral molecules and their parent radical cations.

The spectrum of hexa-1,3,5-triene ion was obtained both at low resolution across the visible/u.v., and also at higher resolution in the visible, where there is well resolved vibrational structure. Identical spectra were obtained for *cis* and *trans* precursor neutrals, in interesting contrast to the observation of distinct *cis* and *trans* spectra in emission. The photodissociation spectrum appears to be that of the *cis* isomer, and a rearrangement of *trans* to *cis* seems a possible explanation of these observations. It was argued that the dissociation of $\tilde{A}$-state *cis* isomer proceeds by a cyclization which is not accessible to the *trans* isomer.[59] Thus,

---

[50] M. S. Kim, R. C. Dunbar, and F. W. McLafferty, *J. Am. Chem. Soc.*, 1978, **100**, 4600.
[51] H. H. Teng and R. C. Dunbar, *J. Chem. Phys.*, 1978, **68**, 3133.
[52] R. C. Dunbar, *J. Chem. Phys.*, 1978, **68**, 3125.
[53] R. C. Dunbar, *J. Phys. Chem.*, 1979, **83**, 2376.
[54] R. C. Dunbar, H. Ho-I. Teng, and E. W. Fu, *J. Am. Chem. Soc.*, 1979, **101**, 6506.
[55] E. W. Fu and R. C. Dunbar, *J. Am. Chem. Soc.*, 1978, **100**, 2279.
[56] E. F. van Dishoek, P. N. Th. van Velzen, and W. J. Van der Hart, *Chem. Phys.*, 1979, **62**, 135.
[57] R. C. Benz and R. C. Dunbar, *J. Am. Chem. Soc.*, 1979, **101**, 6363.
[58] R. C. Dunbar and H. H. Teng, *J. Am. Chem. Soc.*, 1978, **100**, 2279.
[59] M. Allan, J. Dannacher, and J. P. Maier, *J. Chem. Phys.*, in press.

photoexcited *trans* isomers might rearrange to *cis*, but would not dissociate, and no photodissociation features corresponding to the *trans* emission would be seen.

The photodissociation spectrum of the organometallic ion $(C_5H_5)NiNO^+$ in an ICR spectrometer has been obtained.[60] Peaks at 530 and 300 nm are identified with electronic states known from PES, and the photodissociation onset is used to derive an energy for the metal–NO bond of ≤ 43 kcal.

In general, photodissociation of larger ions has been found to yield only the most energetically favoured product. An interesting comparison has been made of the products obtained from a series of protonated aromatic molecules by photodissociation and by collision-induced fragmentation.[61] The differences are striking, with, for example, loss of H being common in CID but not observed in PDS. The differences may be due to the lack of rotational excitation in photo-excitation, or to the excitation of triplet electronic states by collisions.

In a recent *tour de force* of ion trapping, a single Ba^+ ion was trapped in an r.f. quadrupole trap, cooled by an ingenious sideband optical technique, and observed and photographed through a microscope using laser-induced fluorescence.[62–64]

Photodetachment and Photodissociation of Negative Ions.—Photodetachment of electrons from negative ions has been studied for some time, with the primary goal of obtaining accurate electron affinities for atoms and molecules. Among the ions whose photodetachment spectra and electron affinities have been reported recently are a number of enolate ions,[65] acetophenone ions,[66] and alkoxy ions,[67] CH_3S^-,[68] $C_5H_7^-$ and $C_7H_9^-$,[69] HC_2^-,[70] and O_3^-.[71] One development of interest has been the elaboration of a theoretical understanding of the threshold photodetachment lineshape in terms of the electronic symmetry of the ion, which has been used extensively in the determination of electron affinities of polyatomic ions, where the thresholds in the photodetachment spectra are often not clear-cut. A recent elegant application of these theoretical concepts was to the highly symmetric cyclo-octatetraenyl and perinaphthenyl anions,[72] where the high orbital angular momentum of the relevant molecular orbitals provides a particularly stringent test of the theory.

Another recent development in photodetachment spectroscopy has been the observation of various sorts of fine structure in the spectra, which is not expected in the simplest picture of photodetachment lineshapes. A series of sharp peaks

---

[60] R. C. Burnier and B. S. Freiser, *Inorg. Chem.*, 1979, **18**, 906.
[61] V. Franchetti, B. S. Freiser, and R. G. Cook, *Org. Mass. Spectrom.*, 1978, **13**, 106.
[62] W. Neuhauser, M. Hohenstatt, P. Toschek, and H. Dehmelt, *Phys. Rev. Lett.*, 1978, **41**, 233.
[63] W. Neuhauser, M. Hohenstatt, P. E. Toschek, and H. G. Dehmelt, *Appl. Phys.*, 1978, **17**, 123.
[64] W. Neuhauser, M. Hohenstatt, P. E. Toschek, and H. Dehmelt, *Phys. Rev. Lett.*, 1978 **41**, 233.
[65] A. H. Zimmerman, R. L. Jackson, B. K. Janousek, and J. I. Brauman, *J. Am. Chem. Soc.*, 1978, **100**, 4674.
[66] R. L. Jackson, A. H. Zimmerman, and J. I. Brauman, *J. Chem. Phys.*, 1979, **71**, 2088.
[67] B. K. Janousek, A. K. Zimmerman, K. J. Reed, and J. I. Brauman, *J. Am. Chem. Soc.*, 1978, **100**, 6142.
[68] B. K. Janousek and J. I. Brauman, *J. Chem. Phys.*, 1980, **72**, 694.
[69] A. H. Zimmerman, R. Gygax, and J. I. Brauman, *J. Am. Chem. Soc.*, 1978, **100**, 5595.
[70] B. K. Janousek, J. I. Brauman, and J. Simons, *J. Chem. Phys.*, 1979, **71**, 2057.
[71] S. E. Novick, P. C. Engelking, P. L. Jones, J. H. Futrell, and W. C. Lineberger, *J. Chem. Phys.*, 1979, **70**, 2652.
[72] R. Gygax, H. L. Peters, and J. I. Brauman, *J. Am. Chem. Soc.*, 1979, **101**, 2567.

superimposed on the detachment threshold region spectrum observed for some substituted acetophenone enolate ions was attributed to vibrational levels of an autodetaching excited electronic state supported by the dipole field of the neutral product.[66] A resonance in the photodetachment spectrum of sodium halide anions was assigned as an autodetaching electronic state whose nature is essentially that of an excited sodium atom bound to an inert halide anion.[73]

Photoelectron spectroscopy of negative ions, discussed below, has been another fruitful source of new data and understanding, and is complementary to photodetachment spectroscopy.

Spectroscopy as a Structural Tool.—As seen above, the photodissociation spectrum of an ion may for many purposes be considered as equivalent to the electronic absorption spectrum, and is an attractive route to identifying the nature and structure of gas-phase ions in the same way that spectroscopy of neutrals has traditionally been useful. Essentially all work from this point of view has been done in ICR ion-trap instruments.

The idea of spectroscopic ion-structure analysis was suggested some time ago[74] when it was pointed out that the parent radical cations of molecules often contain chromophoric groups which, while they differ markedly from common ideas of chromophores in similar neutral molecules, are nevertheless readily observed by photodissociation and are characteristic of the ion structure. The following are among the ion chromophores which have received recent attention.

Benzenes. While benzene ion itself has a high dissociation threshold and only displays its full spectrum in two-photon spectroscopy (discussed below), many substituted benzene ions display a characteristic spectrum with a strong peak near 400 nm, and another near 270 nm, often accompanied by ring-to-substituent charge-transfer features. The alkylbenzene ions and halogen-substituted benzene ions have been particularly carefully studied.[51,53—55]

Conjugated Dienes and Trienes. The parent ions of conjugated dienes have a peak of modest intensity near 500 nm, and a very strong peak near 330 nm.[74,75] Substituents affect these peak positions substantially, in ways that are still being worked out. Hexa-1,3,5-triene has a similar spectral pattern,[58] but the more extended conjugation shifts the visible peak to near 600 nm, and the u.v. peak to near 360 nm.

Alkanes. For radical ions, the saturated carbon chain is a useful chromophore, giving strong photodissociation in very broad bands across the visible for chains five or more carbons in length.[56,57] (Butane and smaller chains lie at shorter wavelengths, and give apparently weaker peaks.) While for C_5 to C_{10} alkanes and cycloalkanes the photodissociation covers the entire visible region, the intensity maximum shifts progressively to the red for longer chains.

---

[73] S. E. Novick, P. L. Jones, T. J. Mulloney, and W. C. Lineberger, *J. Chem. Phys.*, 1979, **70**, 2210.
[74] R. C. Dunbar, *Anal. Chem.*, 1976, **48**, 723.
[75] R. C. Benz and R. C. Dunbar, 26th Annual Meeting of the American Society for Mass Spectrometry, St. Louis, MO, 1978.

Alkenes. Alkenes, and unconjugated dienes, give parent ion spectra which characteristically have photodissociation rising gradually from the blue into the u.v. with a strong, very broad maximum near or below 250 nm.[74]

The capability of photodissociation spectroscopy for distinguishing ions of different structures has proven to be useful for investigating ion rearrangement chemistry. It is found, for instance,[76] that the parent ions of styrene (1), cyclo-octatetraene (2), and barrelene (3), all $C_8H_8^{+\cdot}$ isomers, do not interconvert on a time scale of seconds, whereas the $C_8H_9^+$ ions obtained by protonating these same neutral molecules all rearrange to the styryl ion structure.

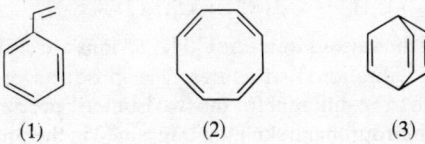

(1)   (2)   (3)

There has been recent attention to the question of migration of double bonds into conjugation. In several ions, for instance 3-phenylpropene[55] and hexa-1,4-diene,[75,77] the double bonds were shown by the photodissociation spectra to migrate to give exclusively ions of conjugated structure, in these cases being β-methylstyrene ion and hexa-2,4-diene ion. However, in several instances where the double bond would have to move by more than one position to reach conjugation, such as hexa-1,5-diene, octa-1,7-diene, and 4-phenylbutene, the spectra showed that no rearrangement takes place. It was also found[77] from laser photodissociation spectra that interconversion of the three *cis–trans* isomers of hexa-2,4-diene ions does not take place on a time scale of seconds.

The electrocyclic ring-opening–ring-closing interconversion between hexa-1,3,5-triene and cyclohexa-1,3-diene ions was investigated by photodissociation spectroscopy,[78] and it was found that no interconversion takes place in seconds. It was noted that this is a case where the concerted process is forbidden by orbital symmetry conservation (and by electronic-state symmetry correlation), so that the interconversion might be expected to be unfavourable. This system has also been studied in a low-temperature solid matrix,[79] where the thermal rearrangement was also found not to occur. In the condensed phase, it was also possible to look for the allowed photochemical interconversion of the ions, and interconversion was observed.

New data have been added to the long-standing question of $C_7H_7^+$ structures.[80] Forming the ion by electron impact on toluene, it was observed that photodissociation occurred in a band near 300 nm, and by comparison with the solution spectrum of tropylium ion it was inferred that at least part of the ion formed has the benzyl structure.

---

[76] R. C. Dunbar, M. S. Kim, and G. A. Olah, *J. Am. Chem. Soc.*, 1979, **101**, 1368.
[77] R. C. Benz and R. C. Dunbar, 27th Annual Meeting of the American Society for Mass Spectrometry, Seattle, WA, 1979.
[78] J. D. Hays and R. C. Dunbar, *J. Phys. Chem.*, 1979, **83**, 3183.
[79] T. Shida, T. Kato, and Y. Nosaka, *J. Phys. Chem.*, 1977, **81**, 1095.
[80] D. A. McCrery and B. S. Freiser, *J. Am. Chem. Soc.*, 1978, **100**, 2902.

A most interesting photodissociation study[81] of the structures of $(C_6H_5NOH)^+$ ions obtained alternatively by protonation of nitrosobenzene or by nitrosation of benzene shows the two structures to be distinct, non-interconverting, and readily distinguished by their spectra.

One of the most promising applications of spectroscopic ion-structure studies is the identification of the structures of product ions from ion–molecule reactions. These possibilities have hardly been touched, but one study[82] used laser photodissociation spectra to identify the $C_3H_5Cl^{+\cdot}$ product ions from reactions of propene ions with 1-chloropropane or 2-chloropropane [reaction (1)].

$$C_3H_6^+ + C_3H_7Cl \rightarrow C_3H_5Cl^+ + C_3H_8 \qquad (1)$$

The laser spectra of the various isomeric $C_3H_5Cl^{+\cdot}$ ions are clearly distinguishable by their distinctive vibrational structures. The products of the ion–molecule reaction were shown to be different for the two isomeric precursor neutrals, and it appeared that the chloropropane skeleton is retained in the product ion through a formal H_2^- abstraction.

3 Photon Emission from Ions

Observing the emission of light from excited gas-phase ions is a method complementary to photodissociation, since in general a given excited electronic state of an ion is likely to choose between emission and fragmentation as its dominant decay mechanism. The three principal ways of observing an emission are, first, the formation of ions in an excited state by an energetic ionization process,[4,83—86] second, formation of excited ions as products of exothermic ion–molecule reactions or energetic ion–molecule collisions,[87,88] and, third, laser-induced fluorescence of ground-state ions.[89—91] All of these are in active use, and ion-emission studies are making major contributions both to the spectroscopy of ions and to elucidating the energy flow among, and competition between, different decay modes of excited ions.

The Basel group has surveyed the extent of electron-impact-induced emission for a very large number of organic molecules. Emission is observed for several classes of ions, among them the substituted benzenes,[92—95] always with at least

[81] W. D. Reents, jun. and B. S. Freiser, *J. Am. Chem. Soc.*, 1980, **102**, 271.
[82] R. G. Orth and R. C. Dunbar, *J. Am. Chem. Soc.*, 1978, **100**, 5949.
[83] J. P. Maier, O. Marthaler, and M. Mohraz, *J. Chim. Phys.*, in press.
[84] S. Leach, M. Devoret, and J. H. D. Eland, *Chem. Phys.*, 1978, **33**, 113.
[85] R. C. Dunbar and D. W. Turner, *J. Chem. Soc., Faraday Trans. 2*, in press.
[86] C. Y. Robert Wu, E. Phillips, L. C. Lee, and D. L. Judge, *J. Chem. Phys.*, 1979, **71**, 769.
[87] R. Marx, in ref. 2.
[88a] C. Ottinger and J. Simonis, *Chem. Phys.*, 1978, **28**, 97.
[88b] I. Kusunoki and C. Ottinger, *J. Chem. Phys.*, 1979, **71**, 4227.
[89] B. H. Mahan, A. O'Keefe, and J. Winn, 'Symposium on Molecular Spectroscopy', Ohio State University, Columbus, OH, 1980.
[90] T. A. Miller and V. E. Bondybey, *Chem. Phys. Lett.*, 1978, **58**, 454.
[91] R. D. Brown, P. D. Godfrey, J. G. Crofts, Z. Ninkov, and S. Vaccani, *Chem. Phys. Lett.*, 1979, **62**, 195.
[92] J. P. Maier and O. Marthaler, *Chem. Phys.*, 1978, **32**, 419.
[93] J. P. Maier, O. Marthaler, M. Mohraz, and R. H. Shiley, *Chem. Phys.*, 1980, **47**, 307.
[94] J. P. Maier, O. Marthaler, M. Mohraz, and R. H. Shiley, *Chem. Phys.*, 1980, **47**, 295.
[95] J. P. Maier, O. Marthaler, M. Mohray, and R. H. Shiley, *J. Electron. Spectrosc. Rel. Phenom.*, 1980, **19**, 11.

two halogen substituents, polyacetylenes and cyanoacetylenes,[96—102] polyenes,[103,104] fluoroborazine,[105] and a variety of triatomic ions.[4] Extensive comparisons have been made with photoelectron and photoion–photoelectron coincidence measurements, and a great deal is known about the ways in which a given ion can choose between radiative, dissociative, and internal-conversion pathways of de-excitation from low-lying excited electronic states. Both direct fluorescence-lifetime measurements and deductions from fluorescence intensity (in competition with non-radiative de-excitation) have yielded insight into the rates of the various processes involved. An excellent review of many aspects of this field has been published.[4] Emission from ions formed by photoionization[85,86] and discharge ionization[84,106—111] and from preformed ions observed by laser-induced fluorescence[89,112—117] has also been reported.

An impressive volume of publication has appeared on emission from fluorine- and chlorine-substituted benzene ions, with work having been active at Basel, Orsay, and the Bell Laboratories. $\tilde{B} \to \tilde{X}$ emission has been observed in the visible spectral region following electron impact for a large number of $C_6F_nH_{6-n}^+$ and $C_6Cl_nH_{6-n}^+$ ions with $n \geq 2$, although not all such ions have detectable emission intensities.[4,84,92—94] Vibrational structure is generally well resolved for this series. Many of these ions have also been observed in both the gas phase and condensed phase by laser-induced fluorescence,[112,90,115,116] and in the gas phase in a discharge-type ion source.[108—111] Sharp vibrational structure has made possible the assignment of a large number of vibrational mode frequencies for both excited- and ground-state ions. One interesting aspect of these ions is the possibility of observable Jahn–Teller effects in the members of the series having three-fold or higher symmetry, since the $\tilde{X}(\pi^{-1})$ ground state is then doubly degenerate. Careful and extensive analyses of Jahn–Teller effects in the cases of $C_6H_3F_3^+$ and $C_6H_3Cl_3^{+\ 108—111,90,115,116}$ and of $C_6F_6^{+\ 108—111}$ have been made. It was concluded that substantial dynamical Jahn–Teller effects are present, with several of the e' vibrational modes showing Jahn–Teller activity. The standard

[96] J. P. Maier, O. Marthlaer, and F. Thommen, *Chem. Phys. Lett.*, 1979, **60**, 193.
[97] M. Allan, E. Kloster-Jensen, J. P. Maier, and O. Marthaler, *J. Electron. Spectrosc. Rel. Phenom.*, 1978, **14**, 359.
[98] M. Allan, J. P. Maier, O. Marthaler, and E. Kloster-Jensen, *Chem. Phys.*, 1978, **29**, 331.
[99] E. Kloster-Jensen, J. P. Maier, O. Marthlaer, and M. Mohray, *J. Chem. Phys.*, 1979, **71**, 3125.
[100] M. Allan, J. P. Maier, O. Marthaler, and J.-P. Stadelman, *J. Chem. Phys.*, 1979, **70**, 5271.
[101] J. P. Maier, O. Marthlaer, and E. Kloster-Jensen, *J. Electron. Spectrosc. Rel. Phenom.*, 1980, **18**, 251.
[102] J. P. Maier, O. Marthaler, and E. Kloster-Jensen, *J. Chem. Phys.*, 1980, **72**, 701.
[103] J. P. Maier, O. Marthaler, and G. Bieri, *Chem. Phys.*, 1979, **44**, 131.
[104] T. B. Jones and J. P. Maier, *Int. J. Mass Spectrom. Ion Phys.*, 1979, **31**, 287.
[105] T. B. Jones, J. P. Maier, and O. Marthaler, *Inorg. Chem.*, 1979, **18**, 2140.
[106] D. Gauyacq and M. Horani, *Can. J. Phys.*, 1978, **56**, 587.
[107] S. Leach, P. R. Stannard, and W. A. Gelbart, *Mol. Phys.*, 1978, **36**, 1119.
[108] C. Cossart-Magos, D. Cossart, and S. Leach, *J. Chem. Phys.*, 1978, **69**, 4313.
[109] C. Cossart-Magos, D. Cossart, and S. Leach, *Chem. Phys.*, 1979, **41**, 363.
[110] C. Cossart-Magos, D. Cossart, and S. Leach, *Chem. Phys.*, 1979, **41**, 345.
[111] C. Cossart-Magos, D. Cossart, and S. Leach, *Mol. Phys.*, 1979, **37**, 793.
[112] T. A. Miller, *J. Chem. Phys.*, 1978, **69**, 2562.
[113] T. A. Miller, V. E. Bondybey, and B. R. Zegarski, *J. Chem. Phys.*, 1979, **70**, 4982.
[114] V. E. Bondybey and T. A. Miller, *J. Chem. Phys.*, 1978, **69**, 3597.
[115] T. A. Miller, V. E. Bondybey, and J. H. English, *J. Chem. Phys.*, 1979, **70**, 2919.
[116] V. E. Bondybey and T. A. Miller, *J. Chem. Phys.*, 1979, **70**, 138.
[117] D. H. Katayama, T. A. Miller, and V. E. Bondybey, *J. Chem. Phys.*, 1980, **72**, 5469.

quantitative model for treating dynamical Jahn–Teller effects proves adequate to account for the spectra in very high detail. However, for $C_6Cl_3H_3^+$, the Jahn–Teller analysis is still controversial: two significantly different choices of Jahn–Teller parameters give plausible fits to the observed spectra, and there is for the moment no agreement between the Orsay and Bell Laboratory groups over the correct choice. Nevertheless, these ions constitute some of the most clear-cut and successfully worked out illustrations of the Jahn–Teller effect.

Early results have been reported from a new instrument in which laser-induced flourescence is observed from ions held in an r.f. quadrupole ion trap.[89] There are interesting advantages to combining the powerful capabilities of LIF with the unique characteristics of trapped-ion experiments, and this is a technique of promise.

A series of experiments examining the ionic emissions observed following charge transfer from Ar^+ or He^+ to neutral molecules has been reported and reviewed.[87] An ICR spectrometer was equipped to observe emission following the near-thermal-energy ion–molecule reactions. A further asset of these experiments is the use of the ICR ion-trapping characteristics to measure the kinetic energy release in the reaction; the combination of optical emission and kinetic energy information reveals a great deal about the charge-transfer process. Optical emissions studied recently include $H_2O^+(\tilde{A} \to \tilde{X})$ from Ar^+ charge transfer to H_2O,[118] $N_2O^+(\tilde{A} \to \tilde{X})$ from He^+ charge exchange (and also electron impact) on N_2O,[119] and $CH^+(CD^+)$ emission resulting from the reaction of He^+ with $C_2H_2(C_2D_2)$.[120] One interesting feature coming from this work is evidence that charge transfer is often a tightly coupled reaction giving substantial kinetic energy release and product ion states not easily populated by vertical ionization from the neutral.

In a quite different type of experiment, emission from reactively excited product ions in ion beams has been studied.[88] A recent investigation of the CH^+ $A^1\Pi$ emissions following C^+ impact on H_2 at 6 Å resolution yielded a large amount of information about the rotational distribution of the products, which could be successfully modelled.

4 Multiphoton Processes

An interesting development of the past few years has been the discovery and increasing understanding of ion two-photon and multiphoton dissociation processes, which have characteristics quite different from multiphoton processes studied in neutral molecules. The basically unique feature of these processes is that the ion is entirely (or nearly) collision free on a time scale of seconds, so that the interesting intramolecular events involved in multiphoton photophysics can be isolated and examined in a leisurely way.

The first two-photon ion dissociation, for benzene ion,[121] was observed in an ICR ion trap. It was suggested correctly that the process was a sequential one, as

---

[118] R. Derai, S. Fenistein, M. Gerard-Ain, T. R. Govers, R. Marx, G. Mauclaire, C. Z. Profous, and C. Sourisseau, *Chem. Phys.*, 1979, **44**, 65.
[119] M. Gerard, T. R. Govers, and R. Marx, *Chem. Phys.*, 1979, **36**, 247.
[120] M. Gerard, T. R. Govers, and R. Marx, *Chem. Phys.*, 1978, **30**, 75.
[121] B. S. Freiser and J. L. Beauchamp, *Chem. Phys. Lett.*, 1975, **35**, 35.

shown in reaction (2), in which first one photon is absorbed to give the electronically excited $C_6H_6^{+*}$.

$$C_6H_6^+ \xrightarrow{h\nu} C_6H_6^{+*} \xrightarrow{\text{fast}} C_6H_6^{+v} \xrightarrow{h\nu} C_6H_5^+ + H \qquad (2)$$

$$\underbrace{\phantom{C_6H_6^+ \xrightarrow{h\nu} C_6H_6^{+*} \xrightarrow{\text{fast}} C_6H_6^{+v}}}_{\text{collisions or i.r. radiation}}$$

Rapid internal conversion gives the vibrationally hot ground-state ion $C_6H_6^{+v}$, which can be deactivated by collision with neutral (M), can relax by infrared radiative cooling, or can be photodissociated by a second photon. Ions for which two-photon dissociation is now known include benzene, chlorobenzene,[54] bromobenzene,[54] iodobenzene,[54] cyanobenzene,[122] naphthalene,[123] methylnaphthalene,[124] and styrene.[55]

In one recent outgrowth from these discoveries,[123] it was found that naphthalene ion, in addition to undergoing a two-photon dissociation in the u.v., also undergoes a four-photon dissociation around 600 nm. The process was presumed to be sequential, and modelling, including both collisional and radiative de-excitation mechanisms, accounted for the intensity and pressure dependences. The cross-section was lower than expected, and some rearrangement of the photoexcited intermediates seems likely.

The fact that the vibrationally excited intermediate $C_6H_6^{+v}$ can be collisionally intercepted was exploited in an investigation of vibrational energy transfer in ion–molecule collisions.[125] The effectiveness of various neutral collision gases (M) in suppressing two-photon photodissociation in bromobenzene ion was measured quantitatively, to give an absolute measure of the rate of collisional quenching of vibrational excitation. For collision partners ranging in size from methane to substituted benzenes, it was found that between three and more than fifty collisions were needed for vibrational quenching, the number correlating fairly well with the size of the collision gas molecules.

Recently, new possibilities in ion multiphoton chemistry have been opened by the observation of i.r. multiphoton dissociation.[126] Using a CW CO_2 laser of a few watts power, it was possible to effect substantial dissociation of the diethyl ether proton-bound dimer ion, and similar multiphoton dissociation has been reported for a variety of other ions.[127–129] It is notable that, for some ions like $C_3F_6^+$, a strong dependence is observed on photon wavelength, and for $C_3F_6^+$ the i.r. spectrum traced by line-tuning the CO_2 laser is very similar to the i.r. spectrum of the neutral molecule. The dissociation process is presumably a sequential one very much like the visible two- and four-photon processes, with a number of i.r. photons of the order of 20 being absorbed sequentially over the course of perhaps

---

[122] T. E. Orlowski, B. S. Freiser, and J. L. Beauchamp, *Chem. Phys.*, 1976, **18**, 439.
[123] M. S. Kim and R. C. Dunbar, *J. Chem. Phys.*, 1980, **72**, 4405.
[124] R. C. Dunbar and Robert Klein, *J. Am. Chem. Soc.*, 1976, **98**, 7994.
[125] M. S. Kim and R. C. Dunbar, *Chem. Phys. Lett.*, 1979, **60**, 247.
[126] D. S. Bomse, R. L. Woodin, and J. L. Beauchamp, *J. Am. Chem. Soc.*, 1979, **101**, 5503.
[127] D. S. Bomse, K. L. Woodin, and J. L. Beauchamp, in 'Advances in Laser Chemistry', ed. A. H. Zewail, Springer Series in Chemical Physics, Springer, 1978.
[128] R. L. Woodin, D. S. Bomse, and J. L. Beauchamp, in 'Chemical and Biochemical Applications of Lasers', ed. C. B. Moore, Academic Press, 1979, Vol. IV.
[129] R. L. Woodin, D. S. Bomse, and J. L. Beauchamp, *J. Am. Chem. Soc.*, 1978, **100**, 3248.

a second. In spite of the many photons involved, the intensity dependence is weak, not much stronger than linear, which may reflect a 'bottleneck' effect in which the first one or few photons are absorbed slowly to elevate the molecule into the 'quasi-continuum,' with subsequent photon absorptions then being much more rapid. One important implication of these developments is that they offer a practical way of obtaining i.r. spectra of ions, with the remaining obstacle being the lack of a sufficiently powerful and tunable i.r. laser source over many parts of the i.r. spectrum.

In another variation of multiphoton ion chemistry, a two-laser experiment was described in which iodobenzene ion was photodissociated by the combined irradiation of both a visible laser and a CO_2 i.r. laser.[130] This process was successfully modelled by computer, taking into account excitation by both lasers, and de-excitation by collisions and by i.r. radiative relaxation. The success of this model in accounting for light intensity dependence and pressure dependence gives reason to be confident that the basic mechanisms involved in multiphoton ion dissociation chemistry have been identified.

Infrared multiphoton dissociation behaviour has been found for CH_3OHF^-, using a pulsed CO_2 laser and an ICR instrument. Since the laser pulse is shorter by several orders of magnitude than the collision time constant, the dissociation process was treated as being collisionless, and the laser pulse was used to probe the collisional cooling history of the ions. It was concluded that, for this ion, collisional thermalization may require as many as 100 collisions with neutrals.

Dissociation of CF_3I^+, CF_3Br^+, and $MeCl^+$ has been effected[132] in a fast beam using a coaxial CW CO_2 laser beam. It is presumed that only one i.r. photon is absorbed and postulated that it is those ions emerging from the ion source with sufficient internal excitation to place them within ~ 0.1 eV of the dissociation threshold which are dissociated by the i.r. laser photons. The i.r. wavelength dependence is pronounced, showing a strong peak at 947 cm^{-1} for CF_3I^+, presumably corresponding to a C—F stretching vibration. This process is of particular interest in that it can be thought of as being equivalent to the *last* photon in a multiphoton i.r. photodissociation process, and thus offers a unique route to studying the characteristics of the last events in i.r. multiphoton dissociation. These experiments have been extended to a variety of fluorine-containing organic ions.[133] For larger ions, the dissociation rates and kinetic energy releases are consistent with RRKM behaviour (although this is not true for CF_3I^+); strong peaks in the wavelength dependence are commonly observed, showing that these highly vibrationally excited molecules have not lost all structure in their i.r. absorptions.

Very recently there was reported the first example of the perturbation of an ion–molecule reaction rate by multiphoton i.r. excitation of one of the reactants.[134] The particular reaction observed was that of dimethyl ether proton-bound dimer ion with H_2O, as shown in reaction (3).

[130] R. C. Dunbar, J. D. Hays, J. Honovich, and N. B. Lev, *J. Am. Chem. Soc.*, 1980, **102**, 3950.
[131] R. N. Rosenfeld, J. M. Jasinski, and J. I. Brauman, *J. Am. Chem. Soc.*, 1979, **101**, 3999.
[132] M. J. Coggiola, P. C. Cosby, and J. R. Peterson, *J. Chem. Phys.*, 1980, **72**, 6507.
[133] R. C. Dunbar, M. Coggiola, P. C. Cosby, and J. R. Peterson, 27th Annual Meeting of the American Society for Mass Spectrometry, New York, 1980.
[134] D. S. Bomse and J. L. Beauchamp, *J. Am. Chem. Soc.*, 1980, **102**, 3967.

$$(\text{MeOH})_2\text{H}^+ + \text{H}_2\text{O} \rightarrow (\text{MeOH})(\text{H}_2\text{O})\text{H}^+ + \text{MeOH} \qquad (3)$$

The reactant ion was known to be a strong absorber of CO_2 laser radiation, and it was shown that the reaction rate was significantly increased by i.r. excitation. This experiment has interest as a rather general way of studying i.r. absorption in ions without the necessity of driving them to dissociation.

In a particularly interesting and surprising experiment, the photodetachment of an electron from the benzyl anion by i.r. laser irradiation was found.[135] Since the i.r. photons certainly excite vibrational degrees of freedom, while the photodetachment is an electronic excitation, this is a process in which the intramolecular conversion of vibrational to electronic excitation occurs. There are so few known examples of this inverse internal conversion process that the observation is of exceptional importance.

Using a coaxial-beams spectrometer there was observed a novel two-photon i.r. dissociation of HD^+, in which the first photon drives a $v = 16 \rightarrow v = 18$ vibrational transition, and the second photon drives the bound → free electronic transition $1s\sigma_g \rightarrow 2p\sigma_u$.[136]

5 Photofragmentation

Experimental results at the level of detail provided by photofragmentation studies have begun to give penetrating understanding of fragmentation processes and the accompanying intramolecular energy flows for a number of fairly small ions. Both ion traps and ion beams are suitable and have been used for photofragmentation studies, although ion-beam experiments are far more numerous. The kinematic compression effect mentioned above as virtually eliminating Doppler-broadening gives a corresponding great advantage to high-energy ion-beam study of kinetic energy release. Kinetic energy releases of the order of milli-electron volts (meV) in the molecule-fixed co-ordinate frame are magnified to appear as energies of the order of electron volts in the laboratory frame, and consequently photofragment energy resolution of the order of 1 meV is routinely achieved for small energy releases.

Photofragmentation of diatomic molecules is particularly easy to interpret, since all of the photon energy in excess of the dissociation endoergicities appears either as kinetic energy or as electronic excitation of the product atoms, and the latter can often be ruled out for lack of low-lying atomic states. The name photofragment spectroscopy has been applied to experiments in which this behaviour is exploited by irradiating at fixed laser wavelength and making a scan of the distribution of fragment ion kinetic energies; this distribution mirrors the distribution of ion excited states leading to fragmentation. Moreover, the fragments must fly apart along the direction of the molecular axis, so that product angular distributions lead readily to relations between molecular orientation and polarization direction of the exciting light. H_2^+ was the first diatomic to receive detailed attention.[137] The photofragment spectrum (the spectrum of product-ion kinetic energies for fixed excitation wavelength) shows well resolved peaks

---
[135] R. N. Rosenfeld, J. M. Jasinski, and J. I. Brauman, *J. Chem. Phys.*, 1979, **71**, 1030.
[136] A. Carrington, J. Buttenshaw, and P. G. Roberts, *Mol. Phys.*, 1979, **38**, 1711.
[137] J. B. Ozenne, D. Pham, and J. Durup, *Chem. Phys. Lett.*, 1972, **17**, 422.

corresponding to many of the populated vibrational levels of the initial $1s\sigma_g$ state, with the transition being polarized along the H—H axis. Although H_2^+ is understood theoretically in such extreme detail that these early experiments added little new information, they did dramatically show the potential utility of ion-beam photofragmentation methods.

Careful study of energy release and angular distribution has been a major part of the great effort spent on O_2^+ quartet states. Comparison of the kinetic energy release spectra for light polarized parallel and perpendicular to the beam direction played a role in assigning the photodissociation at several wavelengths to direct $\tilde{a} \to \tilde{f}\,^4\Pi_g$ transitions, establishing the existence of the $\tilde{f}$ state.[137] More recently, it has been shown[10] that data from a coaxial-beams experiment (where the light is necessarily polarized perpendicular to the flight direction) can, with a detailed analysis, yield angular distributions; angular distributions of individual rovibronic $\tilde{a} \to \tilde{b}$ predissociating transitions were derived and found to agree with theory.

In the difficult process of resolving and assigning the innumerable individual transitions of the $\tilde{a} \to \tilde{b}$ predissociation spectrum of O_2, an important aid has been the ability to select only those product ions having kinetic energy release in a narrow range. This selects only those transitions terminating in vibrational levels of $\tilde{b}\,^4\Sigma_g^-$ lying at a specific energy above the dissociation limit, and has allowed some fairly definitive assignments of many $\tilde{a} \to \tilde{b}$ transitions.[12,138] By observing the lowest-energy levels of the $\tilde{b}$ state which predissociate, it has been possible to derive impressively accurate values of the bond strengths of O_2^+ and O_2.[139] These values can be further improved by accurate determination of the kinetic energy release for specific predissociating levels of $\tilde{b}\,^4\Sigma_g^-$,[10] and now represent the best bond-strength values of O_2 and quartet O_2^+, and among the best bond strengths known for any molecule.

There has been considerable discussion of the nature of the electronic state which predissociates $\tilde{b}\,^4\Sigma_g^-$. The question is best attacked starting from lifetimes of the individual $\tilde{b}$ levels derived from high-resolution Doppler-tuned spectra, and fitting these to calculations of expected linewidths for the possible predissociation mechanism. The conclusion, first reached and subsequently refined and elaborated,[15] that $^4\Sigma_g^+$ is responsible, has stood up well. The suggestion that $\tilde{f}\,^4\Pi_g$ may also contribute to the predissociation is in some doubt.[3] Another topic of current interest is the distribution of the product atoms among the three close-lying spin–orbit components of $O(^3P)$.[10,18,140]

The kinetic energy release in MeI^+ is extraordinarily complex, with each of the innumerable rotation–vibration predissociation peaks showing a distinct energy release.[141] The spectrum can be analysed in terms of progressions in the C—I stretch, and rotational non-equilibration is indicated by the energy release.

The ICR ion trap can be used to obtain photofragmentation information. Kinetic energy release data are available because the trap is of known depth and

[138] M. Tadjeddine, R. Abouaf, P. C. Cosby, B. A. Huber, and J. T. Moseley, *J. Chem. Phys.*, 1978, **69**, 710.
[139] D. L. Albritton, J. T. Moseley, P. C. Cosby, and M. Tadjeddine, *J. Mol. Spectrosc.*, 1978, **70**, 326.
[140] M. Tadjeddine, *J. Chem. Phys.*, 1979, **71**, 3892.
[141] P. C. Cosby, 35th Symposium on Molecular Spectroscopy, Ohio State University, Columbus, Ohio, 1980.

fast ions can escape; angular information comes from the fact that the trap is anisotropic, with escape being easier along the direction of the magnetic field. An ICR study[142] of MeCl$^+$ showed large energy release along with complete parallel polarization of the transition yielding Me$^+$. These results indicate a direct excitation to a repulsive electronic state. Results recently reported for the OCS$^+\tilde{A}^2\Pi$ state[143] show an energy release of 0.55 eV and very small anisotropy, suggesting that predissociation is slow compared with molecular rotation.

6 Other Spectroscopic Methods

Microwave absorption spectra of the ions HCO$^+$, CO$^+$, N$_2$H$^+$, and N$_2$D$^+$ have been obtained.[144] Microwave data, where obtainable, provide by far the best geometry information about ions; these laboratory data are of most interest in providing positive identifications of interstellar species observed in emission in radioastronomy.

Photoelectron spectroscopy of cations, with the necessity for high-energy exciting photons and severe intensity problems, has not yet been achieved, but for anions, where intense visible lasers give sufficient photon energy to observe the highest-lying orbitals, this spectroscopy is feasible. Using an instrument capable of high-quality photoelectron spectroscopy of negative ions, results were reported which have given impressive data for a number of negative ions, and hold promise of bringing negative-ion energetics and spectroscopy to a very high state of development. A mass-analysed beam of negative ions is crossed with an intracavity laser, and the photoelectrons are analysed with a hemispherical energy analyser of 60 meV resolution. The most fundamental use of this capability is the measurement of electron affinities with greater precision and much greater confidence than has been possible before. Among the atoms whose electron affinities have been reported[145,146] are Cu, Ni, and Fe, and molecular species include O$_3$,[71] Fe(CO)$_n$ for $n = 0-4$,[147] MeO,[148] CD$_3$O,[148] MeS,[148] and Me.[149] The observation of the photoelectron spectrum of Me$^-$ is particularly noteworthy, since this ion had not previously been observed and was thought perhaps to be unstable; the electron affinity of 0.08 eV is small, but clearly establishes that the ion is fully stable. The spectrum is interpreted as showing a planar (or quasi-planar) structure for Me, and a pyramidal structure for Me$^-$. For the polyatomic anions, laser photoelectron spectra provide, in addition to electron affinities, information about vibrational frequencies, electronic states, and bond strengths.

The author gratefully acknowledges the support of the National Science Foundation and the donors of the Petroleum Research Fund, administered by the American Chemical Society, during preparation of this report.

---

[142] R. G. Orth and R. C. Dunbar, *J. Chem. Phys.*, 1978, **68**, 3254.
[143] R. G. Orth and R. C. Dunbar, *J. Chem. Phys.*, 1980, **45**, 195.
[144] T. G. Anderson, C. S. Gudeman, T. A. Dixon, and R. C. Woods, *J. Chem. Phys.*, 1980, **72**, 1332.
[145] R. R. Corderman, P. C. Engelking, and W. C. Lineberger, *J. Chem. Phys.*, 1979, **70**, 4474.
[146] P. C. Engelking and W. C. Lineberger, *Phys. Rev. A*, 1979, **19**, 149.
[147] P. C. Engelking and W. C. Lineberger, *J. Am. Chem. Soc.*, 1979, **101**, 5569.
[148] P. C. Engelking, G. B. Ellison, and W. C. Lineberger, *J. Chem. Phys.*, 1978, **69**, 1826.
[149] G. B. Ellison, P. C. Engelking, and W. C. Lineberger, *J. Am. Chem. Soc.*, 1978, **100**, 2556.

5
Aspects of Secondary Ion Emission

BY A. R. KRAUSS AND V. E. KROHN

1 Introduction

The erosion of a surface bombarded by energetic ions or neutral particles is called sputtering and sputtered particles emitted in a charged state are called secondary ions. These ions are of interest because of concern about the physical basis of the emission process, because they are useful in depth profiling and in the analysis of surfaces and of bulk materials, because they are used as a source of negative ions for tandem accelerators, and because they are important in plasma and high-voltage devices.

In recent years, there has been an almost exponential increase in publications concerned with secondary ions. The present survey will assume that the reader is familiar with much of this material and will concentrate on areas where something can be added to the many existing reviews[1-25] on secondary ion

[1] M. Kaminsky, 'Atomic and Ionic Impact Phenomena on Metal Surfaces', Springer-Verlag, Berlin, 1965.
[2] G. Carter and J. S. Colligon, 'Ion Bombardment of Solids', Elsevier, New York, 1968.
[3] P. Sigmund, *Rev. Roum. Phys.*, 1972, **17**, 823, 969, and 1079.
[4] P. Sigmund, in 'Physics of Ionized Gases', ed. M. V. Kurepa, Institute of Physics, Belgrade, 1972, p. 137.
[5] Y. A. Fogel, *Int. J. Mass. Spectrom. Ion Phys.*, 1972, **9**, 109.
[6] C. A. Evans, jun., *Anal. Chem.*, 1972, **44**, 67A.
[7] A. Benninghoven, *Surf. Sci.*, 1973, **35**, 427.
[8] H. Liebl, *Anal. Chem.*, 1974, **46**, 22A.
[9] F. G. Rüdenauer, W. Steiger, and R. Portenschleg, *Mikrochim. Acta*, 1974, Suppl. 5, 421.
[10] H. W. Werner, *Vacuum*, 1974, **24**, 493.
[11] R. E. Honig, *Adv. Mass. Spectrom.*, 1974, **6**, 337.
[12] H. Liebl, *J. Vac. Sci. Technol.*, 1975, **12**, 385.
[13] A. Benninghoven, *Surf. Sci.*, 1975, **53**, 596.
[14] H. H. Andersen, in 'Physics of Ionized Gases 1974', ed. V. Vujnović, Institute of Physics, Zagreb, 1974, p. 361.
[15] H. Liebl, *J. Phys. E*, 1975, **8**, 797.
[16] J. A. McHugh, in 'Methods of Surface Analysis', ed. A. W. Czandera, Elsevier, New York, 1975, p. 223.
[17] R. E. Honig, *Thin Solid Films*, 1976, **31**, 89.
[18] K. Wittmaack, in 'Inelastic Ion-Surface Collisions', ed. N. H. Tolk, J. C. Tully, W. Heiland, and C. W. White, Academic Press, New York, 1977, p. 153.
[19] A. E. Morgan and H. W. Werner, *Adv. Mass. Spectrom.*, 1978, **7A**, 764.
[20] G. Blaise, in 'Material Characterization Using Ion Beams', ed. J. P. Thomas and A. Cachard, Plenum Press, New York, 1978, p. 143.
[21] K. Wittmaack, *Surf. Sci.*, 1979, **89**, 668.
[22] G. Blaise and A. Nourtier, *Surf. Sci.*, 1979, **90**, 495.
[23] P. Williams, *Surf. Sci.*, 1979, **90**, 588.
[24] K. Wittmaack, *Nucl. Instrum. Methods*, 1980, **168**, 143.
[25] A. R. Krauss and R. B. Wright, *J. Nucl. Mater.*, 1980, **89**, 229.

emission and related subjects. We will be concerned with recent advances in the understanding of the emission process, the use of secondary ion mass spectrometry (SIMS) as an analytical technique, secondary ion energy distributions, sputter ion sources, and fusion reactor applications. However, some very important topics of current interest such as single-crystal sputtering and the emission of molecular ions will be neglected entirely because a proper treatment would warrant an entire article on these subjects.

2 The Development of the Equilibrium Treatment

First, it should be noted that, if one assumes that sputtering can be treated as if it were an equilibrium process, the Saha (or Saha-Eggert) equation is not equivalent to the Saha-Langmuir equation. The latter was derived from the former by I. Langmuir[26,27] who recognized that to describe an emission process one must add velocity factors to the Saha equation which describes densities.

The fractional yields (ratios of ions to neutral atoms) given by the Saha-Langmuir equations with the image charge considered (the Dobretsov[28] equations) are given in equations (1) and (2), where the ω's are the partition functions and the ν's the fluxes of neutral atoms and positive and negative ions (designated by 0, +, and −), ϕ is the work function, I is the ionization potential, A is the electron affinity, and x is the critical distance for charge exchange. By considering the image charge, *i.e.* by including the quantity $[e/4x]$, one is assuming that there is thermal equilibrium at the time the sputtered particle is separated from the surface by a critical distance for charge exchange and that, if the particle is an ion, some of its original kinetic energy will be used to separate it from the image charge.

$$\alpha_+ = \nu_+/\nu_0 = (\omega_+/\omega_0) \exp\left[(e/kT)(\phi - I + [e/4x])\right] \quad (1)$$

$$\alpha_- = \nu_-/\nu_0 = (\omega_-/\omega_0) \exp\left[(e/kT)(A - \phi + [e/4x])\right] \quad (2)$$

Veksler and Ben'iaminovich[29] were the first to attempt to account for secondary ion yields by treating sputtering as if it were an equilibrium process. They found that their observed positive ion yields were 10^7—10^9 times larger than the yields calculated using the actual target temperature. They suggested that the high number of secondary ions might be due to the fact that these ions leave the surface with such a high energy that the equilibrium treatment (using the actual temperature) is not valid.

Later[30] it was pointed out that the use of a variable parameter with a value of about 6000 K in place of the actual temperature would lead to calculations in rough agreement with the positive ion yields of Veksler and Ben'iaminovich as well as the negative ion yields from substrates coated with caesium. Of course, given that the discrepancies observed by Veksler and Ben'iaminovich varied by a factor of 100, the single-temperature parameter could hardly be expected to fit the data to better than a factor of 10.

[26] I. Langmuir and K. H. Kingdon, *Proc. R. Soc. London*, 1925, **107**, 61.
[27] I. Langmuir, *J. Am. Chem. Soc.*, 1932, **54**, 2798.
[28] E. N. Dobretsov, *Proc. Phys.-Tech., Inst. Acad. Sci., Uzbek SSR*, 1950, **39**, 3.
[29] V. I. Veksler and M. B. Ben'iaminovich, *Zh. Tekh. Fiz.*, 1956, **26**, 1626.
[30] V. E. Krohn, *J. Appl. Phys.*, 1962, **33**, 3523.

At about this time, Liebl and Herzog[31] expressed considerable enthusiasm in recommending the use of the Saha–Langmuir formula (with image charge and partition functions omitted) to relate secondary ion yields to concentrations, and Smith[32] promptly protested in a letter which included the remark '... the hot-spot model has limited, if any, application'. This was, of course, only the first of many disputes about the validity of the equilibrium treatment.

Schelten[33] (who was aware of the recommendation of Liebl and Herzog) was the first to use a different temperature parameter for each matrix in order to fit the yields of five minor elements per matrix to the Saha–Langmuir equation. He bombarded iron and aluminium matrices with 12 keV Ar^+ and normalized the observed intensity of the (positive) secondary ions of each minor element to the appropriate neutral atom intensities and to the intensity of the secondary ions of the main element of the matrix. Using equation (1) with partition functions set equal to unity and with notation differing somewhat from Schelten's the normalized yields are given by equation (3), where the ν's are intensities, + and 0 designate charged or neutral, i identifies the ith element, m designates the major element of the matrix, and $(\Delta I)_i = I_m - I_i$. Schelten plotted Y_i vs. ΔI_i (Figure 1), but he did not constrain the lines on the semilog plot to intersect the point (0, 1), which we believe to be appropriate because $(\Delta I)_m = 0$ and $Y_m = 1$ by definition. We see that, even with the added constraint, all five minor elements in the aluminium matrix and three out of five in the iron matrix fall within a factor of two of the lines representing equation (3). This agreement is comparable with the agreement achieved in more recent work.

$$Y_i = \frac{(\alpha_+)_i}{(\alpha_+)_m} = \frac{(\nu_+)_i(\nu_0)_m}{(\nu_0)_i(\nu_+)_m} = \exp\left[(e/kT)(\Delta I)_i\right] \qquad (3)$$

It would seem that at this point one had adequate evidence that the equilibrium treatment [equations (1)—(3)] gives an approximate but not an exact description of the probability of sputtered atoms being emitted as ions. However, after a few years, a new group of workers reconsidered the matter without noticing that it had already been examined. The first of these were Andersen and Hinthorne[34,35] who introduced the local thermal equilibrium (LTE) model which is based on the Saha equation and asserts that there is a plasma in the vicinity of the sputtering site. It should be noted that the LTE model works in spite of the missing velocity factors because the equilibrium velocities for neutral atoms and ions of a given mass are equal, and because adjustment of the electron-density parameter compensates for the missing electron-velocity factor. The LTE model with the electron density and the temperature as parameters is exactly equivalent to the Saha–Langmuir equations with the work function and the temperature as parameters (with image charges neglected).

Before considering additional evidence that the predictions of the equilibrium treatment are approximately, if not exactly, correct we will consider comments that are critical of the equilibrium treatment. We have, of course, already heard

---

[31] H. J. Liebl and R. F. K. Herzog, *J. Appl. Phys.*, 1963, **34**, 2893.
[32] H. P. Smith, jun., *J. Appl. Phys.*, 1964, **35**, 3067.
[33] J. Schelten, *Z. Naturforsch.*, Teil A, 1968, **23**, 109.
[34] C. A. Andersen and J. R. Hinthorne, *Science*, 1972, **175**, 853.
[35] C. A. Andersen and J. R. Hinthorne, *Anal. Chem.*, 1973, **45**, 1421.

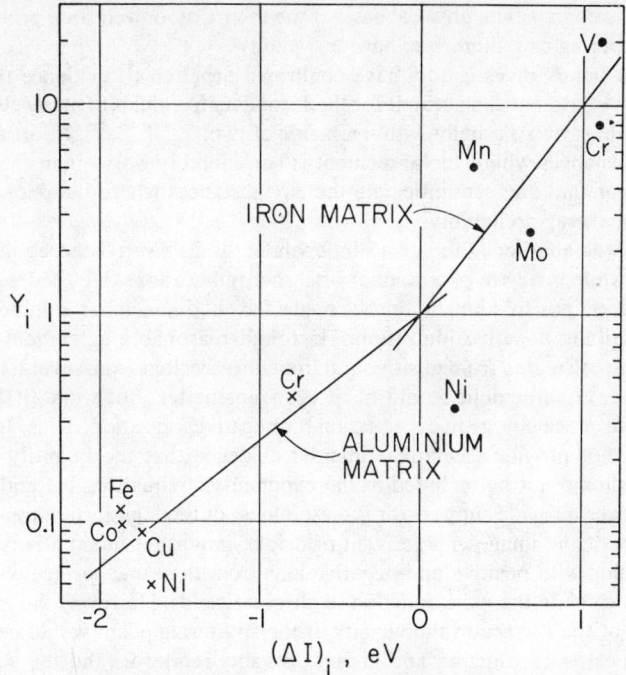

Figure 1 *The yields of five secondary ions from an aluminum matrix (the x's) and five from an iron matrix. The lines representing equation (3) have been constrained to intersect the point (0, 1) so that the only parameters are the slopes of the lines. Eight of the ten points fall within a factor of two of the corresponding line. These data from Schelten are comparable to more recent results. They suggest that the equilibrium treatment gives an approximate, but not an exact description of secondary ion yields*
(Data reproduced by permission from *Z. Naturforsch., Teil A*, 1968, **23**, 109)

from Smith. Later, Benninghoven[36] said that 'the degree of ionization of secondary particles ... cannot be predicted by the Saha–Langmuir equation'; Werner[10] stated, '... there is not thermodynamic equilibrium in the plasma'; Slodzian[37] declared, '... the 'plasma picture' should not be taken too seriously as the temperature and the electron density play only the part of adjustable parameters'; Rüdenauer, Steiger, and Werner[38] suggested that '... thermodynamic equilibrium ... and consequently the existence of a 'temperature' may not be valid'; Krohn[39] protested that '... the sputtered material is not emitted from a plasma'; Simons, Baker, and Evans[40] commented, 'The fitting ... does not by itself prove that local thermal equilibrium exists at the sputtering site'; and

[36] A. Benninghoven, *Z. Phys.*, 1969, **220**, 159.
[37] G. Slodzian, *Surf. Sci.*, 1975, **48**, 161.
[38] F. G. Rüdenauer, W. Steiger, and H. W. Werner, *Surf. Sci.*, 1976, **54**, 553.
[39] V. E. Krohn, *Int. J. Mass Spectrom. Ion Phys.*, 1976, **22**, 43.
[40] D. S. Simons, J. E. Baker, and C. A. Evans, jun., *Anal. Chem.*, 1976, **48**, 1341.

Snowdon[41] said that 'The physical basis of the theory of ionized state production during sputtering can, therefore, have no validity'.

However, many investigators have confirmed Schelten's[33] evidence that the equilibrium treatment can account for the secondary ion yields of many elements from a single surface, usually within a factor of two.[35,38,40,42–50] Unfortunately, there are instances where the agreement is considerably worse than this[51] and it is not clear that one can anticipate the circumstances where the data will be of less than average reliability.

Jurela[52] was another author who independently discovered the equilibrium treatment while unaware of a number of earlier publications. He used equation (1) to describe positive ionization but neglected the image charge in equation (2) in describing negative ionization. He found reasonable agreement in the temperature calculated from positive and from negative ions from several targets, but the overall agreement would have been not better, but worse if he had included the image charge in his equation for negative ionization. Thus, Jureka's work does not provide clear experimental evidence that the quantity $[e/4x]$ should or should not be included in the exponents of equations (1) and (2). It does, however, provide support for the usefulness of the equilibrium treatment.

Inclusion of the image charge is, in principle, important in situations where one is dealing with positive and negative ions from the same surface or where one wants to relate the work function to absolute yields. However, the extreme disruption of the surface in the vicinity of the sputtering event would certainly alter the local work function and in fact probably render invalid the concepts of work function and critical distance for charge exchange. Furthermore, it should be noted that the manner in which the image charge is treated in equations (1) and (2) is a simplification which is valid only in the approximation that x is large compared with interatomic distances. Because of these difficulties one should probably consider treating x and ϕ as parameters along with the apparent temperature T.

It seems worth citing two situations where the equilibrium treatment was found to be particularly bad. In one case,[53] the addition of oxygen caused a lowering of the work function along with an increase in the yield of Mg^+. In

---

[41] K. J. Snowdon, *Radiat. Eff.*, 1979, **40**, 9.
[42] D. E. Newbury, K. F. J. Heinrich, and R. L. Myklebust, in 'Surface Analysis Techniques for Metallurgical Applications', ed. R. S. Carbonara and J. R. Cothill, American Society for Testing Materials, Philadelphia, 1975, p. 101.
[43] P. J. Martin and R. J. MacDonald, *Surf. Sci.*, 1977, **62**, 551.
[44] A. E. Morgan and H. W. Werner, *Anal. Chem.*, 1976, **48**, 699; *ibid.*, 1977, **49**, 927.
[45] D. H. Smith and W. Christie, *Int. J. Mass. Spectrom. Ion Phys.*, 1978, **26**, 61.
[46] P. Vallerand and M. Baril, *Int. J. Mass. Spectrom. Ion Phys.*, 1977, **24**, 241.
[47] A. E. Morgan and H. W. Werner, *J. Chem. Phys.*, 1978, **68**, 3900.
[48] R. J. MacDonald and R. F. Garrett, *Surf. Sci.*, 1978, **78**, 371.
[49] T. Okutani and R. Shimizu, *Surf. Sci.*, 1979, **88**, L51.
[50] M. Bernheim, J. Rebière, and G. Slodzian, in 'Secondary Ion Mass Spectrometry (SIMS II)', ed. A. Benninghoven, C. A. Evans, jun., R. A. Powell, R. Shimizu, and H. A. Storms, Springer-Verlag, New York, 1979, p. 40.
[51] M. A. Rudat and G. H. Morrison, *Anal. Chem.*, 1979, **51**, 1179.
[52] Z. Jurela, *Int. J. Mass. Spectrom. Ion Phys.*, 1973, **12**, 33.
[53] G. Blaise and G. Slodzian, *Surf. Sci.*, 1973, **40**, 708.

the other,[54—56] oxidation caused increases in the yields of both negative and positive ions. In the first case, the variation of the Mg^+ yield was in the opposite direction from the predication of equation (1) and in the second no possible variation of the work function could cause equations (1) and (2) to predict increases in the yields of ions of both signs.

The latter result was interpreted as evidence against the validity of thermodynamic models and the existence of a plasma. However, Williams and Evans[56] have offered a surface polarization model which provides a qualitative explanation of the simultaneous enhancement of positive and negative secondary ion yields and neither supports nor contradicts any fundamental mechanism for secondary ion production. In this model potential changes in an oxygenated surface are considered to occur at highly localized sites of atomic dimensions and to be of variable sign depending on whether the oxygen is adsorbed on the surface or incorporated below it.

Most tests of the equilibrium treatment have been done with noble gas ions on (hopefully) clean surfaces or with oxygen enhancing the yields of positive ions. An alternative is to investigate enhanced negative ion yields from surfaces flooded with an alkali. Although many authors[30,46,50,57—65] have used alkalis to study enhanced negative ion yields, the cases where a variety of elements was emitted from a single surface are limited. In one instance,[35] three ions from one surface were accounted for very well by two parameters; in another,[46] three ions in three phosphor bronze alloys were fitted reasonably well; and in a third instance,[50] six out of seven negative ions from a series of alloys flooded with caesium were well accounted for by an equation of the Saha–Langmuir type.

It should be noted that in the last example the quantity plotted was not $\alpha_- = \nu_-/\nu_0$ but $\eta_- = \nu_-/(\nu_0 + \nu_-)$ and that η is approximately equal to α if $\eta \ll 1$. We call attention to this for two reasons. First, if one substitutes α for η and extrapolates the line in Figure 3 of Bernheim, Rebière, and Slodzian[50] to the electron affinity of gold at 2.3 eV, one finds a value of about 1000 for α_-, *i.e.* one finds that the equilibrium treatment predicts not zero but 0.1% for the neutral fraction (ν_0/ν_-) sputtered from a gold surface flooded with caesium.

Second, we believe that the 'integrated ion counts' cited by the Illinois group[66—71] and reviewed by Wittmaack[24] are proportional to η rather than α

[54] J. M. Morabito, R. K. Lewis, and J. C. C. Tsai, *Appl. Phys. Lett.*, 1973, **23**, 260.
[55] A. Benninghoven, W. Sichtermann, and S. Storp, *Thin Solid Films*, 1975, **28**, 59.
[56] P. Wiliiams and C. A. Evans, jun., *Surf. Sci.*, 1978, **78**, 324.
[57] C. A. Andersen, *Int. J. Mass. Spectrom. Ion Phys.*, 1970, **3**, 413.
[58] M. K. Abdullayeva, A. K. Ayukhanov, and U. B. Shamsiyev, *Radiat. Eff.*, 1973, **19**, 225.
[59] V. E. Krohn and G. R. Ringo, *J. Microsc.*, 1977, **110**, 59.
[60] P. Williams, R. K. Lewis, C. A. Evans, jun., and P. R. Hanley, *Anal. Chem.*, 1977, **49**, 1399.
[61] G. Doucas, *Int. J. Mass. Spectrom. Ion Phys.*, 1977, **25**, 71.
[62] H. A. Storms, K. F. Brown, and J. D. Stein, *Anal. Chem.*, 1977, **49**, 2023.
[63] M. Bernheim and G. Slodzian, *J. Phys.*, 1977, **38**, L325.
[64] M. L. Yu, *Phys. Rev. Lett.*, 1978, **40**, 575.
[65] T. Okutani, K. Shono, and R. Shimizu, in ref. 50, p. 186.
[66] V. R. Deline, C. A. Evans, jun., and P. Williams, *Appl. Phys. Lett.*, 1978, **33**, 58.
[67] V. R. Deline, W. Katz, C. A. Evans, jun., and P. Williams, *Appl. Phys. Lett.*, 1978, **33**, 832.
[68] V. R. Deline, in ref. 50, p. 48.
[69] P. Williams, *IEEE Trans. Nucl. Sci.*, 1979, **26**, 1807.
[70] J. E. Chelgren, W. Katz, V. R. Deline, C. A. Evans, jun., R. J. Blattner, and P. Williams, *J. Vac. Sci. Technol.*, 1979, **16**, 324.
[71] P. Williams, W. Katz, and C. A. Evans, jun., *Nucl. Instrum. Methods*, 1980, **168**, 373.

and that this may have made a difference of more than a factor of ten for some of the F⁻ data which caused Wittmaack[24] to conclude that the negative ion data did not agree with equation (2).

In addition, we believe that a re-examination of some of the results of Deline et al.[67] will provide some additional support for the approximate validity of the equilibrium treatment. The data of interest involved determination of the 'integrated ion counts' of some elements (B, C, F, P, As, and Sb) implanted in matrices (C, Si, Ge, Sn, and GaAs) which were bombarded with caesium to produce negative secondary ions of the six implanted elements or with oxygen to produce positive secondary ions. When the counts for a particular ion were plotted against the inverse of the substrate sputtering rate $(1/S)$ of the various matrices, they were found to be approximately linear on log–log plots.

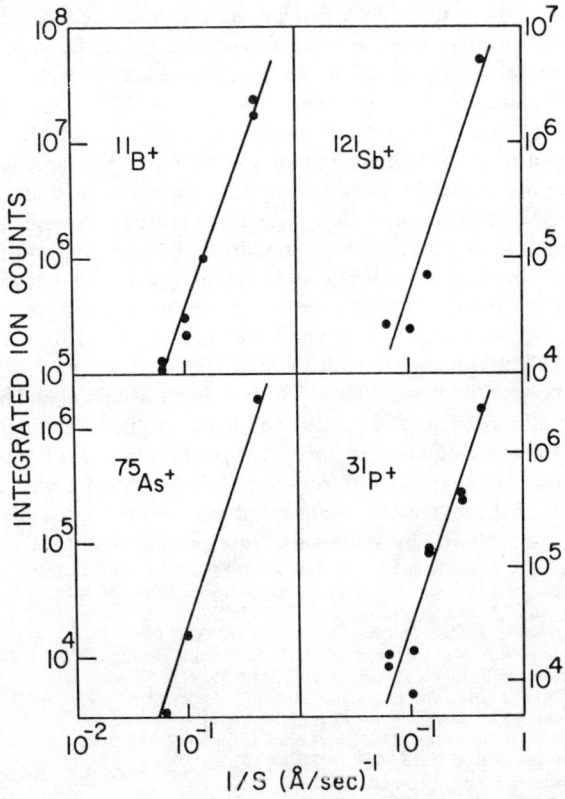

Figure 2 *Secondary ion yields from a series of matrices bombarded by oxygen. The data are from Deline et al. who plotted the yields against the inverse of the sputtering rate of the matrices and found that the log–log plots are linear. The fact that the data for all four ions can be fitted by lines of the same slope is a suggestion of the equilibrium treatment*

(Data reproduced by permission from *Appl. Phys. Lett.*, 1978, **33**, 832)

We believe that it is reasonable to speculate that the exponential increase in counts with increasing abscissa suggests that the abscissa, $\log(1/S)$, is proportional to the apparent work function, ϕ, and that equations (1) and (2) then account for the exponential increase in counts. If this is the case, one expects a single slope to fit all of the negative ion plots and a second slope to fit all four positive ion plots. Figures 2 and 3 show the data of Deline *et al.* with the lines drawn to satisfy this constraint. Except for the F^- plot, the data seem to support our speculation. We have, of course, already suggested that the F^- data reach ionized fractions where α cannot be approximated by η. This would explain the F^- data.

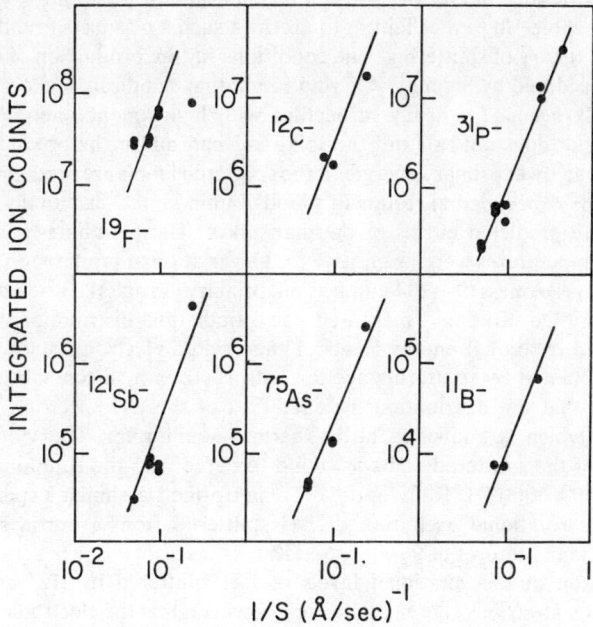

Figure 3 *Secondary ion yields from substrates bombarded by caesium. Again, the data and the form of the plots are from Deline et al. with the slopes of the lines constrained to be identical because of the suggestion of the equilibrium treatment. We believe the F^- data fail to agree with this slope because the published yields are proportional to η rather than α, with F^- the only ion having a yield so high that the depletion of the flux of the corresponding neutral atom is important*
(Data reproduced by permission from *Appl. Phys. Lett.*, 1978, **33**, 832)

3 Further Tests of the Validity of the LTE Model

Because of the overall quantitative success in predicting secondary ion yields, the LTE model and in particular the notion of an equilibrium plasma continue to be considered in the literature in spite of very serious conceptual weaknesses. Clearly the situation will not be resolved until either an alternative model appears which is equally accurate for quantitative analysis while explaining the existence of an apparent thermodynamic equilibrium, or the physical basis of the LTE model can somehow be justified in spite of what appear to be insurmountable difficulties.

An equilibrium plasma is a collection of electrons, atoms, and ions in thermal equilibrium *with each other*. The derivation[26,27] of equations (1) and (2) is based on the assumption of such an equilibrium. Recent experiments investigating both secondary ion emission and the related phenomena of secondary electron emission and electronic excitation of the sputtered material have provided new insights and greater detail concerning the nature and origin of the apparent equilibrium. Specifically, some of the questions addressed are the existence of an equilibrium plasma and its location and mechanism of formation.

An obvious choice for a mechanism to produce such a plasma is found in the thermal-spike theory of sputtering. The conditions for the production of a spike have been considered by Sigmund[72,73] who found that significant spike-induced sputtering only occurs for heavy projectiles with high kinetic energies. This theoretical result does not rule out the spike mechanism for the production of sputtered ions at lower primary energies if the sputtered atoms are predominantly neutral, but the experimental results of Good-Zamin et al.[74] are totally inconsistent with the predicted effects of thermal spikes. These authors found that the derived temperatures were insensitive to the variation of primary ion energy (4—16 keV), target mass (9—204 a.m.u.), and primary ion mass (20—84 a.m.u.).

Thomas and De Koning[75] measured the population distributions in the vibrational and rotational energy levels of the excited electronic states of the CH radical obtained by sputtering acetone adsorbed on a silicon substrate. It was observed that the distribution matched that of the CH spectrum from a flame source which is known to be in thermal equilibrium. The vibrational temperature of the sputtered CH was found to agree with the rotational temperature (4000—5000 K). Kelly and Tolk[76] performed a similar experiment, measuring the rotational excitation of BH sputtered from a boron surface, and obtained temperatures of 2700—3600 K.

A comparison of the rotational levels of CH sputtered by Ne^+ and Kr^+ bombardment[77] at various energies indicates, however, that the electronic excitation of the sputtered molecule sometimes appears to be in accord with LTE and sometimes not. There does not appear to be any correlation with the existence of thermal-spike conditions. It was concluded that neither thermodynamic nor random collision cascade mechanisms could be responsible for the sputtered molecules. Instead, the ejection of electronically excited molecules was attributed to a direct projectile–adsorbate interaction.

Other results arguing against a thermal spike as the source of an equilibrium plasma can be found in the predictions of the LTE model itself. The program

---

[72] P. Sigmund and C. Claussen, in 'Proceedings of a Symposium on Sputtering', ed. P. Varga, G. Betz, and F. P. Viehböck, Institut für Allgemeine Physik, Technische Universität Wien, Austria, 1980, p. 113.
[73] P. Sigmund, *Appl. Phys. Lett.*, 1974, **25**, 169.
[74] C. J. Good-Zamin, M. T. Shehata, D. B. Squires, and R. Kelly, *Radiat. Eff.*, 1978, **35**, 139.
[75] G. E. Thomas and B. R. De Koning, *Chem. Phys. Lett.*, 1978, **55**, 418.
[76] R. Kelly and N. H. Tolk, to be published.
[77] K. J. Snowdon and R. J. MacDonald, *Nucl. Instrum. Methods*, 1980, **170**, 351.

CARISMA has been used by Weyl et al.[78] to measure the composition of series 304 and 316 stainless steels. Measured concentrations were found to be accurate to within a factor of two with typical T, N_e values of 3000—4000 K and 10^{10}—10^{12} cm^{-3}, respectively. Similar values have been obtained by other users of this program.[49] It is extremely difficult to see how a temperature can be meaningfully ascribed to a system in which only one electron out of every 10^{10} present is representative of the thermal 'equilibrium'.

In a similar vein, computer modelling of the collision cascade[79] indicates that, although the collision cascade may be very extensive, almost all of the sputtering occurs as the result of surprisingly few collisions stemming directly from the primary knock-on. This situation is again inconsistent with the establishment of thermal equilibrium.

A strong experimental argument against the thermal spike as the origin of a plasma has been presented by Thum and Hofer.[80] By using a primary ion beam consisting of heavy metal clusters, they were able to ensure thermal spike conditions. They found that the number of electrons ejected per incident atomic particle was a function of the incident velocity and was independent of the number of atoms in the cluster. The absence of enhanced electron emission as the energy density in the spikes increased is evidence against the presence of a plasma. It was noted that, because of the poor momentum transfer between the atoms and electrons and because energetic conduction electrons would distribute themselves over a much larger volume than the energetic atoms, it was not to be expected that the electrons would be in thermal equilibrium with the atoms in the spike.

It can also be shown quite directly that in most cases the atoms themselves are not even approximately in thermal equilibrium. The existence of an equilibrium plasma requires that the kinetic energy distribution of the secondary ions be Maxwellian and identical for all ion species originating within the same plasma. Typical secondary ion kinetic energy distributions are shown for several different matrices in Figure 4.[81] As a first approximation, the kinetic energies are in accord with the Sigmund–Thompson model[82] for sputtering from a random collision cascade. This subject will be dealt with more thoroughly in a later section, but it should be noted here that the secondary ion energies tend to be even higher than the energies of the corresponding sputtered neutrals. The shape of the secondary ion energy distributions is distinctly non-Maxwellian, varying significantly for different species sputtered from the same matrix. The peaks of the distributions correspond roughly to one half of the surface binding energies,[82] which are far in excess of anything ascribable to the apparent temperature deduced from yield measurements.

By using fitting procedures such as the CARISMA program, it is often possible to obtain relatively well defined values of temperature and electron density, N_e. However, when the secondary ions are analysed within a relatively narrow but variably positioned energy window[78] the values obtained vary significantly.

[78] H. Weyl, A. Krauss, and D. Gruen, to be published.
[79] B. J. Garrison, N. Winograd, and D. E. Harrison, jun., *Surf. Sci.*, 1979, **89**, 101.
[80] F. Thum and W. Hofer, *Surf. Sci.*, 1979, **90**, 331.
[81] A. Krauss and D. Gruen, *J. Nucl. Mater.*, 1979, **85/86**, 1179.
[82] M. W. Thompson, *Philos. Mag.*, 1968, **18**, 377.

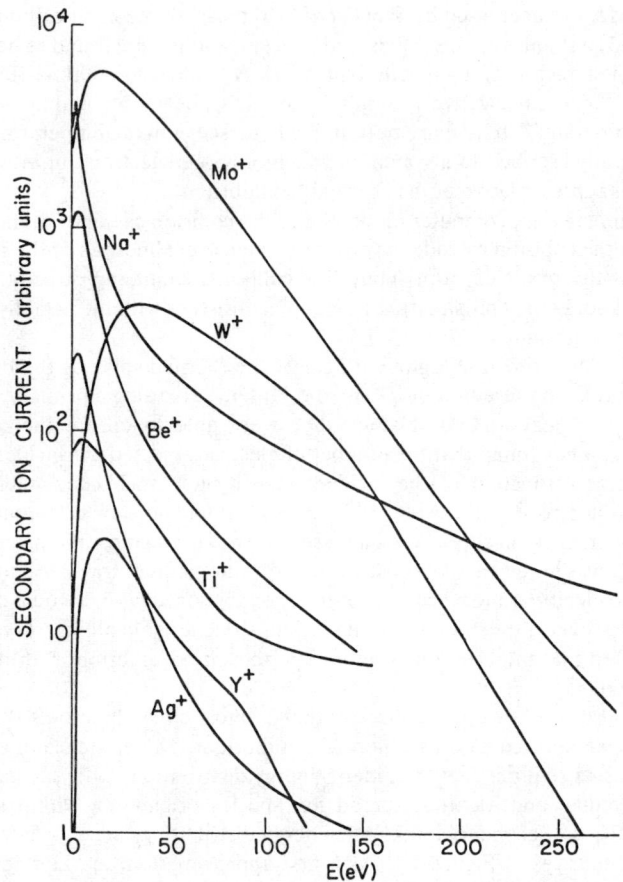

Figure 4 *Secondary ion energy distributions for seven elements. Peak positions range from 1 eV(Na^+) to 35 eV(W^+)*[81]

Figure 5 shows the values derived from the yields from a 304 stainless steel sample as a function of the analyser pass energy. The temperature varies by more than a factor of two and the density varies by almost a factor of 10^5. Some evidence of a similar trend in the value of T vs. pass energy has also been observed by Morgan and Werner.[44] Okutani and Shimizu,[49] on the other hand, found that for a fixed analyser pass energy it was not always possible to obtain unique values for T and N_e. Equally good fits to the elemental concentration were obtained for arbitrary T within the range 2500—7000 K and N_e varying by a factor of 10^{14} as long as the second parameter was adjusted appropriately. Whatever the parameters refer to, they clearly do not represent a well defined plasma in thermal equilibrium within the collision cascade volume.

One is tempted to ask if there is any place other than the collision cascade volume where an equilibrium plasma might exist. One intriguing possibility has

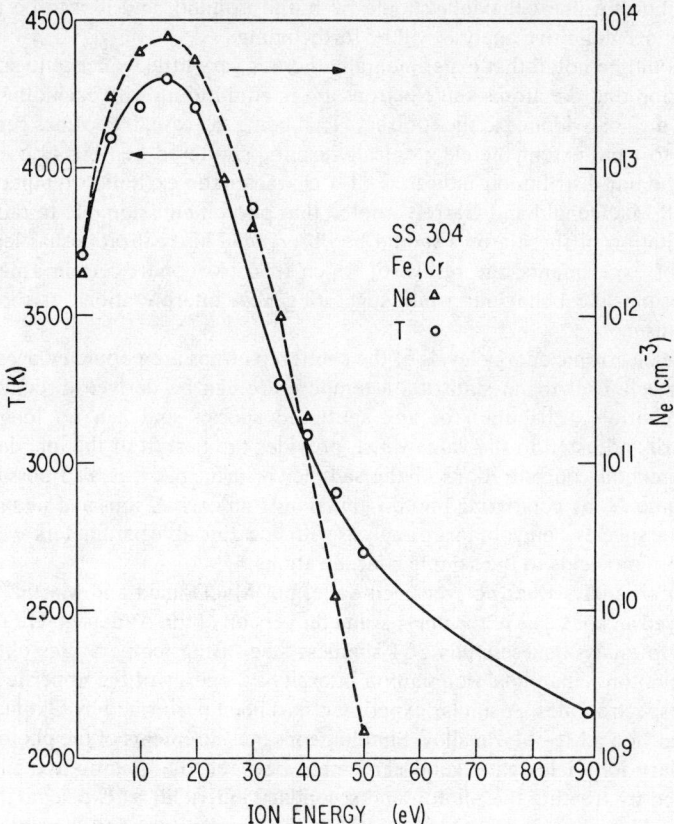

Figure 5 *Effective temperature* (○) *and plasma density* (△) *obtained by CARISMA for SS 304 using known concentrations of Fe and Cr to obtain fitting parameters. An energy window of* ~1 eV *width was used for each evaluation of* T *and* N_e[78]

been presented by Coles.[83] In his model, ionization is considered to be the result of impact with secondary electrons outside the solid. It is arbitrarily assumed that the atom–electron collisions occur in a 1.5 nm hemisphere above the surface. In this region the electron density is calculated to be very high even though the absolute number is only about ten electrons per sputtering event. If the appropriate assumptions are made, Coles finds that this 'micro-plasma' can, on the average, lead to phenomena similar to those characteristic of dense macroscopic plasmas even though the concept of thermal equilibrium is hardly applicable. A similar calculation by Snowdon[41] led to a more pessimistic result, namely that secondary electron yields are a factor of 10^4 to 10^{10} too low to provide the required electron densities. Both calculations require educated guesses to estimate the feasibility of the process. However, the possibility of such an explanation

---

[83] J. N. Coles, *Surf. Sci.*, 1969, **79**, 549.

of equilibrium-like behaviour should be borne in mind, and it is to be hoped that a more definitive analysis will be forthcoming.

It should be noted that experimentally there is very little evidence to support the notion that the atoms and electrons are in equilibrium with each other, and a good deal of evidence to the contrary. That being the case, it becomes germane to ask to what extent the electronically excited sputtered neutrals also exhibit a Boltzmann distribution indicative of a characteristic excitation temperature. In 1978 MacDonald and Garrett[48] noted that photon emission due to radiative de-excitation of these atoms should be observed. This realization has led to a flurry of experiments, the results of which tend to support certain aspects of equilibrium-like behaviour while suggesting new interpretations of the LTE mechanism.

If the electronic energy levels of the sputtered atoms are populated according to Maxwell–Boltzmann statistics, a temperature can be derived directly from the excitation distribution of any sputtered species and can no longer be arbitrarily adjusted to the value which provides the best fit of the ion yields to the elemental concentrations of the sample. In principle it is also possible to determine N_e by comparing photon emission from excited ions and neutrals of the same species[84] and obtain an analysis with no adjustable parameters available to fit the ion yields to the sample concentrations.

Such an analysis has not yet been done, but MacDonald and Martin[84] have employed an analogue of the one-parameter version of the Andersen–Hinthorne model to analyse successfully 304 stainless steel using temperatures obtained from 'photon Saha plots' (log photon intensity $vs.$ energy of the upper level) of the Fe spectral lines. A similar experiment has been performed by Okutani and Shimizu[49] on a Mg–Al–Zn alloy. Simultaneous measurements of the photon and secondary ion yields gave good agreement between the quantitative analyses obtained by treating the photon and secondary ion yields as two separate sets of data. However, it was noted that both the temperature and density values obtained from the photon data were substantially lower than the same quantities obtained from the analysis of the SIMS data, in disagreement with the predictions of LTE. It was further noted that the introduction of oxygen increased the apparent temperature for ion emission[44,49] by approximately a factor of two but had no effect on the temperature obtained from photon emission.

Photon yield measurements by Tsong and Yusuf[85] on a series of oxide glasses failed to give any indication of a Boltzmann population of the energy levels. Least-squares fits to Saha plots yielded temperatures similar to those obtained from SIMS analyses, but with such large error estimates that the authors concluded that the data did not in fact represent thermal equilibrium.

Good-Zamin et al.[74] examined the photon emission for a wide range of projectiles, projectile energies, and elemental targets and found that the excitations could be fitted by thermal distributions. As noted earlier, the derived temperatures proved to be quite insensitive to the bombardment parameters, which is consistent with the interpretation that the temperature is a property of the

---

[84] R. J. MacDonald and P. J. Martin, *Surf. Sci.*, 1977, **66**, 423.
[85] I. S. T. Tsong and N. A. Yusuf, *Surf. Sci.*, 1979, **90**, 417.

electrons only. These authors and others[85a] also noted that an optically excited atom is too big to allow it to exist in the solid, thereby supporting Coles' notion that interactions above the surface are important.

A further test of LTE can be found in an analysis of alloys; all species sputtered from a given sample must have level populations corresponding to the same temperature if LTE is valid. Tsong[86] analysed a Ni–Fe alloy and found a temperature of 2800 K for Ni and 3800 K for sputtered Fe. On this basis he ruled out an electron equilibrium mechanism either above or below the surface. MacDonald et al.[87] noted an error in Tsong's data analysis and presented similar data of their own in which it was claimed that the same temperature was observed for both excited Ni and Fe. However, after correcting the analysis of his own data, Tsong re-analysed the data of MacDonald et al. and came to the conclusion that neither set was consistent with LTE.[88]

It should be noted[89] that the Canadian, Australian, and American groups each used different methods of calculating the initial population of excited states producing photon emission. The extent to which the data were consistent with Boltzmann populations in all three cases indicates that the ability to fit a Saha plot is not a very sensitive test of thermodynamic equilibrium. For this reason we will turn our attention to other criteria for LTE which have been examined recently.

If there is LTE, then the electronic states of all species should have Boltzmann populations corresponding to a *single* temperature. The observed optical emission from excited states of uranium sputtered by 3 keV Kr^+ were found to be characterized by a temperature of ~ 4000 K.[90] Under the same bombardment conditions laser fluorescence spectroscopy was used to probe the low-lying metastable levels.[91] The population fitted a Saha plot, yielding a temperature of only 920 ± 100 K. Almost identical results were obtained by Rusbüldt[92] for sputtered Fe. The implication of these results is that over a restricted energy range the population of excited states is *approximately* Maxwellian, with the effective 'temperature' varying for various degrees of excitation. Unfortunately it has not yet been possible to measure the populations of excited electronic states over a wide and continuous energy range.

One result of the idea that ionization and excitation are caused by the same mechanism is that, if the apparent temperature obtained from the photon data varies, the ion temperature should be the limiting value obtained from Saha plots as one proceeds to the most highly excited states. The lower photon temperatures obtained by Okutani and Shimizu[49] are consistent with this view.

Another prediction of the LTE model is that the secondary ion intensities should be lower than photon yields. Absolute measurements by Williams, Tsong, and Tsuji[93,94] indicate that the converse is true; the secondary ion yield typically

[85a] K. Jensen and E. Veje, *Z. Phys.*, 1974, **269**, 293.
[86] I. S. T. Tsong, *Surf. Sci.*, 1977, **69**, 609.
[87] R. J. MacDonald, R. F. Garrett, and P. J. Martin, *Surf. Sci.*, 1978, **75**, 155L.
[88] I. S. T. Tsong, *Surf. Sci.*, 1978, **75**, 159L.
[89] K. J. Snowdon, B. Andresen, and E. Veje, *Radiat. Eff.*, 1979, **40**, 19.
[90] R. Wright, personal communication.
[91] R. B. Wright, M. J. Pellin, D. M. Gruen, and C. E. Young, *Nucl. Instrum. Methods*, 1980, **170**, 295.
[92] D. Rusbüldt, to be published.
[93] P. Williams, I. S. T. Tsong, and S. Tsuji, *Nucl. Instrum. Methods*, 1980, **170**, 591.
[94] I. S. T. Tsong and N. A. Yusuf, *Appl. Phys. Lett.*, 1979, **33**, 999.

exceeds the photon yield by a factor of 100. This result is inconsistent with the LTE model in general and the Coles theory in particular.

Further difficulties for the LTE model are found in measurements of the kinetic energies of the excited and ionized sputtered material. As discussed earlier and in a subsequent section, the secondary ion energy distributions are similar to those corresponding to the sum of all sputtered particles, but shifted to higher energies. Naïvely, one would expect the kinetic energy distributions of excited atoms to lie somewhere between the total and the secondary ion distributions. However, Tsong and Yusuf,[95] Dzioba et al.,[96] and Heiland et al.[97] have shown that the energy distributions of excited atoms are dramatically different from both ions and neutrals. In particular it is observed that the spectra display threshold kinetic energies E^* below which no excitation occurs. The value of E^* has been found to vary from 31 eV for Na (5890 Å) to 1150 eV for Ba (5535 Å)[95] sputtered from the respective iodides. These threshold energies are far in excess of the typical energies observed in the secondary ion and ground-state neutral energy distributions of these materials. Consequently, it was concluded that both binary collisions and thermal processes could be ruled out as the excitation mechanism.[96]

A rather striking series of experiments has been performed by Snowdon et al.[98] who compared the Zn^I and Zn^{II} level populations for (i) atoms sputtered from a zinc target by inert gas bombardment, (ii) gas-phase single collisions between a Zn^+ ion beam and an argon target, and (iii) a zinc-seeded plasma arc. In the latter case, thermal equilibrium is known to be valid whereas the gas-phase single collisions cannot possibly involve thermal equilibrium. The observed populations from all three processes could not be distinguished from one another. Hence we have direct evidence that non-equilibrium processes can lead to level populations coinciding with those predicted for equilibrium situations.

Experimentally, it is well established that to within a factor of two the secondary ion yields vary exponentially with the ionization potential or electron affinity. The thermodynamic equilibrium models lead naturally to such a dependence but are not unique in this respect. Snowdon et al.[99] have shown theoretically that, to within the experimental error, a Boltzmann-like distribution can be obtained for a variety of non-equilibrium excitation processes, including gas-phase two-body collisions. They concluded that the ability of the LTE model to predict ionization or excitation probability is a necessary but not sufficient condition for the existence of LTE.

4 Sensitivity Factors

Because theory is of negligible help to those interested in quantitative analysis and the equilibrium treatment gives quantitative results of limited reliability, those who are interested in precision have turned to calibrations (sometimes

---

[95] I. S. T. Tsong and N. A. Yusuf, *Nucl. Instrum. Methods*, 1980, **170**, 357.
[96] S. Dzioba, O. Auciello, and R. Kelly, *Radiat. Eff.*, in the press.
[97] W. Heiland, J. Kraus, S. Leung, and N. H. Tolk, *Surf. Sci.*, 1977, **67**, 437.
[98] K. J. Snowdon, G. Carter, D. G. Armour, B. Andresen, and E. Veje, *Surf. Sci.*, 1979, **90**, 429; *Radiat. Eff. Lett.*, 1979, **43**, 201.
[99] K. J. Snowdon, B. Andresen, and E. Veje, *Radiat. Eff. Lett.*, 1979, **43**, 205.

called sensitivity factors) for the ratios of observed ion yields to concentrations. The early work of Beske[100] and Schelten[33] showed that the ion yields of minor constituents of an alloy are proportional to the concentrations. This may hold for all concentrations, but the early work includes examples where concentrations as low as 0.2% caused deviations from linearity, *i.e.* the ratio of the ion yield of the minor constituent to that of the main element ceased to be proportional to the concentration of the minor constituent. (This was true of small concentrations of copper in an aluminium matrix, a system which was later found to be linear for surfaces flooded with caesium.[50]) If one is interested in high precision, linearity cannot be assumed where it has not been demonstrated. In addition, calibrations should only be used for a given instrument and given bombarding conditions. The considerable variation of sensitivity factors for different instruments has been emphasized by Newbury.[101]

Smith and Christie[45] have demonstrated that sensitivity factors can be considerably superior to the equilibrium treatment for purposes of quantitative analysis and Bernheim and Slodzian[63] and Ganjei *et al.*[102] have demonstrated that calibrations can relate ion yields to concentrations with 10—15% precision if bombardment conditions are well controlled. More recently, Leta and Morrison[103] have shown that sensitivity factors for 25 elements in five semiconducting matrices were reliable to a 'conservative' 15%. These results show that the high sensitivity of secondary ion mass spectrometry (SIMS) can be achieved along with high precision if sufficient care is taken.

5 Theory of Secondary Ion Emission

The discussion of the mechanism of secondary ion formation has so far been primarily related to the prediction of secondary ion yields. As we have seen, secondary ion yields are a rather insensitive test of the production process. It has been noted already that the kinetic energy distributions of the secondary ions are incompatible with the LTE model and, in general, it appears that the kinetic energies contain more information than the total secondary ion yields. In this section a variety of models as they relate to energy distributions as well as to yields are considered and in the next section recent kinetic energy measurements are described and consideration is given both to the analytical implications and to their interpretation in terms of models of secondary ion formation.

One school of thought ascribes the ionization mechanism to two-body collisions within the solid, while another relates the ionization process to the interaction between the surface and the sputtered atom. Each point of view has given rise to a number of theoretical treatments, based on different assumptions and analysed within different approximations. The consequent proliferation of predicted secondary ion energy distributions provides the experimenter with a wide range of models to consider.

The energy distribution functions $N^+(E)$, $N^-(E)$, and $N^0(E)$ are related to the ν's (fluxes) discussed earlier by means of equations (4) and (5), and the

---

[100] H. E. Beske, *Z. Naturforsch.*, Teil A, 1967, **22**, 459.
[101] D. E. Newbury, in ref. 50, p. 53.
[102] J. D. Ganjei, D. P. Leta, and G. H. Morrison, *Anal. Chem.*, 1978, **50**, 285.
[103] D. P. Leta and G. H. Morrison, *Anal. Chem.*, 1980, **52**, 514.

ionization probability $R^{\pm}(E)$ is defined by relation (6), where $N(E) = N^{+}(E) + N^{-}(E) + N^{0}(E)$ is termed the total energy distribution. There has been very little experimental or theoretical work on negative secondary ion energy distributions. Consequently, the negative superscript will be dropped in the following discussion.

$$\nu_{\pm} = \int_0^\infty N^{\pm}(E) \, dE \tag{4}$$

$$\nu_0 = \int_0^\infty N^0(E) \, dE \tag{5}$$

$$N^{\pm}(E) = R^{\pm}(E) N(E) \tag{6}$$

For sputtering by random collision cascades,[82] the energy distribution is given by equation (7), where E_b is the surface binding energy, approximately equal to the heat of sublimation, and ϕ_e and ϕ_i are the angles of emission and incidence respectively. This energy spectrum exhibits a peak at $E_b/2$ and has a long high-energy tail which falls as E^{-2}.

$$N(E) \sim \frac{E}{(E + E_b)^3} \frac{\cos \phi_e}{\cos \phi_i} \tag{7}$$

The basic theoretical question can be reduced to a matter of correctly modelling the behaviour of the ionization probability, $R^{+}(E)$. A few years ago, Schroeer[104] summarized the current models, indicating the general form expected for $R^{+}(E)$ on the basis of each model. Since then several new models have appeared. We will consider first the treatments discussed by Schroeer.

The Saha–Langmuir equation, Jurela (after Dobretsov), and Andersen–Hinthorne models predict an ionization probability of the form of expression (8), where $I' = I - \phi, I - \Delta I - \phi$, or $I - \Delta E$ respectively for the three models; ΔI is the image potential at a specified distance from the surface and ΔE is the reduction in the ionization energy due to plasma shielding. This expression [and the specific examples of it, equations (1) and (2)] are independent of the kinetic energy and imply an ion distribution identical to that of the sputtered neutrals. As has already been noted, the neutral and secondary ion energy distributions are not the same experimentally.

$$R^{+}(E) \sim \exp(-I' e/kT) \tag{8}$$

The other models are quantum-mechanical in nature, involving electronic transitions associated with level crossings in two-body collisions or with the surface–sputtered-atom interaction. A number of charge-transfer processes associated with the atom–surface interaction are shown in Figure 6 and have been analysed by Hagstrum.[105] Van der Weg and Bierman[106] assume that the sputtered particle leaves as an ion and is subsequently neutralized by interatomic Auger and resonance processes. They obtain expression (9), where a ($\simeq 2$ Å) is

---

[104] J. M. Schroeer, NBS Spec. Publ. 427, Workshop on Secondary Ion Mass Spectrometry (SIMS) and Ion Microprobe Mass Analysis, Gaithersburg, MD, 1974.
[105] H. Hagstrum, *Phys. Rev.*, 1954, **96**, 336.
[106] W. F. Van der Weg and D. J. Bierman, *Physica*, 1969, **44**, 177 and 206.

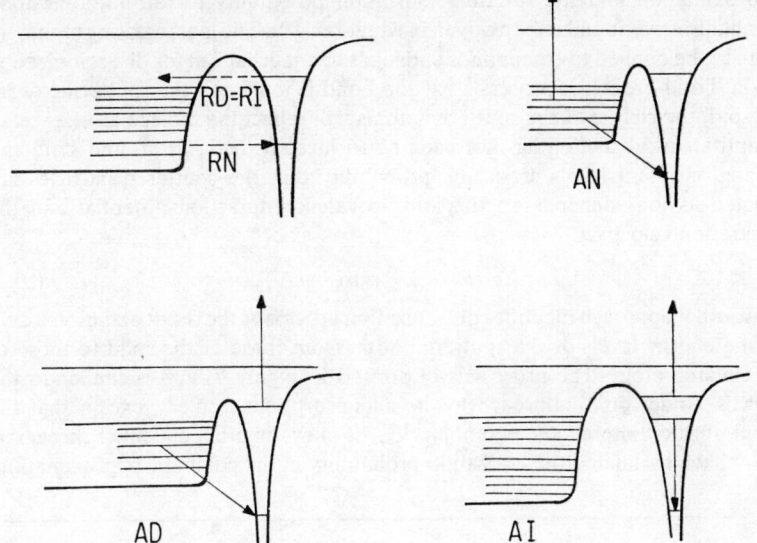

Figure 6 *Charge-exchange and excitation–de-excitation processes associated with the surface–sputtered-atom interactions. A neutral excited atom may de-excite by a resonant charge transfer back to the surface. This process may be thought of as either resonance ionization (RI) or resonance de-excitation (RD). If the sputtered atom was positively ionized the reverse process, resonance neutralization (RN), can occur. A sputtered positive ion may be neutralized by an interatomic Auger transition (AN). Interatomic Auger transitions can also cause de-excitation without changing the charge state (AD). If the sputtered atom escapes to vacuum as an excited neutral, an intra-atomic Auger transition can result in de-excitation by autoionization (AI)*[148]

a parameter representing the atom–surface interaction distance, c is approximately $4 \times 10^{14}\,\text{s}^{-1}$, and v is the velocity.

$$R^+ \sim \exp(-ca/v) \qquad (9)$$

Joyes[107,108] on the other hand assumes that the sputtered atom is neutral with one core electron excited to a valence shell. As long as the atom does not undergo resonance de-excitation while near the surface, it will eventually undergo autoionization by Auger de-excitation. Blaise and Slodzian[109–112] and MacDonald[113] have predicted autoionizing processes for sputtered atoms in which two valence electrons are excited as a result of the energy level rearrangement during surface crossing. Calculations were performed in the one-electron

---

[107] P. Joyes, *J. Phys.*, 1968, **29**, 774; *ibid.*, 1969, **30**, 243, 365.
[108] P. Joyes, *Radiat. Eff.*, 1973, **19**, 235.
[109] G. Blaise, *Radiat. Eff.*, 1973, **18**, 235.
[110] G. Blaise and G. Slodzian, *J. Phys.*, 1974, **35**, 243.
[111] G. Blaise and G. Slodzian, *Rev. Phys. Appl.*, 1973, **8**, 247.
[112] G. Blaise and G. Slodzian, *J. Phys.*, 1970, **31**, 93.
[113] R. J. MacDonald, *Surf. Sci.*, 1974, **43**, 653.

approximation and the resulting ionization probability for an autoionization mechanism was found to be as given in equation (10). The level crossing argument can also be applied to compounds and leads to a mechanism for direct ionization *via* a 'bond-breaking' process.[114] If the bond is ionic and the sputtering event is rapid, the charge of the sputtered atoms will reflect the molecular valence. A quantitative calculation has not been performed. If the ionic ground-state and excited covalent levels cross, the probability that the sputtered particles are ejected as ions depends on the ionic–covalent interaction potential and the separation velocity.

$$R^+(E) \sim 1 - \exp(-c'a/v). \quad (10)$$

Another approach identifies the ionization process as the result of the evolution of the energy levels of the sputtered atom from those of the solid to those of an isolated atom. The process is illustrated in Figure 7, and is similar to the Newns–Anderson[115] approach to the chemisorption problem, except that the atom–surface interaction potential, V_{ak}, is now time-dependent. Schroeer *et al.*[116] have calculated the ionization probability in the adiabatic approximation,

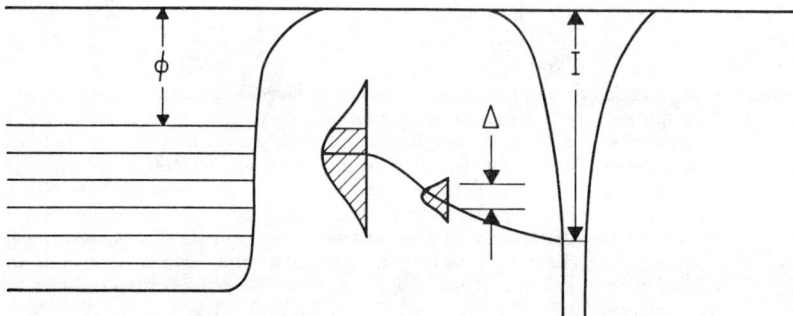

Figure 7 *Evolution of an electronic energy level of a sputtered atom from that of the bulk band structure to that of an isolated atom. The level is described by a resonance half width, Δ, which decreases as the atom leaves the surface*

assuming that the sputtered atom is initially in a neutral, unexcited state. The resulting expression (11) gives an ionization probability in which m (~ 1.25) is

$$R^+(E) = \left[\frac{E_b}{I - \phi}\right]^2 \left[\frac{\hbar v}{a(I - \phi)}\right]^{2m} \quad (11)$$

a fitting parameter, and a is approximately 1.5 Å. Sroubek[117] used the same basic model as Schroeer but employed a molecular orbital approach to evaluate the transition probability. He obtained an expression similar to that of equation (11) but with m varying from $\frac{1}{4}$ to $\frac{3}{2}$ depending on the projected state density in the atomic orbital and the occurrence of level crossings as the atom leaves

---

[114] G. Slodzian, *Surf. Sci.*, 1975, **48**, 161.
[115] D. M. Newns, *Phys. Rev.*, 1969, **178**, 1123.
[116] J. Schroeer, T. Rhodin, and R. Bradley, *Surf. Sci.*, 1973, **34**, 571.
[117] Z. Sroubek, *Surf. Sci.*, 1974, **44**, 47.

the surface. More recently,[118] the calculation has been extended to consider sputtering of a multi-electron atom but was only evaluated in a one-dimensional model. The total ion yield was found to vary as $\exp[-(I - \phi)e/kT]$, in accord with experiment and the predictions of the thermodynamic models. The ionization probability was calculated to vary as given in equation (12).

$$R^+(E) \sim e^{CE}. \tag{12}$$

Further refinements to this model have been made by the French group,[22,119–121] utilizing a many-body theory of time-dependent perturbations to describe ionization produced by the surface–sputtered-atom interaction. This model provides a self-consistent treatment of the fact that the surface interaction produces both excitation and de-excitation. In the limit that the metal states are very broad and have constant density, the problem can be solved, yielding an occupation number of the localized state in terms of the resonance width $\Delta(t)$ of the localized level (Figure 7) obtained from expression (13), where ρ is the metal state density, and $V_{ak}(t)$ is the time-dependent atom-surface interaction potential. The approximation of a broad and constant density of metal states limits the results of the calculation to s-band metals, although the model is in principle applicable to transition metals as well. Depending on the assumptions made for the functional form of $\Delta(t)$, a variety of forms can be found for the energy dependence of the ionization probability. If $\Delta(t)$ is assumed to vary as $\Delta_0 \exp(-qvt)$, then $R^+(E) \simeq \exp(-c'/\hbar qv)$, where c' depends on the value of Δ_0. If, on the other hand, it is assumed that $\Delta(t)$ has values within the intervals shown in expression (14) where $t_c = 1/qv$, then equation (15) follows. This latter expression for $\Delta(t)$ is perhaps not as physically reasonable as the exponential form, but it gives a power law for $R^+(E)$ and, as shown in the next section, this is in better accord with experiment than the expressions for $R^+(E)$ which contain exponentials.

$$\Delta(t) = \pi \rho V_{ak}^2(t) \tag{13}$$

$$\Delta(t) = \begin{cases} \Delta_0 & t < 0 \\ \Delta_0(1 - t/t_c) & 0 < t < t_c \\ 0 & t > t_c \end{cases} \tag{14}$$

$$R^+(E) \simeq \Delta_0 \hbar q E^{1/2} \tag{15}$$

A somewhat unorthodox point of view can be found in the surface excitation model proposed by Williams.[23] The essential feature of this model is that one assumes ionization can be described by an expression having the form of equation (8) with a time-dependent temperature which falls as the collision cascade develops. During the early stages of the cascade when the temperature is high, relatively energetic atoms will be sputtered. These will have a relatively high probability of ionization, R^+, if I' is positive and a relatively low value of R^+ if I' is negative ($I' = I - \phi$ for the Saha–Langmuir equation which is the case considered by Williams[23]). Thus, the model predicts that $R^+(E)$ will increase

[118] Z. Sroubek, J. Zavadil, F. Kubec, and K. Zdansky, *Surf. Sci.*, 1978, **77**, 603.
[119] A. Blandin, D. Hone, and A. Nourtier, *J. Phys. Lett.*, 1975, **36**, L-109.
[120] A. Blandin, A. Nourtier, and D. W. Hone, *J. Phys.*, 1976, **37**, 369.
[121] A. Nourtier, *J. Phys.*, 1977, **38**, 479.

with increasing values of E for atoms of high ionization potential and decrease with increasing E for atoms having a low ionization potential ($<\phi$ if one uses the Saha–Langmuir formulation). This result does not depend on any detailed description of the charge state of an atom when it is first emitted or of the subsequent ionization or neutralization processes.

In general, the surface ionization model predicts that the ionization probability of an atom of low ionization potential sputtered from a surface having a high work function ($\phi > I$) will decrease with increasing kinetic energy and that, if the random collision cascade description [equation (7)] of sputtering is correct, $N^+(E)$ for these cases will fall more rapidly than E^{-2}.

6 Secondary Ion Energies

It seems clear that reliable experimental data on secondary ion energies are necessary as a guide to both the theoretician and the analyst who needs quantitative secondary ion yield data. However, accurate secondary ion energy measurements are notoriously elusive. Results obtained for a given material by workers at different laboratories are often seriously in conflict. This is perhaps to be expected since characterization of the sputtered surface has been usually minimal or non-existent, and much of the data in the literature were obtained under poor vacuum conditions. The transmission vs. energy characteristics of most of the devices used are extremely difficult to specify, particularly if a quadrupole is used in the mass-analyser section of the instrument. The common use of extraction potentials to increase the sensitivity complicates the problem further.[114] In addition, the efficiency of electron multiplier detectors varies with both the mass and kinetic energy of the detected ion.[122]

Both the secondary ion yield[123] and the kinetic energy distribution[124,125] are quite sensitive to the surface topography,[126] which changes quickly under high current density bombardment, and to the degree of surface oxidation. Because of the latter, the dominant secondary ion production may occur at the edge of the sputtered crater where the lower primary current density has allowed the surface to become oxygen contaminated. Ultra-high vacuum (UHV) conditions and high current density are therefore desirable. UHV design is difficult to achieve in an ion microprobe and has usually been obtained only with quadrupole-based instruments. Some instruments have countered this problem either by using collection optics which detect only the central part of the bombarded area[127] or by rastering the beam and gating the counter on only when the beam is near the centre of the rastered area.[128] With these devices, very few data have been obtained from surfaces in which the ion-emitting area has been independently characterized in situ by means of a surface analytical technique such as Auger or X-ray photoelectron spectroscopy.

[122] C. Lao, R. Sander, and R. F. Pottie, *Int. J. Mass Spectrom. Ion Phys.*, 1972/73, **10**, 309.
[123] G. Blaise and M. Bernheim, *Surf. Sci.*, 1975, **47**, 324.
[124] A. R. Krauss and D. M. Gruen, *Appl. Phys.*, 1977, **14**, 89.
[125] A. R. Bayly, P. J. Martin, and R. J. MacDonald, *Nucl. Instrum. Methods*, 1976, **132**, 459.
[126] I. Reid, B. W. Farmery, and M. W. Thompson, *Nucl. Instrum. Methods*, 1976, **132**, 317.
[127] R. Castaing and G. Slodzian, *J. Microsc.*, 1962, **1**, 395.
[128] W. O. Hofer, H. Liebl, G. Roos, and G. Staudenmaier, *Int. J. Mass. Spectrom. Ion Phys.*, 1976, **19**, 327.

Kinetic energies of sputtered neutrals have been measured by a variety of techniques during the past twenty years and it is well established that the average kinetic energy is 10—20 eV. The secondary ions, on the other hand, have average kinetic energies that are much higher. The relevant experiments have been tabulated recently in a review article by Krauss and Wright[25] and will not be listed in detail here. The earliest secondary ion measurements were performed by the French group,[129—132] by Benninghoven,[133] and by Jurela and Perovic.[134] These authors observed a peak at somewhat higher energy than expected for the neutrals and a high-energy tail which fell off approximately as a power law, but more slowly than E^{-2}, typically[20] as $E^{-1.3}$. Blaise and Slodzian[109,135] reported that, although the peak position and relative peak height varied from element to element, the high-energy tails were all rather similar. Jurela and Perovic,[134,136] on the other hand, reported the existence of distinct structure[134] and multiple maxima[136] in the high-energy region which varied from element to element. It should be noted that Slodzian's published data were the result of a deconvolution technique,[114] which corrected for the high extraction field used to collect the secondary ions. The result of this deconvolution is presumably a correct energy distribution, but with much reduced sensitivity at high energies.

In general, the published data[114,124,132,134,135,137—149] report $N^+(E) \simeq E^{-m}$ for $E \gg E_b$, where $m \le 2$, although some of the same authors[134,141,146,150,151] have also observed $N^+(E)$ varying exponentially with energy or velocity for some materials. There have also been several observations of secondary ion energy distributions which exhibit multiple peaks,[125,134,147,152—154] shifting peaks,[152,154—156] or very high-energy peaks.[125,147,154—156] It is perhaps of interest that these reports originated in only three laboratories and that, aside from the

[129] G. Slodzian, *Ann. Phys., Paris*, 1964, **9**, 951.
[130] R. Castaing and J. F. Hennequin, *C. R. Hebd. Seances Acad. Sci.*, 1966, **262**, 1008.
[131] J. F. Hennequin, *J. Phys.*, 1968, **29**, 655.
[132] R. Castaing and J. F. Hennequin, *Adv. Mass Spectrom.*, 1971, **5**, 419.
[133] A. Benninghoven, *Ann. Phys.*, 1965, **15**, 113.
[134] Z. Jurela and B. Perovic, *Can. J. Phys.*, 1968, **46**, 773.
[135] G. Blaise and G. Slodzian, *Rev. Phys. Appl.*, 1973, **8**, 105.
[136] Z. Jurela, *Radiat. Eff.*, 1973, **19**, 175.
[137] E. Dennis and R. J. MacDonald, *Radiat. Eff.*, 1972, **13**, 243.
[138] K. Wittmaack, *Nucl. Instrum. Methods*, 1976, **132**, 381.
[139] S. Miyagawa, *J. Appl. Phys.*, 1973, **44**, 5617.
[140] T. Arikawa, K. Narushima, and M. Inoue, *Jpn. J. Appl. Phys.*, 1976, **15**, 1565.
[141] A. R. Bayly and R. J. MacDonald, *Radiat. Eff.*, 1977, **34**, 169.
[142] K. Komori and J. Okano, *Int. J. Mass Spectrom. Ion Phys.*, 1978, **27**, 379.
[143] T. R. Lundquist, *J. Vac. Sci. Technol.*, 1978, **15**, 684.
[144] T. R. Lundquist, *Appl. Phys.*, 1979, **18**, 221.
[145] T. R. Lundquist, *Surf. Sci.*, 1979, **90**, 548.
[146] M. Rudat and G. H. Morrison, *Surf. Sci.*, 1979, **82**, 549.
[147] A. R. Krauss and D. M. Gruen, *Nucl. Instrum. Methods*, 1978, **149**, 547.
[148] A. R. Krauss and D. M. Gruen, *Surf. Sci.*, 1980, **92**, 14.
[149] R. G. Hart and C. B. Cooper, *Surf. Sci.*, 1980, **94**, 105.
[150] R. J. MacDonald, *Adv. Phys.*, 1970, **19**, 457.
[151] G. Staudenmaier, *Radiat. Eff.*, 1972, **13**, 87.
[152] A. R. Krauss and D. M. Gruen, to be published.
[153] K. J. Snowdon and R. J. MacDonald, *Int. J. Mass Spectrom. Ion Phys.*, 1979, **29**, 101.
[154] K. J. Snowdon, in 'Proceedings of the 7th International Vacuum Congress and 3rd International Conference on Solid Surfaces', ed. R. Dobrozemsky, F. Rüdenauer, F. P. Viehböck, and A. Breth, F. Berger und Söhne, Vienna, 1977, Vol. 3, p. 2557.
[155] K. J. Snowdon, *Radiat. Eff.*, 1978, **38**, 141.
[156] K. J. Snowdon and R. J. MacDonald, *Int. J. Mass Spectrom. Ion Phys.*, 1978, **28**, 233.

Jurela and Perovic[134,136] data, were obtained with probably the only devices specifically designed to counter the experimental difficulties discussed earlier.[124,156,157] One device was a time-of-flight (TOF) energy analyser,[156] while the other two were quadrupole-based mass and energy analysers in which the kinetic energy through the quadrupole stage is fixed independently of the initial kinetic energy by means of a retarding stage in the energy analyser.[124,157] Perhaps the major unresolved difficulty with these two instruments is the velocity dependence of the electron multiplier counting efficiency, leading to a decreased sensitivity for high-energy, high-mass secondary ions. These devices are shown in Figure 8.

Kinetic energy distributions of Cu^+ sputtered by an argon ion beam and measured by means of an electrostatic analyser,[141] a conventional ion microprobe,[135] and the double-sector instrument of Jurela and Perovic[134] are shown in Figure 9. The data were not taken under the same conditions, but the differences in beam-sample-detector geometry, vacuum conditions, current density, and primary beam energy could not account for the variation of the results. Bayly and MacDonald[141] suggested that their relatively high secondary ion energies could be the result of their surface being cleaner (than the others) since it is known that oxygen contamination causes enhanced ion yields at relatively low energies. However, the reported current densities and vacuum conditions of Bayly and MacDonald were not better than those of Blaise and Slodzian[135] and hence it is difficult to understand how their surfaces could have been cleaner.

Snowdon has measured the run-to-run reproducibility[154] of the Ti^+ energy distribution and found that successive scans yielded progressively higher ion energies, as shown in Figure 10. This result supports the correlation of low-energy spectra and contaminated surfaces. Unfortunately, none of the experiments represented in Figures 9 or 10 incorporated any independent means of monitoring the surface oxygen concentration or metal oxidation state. This is especially critical in view of Wittmaack's finding[158] that high-current-density sputtering in poor vacuum can produce bombardment-enhanced oxidation and is not equivalent to a lower current density under correspondingly better vacuum conditions.

Secondary ion energy distributions of Ti^+ and TiO^+ have been measured under UHV conditions in combination with *in situ* X-ray photoelectron spectroscopy (XPS) analysis by Krauss and Gruen.[152] The data (shown in Figure 11) consist of the secondary ion energy spectra and low- and high-resolution XPS data. The Ti^+ energy distributions correspond to the curves labelled as mass 48, while mass 64 corresponds to TiO^+. The low-resolution XPS scan includes both oxygen and titanium peaks; the oxygen concentration at the surface is qualitatively measured by the O:Ti peak ratio. The oxidation state of the titanium may be deduced from the line shape in the high-resolution scan. The data shown in Figure 11a correspond to a moderately oxygen-free surface. After exposure to oxygen at $\sim 10^{-6}$ torr, the data shown in Figure 11b were obtained. The initial kinetic energy distribution (upper trace) shows a low-energy peak plus a

[157] A. R. Bayly and R. J. MacDonald, *J. Phys. E.*, 1977, **10**, 79.
[158] K. Wittmaack, *Surf. Sci.*, 1975, **47**, 358.

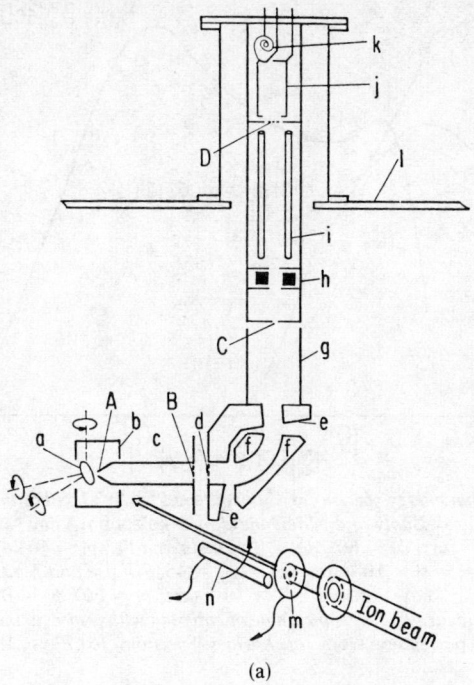

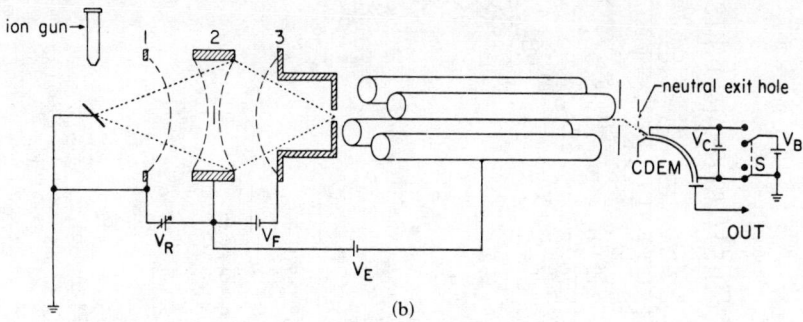

Figure 8 (a) *Electrostatic secondary ion energy analyser of Bayly and MacDonald. The sample is floated and electrodes B–d form a retarding stage followed by a spherical energy analyser*
(Reproduced by permission from *Radiat. Eff.*, 1977, **34**, 169)
(b) *Secondary ion energy analyser of Krauss and Gruen. The sample is grounded and the first pair of grids form a retarding stage while the second pair provides a high-energy cut-off. All potentials including the quadrupole common are referenced to the retardation supply*[124]

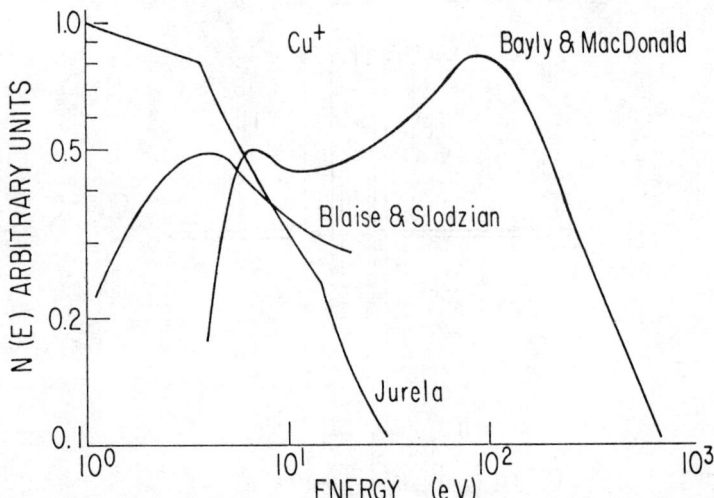

Figure 9 *Secondary ion energy spectra for Cu^+ sputtered by Ar^+ bombardment obtained by several authors: Bayly and MacDonald – 43 keV, 60 $\mu A\ cm^{-2}$, base pressure = 1—4 × 10^{-7} torr, ϕ_i = 60°, ϕ_e = 0°; Jurela and Perovic – 40 keV, 60 $\mu A\ cm^{-2}$, base pressure = 2 × 10^{-6} torr, ϕ_i = ϕ_e = 45°; Blaise and Slodzian – 6.2 keV, 1000 $\mu A\ cm^{-2}$, base pressure = 1 × 10^{-7} torr, ϕ_i = 60°, ϕ_e = 0°, where ϕ_i is the angle of incidence and ϕ_e is the angle of emission with respect to the surface normal*
(Data reproduced by permission from *Int. J. Mass. Spectrom. Ion Phys.*, 1978, **28**, 233)

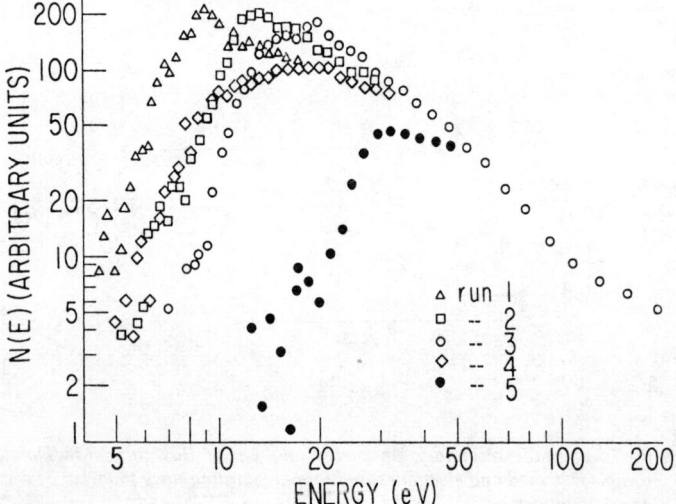

Fig. 10 *A selection of the energy spectra obtained from polycrystalline Ti. Runs 1—4 represent successive scans of Ar^+ bombardment, run 5 Kr^+ bombardment. Analogous behaviour was obtained for both primary ion species. Above 60 eV all spectra were of the same shape*
(Reproduced by permission from 'Proceedings of the 7th International Vacuum Congress and 3rd International Conference on Solid Surfaces', ed. R. Dobrozemsky *et al.*, Vienna, 1977, Vol. 3, p. 2557)

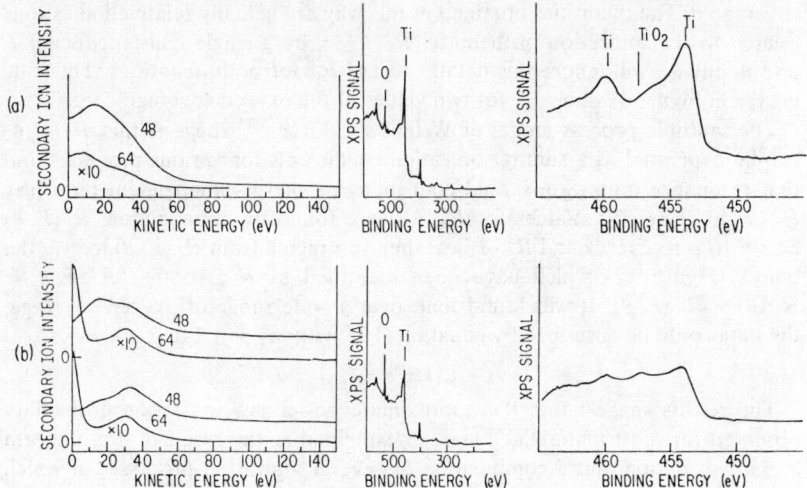

Figure 11 (a) *Secondary ion energy distributions and XPS data for* Ti^+ *and* TiO^+ *sputtered by 3 keV* Ar^+ *from a partially oxidized polycrystalline Ti foil.* (b) *The same data for a more heavily oxidized Ti foil. The upper trace is the first one taken after the ion beam was turned on and the lower trace is the following scan. All subsequent scans were stable. If the ion gun were turned off for a few minutes and then turned back on, the first energy distribution recorded after turn-on matched the upper trace and the next scan again matched the lower trace. The shift in the energy distribution was reproducibly observed for many repetitions. The Auger and XPS lines did not shift, indicating negligible ion-beam-induced surface charging*[152]

higher-energy structure. A subsequent scan (lower trace) shows that the higher energy feature (33 eV) has become the main peak. By varying the primary current the relative intensity of the two peaks could be changed at will. With increased oxygen exposure[124] (not shown) such that the high-resolution XPS spectrum corresponded to that of TiO_2 the ion energy distribution exhibited only a single low-energy peak (~5 eV), in accord with the usual trend to lower secondary ion energies for oxidized surfaces. The behaviour illustrated in Figure 11 has been ascribed to the relative ease of bombardment-induced reduction of the surface oxide phases of titanium.[159] The existence of such complicated behaviour illustrates the need for *in situ* surface analysis in secondary ion experiments.

Additional results have been obtained from the analysis of the Be^+ secondary ion energy spectrum as a function of surface oxidation.[148,160] Beryllium forms a single bulk oxide phase and also appears to have a single surface oxide structure, with a constant sticking coefficient for oxygen. The secondary ion yield for sub-monolayer coverage is linear in oxygen concentration and the kinetic energy distribution undergoes relatively little oxygen-induced peak shift.

In order to evaluate $R^+(E, \theta)$, an E^{-2} dependence was unfolded from the high-energy tail of the energy distribution, $N^+(E, \theta)$, at each value of the oxygen

[159] S. Thomas, *Surf. Sci.*, 1976, **55**, 754.
[160] A. R. Krauss and D. M. Gruen, *Surf. Sci.*, 1979, **90**, 564.

coverage θ. The quantities obtained in this way are actually relative ionizations related to the ionization probability, $R^+(E, \theta)$, by a single constant for all E and θ, but we will ignore this detail and use R^+ for both quantities. The solid curves in Figure 12 show R^+ for two values of the oxygen coverage.

The multiple process model of Wright and Gruen[161] suggests that $R^+(E, \theta)$ can be expressed as a sum of ionization coefficients for various processes and that resonance ionization should contribute an energy-independent term [say $R_s^+(\theta)$ to $R^+(E, \theta)$]. Values for $R_s^+(\theta)$ were found by extrapolating $R^+(E, \theta)$ curves to zero energy and $R_s^+(\theta)$ was then subtracted from $R^+(E, \theta)$ to give the points (Figure 12) which have been identified as R_v^+, so that $R^+(E, \theta) = R_s^+(\theta) + R_v^+(E, \theta)$. It was found that, over a wide range of oxygen coverage, the data could be described by equation (16), with $m_1 = 1.1$ and $m_2 = 1.53$.

$$R_v^+(E, \theta) = C_1 \theta E^{m_1} + C_2(1 - \theta)E^{m_2} \qquad (16)$$

The results suggest that the approximate power law ionization probability observed for most materials[146] may be explained as the result of non-uniform surface oxidation and a combination of several ionization processes of which the dominant one is described by $R^+(E) \sim E^m$, where the value of m depends on whether the surface is locally clean or oxidized. This result is consistent with the surface dipole model of Williams and Evans[56] in which the 'oxidized' region lies within a few atomic radii of the oxygen atom.

Further tests of the effect of the chemical environment on the ionization probability can be made by comparing the secondary ion yields and kinetic energies for a given atomic species sputtered from the bulk elemental solid, from a thin film adsorbed on another element, and from an ionic compound. Reported secondary ion fractions for the alkali halides range up to 89% for Na^+ sputtered from $NaCl$[162] and 93% for K^+ from KBr.[163] Measurements of the alkali secondary ion energy distributions have been made for sodium[139,141] and potassium[140,164] halides. With one exception,[140] these experiments resulted in distributions with high-energy tails following an $E^{-2.0}$ dependence. Arikawa, Narushima, and Inoue observed an $E^{-1.5}$ dependence, although it should be noted that the slower fall-off might have resulted from the high-energy (93 keV) primary beam they used.

Until recently, there have not been any successful energy-distribution experiments on elemental alkali metals. However, Krauss and Gruen[164-166] have measured secondary ion yields and kinetic energy distributions of atoms sputtered from potassium multi-layer and mono-layer films deposited on a transition-metal substrate. Potassium atoms which were in direct contact with the substrate were sputtered with an ion fraction exceeding 60% while atoms sputtered from thicker films were ejected almost entirely (>95%) as neutrals. The K^+ secondary ion energy distribution peaked at ~ 2 eV and the high-energy tail fitted an $E^{-2.0}$ behaviour, as predicted by the random collision cascade model[82] for the total emission.

[161] R. B. Wright and D. M. Gruen, *J. Chem. Phys.*, 1980, **72**, 147.
[162] B. Navinsek, *J. Appl. Phys.*, 1965, **36**, 1678.
[163] I. A. Abroyan, V. P. Laurov, and I. G. Fedorova, *Sov. Phys. – Sol. State*, 1966, **7**, 2954.
[164] A. Krauss and D. Gruen, in preparation.
[165] A. Krauss and D. Gruen, *J. Nucl. Mater.*, 1980, **93/94**, 686.
[166] A. Krauss and D. Gruen, in ref. 72, p. 484.

(a)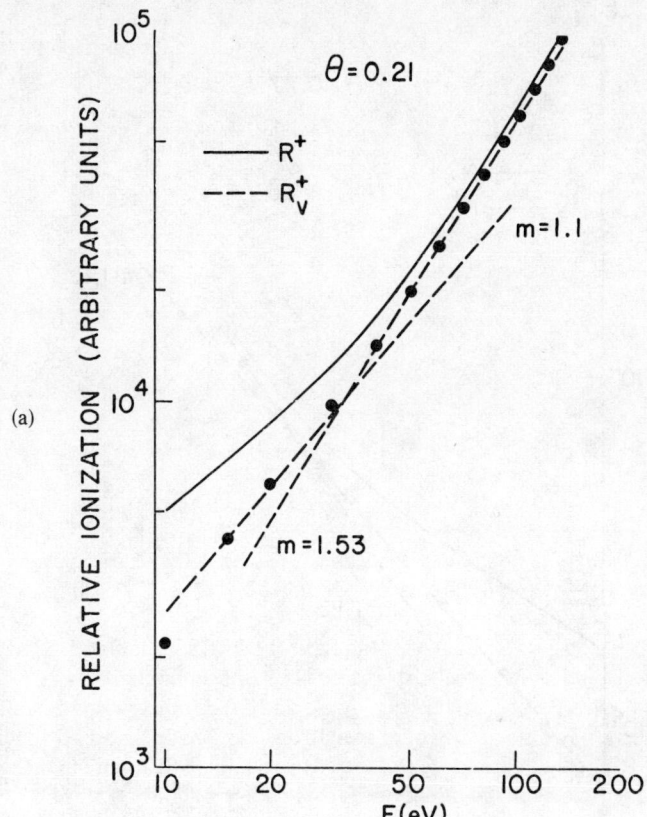

Figure 12 (a) *The relative ionization, $R^+(E)$, for a moderately heavily ($\theta = 0.21$) oxygen-covered Be surface.* (b) *The relative ionization for a more heavily oxygen-covered ($\theta = 0.77$) Be surface. For $\theta = 0.03$ (not shown) a single slope results, characterized by $m = 1.53$. For high oxygen coverage (not shown) a single slope with $m = 1.1$ results*[148]

It will be recalled that the surface excitation model[23] predicted that, if $I < \phi$, $R^+(E)$ should decrease with increasing energy and that the secondary ion energy distribution for the ions would therefore fall more rapidly than E^{-2}. Experimentally, one would like to consider a case like caesium sputtered from a tungsten substrate where $I < \phi$ by a substantial margin, but at the present time the best test of this sort involves potassium sputtered from a molybdenum substrate where $I \sim \phi$. The fact that in this instance[164] the secondary ion energy distribution fell approximately as E^{-2} tends to support the surface excitation model.

In addition to chemical effects, the observed secondary ion energies are also sensitive to the collection angle, primary energy, and angle of incidence. Angular effects have already been noted at high emission energies in the neutral sputtering of polycrystalline gold as the result of surface recoils,[126] and these effects are

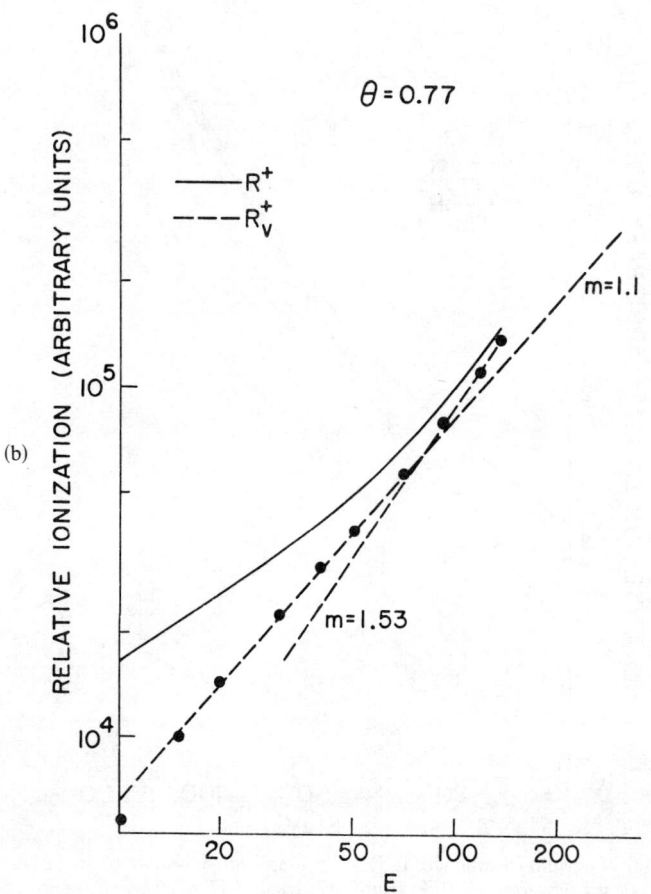

reflected in the secondary ion energy distribution.[147,167] However, shifts of the low-energy peak as a function of angle of emission[81,132,142,147,168] have also been observed. Typical results for aluminium are shown in Figure 13. The position of the peak of the energy distribution shifts about 10 eV as the angle of emission goes from 55° to 82.5°. This shift has been interpreted[81] as the result of the fact that the amount of time the sputtered ion spends in the near-surface region where charge-transfer processes (Figure 6) occur depends on both the velocity and the angle of emission. The experimental energy distribution fits expression (18), where $m = 0.6$ was obtained from experiment,[147] and $v_\perp$ is the secondary ion velocity component normal to the surface. The arrows in Figure 13 indicate the calculated peak positions, assuming $A = 2.5 \times 10^3$ m s^{-1}, the value obtained

---

[167] G. A. V. D. Schootbrugge, A. G. J. DeWit, and J. M. Fluit, *Nucl. Instrum. Methods*, 1976, **132**, 321.
[168] J. N. Smith, *IEEE Trans. Nucl. Sci.*, 1979, **26**, 1292.

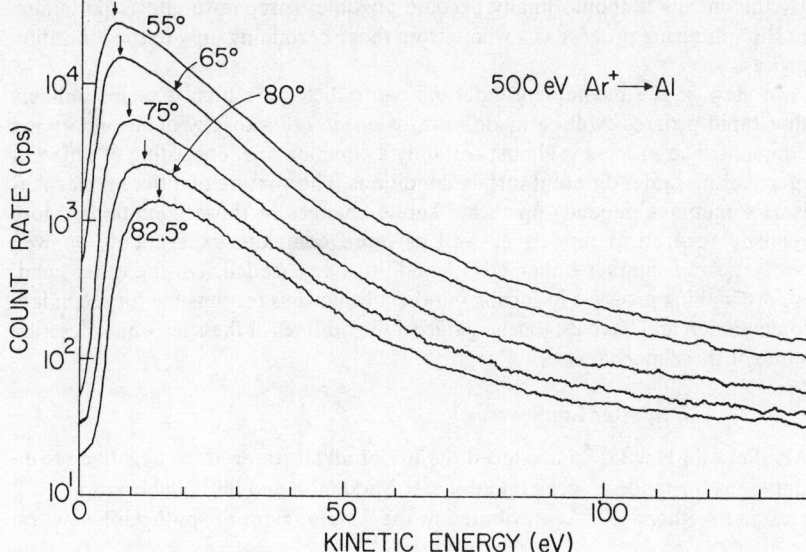

Figure 13 *Secondary ion energy distributions for Al^+ sputtered by a 500 eV Ar^+ beam from a polycrystalline foil at various angles of emission*[81]

from ion-induced photon emission data.[169]

$$R^+(E) \sim E^m \exp(-A/v_\perp) \qquad (18)$$

Several authors[147,149,170] have investigated the effect of the primary energy on the secondary ion energy distribution and concluded that the energies of the sputtered atoms are not independent of the primary ion mass or energy as predicted by the random collision cascade model, but decrease with decreasing primary mass and energy. Presumably, the secondary ions are merely reflecting changes in the total energy. However, the shift of the peak energy observed by Sroubek[170] and by Krauss and Gruen[147] for Ni^+ and Al^+, as well as the variation of the Cu^+ high-energy tail observed by Hart and Cooper,[149] greatly exceed changes in the Cu neutral energy spectra observed by Stuart and Wehner.[171,172]

It should be noted that the conditions of these experiments are not comparable, that the secondary ion data are rather incomplete, and that the neutral data, although more complete, were taken under less than desirable conditions and are not to be considered authoritative. Currently, the technique of laser fluorescence spectroscopy is being developed into a powerful tool for the velocity measurement of neutral and (in principle) ionized sputtered particles.[91,173,174]

[169] R. Hippler, W. Kruger, A. Scharmann, and K. H. Scharfner, *Nucl. Instrum. Methods*, 1976, **132**, 439.
[170] Z. Sroubek, *Surf. Sci.*, 1974, **44**, 47.
[171] R. V. Stuart and G. K. Wehner, *J. Appl. Phys.*, 1964, **35**, 1819.
[172] R. V. Stuart, G. K. Wehner, and G. S. Anderson, *J. Appl. Phys.*, 1969, **40**, 803.
[173] W. Husinsky, R. Bruckmuller, P. Blum, F. Viehbock, D. Hammer, and E. Benes, *J. Appl. Phys.*, 1977, **48**, 4734.
[174] W. Husinsky and R. Bruckmuller, *Surf. Sci.*, 1979, **80**, 637.

By this means it should finally become possible to separate effects pertaining to the sputtering process as a whole from those pertaining only to the ionization process.

In view of the incomplete data and contradictory results, it seems unlikely that rapid progress will be made towards an *ab initio* theory of secondary ion emission. The process is almost certainly a complex one, consisting of different mechanisms under different surface conditions. The mixture of different mechanisms sometimes depends on rather subtle changes in those conditions. More realistic theoretical models as well as more complete experiments on well characterized samples under UHV conditions are needed. On the other hand, we are making progress in sorting out the mechanisms responsible for secondary ion emission and have established at least the outlines of theories which describe some of these mechanisms.

7 Sputter Ion Sources

Mueller and Hortig[175] introduced the use of alkalis to produce negative secondary ions for tandem accelerators. Later Middleton and his collaborators[176–180] and many others[181–205] contributed to the development of sputter ion sources.

[175] M. Mueller and G. Hortig, *IEEE Trans. Nucl. Sci.*, 1969, **NS-16**, 38.
[176] R. Middleton and C. T. Adams, *Nucl. Instrum. Methods*, 1974, **118**, 329.
[177] R. Middleton, *Nucl. Instrum. Methods*, 1974, **122**, 35.
[178] R. Middleton, *IEEE Trans. Nucl. Sci.*, 1976, **NS-23**, 1098.
[179] R. Middleton, *Nucl. Instrum. Methods*, 1977, **144**, 373.
[180] R. Middleton, in 'Proceedings of the Symposium of Northeastern Accelerator Personnel (SNEAP 78)', compiled by J. K. Bair and C. M. Jones, ed. G. D. Alton, J. K. Bair, J. A. Benjamin, C. M. Jones, R. C. Juras, J. E. Mann, W. T. Milner, C. D. Moak, R. O. Sayer, and G. F. Wells, Oak Ridge National Laboratory, CONF-781051, June, 1979, p. 114.
[181] H. H. Andersen and P. Tykesson, *IEEE Trans. Nucl. Sci.*, 1975, **NS-22**, 1632.
[182] P. Tykesson, H. H. Andersen, and J. Heinemeier, *IEEE Trans. Nucl. Sci.*, 1976, **NS-23**, 1104.
[183] H. V. Smith, jun. and H. T. Richards, *Nucl. Instrum. Methods*, 1975, **125**, 497.
[184] H. V. Smith, jun., *IEEE Trans. Nucl. Sci.*, **NS-23**, 1118.
[185] H. V. Smith, jun., *Nucl. Instrum. Methods*, 1978, **153**, 605.
[186] H. V. Smith, jun., *Nucl. Instrum. Methods*, 1979, **164**, 1.
[187] G. T. Casky, R. A. Douglas, H. T. Richards, and H. V. Smith, jun., *Nucl. Instrum. Methods*, 1978, **157**, 1.
[188] H. V. Smith, jun., *Nucl. Instrum. Methods*, **153**, 605.
[189] G. D. Alton, *IEEE Trans. Nucl. Sci.*, 1976, **NS-23**, 1113.
[190] G. D. Alton, *IEEE Trans. Nucl. Sci.*, 1979, **NS-26**, 1542.
[191] G. D. Alton, *IEEE Trans. Nucl. Sci.*, 1979, **NS-26**, 3708.
[192] G. Doucas and H. R. McK. Hyder, *Nucl. Instrum. Methods*, 1974, **119**, 413.
[193] G. Doucas, H. R. McK. Hyder, and A. B. Knox, *Nucl. Instrum. Methods*, 1975, **124**, 11.
[194] G. Doucas, T. J. L. Greenway, H. R. McK. Hyder, and A. B. Knox, *IEEE Trans. Nucl. Sci.*, 1976, **NS-23**, 1155.
[195] G. Doucas, *Rev. Phys. Appl.*, 1977, **12**, 1465.
[196] K. R. Chapman, *IEEE Trans. Nucl. Sci.*, 1976, **NS-23**, 1109.
[197] G. Braun-Elwert, J. Huber, G. Korschinek, W. Kutschera, W. Goldstein, and R. L. Herschberger, *Nucl. Instrum. Methods*, 1977, **146**, 121.
[198] G. Korschinek and W. Kutschera, *Nucl. Instrum. Methods*, 1977, **144**, 343.
[199] G. Korschinek and W. Kutschera, *Rev. Phys. Appl.*, 1977, **12**, 1459.
[200] R. Maier, G. Korschinek, P. Spolaore, W. Kutschera, H. J. Maier, and W. Goldstein, *Nucl. Instrum. Methods*, 1978, **155**, 55.
[201] K. Brand, *Nucl. Instrum. Methods*, 1977, **141**, 519.
[202] K. Brand, *Nucl. Instrum. Methods*, 1978, **154**, 595.
[203] M. Dumail, *Nucl. Instrum. Methods*, 1979, **163**, 61.
[204] R. I. Cutler, K. W. Kemper, and K. R. Chapman, *Nucl. Instrum. Methods*, 1979, **164**, 605.
[205] G. D. Alton and G. C. Blazey, *Nucl. Instrum. Methods*, 1979, **166**, 105.

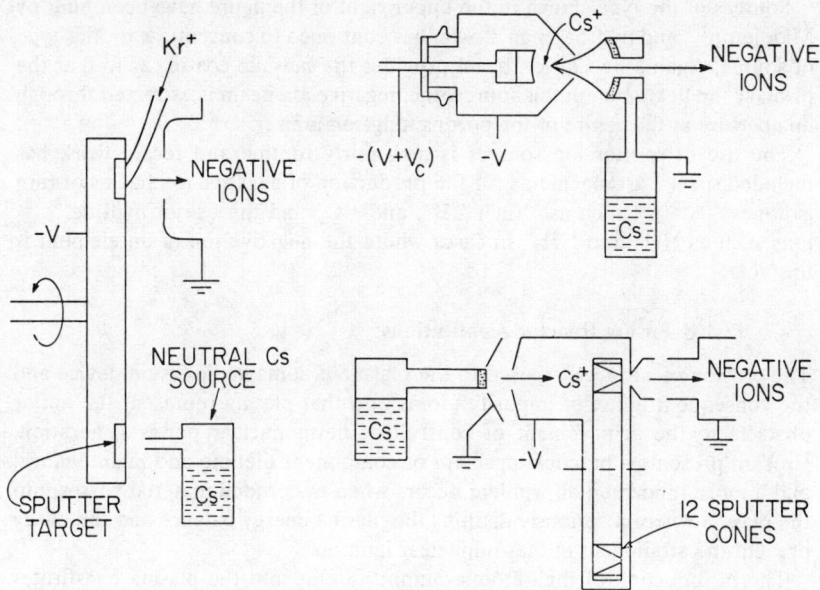

Figure 14 *Schematic drawings of the sputter ion sources of Mueller and Hortig[175] (left), Middleton and Adams[176] (lower right), and Chapman[196] and Middleton[178] (upper right). The schematic drawings are from Middleton[180] with a few modifications*
(Reproduced with the permission of Roy Middleton and the Oak Ridge National Laboratory operated by the Union Carbide Corporation for the United States Department of Energy)

The large number of publications cited in this section are indicative of the importance of efforts to make secondary ions available as a source of negative ions for accelerators.

Figure 14 shows schematic drawings of three sputter ion sources. These were selected, in part, because the negative ions from them are probably almost 100% secondary ions whereas the negative ions from some of the other sources include substantial numbers of atoms sputtered as neutrals and subsequently ionized in caesium vapour or plasma.

On the left of the figure is the rotating-wheel source of Mueller and Hortig.[175] Here the caesium coating needed for enhanced negative ions yields is applied from a caesium boiler and the secondary ions are produced by bombardment by a primary ion beam of Kr^+ at a different location.

The source at the lower right of the figure is that of Middleton and Adams.[176] In this source the Cs^+ beam from a heated porous tungsten ionizer provides a caesium coating on the inside of a cone of the target material and also is the primary ion beam which causes the emission of secondary ions. The latter are extracted through a hole at the apex of the cone. The target material can be changed by rotating twelve different cones into the primary ion beam.

Sources of the type shown in the upper right of the figure have been built by Middleton[178] and by Chapman[196] who has continued to concentrate on this type of source. Again, the Cs^+ ion beam provides the caesium coating as well as the primary ion beam but, in this source, the negative ion beam is extracted through an aperture at the centre of the porous tungsten ionizer.

The use of sputter ion sources is now fairly routine and recent work has included special arrangements for the production of negative ion beams of rare isotopes[198—200,204] such as $^{48}Ca^-$, $^{10}B^-$, and $^{14}C^-$ and the use of hydride[180,197] ions such as NH^- and NH_2^- in cases where the negative ion of an element is unstable.

8 Fusion Reactor Applications

The erosion of surfaces exposed to the plasma of a magnetic fusion device and the consequent influx of impurity atoms into that plasma represent the major obstacle to the achievement of controlled thermonuclear power generation. Erosion presents a practical problem of component lifetime and maintenance, and a more fundamental problem occurs when the eroded material passes into the plasma where it seriously disrupts the plasma energy balance and can easily prevent the attainment of thermonuclear ignition.

The introduction of high-atomic-number atoms into the plasma constitutes an especially serious problem since plasma energy losses due to atomic line radiation increase approximately as Z^3. However, durability requirements have traditionally dictated the use of high-Z refractory materials such as tungsten for limiters and neutral-beam dump areas. Consequently, all countries engaged in magnetic-confinement fusion research have placed very high priority on understanding the interactions between the plasma and first wall.

Since a major portion of the erosion is caused by sputtering, it seems appropriate here to address the effect of secondary ion emission on the operation of these devices. All atoms sputtered from the device components are of course ionized by the time they enter the plasma. Because of the intense magnetic and electric fields encountered in a device such as a Tokamak, the impurity atom transport properties are strongly affected by their *initial* velocity and charge state. To put it simply, sputtered neutral atoms with high velocities have a high probability of penetrating into the plasma before ionization occurs. A sputtered ion on the other hand would be immediately returned to the surface by the electrostatic sheath potential (for positive ions) and by the toroidal magnetic field, and have little likelihood of penetrating sufficiently far into the plasma to encounter a region of sufficiently high density to allow the collisions required for transport across magnetic field lines.[206]

The desired reduction of erosion and impurity influx is not significant unless the ion fraction approaches unity. Figure 15 shows the published values of the secondary ion fraction for forty-two clean and oxidized elements, suggesting that high ion fractions are favoured for the alkali metals and metallic oxides.[81] Additionally, it has been found that, for some hydride-forming metals such as

---

[206] G. A. Emmert, in 'Proceedings of the Workshop on Sputtering by Plasma (Neutral Beam) Surface Interaction', Office of Fusion Energy, USDOE, CONF-790775.

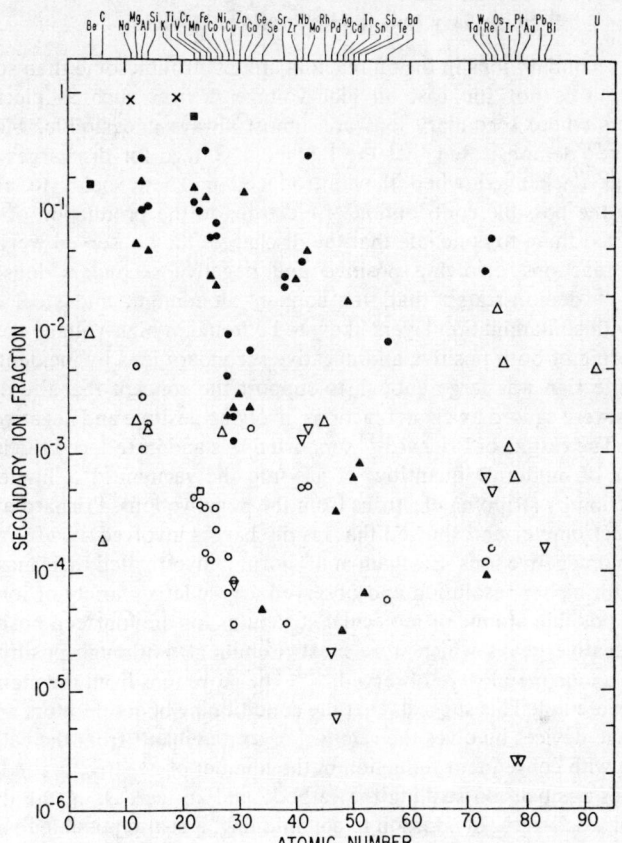

Figure 15 Plot of the secondary ion fraction for forty-two elements. Open symbols are for nominally clean surfaces and filled symbols are for oxygen-covered surfaces. References are: ▲ Andersen (57), ●, ○ Benninghoven (13), △ Beske (100), × Blaise (20), □, ■ Krauss and Gruen (147, 160), and ▽ Schroeer (116)[81]

Ti, the presence of hydrogen is sufficient to produce a substantial increase in the secondary ion fraction.[207]

A number of techniques have been proposed[81,165,207–210] for producing reactor surfaces with high secondary ion yields. These surfaces should preferably be self-sustaining[81,165] and it should at least be possible to regenerate them *in situ*. However, they must also meet a number of other constraints to be useful in a reactor environment. Work is in progress but it is not yet possible to evaluate the practicality of using relatively high secondary ion yields as a means of reducing plasma–wall interaction problems.

[207] A. Krauss and D. Gruen, *J. Nucl. Mater.*, 1976, **63**, 380.
[208] J. N. Smith, jun., *J. Nucl. Mater.*, 1979, **82**, 179.
[209] W. O. Hofer, H. L. Bay, and P. J. Martin, *J. Nucl. Mater.*, 1978, **76/77**, 15.
[210] W. O. Hofer and P. J. Martin, *Appl. Phys.*, 1978, **16**, 271.

9 Secondary Ions in High-voltage Devices

Although secondary ions in fusion reactors are less troublesome than sputtered neutrals, this is not the case in high-voltage devices such as electrostatic accelerators where secondary ions are almost always undesirable. McKibben and Boyer[211] demonstrated that the threshold voltage for discharges in their system was unchanged when they introduced magnetic fields to eliminate effectively the possible contribution of electrons to the production of positive ions. This led them to conclude that the discharges they observed were caused by chain reactions involving positive and negative secondary ions. Later, Mansfield[212] demonstrated that, for copper, aluminium, and steel surfaces covered with contamination layers likely to be found in high-voltage apparatus, the sputtering of both positive and negative secondary ions by incident ions of the opposite sign was large enough to support the concept that the observed discharges were caused by chain reactions involving positive and negative secondary ions. The results of Prichard[213] supported this and introduced the idea that the release of moderate quantities of gas into the vacuum[213,214] hindered the chain reaction by stripping electrons from the negative ions. Prichard also used a mass spectrometer and showed that his discharges involved a wide variety of positive and negative ions. Eastham and Thorn[215] investigated their mass distributions with higher resolution and observed a very large variety of ions. They suggested possible atomic or molecular structures for the fourteen positive and sixteen negative peaks which were most prominent. Although positive metal ions from anode metal were observed,[214,215] negative ions from the metal of the cathodes were not. This suggests that the conditioning of accelerators and other high-voltage devices involves the removal of contaminants from the cathode by sputtering with consequent reduction in the number of negative secondary ions emitted per positive ion striking the cathode and an increase in the threshold voltage for discharges. In addition to conditioning, it is also possible to suppress partially the secondary ion chain reaction by introducing 'back-biasing' electrodes or by arranging an accelerator's electrodes so that the electric field will inhibit the escape of secondary ions of one sign from electrodes which are subject to bombardment by ions of the opposite sign.[216]

Finally, it should be mentioned that high-voltage breakdown is a complicated subject and that electrons and charged microparticles ($\sim 10~\mu$m)[217,218] sometimes play important roles as do flashovers on the surfaces of insulators and gas discharges in neutral material emitted under bombardment. However, current loading and microdischarges in ion accelerator tubes are believed to be almost entirely secondary ion chain reactions and secondary electrons from these ions.

---

[211] J. L. McKibben and K. Boyer, *Phys. Rev.*, 1951, **82**, 315.
[212] W. K. Mansfield, *Br. J. Appl. Phys.*, 1960, **11**, 454.
[213] B. A. Prichard, jun., *J. Appl. Phys.*, 1973, **44**, 4548.
[214] C. M. Turner, *Phys. Rev.*, 1951, **81**, 305.
[215] D. A. Eastham and R. Thorn, *J. Phys. D*, 1978, **11**, 1149.
[216] A. Langsdorf, in Argonne National Laboratory Report ANL-76-96, 105, unpublished.
[217] L. Cranberg, *J. Appl. Phys.*, 1952, **23**, 518.
[218] G. L. Griffith and D. A. Eastham, *J. Phys. D*, 1979, **12**, L105.

6
Development and Trends in Instrumentation in Mass Spectrometry

BY A. McCORMICK

1 Introduction

Although the previous article in this series has been called[1] 'a comprehensive review of trends in instrumentation' the emphasis was on the 'narrow limits' part of the definition of 'comprehensive' (Collins National Dictionary) and it was a 'review' in the limited sense that only some of the publications cited were subjected to critical appraisal. The same format has been adopted for this report. Coverage has been restricted to publications in which the application or potential application of mass spectrometry instrumentation in organic chemistry, and particularly analytical organic chemistry, is described. As before, some of the subject matter has been selected for more detailed discussion and this selection has been governed firstly by the Reporter's personal interest and experience and secondly by his assessment of the need for further consideration of these topics. A more comprehensive list of publications dealing with instrumentation related to the general subject matter of this volume is to be found in reference 1 and a complete bibliography on mass spectrometry instrumentation is contained in Section One of the 'Mass Spectrometry Bulletin'.[2]

The triennial International Mass Spectrometry Conference is one of the most important venues for the announcement of new instrumental methods. Unfortunately publication of the proceedings of the eighth meeting held in Oslo in August 1979, 'Advances in Mass Spectrometry' Vol. 8, has been delayed. The reporter has therefore had to make do with the abstracts only of the presentations mentioned below. However, thanks to the editor, Mr. A. Quayle, it has been possible to cite the page numbers as they will appear in the final publication.

From the many publications on organic mass spectrometry which have been issued in the past few years it is clear that one of the 'trends in instrumentation' is a growing awareness of the importance of instrumental factors. As pointed out in ref. 1, mass spectrometry is as yet unmatched in sensitivity and specificity or in the scope of its applications. Biochemistry, forensic science, and environmental research are concerned with increasingly smaller quantities of sample. Mass spectrometry is often the only method of identification available. For identification of known compounds, a low-resolution spectrum which can be compared with a reference library will usually suffice but, where the sample is of an unknown compound or a more rigorous proof of identity is required, there is a

---

[1] A. L. Burlingame, T. A. Baillie, P. J. Derrick, and O. S. Chizhov, *Anal. Chem.*, 1980, **52**, 214R.
[2] 'Mass Spectrometry Bulletin', Mass Spectrometry Data Centre, U.K.C.I.S., The University, Nottingham, NG7 2RD.

demand for obtaining more structurally related information from the mass spectrometer or an even higher selectivity. In addition, there is a need to extend the scope of the method to include compounds of higher molecular weight or those which are thermally too sensitive to be analysed by current techniques. A mass spectrometer is a costly investment and for some laboratories an important requirement is a high rate of sample throughput, preferably in a fully automated system. In addition to these analytical requirements there is the intellectual compulsion to uncover the processes by which the analyst's result is reached and to try to understand why one method should be successful where others have failed. It is probably true to say that instrumental methods developed to provide the answers to these theoretical problems are the forerunners of analytical techniques rather than the reverse. It is the wealth of publications describing different ways of tackling the above problems that has led to the much better understanding of the instrumental aspects of mass spectrometry which exists today even amongst researchers who would regard themselves as being biochemists or botanists rather than mass spectroscopists.

However knowledgeable they may be, the majority of users will probably prefer to purchase 'off-the-shelf' or custom-built equipment rather than become involved in home construction. Fortunately in this respect there is healthy competition amongst manufacturers and new developments now seem to be available much more quickly than in the past, helped no doubt by the widespread use of modular construction and electronics. A review by Craig[3] gives a well balanced account of the relative merits of single-focusing and double-focusing sector mass spectrometers and quadrupoles. In spite of what the manufacturers may claim, commercial instrumentation available to-day tends to be the result of an evolutionary development of well tried principles rather than being revolutionary in character. For example, in spite of the advantages claimed for reversed geometry instruments one well known manufacturer has stuck firmly to conventional design even for its latest product[4] and a recent survey[5] of currently available instrumentation for organic mass spectrometry makes no mention of ICR spectrometers, whereas in the previous report in this series it was suggested that these could soon be available as general purpose instruments.

Concomitantly with this growing interest in the instrumental aspects of organic mass spectrometry, text books on the subject now tend to have much more space devoted to practical procedures than was the case in the past when a sketchy (and often erroneous) account of the physical principles only was given. A few recent books covering instrumentation and procedures in some detail are listed in reference 6.

[3] R. D. Craig, R. H. Batemann, B. N. Green, and D. S. Millington, *Philos. Trans. R. Soc. London, Ser. A*, 1979, **293**, 135.
[4] D. R. Denne, G. Warburton, and M. L. Aspinal, in 'Advances in Mass Spectrometry', ed. A. Quayle, Heyden, London, 1980, Vol. 8, p. 1786.
[5] C. Tayler, *Lab. Equip. Dig.*, June 1980, 75.
[6] (*a*) 'Practical Mass Spectrometry', ed. B. S. Middleditch, Plenum Press, New York, 1979; (*b*) 'Biochemical Applications of Mass Spectrometry', ed. G. R. Waller and O. C. Dermer, Wiley-Interscience, New York, 1980; (*c*) 'GLC and HPLC Determination of Therapeutic Agents', ed. K. Tsuji and W. Morozowich, Marcel Dekker, New York, 1978; (*d*) 'The Medical and Biological Application of Mass Spectrometry', ed. J. P. Payne, J. A. Bushman, and D. W. Hill, Academic Press, London, 1979; (*e*) 'Organics Analysis Using Gas Chromatography/Mass Spectrometry. A Techniques and Procedures Manual', W. L. Budde and J. W. Eichelberger, Ann Arbor Sci. Publ. Inc., Ann Arbor, 1979.

2 'In-beam' Sample Introduction

A number of publications have appeared over the past two years describing instrumental variations on the 'new' method of 'in-beam' analysis. When used with CI this method has been termed 'plasma desorption' by some authors in the belief – mistaken in the opinion of the Reporter for reasons discussed below – that ionization takes place by the action of reagent gas ions on the sample molecules in the condensed phase. Activity in this area was inspired by a publication[7] which described the successful analysis of underivatized oligopeptides by coating of the samples on the tip of a glass rod which could be inserted directly into the ion chamber of a CI mass spectrometer to a position close to the electron beam axis. Good CI spectra could be obtained at temperatures 150 °C below that required by the normal probe supplied for this ion source. CI spectra obtained by the latter method were of thermally degraded molecules and EI spectra of these compounds could not be obtained at all, even using the modified sample introduction method. Later, FD emitters were used with similar success as direct insertion probes for CI and EI respectively[8,9] and the terms CI/D and EI/D were coined. It was noted[8] that better results could be obtained for thermally labile materials if rapid heating of the emitter, as advocated earlier,[10] was used. Other workers[11] initially used a glass probe tip for 'in-beam' sample introduction and emphasized the effectiveness of an inert probe tip surface for analysis of labile compounds. Coating of the probe tip by dipping it into a solution of SE 30 was effective but the coating could be made more resistant to removal by non-polar samples if a surface bonding technique was used.[12] Later, it was reported that use of a glass probe tip tended to cause erratic source behaviour and a metal probe[13] which can be floated at source potential is now favoured. Reports have appeared of sensitivity increases of up to six times for polar compounds when the probe tip is silanized.[14] The use of a 'Vespel' probe tip for direct exposure of the sample to CI reagent gas ions has been used;[15] it was noted that the best probe position is 2 mm from the electron beam axis and, significantly, that using this method molecular ion abundance in EI mode is greater than that observed when using the normal probe of the instrument. Coiled non-activated rhenium wires as probe tips were used for 'desorption chemical ionization'[16] and gave good spectra from higher glycosides and trisaccharides which could not be obtained by flash evaporation from the moving belt l.c.–m.s. interface although this has been found[17] to be a convenient direct introduction method for many types of

---

[7] M. A. Baldwin and F. W. McLafferty, *Org. Mass Spectrom.*, 1973, **7**, 1353.
[8] D. F. Hunt, J. Shabanowitz, F. K. Both, and D. A. Brent, *Anal. Chem.*, 1977, **49**, 1160.
[9] B. Soltmann, C. C. Sweeley, and J. F. Holland, *Anal. Chem.*, 1977, **49**, 1164.
[10] R. J. Beuhler, E. Flanigan, L. J. Greene, and L. Friedman, *J. Am. Chem. Soc.*, 1974, **96**, 3990.
[11] J.-P. Thenot, J. Nowlin, D. I. Carroll, F. E. Montgomery, and E. C. Horning, *Anal. Chem.*, 1979, **51**, 1101.
[12] C. Madani, E. M. Chambaz, M. Rigaud, J. Durand, and P. Chebroux, *J. Chromatogr.*, 1976, **126**, 161.
[13] D. I. Carroll, I. Dzidic, M. G. Horning, F. E. Montgomery, J. G. Nowlin, R. N. Stillwell, J.-P. Thenot, and E. C. Horning, *Anal. Chem.*, 1979, **51**, 1858.
[14] K. S. Webb, B. J. Wood, and R. Davis, in ref. 4, p. 1921.
[15] R. J. Cotter, *Anal. Chem.*, 1979, **51**, 317.
[16] U. Rapp, G. Dielmann, D. E. Games, J. L. Gower, and E. Lewis, in ref. 4, p. 1660.
[17] D. E. Games, C. Eckers, P. Hirter, E. Lewis, and K. R. N. Rao, 27th Annual Conference on Mass Spectrometry and Allied Topics, Seattle, 1979, p. 627.

compound. There has also been described a metal filament system,[18] in this case platinum, for which advantages are claimed, *e.g.* ease of cleaning and simplicity of construction as platinum can be fused in to a soda glass holder/insulator. Fortunately for the unwary, it was noted that this platinum probe is suitable for a low-voltage quadrupole instrument only, and careful consideration would have to be given to insulation problems if it were considered for use on a high-voltage sector instrument. Modification to the EI source of a Kratos MS 50 spectrometer has enabled the FD emitter system to be used as an 'in-beam' solids probe.[19] An extensive study of the 'in-beam' method of sample introduction using a glass probe tip in a CI instrument with an in-line arrangement of the electron entrance aperture and ion exit slit has been made.[20] For compounds such as arginine or creatine, it was found that the optimum probe tip position was 2.5 mm from the electron beam axis, 'optimum' conditions being those which gave the maximum absolute and relative yield of $(M + H)^+$ ions. Other factors contributing to this optimum were the use of rapid heating (60 °C per second) and a 'Teflon'-covered probe tip. The probe temperature control in addition to the ballistic heating rate of 60 °C per second allowed a slower, variable heating rate of as little as 30 °C per minute to be used. Whilst this was more convenient for routine analyses it did give an overall reduction in sensitivity of about a factor of two and a smaller ratio of $(M + H)^+$ to $(M - OH)^+$ compared with the rapid-heating method for compounds such as creatine. Another point noted by these authors is that, in addition to the probe surface condition, other compounds co-deposited with the sample of interest may have an effect on the success or otherwise of the analysis. For example, in the presence of oxalic acid, creatine gave no $(M + H)^+$ ions but there were no matrix effects with urea or ammonium chloride. It was noted[21] that, for some compounds such as aliphatic alcohols, 'in-beam' electron impact spectra contain evidence for CI ionization giving rise to protonated and dimeric species.

In considering this 'in-beam' method of sample introduction it is worth having a brief historical look at the development of the direct insertion probe for electron bombardment ionization. This was first used in the 1950's and the success of the method in producing spectra of hitherto intractable compounds was clearly attributed to the direct evaporation of the sample *within the ionization chamber and close to the electron beam.* One of the earliest probes[22] passed through a hole in the ion repeller electrode of a Metropolitan-Vickers MS2 mass spectrometer and it was found that the optimum position for the probe tip (*i.e.* as close as possible to the electron beam without disturbing the source operation) was just level with the repeller plate (about 2—3 mm from the electron beam; *cf.* refs. 15 and 20). The first 'direct' probe on one of the earliest commercial instruments produced specifically for qualitative analysis of high molecular weight organics (AEI MS9) was a heated tungsten filament (*cf.* refs 16 and 18). This was not very successful as the filament was positioned outside the enclosed ionization region. It

[18] A. P. Bruins, *Anal. Chem.*, 1980, **52**, 605.
[19] G. K. Eigendorf, in ref. 17, p. 717.
[20] (*a*) G. Hansen and B. Munson, *Anal. Chem.*, 1978, **50**, 1130; (*b*) *ibid.*, 1980, **52**, 245.
[21] M. Ohashi and N. Nakayama, *Org. Mass Spectrom.*, 1978, **13**, 642.
[22] R. I. Reed, *J. Chem. Soc.*, 1958, 3432.

was replaced by a pyrophyllite rod with two longitudinal holes into which a platinum heating wire was inserted. The pyrophyllite probe was thin enough to pass through the sample inlet channel in the source block. As the probe assembly did not incorporate a vacuum lock, but was attached to a small flange on the source housing and was non-adjustable, it was made just long enough for the tip to be positioned about 2—3 mm from the electron beam axis. Operation of this probe meant cooling and venting the source for each sample and the procedure was that after attaching the probe and obtaining a working vacuum the source was allowed to warm up as the mass range was repetitively scanned using the oscilloscope display as a spectrum monitor. Spectra were recorded when sufficient ion beam intensity was obtained or changes in the spectral pattern were observed. The platinum wire heater rarely had to be used except as a means of attaching the probe and of floating it at source potential (*cf.* ref. 13). Spectra were recorded in every instance at the lowest temperature consistent with obtaining a reasonable signal-to-noise ratio. This fixed probe was later changed for one with a vacuum lock, the probe tip distance from the electron beam being adjustable by use of a soft iron magnet follower as in the earlier design.[22] Quartz or pyrophyllite probe tips were used in this unheatable probe assembly and the standard procedure used in the Reporter's and other laboratories was to insert the probe until it just intercepted the electron beam (observable by the response of the filament emission or total ion beam monitor meters) and then to withdraw it to the minimum distance consistent with obtaining a suitable ion current intensity. Control of the evaporation/ionization rate was effected by adjusting the source block temperature and the distance of the probe tip from the electron beam. The minimum temperature at which a sample could be run was achieved by placing the probe tip as close as possible to the electron beam (2—3 mm) and adjusting the source temperature (*cf.* ref. 7). Although later sources for this instrument had more efficient heaters and a fast cooling system it was generally more convenient and quicker for routine analyses to maintain the source at a reasonably high (200—250 °C) temperature. Sample throughput was greatly increased with this new probe system but the best emission temperature (*cf.* ref. 8) was not necessarily automatically obtained as it was with the older type of fixed probe. For some compounds which would otherwise have been too volatile for probe analysis, results could be obtained by dipping the loaded probe tip into liquid nitrogen before insertion into the preheated source. It was also accidentally discovered that by this method spectra could be obtained of some substances which by the more routine probe method gave spectra of only decomposition products. For example, a molecular ion could be obtained for the sesquiterpene caparrapidiol[23] by this method but only for a few minutes after probe insertion after which only $(M - 2H_2O)^+$ ions were observed. Efforts to obtain the spectrum of this very labile substance by using a cooled ion source also resulted only in the spectrum of the decomposition product. In other words, this liquid nitrogen dipping technique appears to have been an early example of the rapid-heating method[10] though it was not recognized as such at the time.

---

[23] J. Borges del Castillo, C. J. W. Brooks, and M. M. Campbell, *Tetrahedron Lett.*, 1966, 3731.

Later developments in direct insertion probe design on almost all commercial mass spectrometers included methods of independent probe heating and cooling. This type of probe is very convenient in that the rate of sample evaporation can be finely controlled or the heating rate can be programmed. However, these devices are bulkier than their predecessors and cannot be inserted into the ionization region of the mass spectrometer source. Sample molecules evaporated from such a probe then have to pass through a hole in the source block before ionization (see, for example, the diagram on p. 28 of ref. 6a) and this channel in the hot metal block may be as long as 20—30 mm and the surface may become increasingly more active with use. Therefore, these devices are *not* direct insertion probes according to the original conception. It would appear to this Reporter that the success of the 'in-beam' method for EI analysis[15] is due mainly to reinvention of the direct insertion probe but partly to the use of inert probe materials and partly to the use of rapid-heating methods. The Reporter's laboratory certainly has been able to obtain spectra of many compounds which had failed elsewhere, thanks probably to the fact that the same 'old-fashioned' probe has been retained through several changes of mass spectrometer over the years. The question that has not been satisfactorily answered is why the 'in-beam' method is so successful for CI. Most of the authors cited above state or imply that ionization occurs by virtue of direct exposure of the solid sample to the ions of the reactant gas, presumably to produce product ions on the probe surface. In a highly critical assessment[24] of some theories of FD[25] it was convincingly demonstrated that ions cannot be desorbed from the condensed phase in the absence of strong electric fields. There are no such fields in a CI source so how do the ions leave the probe surface, if indeed they are produced there at all? The fact that 'CI/D' is assisted by rapid heating[8,20b] may provide the clue. In a 'conventional' CI source, sample molecules enter the gas phase at a position remote from the point of ion formation and extraction. These molecules may then suffer many wall collisions, and hence undergo thermal degradation, before they can be ionized, whereas in the 'in-beam' configuration, particularly when rapid heating is used, a dense pocket of sample gas may be formed at the optimum position with a higher probability of ionization before wall collisions can occur. In other words CI/D is no different from CI except in the optimization of presentation of *gaseous* sample molecules to the ionizing plasma. Some support for this suggestion comes from where it was reported that attempts to obtain EI spectra of certain compounds (*e.g.* aliphatic alcohols) resulted in self chemical ionization, *i.e.* the sample pressure in the ionization region must have been much higher than the average pressure in the enclosure.[21] It seems also to have been forgotten that Reed[26] in 1961, before CI was invented, reported obtaining the spectra of sugars by the direct insertion probe/EI technique and that these spectra exhibited $(M + H)^+$ peaks but no $M^{+\cdot}$. The rapid-heating methods[10] which have been so successful in many applications[27]

[24] H. D. Beckey and F. W. Röllgen, *Org. Mass Spectrom.*, 1979, **14**, 188.
[25] (a) J. F. Holland, B. Soltmann, and C. C. Sweeley, *Biomed. Mass Spectrom.*, 1976, **3**, 340; (b) J. F. Holland, B. Soltmann, R. Olinger, and J. Dye, ref. 17, p. 174.
[26] R. I. Reed, W. K. Reid, and J. M. Wilson, in 'Advances in Mass Spectrometry', ed. R. M. Elliott, Pergamon, London, 1963, Vol. 2, p. 416.
[27] See for example W. R. Anderson, jun., W. Frick, and G. D. Daves, jun., *J. Am. Chem. Soc.*, 1978, **100**, 1974.

are in fact also 'in-beam' methods and, although this success has been attributed solely to the evaporation mechanism, one wonders whether in fact direct ionization without further wall collisions on account of the density of sample vapour produced initially at the ionization point is not an important contributary factor. In summary, a successful 'in-beam' system requires (i) an inert probe material, (ii) placement of the sample at the extreme tip of the probe[28] or on the surface of a filament, (iii) positioning of the probe close to the electron beam axis, and (iv) rapid heating of the probe. Additionally, it might be worthwhile using a cooled source block so that decomposition products formed at the walls are trapped and cannot contribute to the observed spectrum.

3 Laser Applications

The use of a pulsed laser as a means of ionizing solid samples for mass spectrometric analysis was reported as long ago as 1963 and a number of publications on this topic have appeared since that time. An excellent review[29] of this method, which is useful mainly for elemental analysis of small samples, has been published. The first use of lasers in organic mass spectrometry was described in 1966 and a few other reports appeared in later years.[30] Now that tunable lasers with high pulse repetition rates are becoming more generally available and at reasonable cost, interest in their application to organic mass spectrometry is becoming more widespread. These devices have been used purely as sources of thermal energy for sample vaporization, ionization being accomplished by one of the long established methods. The attraction of the laser is that it can be used to heat a small precisely defined area, and at a high rate if required, without the electrical insulation and mechanical problems associated with conventional heating methods. Tunable lasers have been used to heat emitters in FD experiments[31] and it is claimed that this is not only convenient but gives better instrument performance than the traditional ohmic heating method. Another obvious use is in pyrolysis mass spectrometry and a number of such applications have been reported.[32] The use of pulsed lasers for flash evaporation of the solvent in on-line l.c.–m.s. has been described.[33,34] However, it is likely that the greatest area of growth in laser applications will be in their use as ionization devices. A laser/TOF spectrometer system for the study of polymer pyrolysis produced organic ions as well as neutral species. Further investigation[35] of lower molecular weight compounds showed that molecular ions of polycyclic aromatics and even of leucine could be obtained by laser ionization. Recent investigations have tended to concentrate on the types of compound which are difficult to analyse by EI

[28] A. Dell, D. H. Williams, H. R. Morris, G. A. Smith, J. Feeney, and G. C. K. Roberts, *J. Am. Chem. Soc.*, 1975, **97**, 2497.
[29] I. D. Kovalev, G. A. Maksimov, A. I. Suchkov, and N. V. Larin, *Int. J. Mass Spectrom., Ion Phys.*, 1978, **27**, 101.
[30] See for example R. J. Conzemius and J. M. Capellen, *Int. J. Mass Spectrom., Ion Phys.*, 1980, **34**, 197.
[31] H.-R. Schulten and W. D. Lehmann, *Org. Mass Spectrom.*, 1978, **13**, 36.
[32] See for example 'Analytical Pyrolysis', ed. C. E. R. Jones and C. A. Cramers, Elsevier, Amsterdam, 1977.
[33] C. R. Blakley, M. J. McAdams, and M. L. Vestal, in ref. 4, p. 1616.
[34] E. R. Lory and F. W. McLafferty, in ref. 4, p. 954.
[35] F. J. Vastola and A. J. Pirone, in 'Advances in Mass Spectrometry', ed. E. Kendrick, Institute of Petroleum, London, 1968, p. 107.

methods. The successful analysis of underivatized oligosaccharides, cardiac glycosides, and nucleotides was described using two different lasers in a sector mass spectrometer with simultaneous electrical ion detection and in a double-focusing instrument with photoplate detection.[36] Ions observed were protonated or cationized molecular ions. Deliberate 'doping' of the sample with alkali halides was found to be helpful in producing more efficient ionization but, although trace amounts of potassium had been found to produce disproportionately large amounts of $(M + K)^+$ ions compared with $(M + Na)^+$, deliberate doping with lithium halides did not give the expected increase in efficiency. As essentially the same results were obtained with the two lasers of different wavelengths it was concluded that the ion desorption process is not strongly influenced by the wavelength. In fact organic compounds absorb only very weakly at 1.06 μm, the wavelength of the neodimium laser used in this study, and from this it was concluded that it is fast heating of the substrate which effects ionization/desorption in a manner similar to that proposed for fission fragment ionization.[37] A pulsed laser–TOF system was used for organic analysis in which quasimolecular ions were observed, but matching of the wavelength to an absorption band of the sample appeared to be important.[38,39] This has also been the conclusion for a CW carbon dioxide laser at low-power density.[40] However, no quasimolecular positive ions were observed at all, even for citric acid. It was essential to have the organic sample admixed with alkali halides, when cationized species could be produced. Using the low-power CW laser irradiation of sucrose plus sodium iodide an ion corresponding to $(M - H_2O + Na)^+$ is observed which is absent in the spectrum produced by the pulsed laser.[36] Another significant point is that in these later experiments cationized organic ions were seen for only one or two seconds, after which thermionic emission of alkali-metal ions only was observed and poor results were obtained for amino-acids compared with those achieved by a pulsed laser method.[41] Other similar results have been reported,[42] *i.e.* organic compounds alone produce no positive ions but, when alkali halide is present, cationized species can be observed.

A possible explanation for the discrepancies in results produced by different systems is that the low-power CW laser acts purely as a heating device, *i.e.* the cationized organic ions observed are formed in the gas phase from neutrals flash evaporated from the surface and alkali-metal ions produced by thermal ionization ('organic' background is one of the most troublesome features in thermal ionization mass spectrometry and could also be attributed to cationization of residual vapour in the region of the source slit). In other words this laser experiment is not analogous to FD as suggested in ref. 1. The production of cationized species by rapid heating of mixtures of alkali halides and polar organics

---

[36] M. A. Posthumus, P. G. Kistemaker, H. L. C. Meuzelaar, and M. C. Ten Noever de Brauw, *Anal. Chem.*, 1978, **50**, 985.
[37] R. D. Macfarlane and D. F. Torgerson, *Science*, 1976, **191**, 920.
[38] H. J. Heinen, S. Meier, H. Vogt, and R. Wechsung, in ref. 4, p. 942.
[39] K.-D. Kupka, F. Hillenkamp, and Ch. Schiller, in ref. 4, p. 935.
[40] R. Stoll and F. W. Röllgen, *Org. Mass Spectrom.*, 1979, **14**, 642.
[41] K.-D. Kupka, F. Hillenkamp, and Ch. Schiller, *Verh. Dtsch. Phys. Ges.*, 1979, **14**, 477.
[42] G. D. Daves, jun., T. D. Lee, W. R. Anderson, jun., D. F. Barofsky, G. A. Massey, J. C. Johnson, and P. A. Pincus, in ref. 4, p. 1012.

from a rhenium filament into an electron beam has been described recently.[43] It would be interesting to repeat this experiment with the electron beam switched off! This was tried in a laser source instrument but no cationized species were observed; it is possible that the direct heating of the tungsten filament was not fast enough to produce a sufficient density of neutrals. On the other hand, the high-power pulsed laser may act as a combination of heating device and field desorber, i.e. this process is exactly analogous to conventional FD and not as suggested in ref. 36. The electric field of a laser beam is given by $E = 27I^{1/2}$ where I is the power density in Wcm^{-2}. The peak power density of the neodimium laser used in the work described above[36] would be about $10^8\ Wcm^{-2}$. This would give a field strength of approximately $3 \times 10^5\ Vcm^{-1}$ which is probably just sufficient to field desorb ions already formed on the surface of the emitter (see, for example, ref. 24). The observation[36] that some substances produce negative ions whereas others do not and that these negative ions occur as $(M - H)^-$ and not as, e.g., $(M + Cl)^-$ is another indication that laser desorption is a field process. In fact it has been suggested[44] that, for solids vaporized by a pulsed laser, neutral species in the vaporized material could be ionized by electrons accelerated in the field of the incident light wave. These authors[44] also point out that a non-Q-switched laser pulse actually consists of a train of narrower pulses. This would mean that at the peak value of these narrow pulses the power density, and hence the field, is instantaneously much higher than that calculated for an integrated pulse. (The field effect of the laser is also suggested as a mechanism of ion desorption in the 'Field Ionization and Field Desorption' section of ref. 1.) Whatever the mechanism for laser ionization of solid organic compounds the technique appears to rival FD in its range of applications and with a little more development could probably be simpler to apply.

Lasers used for ionization of solid samples cannot yet be claimed to have broadened the scope of organic mass spectrometry as the results reported to date have shown only that the method could be considered as an alternative to FD. Photoionization by a tunable laser source does offer the distinct possibility of increased specificity of ionization, which of course also means better sensitivity because of background suppression. Currently available tunable lasers cannot photoionize organic molecules directly, although a very short wavelength device is under development at Stanford University.[45] Ionization must then take place by the absorption of two or more photons. The two-photon ionization of benzene has been described[46] using a frequency doubled dye laser which gives a photon energy of 4.786 eV, corresponding to a real intermediate state of the molecule. Benzene vapour was introduced into the ion source of a quadrupole mass spectrometer as a molecular beam, and ion currents comparable in magnitude with conventional EI were obtained. Only molecular ions were observed with no fragmentation and the point was made that, since the production of the intermediate state is a resonance process, the ionization is compound specific and would enable the detection of small traces in mixtures to be accomplished. The first multiphoton ionization

---

[43] G. D. Daves, jun. and W. R. Anderson, jun., *Int. J. Mass Spectrom., Ion Phys.*, 1979, **30**, 385.
[44] K. A. Lincoln and M. A. Covington, *Int. J. Mass Spectrom., Ion Phys.*, 1975, **16**, 191.
[45] *Laser Focus*, June 1980, p. 40.
[46] U. Boesl, H. J. Neusser, and E. W. Schlag, *Z. Naturforsch., Teil A*, 1978, **33**, 1546.

process with mass analysis is claimed,[47] using iodine as sample vapour. A sensitivity of less than 1 ion count per second could be achieved by gating the detector to the laser pulse frequency. Multiphoton ionization of hydrocarbons is also mentioned but no data are given. Using a rare-gas halide laser in a time-of-flight mass spectrometer, multiphoton ionization of a range of organic compounds was achieved.[48] Certain molecules could not be ionized and others yielded wavelength-dependent mass spectra which differ substantially from their EI spectra. A very recent report[49] describes use of an ArF excimer laser with a quadrupole mass spectrometer. Twenty seven compounds of various types were successfully analysed and it was found that, with two exceptions, the spectra obtained were similar to the 10 eV EI spectra. From this it was concluded that the ionization was a non-resonant two photon process. Efficiency of ionization in this experiment was less than for 70 eV EI but again the laser method results in essentially nil ion background so that excellent signal-to-noise can be achieved. The two exceptions to the two-photon non-resonant absorption noted above were carbon disulphide and benzaldehyde. In the spectrum of the former, fragmentation processes are observed which require three-photon absorption. The benzaldehyde ionization efficiency was found to be about one hundred times greater than that of the other compounds examined and the spectrum showed much more extensive fragmentation than even the 70 eV EI spectrum; this was explained as due to resonant absorption of the first photon. Resonance ionization seems to be a promising method of increasing the specificity (and hence sensitivity) of mass spectrometry which is increasingly being demanded for some applications, as mentioned in the Introduction to this report. An excellent review[50] of the theory and practical applications of resonance ionization methods, including a method for *single-atom detection*, has recently appeared.

Laser irradiation can also be conveniently used for dissociation of ions produced by other methods. Advantages over the collision gas method are simplicity (no collision cell with its associated pumping, *etc.*), choice of photon energy, and detection of photo-dissociated ions only by use of beam chopping methods. Applications in ICR[51], double-focusing mass spectrometers,[52] and tandem quadrupole systems[53] have been described.

4 Field Ionization and Field Desorption

In a lengthy review of FD,[54] there was concentration on chemical applications but some description of instrumentation and techniques is given. Reference was made to 302 publications covering almost all aspects of FD mass spectrometry. Beckey[24] has severely criticized Holland's[25] theories about the mechanism of FD ionization and has clearly demonstrated that high electric fields are essential for

---

[47] L. Zandee, R. B. Bernstein, and D. A. Lichtin, *J. Chem. Phys.*, 1978, **69**, 3427.
[48] J. P. Reilly and K. L. Kompa, in ref. 4, p. 1800.
[49] M. Seaver, J. W. Hudgens, and J. J. Decorpo, *Int. J. Mass Spectrom., Ion Phys.*, 1980, **34**, 159.
[50] G. S. Hurst, M. G. Payne, S. D. Kramer, and J. P. Young, *Rev. Mod. Phys.*, 1979, **51**, 767.
[51] D. S. Bomse, R. L. Woodin, and J. L. Beauchamp, *J. Am. Chem. Soc.*, 1979, **101**, 5503.
[52] (*a*) E. S. Mukhtar, I. W. Griffiths, F. M. Harris, and J. H. Beynon, in ref. 4, p. 1685; (*b*) E. S. Mukhtar, I. W. Griffiths, F. M. Harris, and J. H. Beynon, *Org. Mass Spectrom.*, 1980, **15**, 51.
[53] D. C. McGilvery and J. D. Morrison, *Int. J. Mass Spectrom., Ion Phys.*, 1978, **28**, 81.
[54] H.-R. Schulten, *Int. J. Mass Spectrom., Ion Phys.*, 1979, **32**, 97.

this to operate. This has also been emphasized by Giessmann[55] who points out that, for some types of organic compounds, ions may be formed on the emitter by field-independent electrolytic processes and may require lower field strengths for desorption than other types which do not form solvated ions; the latter in fact are ionized by FI which requires fields of the order of 10^{10} V m^{-1}. In a study of cluster ion formation in alkali halides, it was likewise concluded that FD is essentially a surface process[56] and replacement of the usual alkali halide additive by polyphosphoric acid coated on the emitter has shown that $(M + H)^+$ ions can be produced at relatively low applied potentials for substances which protonate readily whereas others, such as hexane, do not ionize at all.[57] From this it was concluded that FD and electrohydrodynamic ionization are one and the same process, i.e. extraction of ions already formed in the condensed phase by the applied field.

Depending on one's point of view the absence of a universally accepted methodology for FI/FD analysis can be considered either as a serious problem or as providing freedom for experimentation. Invention of new types of emitter or new techniques for making established types continues to be a flourishing business. The use of activated cobalt dendrites prepared as described previously[59] for lithium cationization of organic molecules has been advocated.[58] Advantages claimed are that lower temperature can be used and that, for this particular FD method, ionization efficiency can be as much as ten times·greater than that obtained by using carbon microneedles. On the other hand, for normal FD analysis, the use of platinum emitters made by electrochemical deposition is suggested.[60] These require no heat treatment, unlike cobalt and nickel dendrites,[59] give ionization efficiencies comparable to those obtained with the carbon type, and can be cleaned, even in chromic acid, for re-use. The preparation and performance of graphite filaments for FI and FD have been described.[61] With these, the requisite field strengths can be obtained at applied potentials as low as 2000 V. Multiple FI emitters of tungsten prepared by a vapour deposition process[62] have been carefully compared[63] with carbon emitters and have been found inferior in ionization efficiency. They can, however, be improved by treatment in a benzene atmosphere at about 1300 K and an applied voltage of 15 kV for 1 minute. A possible advantage of the dendritic tungsten emitter over the carbon one is in its use for the study of surface reactions. A disadvantage of the dendritic carbon emitter is the time taken in its preparation, although a fast method has been described using a tungsten wire subjected to a *negative* discharge in a benzonitrile–tungsten hexacarbonyl atmosphere.[64] The latter

---

[55] U. Giessmann, H. J. Heinen, and F. W. Röllgen, *Org. Mass Spectrom.*, 1979, **14**, 177.
[56] C. E. Rechsteiner, T. L. Youngless, M. M. Bursey, and R. P. Buck, *Int. J. Mass Spectrom., Ion Phys.*, 1978, **28**, 401.
[57] W. V. Ligon, *Science*, 1979, **204**, 198.
[58] T. L. Youngless, M. M. Bursey, and C. E. Rechsteiner, *Anal. Chem.*, 1978, **51**, 1951.
[59] M. M. Bursey, C. E. Rechsteiner, T. L. Youngless, D. F. Fraley, M. C. Sammons, and J. R. Mars, in 'Advances in Mass Spectrometry', ed. N. R. Daly, Heyden, London, 1978, Vol. 7, p. 932.
[60] G. Semrau and J. Heitbaum, *Anal. Chem.*, 1979, **51**, 1998.
[61] D. J. Allman, P. L. Gutshall, and P. J. Bryant, in ref. 17, p. 180.
[62] F. Okuyama, *J. Vac. Sci. Technol.*, 1975, **12**, 1399.
[63] F. Okuyama and H. D. Beckey, *Int. J. Mass Spectrom., Ion Phys.*, 1978, **27**, 391.
[64] H. B. Linden, E. Hilt, and H. D. Beckey, *J. Phys. E.*, 1978, **11**, 1033.

compound was first used to stabilize the discharge but it was also discovered that it could be used alone to produce tungsten microneedles with good adhesion to the support wire and good FI/FD sensitivity. Similar results could be obtained using other metal carbonyls. 'Short' carbon microneedles grown quickly under operating conditions in the mass spectrometer[55] have been found particularly effective for FD of non-electrolytes. Microfabrication technology developed for the semiconductor industry can be used[65] for the preparation of multipoint emitter arrays or volcano ion sources. A method[66] of making dendritic silicon emitters of high FI/FD ionization efficiency looks particularly attractive. A 60 μm gold-plated wire is simply heated by a d.c. current in an argon–silane atmosphere for 1 minute. No high voltages are required and as many as two dozen can be made simultaneously. The silicon whiskers and the thick supporting wire are particularly resistant to electrical breakdown damage in the source. Pretreatment of these silicon emitters[67] with tartaric acid has been found to enhance ionization efficiency, particularly for peptides. A review[68] of experimental techniques in FI/FD concentrates mainly on the fabrication and properties of the emitter and, with the exception of the most recent publications mentioned above, critically assesses all of the published methods using the criteria of simplicity in manufacture and sample loading, ionization efficiency for FI and for FD of different types of compound, and durability in use. It is concluded that different problems require different types of emitter and the best types to use for these different applications are suggested.

The volcano type ionizer has been found to be not only more efficient than other types for FI but also to be usable as a selective detector.[69] Using this type of source for g.c.–m.s. as little as 10 pg of anthracene could be quantitatively measured.[70] Discrimination against the saturated hydrocarbon solvent (of higher ionization energy) was as high as a factor of 10^8. The relative sensitivity at high temperature (270—300 °C) of FI for alkanes and cycloalkanes has been found[71] to increase linearly with increasing carbon number, in contrast to the variations found at low temperature, and this has proved invaluable for quantitative measurements. Fingerprinting of urine samples using a Wien filter with a volcano FI source has been described.[72] Particular advantages of the Wien filter are fast scanning and high accelerating voltage which gives good transmission and resolving power. The volcano source has been found to have a lifetime of greater than 130 hours of continuous operation for this type of analysis.

A quadrupole with a specially designed ionization chamber, sharp-edge emitter, and retardation lens system has been found[73] to have exceptionally high FI sensitivity. As deduced earlier,[74] FI background is shown to be due to fast

---

[65] C. A. Spindt, in ref. 17, p. 39.
[66] T. Matsuo and H. Matsuda, *Anal. Chem.*, 1979, **51**, 69.
[67] I. Katakuse, T. Matsuo, H. Matsuda, Y. Shimonishi, and Y. Izumi, *Shitsuryo Bunseki*, 1979, **27**, 127.
[68] H. D. Beckey, *J. Phys. E.*, 1979, **12**, 72.
[69] S. E. Buttrill, G. A. St. John, and C. A. Spindt, in ref. 17, p. 42.
[70] S. E. Young, S. E. Buttrill, and G. A. St. John, in ref. 17, p. 353.
[71] S. E. Scheppele, C. S. Hsu, T. D. Marriott, P. A. Benson, K. N. Detwiler, and N. B. Perreira, *Int. J. Mass Spectrom., Ion Phys.*, 1978, **28**, 335.
[72] W. Aberth, in ref. 17, p. 458.
[73] M. G. Darcy, D. E. Rogers, and P. J. Derrick, *Int. J. Mass Spectrom., Ion Phys.*, 1978, **27**, 335.
[74] H. J. Heinen, Ch. Hötzel, and H. D. Beckey, *Int. J. Mass Spectrom., Ion Phys.*, 1974, **13**, 55.

electrons sputtered from the counter-electrode and not to superexcited neutrals.[75] The same publication[73] describes the design of a very large double-focusing instrument for FI/FD studies and this is further discussed in relation to its application for analysis of large molecules.[76] FI can produce negative ions without intolerable production of electrons provided the emitter location with respect to the counter electrode and the field are suitably adjusted.[77] A pulsed FD source with time-of-flight mass analysis is claimed[78] to have several advantages over the conventional continuous process, *viz* simplicity, high transmission, and the ability to distinguish FD from FI processes. A technique[79] for transfer of air-sensitive transition-metal complexes to the emitter of a FD source has been described in detail.

5 Chemical and Other High-pressure Ionization

A general review[80] of CI includes a section dealing with methodology and instrumentation whilst the latter has been more comprehensively reviewed.[81] It has been stressed[82] that a major factor in the lack of comparability of CI spectra recorded on different instruments is the uncertainty in the quoted reagent gas pressure. Since commercial instruments are usually not fitted with accurate gauges for this purpose, a method of pressure measurement based on kinetics has been developed[82] and tested for methane and isobutane. The effect of reagent gas pressures in the ion source on sensitivity has been investigated[83] when it was found there is an optimum, *i.e.* a peak in the measured total ion current, for each gas. It is suggested that CI spectra should be run under these conditions for both maximum sensitivity and minimization of source contamination caused by the high reagent gas pressure. This shortened source life compared with that of the EI method is perhaps too often ignored or understated by advocates of CI. In this respect a low-pressure CI source looks promising.[84] This uses a d.c. discharge instead of the more usual hot filament, is claimed to have a sensitivity 1—2 orders higher than 'normal' CI sources, requires smaller capacity pumps, and has worked for over 1000 hours without measurable change. ICR methods of course offer the ultimate in low-pressure CI and some potential applications of a Fourier transform system have been described.[85] One distinct advantage of ICR is the unambiguous distinction between isomers,[86] *e.g.* as with primary, secondary, and tertiary alcohols. The Quadrupole Ion Storage Trap in theory could also be used as a low-pressure CI source and some experimental work on this aspect has been

---

[75] H. L. Brown and M. Anbar, *Int. J. Radiat. Phys. Chem.*, 1975, **7**, 281.
[76] P. G. Cullis, G. M. Neumann, D. E. Rogers, and P. J. Derrick, in ref. 4, p. 1729.
[77] (*a*) J. Van der Greef and N. M. M. Nibbering, *Int. J. Mass Spectrom., Ion Phys.*, 1979, **31**, 71; (*b*) G. F. Mes, J. Van der Greef, N. M. M. Nibbering, K. H. Ott, and F. W. Röllgen, *ibid.*, 1980, **34**, 295.
[78] R. W. Odom and S. E. Buttrill, in ref. 17, p. 182.
[79] C. N. McEwen and S. D. Ittel, *Org. Mass Spectrom.*, 1980, **15**, 35.
[80] J. Wilhelm, J. Richter, and H. Schwarz, *Angew. Chem., Int. Ed. Engl.*, 1978, **17**, 424.
[81] R. E. Mather and J. F. J. Todd, *Int. J. Mass Spectrom., Ion Phys.*, 1979, **30**, 1.
[82] R. A. Hancock, R. Walder, and H. Weigel, *Org. Mass Spectrom.*, 1979, **14**, 507.
[83] R. P. Morgan, E. J. Hayward, and G. Steel, *Org. Mass Spectrom.*, 1979, **14**, 627.
[84] J. Eyem, G. Weiss, and J. Franzen, 8th International Mass Spectrometry Conference, Oslo, August 1979, unpublished.
[85] S. Ghaderi, P. S. Kulkarni, R. B. Spencer, E. Ledford, C. L. Wilkins, and M. L. Gross, in ref. 17, p. 235.
[86] G. Boand, R. Houriet, and T. Gläumann, in ref. 4, p. 238.

done[87] but, using methane, the maximum m/z value of ions which could be transmitted to the mass spectrometer was 80 and higher molecular weight reagent gases would have to be used to extend the mass range. One distinct advantage of the low-pressure CI methods is the lack of cluster ion formation[88] but these can be destroyed by interposing a drift tube system between the ion source and mass analyser. One commerical atmospheric-pressure mass spectrometer incorporates such a device[89] which, by alteration of the potential applied to the drift tube, can be used to control fragmentation. This effect has also been studied for the selected fragmentation of esters ionized by CI by alteration of the ratio of applied potential to pressure of collision gas in the drift tube.[90] The relationship between cluster bond strength and the voltage required for dissociation in a drift field at fixed pressure (1 torr) has been investigated.[91] Ion drift times as measured in a conventional CI source derived from an EI design can be seriously in error and a source incorporating a number of refinements has been constructed especially for this purpose[92] but the high sensitivity achieved (1×10^{-10} C μg^{-1} for methyl stearate) would also recommend its use for general analytical work.

High-pressure negative ionization continues to extend the range of mass spectrometry and has been applied to underivatized steroids,[93] aliphatic alcohols,[86] steroid glycosides,[94] and essential oils[95] to name but a few compound types. Since it has been found that for compounds with electronegative atoms the negative ionization efficiency may be as much as one thousand fold that of the positive ionization,[96] derivatization with reagents containing such atoms or groups has been advocated.[97] For example, trace metal analysis with a precision equalling that of neutron activation was achieved by formation of the trifluoropentanedionates.[98] A judicious choice of reagent gases and use of both positive and negative CI does, however, have much to offer for qualitative applications,[99] but positive/negative CI still does not match a combination of EI and CI for structure elucidation.[100] Even higher switching rates can be achieved with a tandem CI/EI ion source[101] in which beam selection for mass analysis is controlled by ion optical shuttering, with no requirement for reagent gas removal or introduction. A new type of CI source employing a microwave discharge is under development.[102] Parts per billion sensitivity for impurities in helium has been achieved but it is expected that this source will also be useable for organic

[87] R. E. Mather, G. Lawson, and J. F. J. Todd, *Int. J. Mass Spectrom., Ion Phys.*, 1978, **28**, 347.
[88] See for example R. A. Hancock, B. S. Thyagarajan, and R. Walder, *Org. Mass Spectrom.*, 1980, **15**, 101.
[89] (a) N. M. Reid, B. A. Thomson, W. R. Davidson, J. B. French, and J. A. Buckley, in ref. 17, p. 237; (b) N. M. Reid, J. A. Buckley, C. C. Poon, and J. B. French, in ref. 4, p. 1843.
[90] P. C. Price, H. S. Swofford, and S. E. Buttrill, *Anal. Chem.*, 1978, **50**, 1127.
[91] H. Kambara, Y. Mitsui, and I. Kanomata, *Anal. Chem.*, 1979, **51**, 1447.
[92] A. J. Illies and G. G. Meisels, *Anal. Chem.*, 1980, **52**, 325.
[93] T. A. Roy, F. H. Field, Y. Y. Lin, and L. L. Smith, *Anal. Chem.*, 1979, **51**, 272.
[94] A. P. Bruins, in ref. 4, p. 246.
[95] A. P. Bruins, *Anal. Chem.*, 1979, **51**, 967.
[96] J. R. Hass, M. D. Friesen, D. J. Harvan, and C. E. Parker, *Anal. Chem.*, 1978, **50**, 1474.
[97] D. F. Hunt and F. W. Crow, *Anal. Chem.*, 1978, **50**, 1781.
[98] S. R. Prescott and T. H. Risby, *Anal. Chem.*, 1978, **50**, 562.
[99] D. F. Hunt and S. K. Sethi, in ref. 17, p. 737.
[100] K. T. Taylor and C. J. Wakefield, in ref. 4, p. 1650.
[101] M. W. Siegel, in ref. 4, p. 1812.
[102] M. W. Siegel, in ref. 4, p. 1655.

analysis. Atmospheric-pressure ionization methods have been devised for investigation of ions formed in a pulsed electron capture g.c. detector.[103] Enhanced relative abundance of negative molecular ions produced by electron impact can be achieved[104] by introducing a pressure of about 0.1—0.5 torr of argon into the ion source. Contrary to what might be expected, CI can be used to give a precise determination of stable isotope enrichment in small samples.[105] Use of photoionization as a means of primary ionization of the solvent (cyclohexane) is a useful means of 'fingerprinting' complex hydrocarbon mixtures[106] but care must be taken to ensure that the sample is very dilute otherwise the aliphatic components will not be detected because of secondary reactions with molecules of lower ionization energy.

Sample introduction for CI of 'difficult' compounds has been largely dealt with in Section 2. A new method of introduction involving atomization by a helium gas stream has been described[107] and enabled free amino-acids, deoxycortisone, and ethylene glycol to be analysed with high sensitivity.

6 Other Ionization Methods

A collimated electron beam ion source[108] for a time-of-flight mass spectrometer has given a two-fold increase in resolving power but with a two-fold reduction in sensitivity, but the product of resolving power and sensitivity is still claimed to be comparable to that of sector instruments. Photoionization by a vacuum ultraviolet lamp, used[109] for g.c.–m.s., is less sensitive than EI but gives no fragmentation and does not ionize water. A new method of ionization of organic compounds in the liquid phase has been introduced.[110] In this, a gas such as argon is ionized at atmospheric pressure by corona discharge and the resultant ions are accelerated through a grid to interact with the liquid sample. Mass spectra of involatile compounds can be obtained even at ambient temperature and this looks to be a very promising device for l.c.–m.s. applications. A specially designed ion source incorporating cryopumping has made possible the electrohydrodynamic ionization of aqueous solutions.[111] Two slightly different designs of a very high-efficiency electron impact source for a quadrupole mass filter have been described.[112]

Secondary ion mass spectrometry (SIMS) is dealt with more fully in Chapter 5, so only a brief coverage is included here. As no recent publications appear to mention the work of Dillon and co-workers it is worth noting that, with a time-of-flight instrument, bombardment of polymers with an argon ion beam

---

[103] E. P. Grimsrud, S. H. Kim, and P. L. Gobby, *Anal. Chem.*, 1979, **52**, 223.
[104] J. E. Szulejko, I. Howe, J. H. Beynon, and U. P. Schlunegger, *Org. Mass Spectrom.*, 1980, **15**, 263.
[105] D. E. Matthews, E. Ben-Galim, and D. M. Bier, *Anal. Chem.*, 1979, **51**, 80.
[106] L. W. Sieck, *Anal. Chem.*, 1979, **51**, 128.
[107] Y. Hirata, T. Takeuchi, S. Tsuge, and Y. Yoshida, *Org. Mass Spectrom.*, 1979, **14**, 126.
[108] J.-P. Lehmann and E. J. Younginger, *Int. J. Mass Spectrom., Ion Phys.*, 1980, **33**, 95.
[109] N. Washida, H. Akimoto, H. Takagi, and M. Okuda, *Anal. Chem.*, 1978, **50**, 910.
[110] M. Tsuchiya and T. Taira, *Int. J. Mass Spectrom., Ion Phys.*, 1980, **34**, 351.
[111] V. E. Skurat, N. B. Zoloty, G. V. Karpov, V. L. Talrose, Yu. V. Vasjuta, and G. L. Ramendik, in ref. 4, p. 1054.
[112] Yu. T. Kalinin, A. A. Sysoev, V. N. Vilatov, and R. S. Khafizov, *Instrum. Exp. Technol.*, 1979, **22**, 177.

produced a spectrum of secondary ions up to about m/z 400 which was characteristic of the polymer structure.[113] Current interest in SIMS for organic applications is attributable mainly to Benninghoven who introduced the method of 'static' SIMS,[114] i.e. the use of very low-current density primary beams, which minimizes radiation damage. This was announced[115] as a new method for organic mass spectrometry, particularly for sensitive biological compounds, in 1977. A more detailed account[116] describes improvements in techniques and instrumentation. Positive and negative ion spectra of forty compounds such as amino-acids and peptides were obtained. Although these showed $(M + H)^+$ or $(M - H)^-$ the spectra are also characterized by complex ions formed by reaction between the metal support (Ag foil) and the organic compound and by metal and metal halide ions. The effects of sample preparation and loading on the absolute and relative amounts of these ions have been studied when it was concluded that a thin sample film giving close contact with the metal substrate is essential for a successful SIMS analysis.[117] SIMS appears to offer a wider range of cationization[118] for mass analysis than other techniques such as FD but the efficiency depends to some extent on the metal and the structure of the organic ligand.[119] A recent study[120] of the application of static SIMS to polymers has shown some success both in structure differentiation and in impurity analysis. No definitive mechanism for the production of organic ions by SIMS has been agreed on. Some authors[116] propose a mechanism similar to that advocated for fission fragment ionization. However, cationization is observed[118] only for those metals which also produce intense metal ion beams and from this it is suggested that the complexes are produced in the gas phase between sputtered metal ions and evaporated neutrals (cf. comments made above in Section 3 on cationization by fast evaporation from metal filaments and low power CW laser irradiation). Although it is implied[118] that cationization in static SIMS offers a method of quantitative analysis of metals the Reporter is not aware of any detailed publications as yet on this subject.

7 Ion Detection and Measurement

Factors affecting the response of an ion detector to an incident ion beam were discussed at some length in the previous report. A practical study of various ways of producing surfaces with high secondary electron coefficients has been described[121] as have the construction and performance of a compact high-gain multiplier.[122] Detector discrimination against positive[123a] and negative[123b] ions in SIMS has been investigated for a large number of elements. Ultra-sensitive

---

[113] A. F. Dillon, R. S. Lehrle, J. C. Robb, and D. W. Thomas, in ref. 35, p. 477.
[114] A. Benninghoven, *Surface Sci.*, 1975, **53**, 596.
[115] (a) A. Benninghoven, D. Jaspers, and W. Sichterman, in ref. 59, p. 1433; (b) A. Benninghoven and W. Sichterman, *Org. Mass Spectrom.*, 1977, **12**, 595.
[116] A. Benninghoven and W. K. Sichtermann, *Anal. Chem.*, 1978, **50**, 1180.
[117] R. J. Colton, J. S. Murday, J. R. Wyatt, and J. J. DeCorpo, *Surface Sci.*, 1979, **84**, 235.
[118] R. J. Day, S. E. Unger, and R. G. Cooks, *Anal. Chem.*, 1980, **52**, 353.
[119] R. J. Day, S. E. Unger, and R. G. Cooks, *J. Am. Chem. Soc.*, 1979, **101**, 499.
[120] J. A. Gardella, jun. and D. M. Hercules, *Anal. Chem.*, 1980, **52**, 226.
[121] D. L. Swingler, *Int. J. Mass Spectrom., Ion Phys.*, 1978, **27**, 359.
[122] D. L. Swingler, *Int. J. Mass Spectrom., Ion Phys.*, 1978, **27**, 367.
[123] (a) M. A. Rudat and G. H. Morrison, *Int. J. Mass Spectrom., Ion Phys.*, 1978, **27**, 249; (b) M. A. Rudat and G. H. Morrison, *ibid.*, 1979, **29**, 1.

detection over a wide mass range by ion counting and signal averaging using commercially available equipment has been described.[124] Substitution of the detector on an unmodified single-stage mass spectrometer[125] by an electro-optical system seems not to have been very successful as the mass range covered was limited, signal-to-noise was poor, and the dynamic range was only about one hundred to one. Although it is pointed out in the 'Field Ionization and Field Desorption' section of ref. 1 that ions up to m/z 11 000 have been detected using 'standard electron multipliers operated at standard dynode potentials' the detection efficiency is not quantified and this statement cannot be taken as a refutation of the arguments put forward in Section 7 of the previous report in which it was suggested that detection efficiency for an ion of m/z 2500 could only have been about 50%. Efficient detection and measurement of ions is important and new methods will have to be introduced, particularly as interest in negative ions grows and pulse-counting methods of measurement are beginning to be used[126] in organic mass spectrometry. Although negative ions can be measured by use of a negative ion/positive ion conversion dynode[127] a different type of detector looks more promising for precise quantitative measurements.[128] This comprises a conventional Allen-type multiplier which can be floated at different potentials depending on whether it is desired to detect positive or negative ions, but instead of the more usual capacitative decoupling of the anode[129] an electro-optical coupling is used so that normal measurement systems at ground potential can be employed.

Instead of a photodiode array[125] a row of metal strips interconnected by a resistor/capacitor network has been used[130] as the anode for a channel electron multiplier array (CEMA) detection system in an apparatus for photodissociation studies. This can give good spatial resolution of simultaneously detected ion beams but the time constants of the system and the method of image position determination limit the total spectrum count rate to about 10^4 s^{-1} so it would have a very limited application in organic mass spectrometry. A detailed discussion[131] on the use of CEMAs for simultaneous ion detection and measurement in spark source mass spectrometry, but which is also relevant to organic mass spectrometry, has been published. Background is the real limitation to mass spectrometer sensitivity. One source of this is ions scattered from other ions, neutrals, or spectrometer surfaces. Fortunately, these scattered ions will usually have different trajectories and/or translational energies from those which it is desired to measure and can be prevented from reaching the detector by the judicious use of baffles and energy selectors. Incorporating such devices into a reversed geometry double-focusing mass spectrometer[132] has reduced the background to

---

[124] W. V. Ligon, jun., in ref. 17, p. 481.
[125] B. Hedfjäll and R. Ryhage, *Anal. Chem.*, 1979, **51**, 1687.
[126] J. M. Hayes, D. E. Matthews, and D. A. Schoeller, *Anal. Chem.*, 1978, **50**, 25.
[127] D. F. Hunt and F. W. Crow, *Anal. Chem.*, 1978, **50**, 1781.
[128] L. A. Dietz and L. R. Hanrahan, *Rev. Sci. Instrum.*, 1978, **49**, 1250.
[129] See for example A. L. C. Smit and F. H. Field, *Anal. Chem.*, 1976, **48**, 2042.
[130] P. J. C. M. Nowak, H. H. Holsboer, W. Heubers, R. W. Wignaendts Van Resandt, and J. Los, *Int. J. Mass Spectrom., Ion Phys.*, 1980, **34**, 375.
[131] D. L. Donohue and J. A. Carter, *Int. J. Mass Spectrom., Ion Phys.*, 1980, **33**, 45.
[132] A. G. Brenton, J. H. Beynon, and R. P. Morgan, *Int. J. Mass Spectrom., Ion Phys.*, 1979, **31**, 51.

such a level that a limit of detection of about 1 ion s^{-1} has been achieved. A combination of two channeltron multipliers (CEM) in series[133] gives a detection system with a dynamic range of about 10^{10}. The first CEM is used as an analogue device for measuring large in-put ion currents. For small ion beams the out-put of the first CEM is further multiplied in the second CEM, ion current measurement in this case being done by pulse counting. This detector is very similar in concept to one described earlier[134] and mentioned in the previous report. A particle guide and integrating amplifier system[135] gives a detection limit of about 200 ions s^{-1} for a quadrupole mass spectrometer and also allows fast switching to a low-gain multiplier for total ion current measurements.

8 G.C.–M.S. and L.C.–M.S.

In an article entitled 'The Hyphenated Methods', various combined methods of separation and analysis were examined[136] and it was concluded from a statistical argument, perhaps not very surprisingly, that the combination is more powerful than the sum of its parts. Recognition of this is perhaps the most important contributory factor to the awakening of interest in mass spectrometry instrumentation amongst users from all disciplines referred to in the Introduction. The two most important 'hyphenated methods' in analytical mass spectrometry are the well established g.c.–m.s. and the newer but rapidly growing l.c.–m.s. techniques and these are comprehensively covered in Chapter 8.

An ingenious method[137] permits the automatic on-line adjustment of the quantity of individual components in complex mixtures for capillary column g.c.–m.s. Glass is superior to platinum for open-split g.c.–m.s. interfaces[138] and an all-glass solvent venting system has been designed.[139] Vacuum gas chromatography is more efficient at higher flow velocities than conventional g.c.[140] and short glass capillary columns are useful for fast analyses at low temperature, with the further advantage of not requiring separators or splitters for g.c.–m.s. Organic compounds can be separated by g.c. and converted into CO_2 or N_2 on-line to the mass spectrometer[141] for measurement of carbon or nitrogen isotope ratios with a precision of 0.5% or better. The problems associated with adsorption and sample decomposition in g.c.–m.s. and the methods developed to overcome these have been reviewed[142] in a publication describing the successful analysis of alkyl thiols and sulphides using a PTFE g.c.–m.s. system. Dense gas chromatography can be coupled to mass spectrometry using a supersonic molecular beam separator[143] and may offer a compromise between l.c.–m.s. and conventional g.c.–m.s. but additionally has some unique advantages.

---

[133] E. A. Kurz and R. L. Roy, in ref. 17, p. 479.
[134] J. J. Stoffels, C. R. Lagergren, and P. J. Hof, *Int. J. Mass Spectrom., Ion Phys.*, 1978, **28**, 159.
[135] G. Weiss, in ref. 4, p. 1624.
[136] T. Hirschfeld, *Anal. Chem.*, 1980, **52**, 297A.
[137] R. J. May and W. V. Ligon, jun., in ref. 17, p. 356.
[138] H.-J. Stan and B. Abraham, *Anal. Chem.*, 1978, **50**, 2161.
[139] D. L. Winmill and R. L. Williams, in ref. 17, p. 467.
[140] F. W. Hatch and M. E. Parrish, *Anal. Chem.*, 1978, **50**, 1465.
[141] D. E. Matthews and J. M. Hayes, *Anal. Chem.*, 1978, **50**, 1465.
[14] M. Thompson and M. Stanisavljevic, *Talanta*, 1980, **27**, 477.
[14] L. G. Randall and A. L. Wahrhaftig, in ref. 17, p. 451.

In a review[144] of development and trends in l.c.–m.s. technology it was concluded that only the moving belt and direct liquid injection (DLI) systems hold any promise for the future. The effects of interface dead volume and pumping requirements for the DLI method favour the use of a membrane pin-hole system with cryopumping.[145] Using this combination[146] solvent removal capability of the interface was increased by about a factor of ten compared with that achieved by diffusion pumps alone but scanning of a complete spectrum still required 50—500 ng of sample and about 2—20 ng was required for selected ion monitoring. Similar results have been obtained using a glass probe inlet on an unmodified CI quadrupole.[147] For the molecular beam interface with laser heating[33] for solvent evaporation, nanogram detectability is claimed. A new type of moving belt interface for use on either magnetic sector or quadrupole instruments is now available.[148] By inserting a modified segmented-flow extractor[149] between the column and a moving belt interface the solute can be extracted into a volatile organic solvent; this gives complete removal of salts and allows reversed phase chromatography to be done without difficulty. Sensitivity claimed for this system is in the picogram range. An alternative to the moving belt interface is a novel static wire concentrator with automatic flow control.[150]

9 Linked Scanning Methods

The various ways of adjustment or scanning of the energy and momentum defining parameters for ion transmission in double-focusing instruments in order to determine the pathways and energetics of selected ion decompositions were discussed in the previous report. Developed initially as research tools for mass spectrometry these methods when coupled with some means of stimulating fragmentation of the selected ions are beginning to show promise as routine means of compound identification or structure elucidation. Practical examples of this are given in a clear and concise account[151] of the reversed geometry or MIKES method. These linked scanning methods promise to be particularly useful for analysis of mixtures when CI[152] or FI/FD[153] are used as the means of primary ionization. Computer control of the spectrometer[154] gives a system fast enough even for g.c.–m.s. applications. Useful results can be obtained from slightly modified instruments constructed primarily as straightforward double-focusing mass spectrometers[155] but specially designed systems offer many advantages.[156]

[144] P. J. Arpine and G. Guichon, *Anal. Chem.*, 1979, **51**, 682A.
[145] P. J. Arpino and G. Guichon, in ref. 17, p. 366.
[146] (a) A. Melera, in ref. 17, p. 363; (b) A. Melera, in ref. 4, p. 1597.
[147] J. D. Henlon, *Anal. Chem.*, 1978, **50**, 1687.
[148] D. S. Millington, D. A. Yorke, and P. Burns, in ref. 4, p. 1819.
[149] B. L. Karger, D. P. Kirby, P. Vouros, R. L. Foltz, and B. Hidy, *Anal. Chem.*, 1979, **51**, 2324.
[150] R. G. Christensen, H. S. Hertz, S. Meiselman, and E. White, in ref. 17, p. 620.
[151] M. H. Bozorgzadeh, R. P. Morgan, and J. H. Beynon, *Analyst*, 1978, **103**, 613.
[152] G. L. Glish, V. M. Shaddock, K. Harmon, and R. G. Cooks, *Anal. Chem.*, 1980, **52**, 165.
[153] H. Kambara, F. C. Walls, R. McPherron, K. Straub, and A. L. Burlingame, in ref. 17, p. 184.
[154] (a) W. F. Haddon, *Anal. Chem.*, 1979, **51**, 983; (b) M. H. Bozorgzadeh, J. H. Beynon, R. P. Morgan, and A. G. Brenton, *Int. J. Mass Spectrom., Ion Phys.*, 1979, **29**, 191.
[155] (a) V. Vrscaj, V. Kramer, M. Medved, B. Kralj, J. Marsel, J. H. Beynon, and T. Ast, *Int. J. Mass Spectrom., Ion Phys.*, 1980, **33**, 409; (b) U. Herzig and P. Krenmayr, *Org. Mass Spectrom.*, 1979, **14**, 75.
[156] R. P. Morgan, J. H. Beynon, R. H. Bateman, and B. N. Green, *Int. J. Mass Spectrom., Ion Phys.*, 1978, **28**, 171.

Additional stages of analysis can be useful[157] and, for collisional activation experiments, high resolution in the selection of the primary ion beam and fragments by use of two double-focusing instruments in tandem is being developed.[158] Higher detection sensitivity for daughter ions and extremely fast analyses can be achieved[159] by using a specially designed instrument having high post-acceleration of the fragments and a simultaneous electro-optical detection system. Angle-resolved mass spectrometry[160] adds yet another dimension to the linked scanning methods. This is applicable to both conventional[161] and reversed geometry[162] instruments though the latter[162b] are easier to use in practice. Artifacts may arise in linked scanning methods either from purely instrumental factors[163] or from a combination of these and the ion chemistry of the sample being examined.[164] Methods of recognizing various artifact peaks are detailed in ref. 164c. Ambiguous results can be avoided by using more than one linked scanning mode[165] and artifacts can be suppressed by field modulation coupled with phase-sensitive detection.[166] Tandem quadrupole systems[167] may not have the energy or angular resolution capabilities of the large sector instruments but, when generally available, they must surely be cheaper. They have already been shown to be useful for chemical applications[168] and the low-energy collision-induced method has potential advantages[53,169] over the high-energy method used in sector instruments.

10 Miscellaneous

Electronic circuits have been described which give faster scan speeds[170] and better stability for multiple ion detection[171] on the AEI MS902. A rapid scan circuit[172] for the AEI MS30 allows monitoring of 4 peak positions at 13 000 resolving power, and 10 p.p.m. mass measurement accuracy at low resolving power for up to

---

[157] M. Medved, B. Kralj, J. Marsel, V. Kramer, T. Ast, and J. H. Beynon, *Org. Mass Spectrom.*, 1979, **14**, 307.
[158] F. W. McLafferty, *Acc. Chem. Res.*, 1980, **13**, 33.
[159] G. J. Louter, A. J. H. Boerboom, P. F. M. Stalmeier, H. H. Tuithof, and J. Kistemaker, *Int. J. Mass Spectrom., Ion Phys.*, 1980, **33**, 335.
[160] J. A. Laramée, J. J. Carmody, and R. G. Cooks, *Int. J. Mass Spectrom., Ion Phys.*, 1979, **31**, 333.
[161] J. A. Laramée, P. H. Hemberger, and R. G. Cooks, *Int. J. Mass Spectrom., Ion Phys.*, 1980, **33**, 231.
[162] (a) A. G. Brenton, C. J. Proctor, and J. H. Beynon, in ref. 4, p. 1689; (b) D. M. Fedor and R. G. Cooks, *Anal. Chem.*, 1980, **52**, 679.
[163] S. Howells, A. G. Brenton, J. H. Beynon, and R. P. Morgan, *Int. J. Mass Spectrom., Ion Phys.*, 1979, **32**, 35.
[164] (a) J. N. Bilton, N. Kyriakidis, and E. S. Waight, *Org. Mass Spectrom.*, 1978, **13**, 489; (b) B. Schaldach and H. F. Grützmacher, *Org. Mass Spectrom.*, 1980, **15**, 182; (c) T. Ast, M. H. Bozorgzadeh, J. L. Wiebers, J. H. Beynon, and A. G. Brenton, *Org. Mass Spectrom.*, 1979, **14**, 313.
[165] (a) M. J. Lacey and C. G. Macdonald, *Int. J. Mass Spectrom., Ion Phys.*, 1979, **30**, 359; (b) M. J. Lacey and C. G. Macdonald, *Org. Mass Spectrom.*, 1979, **14**, 465; (c) M. J. Lacey and C. G. Macdonald, *Org. Mass Spectrom.*, 1980, **15**, 135.
[166] B. Shushan and R. K. Boyd, *Int. J. Mass Spectrom., Ion Phys.*, 1980, **34**, 37.
[167] R. A. Yost and C. G. Enke, *Anal. Chem.*, 1979, **51**, 1251A.
[168] R. A. Yost and C. G. Enke, in ref. 17, p. 461.
[169] R. A. Yost, C. G. Enke, D. C. McGilvery, D. Smith, and J. D. Morrison, *Int. J. Mass Spectrom., Ion Phys.*, 1979, **30**, 127.
[170] V. P. Williams, A. Moore, C. Lichtenstein, and J. R. Kruse, *Int. J. Mass Spectrom., Ion Phys.*, 1978, **28**, 427.
[171] D. A. Durden, in ref. 17, p. 453.
[172] J. G. Solch, C. D. Miller, R. O. Yelton, M. L. Taylor, and T. O. Tiernan, in ref. 17, p. 545.

6 mass positions is achieved[173] by the use of circuitry developed for computerized multiple ion monitoring. Digital-type data smoothing in real time can be obtained[174] from an analogue delay device. New mass reference compounds, up to m/z 2000[175] and even m/z 10 000,[176] have been suggested.

The development of ICR in recent years has been described[177] and it is suggested that fast scanning or Fourier transform ICR will be the high-performance mass spectrometer of the future. Progress made in chemical applications of Fourier transform mass spectrometry in one laboratory[178] has been described and the advantages over conventional ICR, particularly for the study of ion/molecule chemistry, have been discussed and illustrated.[179]

Further examples of the methods and results obtained for polymer analysis by direct pyrolysis–m.s.[180] or pyrolysis–g.c.–ms.[181] have been given.

---

[173] R. J. Weinken and J. L. D'Angona, *Anal. Chem.*, 1979, **51**, 1074.
[174] D. L. Doerfler and I. M. Campbell, *Anal. Chem.*, 1978, **50**, 1018.
[175] W. V. Ligon, jun., *Anal. Chem.*, 1978, **50**, 1228.
[176] T. Matsuo, H. Matsuda, and I. Katakuse, *Anal. Chem.*, 1979, **51**, 1331.
[177] R. L. Hunter and R. T. McIver, jun., *Anal. Chem.*, 1979, **51**, 699.
[178] E. B. Ledford, jun., S. Ghaderi, C. L. Wilkins, and M. L. Gross, in ref. 4, p. 1707.
[179] M. B. Comisarow, in ref. 4, p. 1698.
[180] (a) W. E. Franklin, *Anal. Chem.*, 1979, **51**, 992; (b) N. Evans and J. E. Williamson, in ref. 4, p. 1023.
[181] J. L. Wuepper, *Anal. Chem.*, 1979, **51**, 997.

7
Applications of Computers and Microprocessors in Mass Spectrometry

BY R. D. SEDGWICK

1 Introduction

The format adopted in this report has been modified slightly from that previously used[1] in order to reflect the changed emphasis in interest during the period reviewed. As predicted in the last report in this series there has been a significant increase in the applications of microprocessors to most areas and in particular to automatic control of instrument functions. This is due mainly to their low cost when compared with minicomputer systems. The cost advantage is offset by difficulties associated with machine code programming, which is the only method available with many current microprocessors. Whilst the speed of operation of many instrument control and data acquisition functions demands the use of a low-level language, this is certainly not the case for data processing of mass spectral data. Thus, complex programs, associated with spectral manipulations and as aids to the interpretation of mass spectra, can only be written with facility in high-level languages which remain largely in the province of minicomputers. These have dominated the mass spectrometry scene for the past decade, a state of affairs likely to continue until microcomputer systems are introduced which are compatible with the operating systems and software of existing minicomputers. This step, which has already been taken by some computer manufacturers, must be rate determining in introducing cheaper hardware systems with the capability of exploiting existing software, with minimal modifications. The microprocessor has also been adopted as an adjunct to many computer peripherals thus increasing their power and versatility. The most striking of these is the production of 'intelligent' terminals and printers with powerful graphics functions which can add a new dimension to data presentation, either interacting with or operating independently of the host computer.

The microprocessor-based computer and the mini-computer differ only in physical size and cost and not in any other really significant way. Consequently the introduction of the microprocessor has not resulted in the introduction of any fundamentally new techniques or applications which were beyond the capabilities of previous devices. They have been responsible for the consolidation and refinement of many existing techniques and methods and the prospect of easier and more reproducible instrumental operation has been opened to a much wider audience.

---

[1] F. A. Mellon, in 'Mass Spectrometry', ed. R. A. W. Johnstone, (Specialist Periodical Reports) The Chemical Society, London, 1978, Vol. 5, p. 100.

It is appropriate that distinction is made between the minicomputer which has been the principal subject of previous reports in this series and the microprocessor as it presently exists.

A minicomputer consists of a central processor unit (CPU), which can execute instructions in a program which is stored in a memory unit. The instructions operate on data which are read from the memory, and results are written back into the memory as the program proceeds. In a typical installation the memory may be able to hold only 4000 words but may be extendable up to 128 000 words or more. Word lengths of 12 and 16 bits are the most common. The total system will also have input/output devices for communication purposes, *e.g.* keyboard, visual display unit (VDU), printer, *etc.*, and a backing store of magnetic disc or tape. The system is connected to a mass spectrometer through an interface which may use digital/analogue conversion for control and analogue/digital conversion for data acquisition operations.

A microcomputer is made up from a number of microcircuits (chips) usually mounted on a single printed circuit board. The most complex of these is the microprocessor unit (MPU) which can perform similar functions to a CPU. Two types of memory unit are used: (i) a random access memory (RAM), which is similar to that used in minicomputers but may be of much smaller capacity, containing only 500 8-bit words, and (ii) a programmable read-only memory (PROM), which can transmit but not receive information. A PROM is a low-cost device which may be used to store a short set of instructions to perform a well defined task and as such is suited to mass-produced digital control systems. A PROM cannot be reprogrammed in general, unless it is an erasable unit (EPROM) which is reprogrammable after a fairly drastic full erasure procedure outside the microcomputer.

Most systems use both RAM and ROM. Control systems would generally use PROM to store instrument control operations, the digital equivalents of turning dials and closing switches. For a flexible system the control sequences and the limits of operations such as scan times and rates would be under operator control and so would require some RAM for their storage and for assembling the control program.

For data acquisition and data processing where large amounts of data are transferred between devices, it is necessary to use RAM as the transfer device.

Program development is only possible using RAM since PROM cannot be edited. Furthermore because of the small storage capacity of many microprocessor-based systems, it is necessary to write all programs in a machine code. This may result in a penalty in program development effort which may not be offset by the low cost of the device compared with a minicomputer with good software support. Additionally both types of computer require expensive peripherals whose costs are a constant whichever system is used and may be a major proportion of any total system cost.

A full description of the practical problems associated with interfacing mini- and micro-computers to scientific instruments, including mass spectrometers, has appeared in a book[2] covering both control and measurement applications.

---

[2] A. Carrick, 'Computers and Instrumentation', Heyden and Son Ltd., London, 1979.

A timely and useful book which brings together all the applications of computers and computer techniques in mass spectrometry has also been produced.[3] A procedural manual with the objective of standardizing gas chromatograph–mass spectrometer (g.c.–m.s.) analysis in the environmental field has limited appeal since it is written around the operations of a specific computerized data system.[4] A general review of all aspects of computerized mass spectrometry has appeared[5] while critical reviews have covered the methods available for the classification of mass spectra by pattern recognition techniques[6] and of mass spectral library search algorithms.[7] An evaluation of chemical information systems in which mass spectra are combined with other types of spectroscopic data[8] has been used in the development of a specific system for use with a minicomputer.[9]

The role of the computer in mass spectral methods of environmental analysis has been reviewed,[10] as well as computer techniques for mass spectral identification.[11]

There have been reports on developments in systems for computer-assisted mass spectral interpretation which are available over computer networks, namely Probability Based Matching (PBM),[12–14] Self-Training Interpretative and Retrieval System (STIRS),[12–14] Artificial Intelligence (CONGEN),[15] and Mass Spectral Search System (MSSS).[16] The operators of the last system, MSDC-NIH-EPA, have announced that both PBM and STIRS will become available alongside MSSS.

2 Computer Control

Microprocessor Control.—Two remotely controlled and therefore completely automated mass spectrometers have been described which serve to emphasize the remarkable reduction in size which has been achieved in the production of micro-electronic circuitry. The first was a balloon-borne mass spectrometer used for the automatic sampling of ambient positive and negative ions in the stratosphere.[17] The spectrometer was developed from a commercial quadrupole mass

---

[3] J. R. Chapman, 'Computers in Mass Spectrometry', Academic Press, London, 1978.
[4] W. L. Budde and J. W. Eichelberger, 'Organic Analysis using Gas Chromatography–Mass Spectrometry', Ann Arbor Sci. Publ. Inc., Michigan, 1979.
[5] J. R. Chapman, *J. Phys. E*, 1980, **13**, 365.
[6] J. R. McGill and B. R. Kowalski, *J. Chem. Inf. Comput. Sci.*, 1978, **18**, 52.
[7] G. T. Rasmussen and T. L. Isenhour, *J. Chem. Inf. Comput. Sci.*, 1979, **19**, 179.
[8] J. Zupan, *Anal. Chim. Acta*, 1978, **103**, 273.
[9] J. Zupan, M. Penca, M. Razinger, B. Barlic, and D. Hadzi, *Anal. Lett.*, 1979, **12**, 109.
[10] L. H. Keith, *J. Chromatogr. Sci.*, 1979, **17**, 48.
[11] F. W. McLafferty and R. Venkataraghavan, *J. Chromatogr. Sci.*, 1979, **17**, 24.
[12] R. Venkataraghavan, H. E. Dayriger, B. L. Atwater, G. M. Pesyna, and F. W. McLafferty, in 'Advances in Mass Spectrometry', ed. N. R. Daly, Heyden and Son Ltd., London, 1978, Vol. 7, p. 989.
[13] F. W. McLafferty, *Pure Appl. Chem.*, 1978, **50**, 197.
[14] F. W. McLafferty and R. Venkataraghavan, 'High Performance Mass Spectrometry: Chemical Applications' (American Chemical Society Symposium Series, Series 70), American Chemical Society, Washington, 1978, p. 310.
[15] D. H. Smith and R. E. Cathcart, in ref. 14, p. 325.
[16] S. R. Heller, R. S. Heller, A. McCormack, D. C. Maxwell, and G. W. A. Milne, in ref. 12, p. 985.
[17] J. R. Olson, R. C. Amme, J. N. Brooks, D. G. Murcray, D. A. Steffen, R. E. Sturm, and G. E. Keller, *Rev. Sci. Instrum.*, 1978, **49**, 643.

filter adapted to respond to on-board command electronics. Mass spectra with a mass range from 1—150 a.m.u. were produced using an ion-counting detector interfaced to a microcomputer having 500 words of RAM. This used processing and encoding instructions stored in a PROM, to produce digitally coded mass spectra that were stored on magnetic tape prior to transmission to a ground station.

The second automated g.c.–m.s. system was used in the Viking program which made two successful landings on the planet Mars.[18] The objective was to determine the atmospheric composition of the planet and to detect organic materials which could be vaporized or pyrolysed from its soil. The mass spectrometer was a small double-focusing machine of the Nier–Johnson type. This was fitted with an atmospheric gas sampling and inlet system and also interfaced to a temperature-programmable gas chromatograph with a palladium alloy molecular separator. Soil samples were collected, pulverized, and then pyrolysed at controlled temperatures to produce samples for analysis on the g.c.–m.s. The complete system was controlled by commands transmitted from earth and stored in the Viking lander computer before being transferred to the g.c.–m.s. computer for execution. The computer had two distinct functions, *viz.* command processing and control of operation sequences, and data acquisition and output control logic. Individual control instructions and timing sequences were stored in 512-word (12-bit word) PROM units which were prepared before the launch. Thus a variety of experimental operations and sequences could be initiated through a short command that selected the required PROM units. Control of the mass spectrometer was simplified by using a permanent magnet and electrostatic scanning over a mass range of 12—215 a.m.u. Mass spectra acquired by the g.c.–m.s. computer were transferred to the lander computer and were stored on magnetic tape ready for transmission to earth daily. The instruments on both landings performed for the full duration of the mission without serious malfunction.

Microprocessor control has been applied most widely to automation of quadrupole mass spectrometers which lend themselves more easily to precise electronic control of resolving power, focusing, and scanning operations. Two separate microprocessors have been used on a single quadrupole instrument, one being dedicated to instrument control and the second to data acquisition.[19]

The relative merits of three commercially available automated mass spectrometers have been evaluated with reference to their applications in clinical chemistry.[20] All are g.c.–m.s. instruments and two use microprocessor-controlled gas chromatographs linked to quadrupole mass spectrometers with different degrees of microcomputer control. Only one of these systems uses the microprocessor for data acquisition and spectral analysis.[21] The others adopt the general philosophy that a combination of micro- and mini-computers provides the best

[18] D. R. Rushnek, A. V. Diaz, D. W. Howarth, J. Rampacek, K. W. Olson, W. D. Dencker, P. Smith, L. McDavid, A. Tomassian, M. Harris, K. Bulota, K. Biemann, A. L. LaFleur, J. E. Biller, and T. Owen, *Rev. Sci. Instrum.*, 1978, **49**, 817.
[19] R. M. Lum, '30th Pittsburgh Conference on Analytical Chemistry and Applied Spectroscopy', Cleveland, 1979, p. 674.
[20] N. Gochman, L. J. Bowie, and D. N. Bailey, *Anal. Chem.*, 1979, **51**, 525.
[21] E. M. Chait and P. A. Strauss, *Int. Lab.*, 1979, **9**, 71.

answer to the problems of extending the scope of mass spectrometer control systems whilst preserving the complex and extensive software which already exists for data processing. A similar approach is apparent in further development of a commercially produced, fully automated g.c.–m.s. system for organic analysis which also uses a combination of micro- and mini-computers.[22]

A fully automated mass spectrometer that can process up to thirty organic solids using an automatic direct inlet probe[23] uses a similar approach to those above. A process control unit determines the sample evaporation rate and operation sequences, including sample change-over and spectrometer scan control, while data acquisition is left to a conventional minicomputer.

The advantages of using a microcomputer to upgrade the performance of an existing fully automated mass spectrometer for gas analysis have been assessed and the associated hardware[24] and software[25] described. The analysis of gases in a nuclear reactor has been described for a situation requiring a remotely controlled mass spectrometer.[26] The control system was based on a programmable calculator to control the mass spectrometer and its inlet system.

A fully digitally controlled quadrupole mass spectrometer for use with an atmospheric pressure chemical ionization source has been described.[27] It is claimed to be capable of detecting trace compounds present only in femtogram quantities and can be used to monitor atmospheric pollutants.

The ease of operation afforded by microprocessor control of a low-resolution quadrupole mass spectrometer has been exploited to the full in the design of residual gas analysers[28,29] and respiratory gas monitors.[30,31]

In addition to full instrument control, the microprocessor has been used in the development of a number of accessories for mass spectrometers.

A full description has appeared of the construction and programming of a low-cost microcomputer electric current programmer for control of a direct insertion probe for use in an electron impact ion source but having applications in field desorption. The current program steps can be advantageously synchronized with the spectrometer scan.[32] Programmed temperature control is also a feature of a high-temperature Knudsen cell effusion source that has been operated with a computer-controlled quadrupole mass spectrometer and data acquisition system to produce quantitative mass pyrograms.[33]

A mass spectrometer data system has been described which features a wide dynamic range (250 000:1) due to the introduction of a microprocessor-based preprocessor.[34]

[22] R. G. Beimer, E. M. Chait, C. W. Hull, and P. Burroughs, in ref. 12, p. 892.
[23] H. Hillig, H. Kuper, W. Riepe, and H. P. Ritter, *Anal. Chim. Acta*, 1979, **112**, 123.
[24] V. Barton, R. Downey, and C. Pomernacki, in ref. 19, p. 672.
[25] R. G. Bedford, R. W. Crawford, H. R. Brand, V. Barton, and R. Downey, in ref. 19, p. 673.
[26] L. F. Herzog, in ref. 19, p. 88.
[27] N. M. Reid, J. A. Buckley, J. B. French, and C. Poon, in ref. 19, p. 231.
[28] F. W. Karasek, *Ind. Res. Dev.*, 1978, **20**, 123.
[29] D. H. Holkeboer, W. Parfitt, F. H. Schlereth, and P. F. McGinnis, *J. Vac. Sci. Technol.*, 1978, **15**, 787.
[30] I. Nishi, G. Tomizawa, and L. Tatsuta, *Shitsuryo Bunseki*, 1978, **26**, 67.
[31] I. Nishi, G. Tomizawa, S. Sugai, and K. Sasaki, in ref. 12, p. 1103.
[32] R. J. Cotter, *Biomed. Mass Spectrom.*, 1979, **6**, 508.
[33] N. W. K. Liu and D. W. Muenow, *High Temp. Sci.*, 1978, **10**, 145.
[34] E. Muller and U. Krienen, *Fresenius Z. Anal. Chem.*, 1979, **294**, 241.

Minicomputer Control.—The modular nature of a computer-aided measurement and control (CAMAC) system, designed to operate with a minicomputer, has been exploited in the design of a time-of-arrival spectrometer[35] that can be operated automatically. The apparatus was used to measure the mobilities of ions through a drift space followed by mass analysis using a quadrupole mass spectrometer.

A modular approach has also been adopted in the design of the digital control for the close-loop operation of a commercial quadrupole mass spectrometer/minicomputer system for organic analysis.[36,37] The same type of instrument has been integrated into the control system of a steel-making plant where compositions of waste gases were monitored.[38]

A system for controlling a quadrupole mass spectrometer from a remote computer system has been described.[39] The applications of the g.c.–m.s. system orientated to research are enhanced by comprehensive data manipulation programs and display facilities, available on a central computing facility.

Several mass spectrometers have been described which have conventional data acquisition systems based on the minicomputers but have extended their range of operations by direct digital control of major instrumental features such as rapid switching between electron impact and chemical ionization modes[40,41] and temperature control of a solid-sample evaporation probe.[42] The automatic optimization of the resolving power of an ultra-high-resolution mass spectrometer has been achieved at minimal cost using stepping motors acting under control from a modified commercial data system.[43]

Further development of a minicomputer-controlled secondary ion mass spectrometer has extended its use to allow recording of the kinetic energy spectra of sputtered ions.[44] After mass analysis in a magnetic system the secondary ions are energy analysed in a scanning spherical electrostatic analyser operated under digital control. The system has shown that markedly different energy profiles correspond with different fragment ions sputtered from the same sample.

Minicomputer systems continue to be used to analyse high-resolution mass measurement data from scanning instruments. The rate of computerized data acquisition is one of the design factors that determine accuracy of measurement and scan speed. These factors have been discussed in relation to the operation of a high-resolution gas chromatograph coupled to a high-resolution mass spectrometer.[45]

A minicomputer has been used to control a stepped-voltage scan of a mass spectrometer over a limited mass range. Analyses of the spectra using a conventional data acquisition system have been used to produce useful accurate mass

---

[35] D. Verejaus, E. Arijis, and J. Ingels, *J. Phys. E*, 1978, **11**, 955.
[36] R. Fluckiger and M. Schalcher, in ref. 12, p. 1034.
[37] P. Nageli and H. P. Egli, in ref. 12, p. 1713.
[38] G. G. Cameresi and B. Costa, in ref. 12, p. 1062.
[39] J. E. Campana, T. H. Risby, and P. C. Jurs, *Anal. Chim. Acta*, 1979, **112**, 321.
[40] K. T. Taylor, C. J. Wakefield, J. R. Chapman, and K. R. Compson, in ref. 19, p. 82.
[41] D. R. Denne, A. Taylor, L. Rutherford, and K. R. Compson, in ref. 19, p. 89.
[42] E. M. Chait, in ref. 19, p. 87.
[43] S. L. Kaberline and C. L. Wilkins, in ref. 19, p. 15.
[44] M. A. Rudat and G. H. Morrison, *Anal. Chim. Acta*, 1979, **112**, 1.
[45] J. Meili, F. C. Walls, R. McPherron, and A. L. Burlingame, *J. Chromatogr. Sci.*, 1979, **17**, 29.

data at resolving powers of less than 2000.[46] An alternative system,[47] which is capable of scanning over a wider mass range, uses magnetic field sensing to give improved reproducibility of the scan law for a magnetic scan. Additionally, the computer-acquired magnetic field values can be related to a scale of accurate masses using suitable software. Mass measurements on molecular ions bracketed by two widely spaced reference peaks on the mass scale have been used to calculate plausible atomic compositions for g.c.-separated peaks containing only nanograms of material.

A novel design for a control system to scan a quadrupole mass spectrometer uses a combination of a step voltage with a saw-tooth voltage of the same frequency to produce the linear ramp used in normal scanning of the mass scale.[48] Each of the steps corresponds to one mass number so that equal times are used for ion collection at each mass. By digital selection of the step numbers, it is possible to switch between mass numbers, in either direction. More impressively, the step number may be placed under program control using a minicomputer. In this case, the mass scale can be scanned in any order, *i.e.* the scale could consist of ion collections at m/z 18, 54, 55, 56, 18, 44, 43, 42, 15, 64, *etc*.

An automated pyrolysis mass spectrometer has been developed in which a minicomputer controls a solid sample selector, a Curie point pyrolyser, and a quadrupole mass spectrometer, in addition to performing the usual data acquisition functions.[49] The reproducible nature of the operations produces mass spectra which are 'typical' of such complex samples as micro-organisms. This point has been further emphasized by examining the cluster behaviour in hyperspace (see the later section on pattern recognition in this chapter) of the mass spectra. This showed interesting and close relationships between the mass spectrum and the strain of bacterium from which it came. A related approach to the same problem used a double-focusing mass spectrometer coupled to a gas chromatograph.[50] The system showed that retention times, peak intensities, and peak identities can be used as alternative methods of demonstrating that normal human tissues exhibit common properties.

Multiple Instrument Systems.—A minicomputer has been used for instrument control and data acquisition in a laboratory devoted to surface analysis where there are seven surface-science instruments, including a quadrupole mass spectrometer, under computer control.[51] The digital control circuits of a quadrupole mass spectrometer with applications in medical research have been described.[52]

A further description has appeared of a time-sharing computer to which nineteen laboratory instruments were interfaced. These included three magnetic mass spectrometers that could be operated simultaneously. The laboratory

---

[46] R. J. Weinkam and J. L. D'Angona, *Anal. Chem.*, 1979, **51**, 1074.
[47] P. Powers, P. H. D'Arcy, J. C. Bill, and M. J. Wallington, in 'Proceedings of the 26th Annual Conference on Mass Spectrometry and Allied Topics', St. Louis, 1978, p. 480.
[48] K. Maeda, I. H. Suzuki, and Y. Koyama, in ref. 12, p. 898.
[49] H. L. C. Meuzelaar, P. G. Kistemaker, W. Eshuis, and A. J. H. Boerboom, in ref. 12, p. 1452.
[50] E. Reiner, L. E. Abbey, T. F. Morgan, P. Papamichalis, and R. W. Schafer, *Biomed. Mass Spectrom.*, 1979, **6**, 491.
[51.] S. H. McFarlane, in ref. 19, p. 446.
[52] S. Marlowe, in 'Medical and Biological Applications of Mass Spectrometry', ed. J. P. Payne, J. A. Blackman, and D. W. Hill, Academic Press, London, 1979, p. 21.

computer is connected by a high-speed digital interface to a larger system capable of manipulating large sets of spectra including library searching.[53]

Two mass spectrometers have been interfaced to a minicomputer by means of its memory management facility.[54] Full control and data acquisition from a quadrupole mass spectrometer were handled as background operations whilst data acquisition from a double-focusing magnetic mass spectrometer was carried out in the foreground.

The complex procedures needed for operation of three quadrupole mass filters in tandem have been achieved using a minicomputer for both control and data acquisition.[55] A tunable dye laser was scanned under program control in 1400 steps across its photon energy range. Several hundred flashes of the laser were directed along the axis of the triple quadrupole analyser which was used to detect the photodissociation of selected ions.

Fourier Transform Ion Cyclotron Resonance.—The power and potential of the recently developed, computer-dependent technique of Fourier Transform Ion Cyclotron Resonance (FT–ICR) spectroscopy have been further demonstrated. Examples of mass spectra of compounds with molecular weights greater than 1000 have been reported as well as the ability to achieve ultra-high resolving power. A mass resolving power of 250 000 has been obtained using the $N_2:C_2H_4$ doublet without the need for any increase in magnetic field homogeneity over that used in conventional ICR.[56]

The factors affecting the accuracy of mass measurements in Fourier transform mass spectrometry have been discussed,[57] and a Fourier transform mass spectrometer designed to make rapid measurements on electron impact and chemical ionization mass spectra at low pressures has been described.[58]

A digitally based control system for a pulsed ion cyclotron resonance (ICR) mass spectrometer has been used as an integral part of a microprocessor-based Fourier transform ICR system.[59]

3 Data Acquisition and Processing

Pulse Counting.—Interest has continued in the use of pulse counting methods and the technique has been applied in a peak matching system suitable for use at low signal levels[60] which must be less than 10^5 ions s^{-1}. Peak profiles were accumulated in a multichannel analyser. The system has been used to produce accurate mass measurements during elution times for small peaks from an on-line gas chromatograph.

---

[53] J. F. Ziersel, M. D. Kenny, J. B. Aldrich, L. Baczynaskyj, D. J. Duchamp, and J. E. Herman, in ref. 12, p. 1038.
[54] P. D. Bolton and G. W. Trott, *Int. J. Mass Spectrom. Ion Phys.*, 1978, **27**, 419.
[55] D. C. McGilveny and J. D. Morrison, *Int. J. Mass Spectrom. Ion Phys.*, 1978, **28**, 81.
[56] M. Comisarow, in ref. 12, p. 1042.
[57] E. B. Ledford, jun., R. L. White, P. S. Kulkarni, R. B. Spencer, S. Ghaderi, C. L. Wilkins, and M. L. Gross, in ref. 19, p. 117.
[58] S. Ghaderi, P. S. Kulkarni, M. L. Gross, C. L. Wilkins, R. B. Spencer, and E. B. Ledford, jun., in ref. 19, p. 118.
[59] R. J. Dugan, L. N. Morgenthaler, R. O. Daubach, and J. R. Eyler, *Rev. Sci. Instrum.*, 1979, **50**, 691.
[60] J. Freudenthal and L. G. Gramberg, *Anal. Chem.*, 1977, **49**, 2205.

An ion counting system has been developed for the detection of low-intensity peaks in negative ion quadrupole mass spectrometry.[61] A multichannel analyser was again used to accumulate multiple, digitally controlled mass scans. A similar system, aimed at improving the signal-to-noise ratio, has been used to overcome some of the problems associated with the unstable and inherently weak signals produced by a field desorption ion source.[62]

Focal Plane Detection.—Several systems of focal plane detector are being developed as an alternative to the photoplates currently used in mass spectrographs. The attributes of focal plane detectors have long been recognized for their advantageous integrating properties when unstable systems such as spark sources are used. Similar difficulties, coupled with low sensitivity, are encountered in field desorption experiments. However, some of the sensitivity gain associated with total ion collection is lost when photoplates are used because they lack the single ion sensitivity of the electron multiplier detector. The development of multichannel array multiplier detectors has stimulated attempts to develop what may be considered the electronic equivalent of the photoplate as a mass spectrographic detector.

One arrangement uses the multichannel array multiplier, which is a collection of microscopic electron multipliers, to detect the ion beam. The electron beam output from the array is used to activate a phosphor which in turn is viewed through a window by a photodiode array that can be outside the vacuum system.[63] The diode array thus reacts in a geometrical pattern which represents the ion image at the detector. The diode array can be interfaced to a computer and provides a rapid and direct link between the focal plane detector and the mass spectrometer data system. The system is equivalent to photoplate detection but with lower dynamic range and lower resolution, since the latter is determined by the dimensions of the array multiplier channels and the dispersion of the mass spectrograph. The system does have single ion detection capability and the dynamic range can be increased by accumulating multiple scans of the diode array into the computer and performing signal averaging.

An alternative system[64] employs a degree of magnetic focusing of the array multiplier output onto the phosphorescent screen. The optical signal is then channeled using fibre optics in an ordered array to a VIDECON television scan system, the electrical output of which is interfaced to a computer data acquisition system. Overall the system has a gain in mass resolving power over other similar systems.[65]

As an alternative to photoplate recording, a minicomputer system has been developed to acquire low-resolution mass spectra from a scanning mass spectrometer.[66] A multichannel analyser was used to acquire data from repetitive low-resolution scans and then computer analysis was used for signal smoothing and integration to give quantitative analyses.

---

[61] T. Fujii, *Anal. Chim. Acta*, 1979, **104**, 167.
[62] W. D. Lehmann and H. R. Schulten, *Biomed. Mass Spectrom.*, 1978, **5**, 208.
[63] B. Hedfjall and R. Ryhage, *Anal. Chem.*, 1979, **51**, 1687.
[64] H. G. Boettger, C. E. Giffin, and D. D. Norris, 'Multichannel Image Detection' (American Chemical Society Symposium Series, Series 102), American Chemical Society, Washington, 1979, p. 291.
[65] H. H. Tuithof, A. J. H. Boerboom, P. G. Kistemaker, and H. L. C. Meuzelaar, in ref. 12, p. 838.
[66] F. De Ceuninck, E. Van Hoye, and F. Adams, *Proc. Anal. Div. Chem. Soc.*, 1978, **15**, 104.

A theory describing the gain of quantitative data from spark source mass spectra by automatic photoplate evaluation, without the use of calibration standards, has been developed further.[67] A parabolic relationship was deduced between ion intensities and ionization state of each element, modified by effects of the matrix. Useful quantitative results can be obtained for conducting samples containing elements for which three ionization states can be observed.

Reference Materials.—The lack of mass calibration compounds for negative ion electron-impact mass spectrometry is of interest. While positive ion calibration data may work after instrument polarity reversal at low resolving power, the need for an internal mass standard for computerized scanning and peak matching at high resolving powers is evident. The use of polymeric phosphonitrile chloride (1), where $n = 9$, has been shown to give useful negative ion mass spectra for these purposes in the mass range 310—1035.[68]

$$\left[\begin{array}{c} Cl \\ | \\ P=N \\ | \\ Cl \end{array} \right]_n$$

(1)

A new high molecular weight fluorine-containing oil has been proposed as an additional wide-range mass calibration standard and has been used for positive ion electron-impact work.[69]

Selected Ion Monitoring.—Detection limits in single ion monitoring experiments at low resolving powers are determined by background levels arising from the sample and the mass spectrometer itself. Often the situation can be improved by utilizing the increased selectivity afforded by operation at high resolving power. It has been claimed that the decrease in absolute sensitivity, which is the penalty incurred by operating at high resolving power, is more than compensated for by the gain in sample-to-background signal ratio.[70]

In a study of the methods of determining the levels of insecticides in fish, a comparison has been made between single ion monitoring methods, including the use of reconstructed data from g.c.–m.s. runs, and a number of non-mass spectrometric methods of analysis. The preferred method is single ion monitoring although this is shown to be reliable only when the molecular ion is used.[71]

In a study of the accuracy of conducting computerized single ion monitoring experiments to detect compounds at the 100 pg level in g.c.–m.s. experiments, it was found that significant variations could arise from the methods used to calculate peak areas.[72] The technique of single ion monitoring at high resolution during g.c.–m.s. operations has been used to good effect in the analysis of pollutants in air samples.[73]

---

[67] L. Radermacher and H. E. Beske, *Spectrochim. Acta*, 1979, **34b**, 105.
[68] Y. Hirata, K. Matsumoto, and T. Tabenchi, *Org. Mass Spectrom.*, 1978, **13**, 111.
[69] G. A. Warburton, R. A. McDowell, and J. R. Chapman, Data Sheet No. 113, Kratos Ltd., Manchester, 1979.
[70] T. A. Gough, K. S. Webb, A. Carrick, and D. Hazelby, *Anal. Chem.*, 1979, **51**, 989.
[71] J. L. Laseter, *Anal. Chem.*, 1978, **50**, 1169.
[72] V. DaGragnano and J. H. Harris, in ref. 12, p. 1069.
[73] D. Schuetzle, T. J. Prater, T. M. Harvey, and D. E. Hagge, in ref. 12, p. 1082.

A system to monitor the dynamic performance of multiple ion monitoring systems is suitable for both single- and double-focusing magnetic mass spectrometers. It uses a programmable power supply acting under computer control to effect peak-switching operations. Attention was drawn to the sources of error in this type of experiment.[74]

Isotopic Analyses.—The value of the reproducibility which results from the computer control of all instrumental parameters is stressed in the description of a thermal ionization mass spectrometer for the isotopic analysis of solids.[75] In addition to control of the instrument electronics and data acquisition system, there is also an automatically operated turret magazine for loading samples into the ion source, making the whole operation automatic. Similar conclusions were reached using a system developed for the isotopic analysis of nuclear fuels.[76]

A statistical analysis program has been developed for the computerized evaluation of isotope ratio measurements for solids produced by a thermal ionization, minicomputer system.[77]

A control system for a magnetic sector mass spectrometer has been developed for the isotopic analysis of carbon dioxide in breath.[78] The breath samples need to be purified before analysis in the single collector isotope mass spectrometer which was operated in a peak switching mode. This mode of operation limits the precision of the measurements to 0.3%. The gas purifier, dual gas inlets, peak switching, and data acquisition were all controlled with a microcomputer having 12K of PROM, 8K of RAM, and additional cassette tape storage facilities.

A re-evaluation of the statistical methods used for calculating high-precision isotopic enrichments for double collector mass spectrometers has indicated a method by which the run length can be related to the desired confidence interval of the measurements.[79]

An algorithm for the interpolation of data points between repetitive scans to improve the analysis of accurate isotope ratio measurements has been described.[80] A linear method with automatic correction for second differences is used which is equivalent to a second-order polynomial fit.

A g.c.–m.s. method with computerized data acquisition and analysis has been developed for protein amino-acid analysis using an isotope dilution technique for quantification.[81] A re-evaluation of the calculation procedures used in this type of experiment has been reported.[82]

G.C.–M.S.—Modified algorithms have been developed for the extraction of significant mass spectra from series of spectra obtained in g.c.–m.s. or repetitive

---

[74] D. E. Matthews, K. B. Denson, and J. M. Hayes, *Anal. Chem.*, 1978, **50**, 681.
[75] C. Brunnee, G. Kappus, H. Rache, E. U. Seiler, and B. Windel, *Int. Lab.*, 1978, **8**, 89.
[76] L. Koch, B. Brandalise, G. Rijkeboer, M. Romkowski, M. Wilhelmi, K. Brachmann, and G. Heinen, in ref. 12, p. 1052.
[77] J. G. Van Raaphorst, J. E. Ordelman, and A. Brockman, *Int. J. Mass Spectrom. Ion Phys.*, 1979, **31**, 65.
[78] D. A. Schoeller and P. D. Klein, *Biomed. Mass Spectrom.*, 1979, **6**, 350.
[79] K. W. Steele, M. J. Grinsted, A. T. Wilson, and C. B. Dyson, *Int. J. Mass Spectrom. Ion Phys.*, 1978, **28**, 313.
[80] M. H. Dodson, *J. Phys. E*, 1978, **11**, 296.
[81] J. J. Rafter, M. Ingelman-Sundberg, and J. A. Gustafsson, *Biomed. Mass Spectrom.*, 1979, **6**, 31.
[82] B. N. Colby and M. W. McCaman, *Biomed. Mass Spectrom.*, 1979, **6**, 225.

scanning during solid sample evaporation[83] and have been combined with a fast on-line reverse library search programme which can operate in real time.[84]

The accuracy of retention times for peaks in gas chromatographs derived from computer-acquired mass spectral data has been shown to depend significantly on the mass spectrometer's scan time.[85a] To avoid degrading the gas chromatographic data while maximizing the mass spectral information, a system of scanning a preselected set of ions is recommended. An automatic calculation of Kovat's retention indices during g.c.-m.s. has been described.[85b]

Ionization Efficiency.—A fully automatic system for the collection of ionization efficiency data from a conventional electron impact ion source has been used to determine ionization energies for some small organic molecules.[86] A saw-tooth voltage at a frequency of 50 Hz was used to scan the energy scale, and computer averaging of multiple scans gave data with a good signal-to-noise ratio. The data were processed by the double EDD method.[87] Ionization and appearance potentials were reproducible to ±0.05 eV and were slightly higher than comparative photoionization values.

A similar inverse convolution method of calculation has been adopted for the analysis of experimental data from studies of the ionization of zinc. The ionization efficiency curves were obtained using a computer-controlled instrument with digital control of the electron beam energy and an ion counting detection system.[88] After computer processing the results showed evidence for several electronic states.

Metastable Ion Spectra.—A system for the three-dimensional graphical display of the sequential transitions which give rise to metastable ions in the first field-free region of a double-focusing magnetic mass spectrometer has been illustrated.[89] The data are presented using two parameters, ρ and μ. For an instrument in which ions are accelerated by a voltage V, pass through an electric sector at a voltage E, and are analysed in a magnetic sector which transmits ions of mass m_B, when $E = E_0$ and $V = V_0$, then

and

$$\rho = \frac{V_0}{V} \cdot \frac{E}{E_0}$$

$$\mu = \frac{E_0}{E} \cdot m_B$$

Three-dimensional plots of transmitted ion intensity are plotted on the $\mu:\rho$ plane. The interpretation of data from metastable scans from toluene and chlorophenol were used to illustrate the applications of the method. More

---

[83] D. Henneberg, H. Damen, and B. Weimann, in ref. 12, p. 975.
[84] P. Powers, M. J. Wallington, and J. A. V. Hopkinson, in ref. 12, p. 1029.
[85] (a) R. C. Murphy, S. E. Hattox, and H. R. Helbig, *Biomed. Mass Spectrom.*, 1978, **5**, 444; (b) K. R. Betty and F. W. Karasek, *J. Chromatogr.*, 1978, **166**, 111.
[86] R. A. W. Johnstone and B. N. McMaster, in ref. 12, p. 1047.
[87] J. Vogt and C. Pascuel, *Int. J. Mass Spectrom. Ion Phys.*, 1972, **9**, 441.
[88] A. Hashizume and N. Wasada, *Jpn. J. Appl. Phys.*, 1979, **18**, 429.
[89] M. J. Lacey and C. G. Macdonald, *Org. Mass Spectrom.*, 1978, **13**, 284.

conventionally presented metastable scans can be extracted from relevant sections of the plot.

A linked-scan system, in which a constant ratio is maintained between the magnetic field and the electrostatic analyser voltages (B/E scan) in a double-focusing mass spectrometer, has been described.[90] Daughter ions, originating from a given precursor ion, are selectively recorded at good resolution and signal-to-noise ratio. Sensing of the magnetic field is used to generate the control necessary to ensure that the electrostatic analyser follows the magnet scan.

A further B/E-scan study has used a system which dispenses with direct measurement of the magnetic field to produce the linked scan.[91] A data system is used to calculate a table of effective values for the magnetic field at times during a normal magnetic scan using conventionally acquired mass:time files. The same magnetic scan law is subsequently used for the metastable scan but with the electrostatic analyser being stepped under program control to maintain a constant ratio for B/E and assuming no change in the calculated magnet scan law. Reproducibility is claimed to be at least as good as systems which use field probes to monitor the magnetic field.

Computer-controlled scanning of the ion-accelerating voltage and averaging of repetitive scans in the same minicomputer have been used for acquiring accurate peak profile data for metastable peaks arising in the first field-free region of a double-focusing mass spectrometer.[92] A different objective, namely the recording of all the metastable peaks arising from one compound, has been achieved by rapid scanning of the electric sector voltage and simultaneous slow scanning of the magnetic field.[93] The electric sector was scanned by the application of a linear ramp voltage which was sampled alternately with the spectrometer detector signal by a multiplexer interposed before a single channel input data acquisition system. The combination of a fast electric sector scan superimposed on a slow magnetic scan produces data for both normal and metastable peaks which the system recognizes to produce comprehensive lists of detected transitions.

The technique of mass analysed ion kinetic energy spectrometry (MIKES) has been developed using double-focusing mass spectrometers of reversed geometry. The development of a computer-control system for such an instrument has been described.[94] In a typical application, a particular species is first mass analysed by the magnetic sector which is accurately field locked. The electric sector, which is the second element in the ion–optical system, is then scanned under computer control and the ion signal at the detector, the IKES spectrum, is acquired by the same minicomputer. The electric sector scan is under digital control and energy steps as small as 0.1 eV may be used. The scan rates of the system are fast enough to produce MIKES spectra from an on-line gas chromatograph.[95] The data acquisition system has some specialized features such as signal

[90] A. P. Bruins, K. R. Jennings, and S. Evans, *Int. J. Mass Spectrom. Ion Phys.*, 1978, **26**, 395.
[91] W. F. Haddon, *Anal. Chem.*, 1979, **51**, 983.
[92] H. Mandli, R. Robbiani, T. Kuster, and J. Seibl, *Int. J. Mass Spectrom. Ion Phys.*, 1979, **31**, 57.
[93] C.-S. Hwang and R. W. Kiser, *Int. J. Mass Spectrom. Ion Phys.*, 1978, **27**, 209.
[94] P. D. Bolton, G. W. Trott, R. P. Morgan, A. G. Brenton, and J. H. Beynon, *Int. J. Mass Spectrom. Ion Phys.*, 1979, **29**, 179.
[95] M. H. Bozorgzadeh, J. H. Beynon, R. P. Morgan, and A. G. Brenton, *Int. J. Mass Spectrom. Ion Phys.*, 1979, **29**, 191.

averaging, signal smoothing,[96] and peak subtraction, in addition to more commonplace facilities.

Miscellaneous Applications.—A data acquisition system based on a programmable calculator has been developed to acquire data from a slow scanning magnetic sector mass spectrometer used for gas analysis and to accept and analyse data from a microdensitometer used to examine photoplates from a mass spectrograph.[97]

A quadrupole g.c.–m.s. system has been used to monitor the output from a pulsed chemical reactor using conventional data acquisition methods.[98] The data were processed further to give chemical equilibrium constants from small amounts of materials.

Algorithms for processing secondary ion mass spectral data (SIMS) have been described;[99] these produce qualitative information about the presence of metallic elements, their cluster ions, oxides, and hydroxides from isotopic analysis of the spectra. The information can be combined with calibration data from an internal standard to produce semi-quantitative analyses.

A computer program has been used to generate all the possible fragment ion sequences which could arise from small polypeptides. The degeneracy of mass values is high and the effect of this on sequencing experiments has been discussed.[100]

4 Interpretation of Mass Spectra

Library Search Methods.—The computer operations involved in compiling libraries of mass spectral and other spectroscopic data have been discussed with reference to selection, digitization, completion, formatting, and verification of the data.[101] The importance of using good-quality mass spectra in library collections has been emphasized and a quality index has been used to examine the spectra in three commonly used data bases and also a collection compiled from the recent literature.[102] The seven quality factors examined were: QF1. Source of the spectrum, QF2. Ionization conditions, QF3. Higher molecular weight impurities, QF4. Illogical neutral losses, QF5. Isotopic abundance accuracy, QF6. Number of peaks, and QF7. Lower mass limit. The number of peaks (N) in the spectrum was found to be the most sensitive quality factor and, in good-quality mass spectra, was found to correlate with the number of atoms in the molecule (Z) plus fifteen when low molecular weight compounds had been excluded as given in equation (1). Examples taken from the literature generally are of lower quality than data base spectra owing to the practice of publishing only those ions considered to be significant.

$$\text{QF6} = \frac{N}{(Z + 15)} \qquad (1)$$

[96] G. W. Trott and J. H. Beynon, *Int. J. Mass Spectrom. Ion Phys.*, 1979, **31**, 37.
[97] L. J. Rigby, in ref. 12, p. 1074.
[98] D. Schweich and J. Villermaux, *Anal. Chem.*, 1979, **51**, 77.
[99] W. Steiger and F. G. Rudenauer, in ref. 12, p. 770.
[100] M. Mudgett, J. A. Sogn, D. V. Bowen, and F. H. Field, in ref. 12, p. 1506.
[101] R. Buchi, J. T. Clerc, C. Jost, H. Koenitzer, and D. Wegmann, *Anal. Chim. Acta*, 1978, **103**, 21.
[102] D. D. Speck, R. Venkataraghavan, and F. W. McLafferty, *Org. Mass Spectrom.*, 1978, **13**, 209.

A general text-searching algorithm has been used to conduct searches of a mass spectral library on the basis of data not necessarily extracted from the spectra.[103] The error-checking features of the system proved particularly useful and serve to underline a recurrent theme, namely the ability of library search systems to cope with variations between mass spectra of the same compound. It has been proposed that as a preliminary to the computer-matching of mass spectra the spectra should be subjected to a scaling procedure designed to minimize the effects of mass discrimination and other instrumental distortions of the intensity values.[104] Relative peak intensities are conventionally scaled to the intensity of the most abundant ion or to the total ionization intensity of all the observed ions. It has been suggested that these methods be replaced by pairs of spectra whose intensities have been systematically scaled to be at their minimum distance apart as measured by the sum of the squared differences of peak intensities in the two spectra. The distance D between the unknown spectrum U and the reference spectrum R depends on two scaling factors C and d, according to equation (2), where the sum is over all the mass values m in both spectra. The scaling factor d allows, in this case, for linear mass-dependent discrimination effects. Minimization of D gives a formal basis for normalizing intensity data for comparison purposes which takes account of the commonest intensity distortions.

$$D = \sum [U_m - (C + dm)R_m]^2 \qquad (2)$$

Variations in ion abundance data can be particularly serious when spectra are abbreviated by the use of binary codes or difference codes. Thus, difficulties will arise in cases where the spectra are abbreviated by retaining only the most intense peak in each interval of fourteen masses. Two peaks of similar intensity in one mass interval may have reversed intensity values in spectra from different sources and will result in different codes. This effect has been taken into account in a new method of creating an ordered library of abbreviated mass spectra which is particularly suited for rapid searching on a small computer system.[105] Each mass spectrum is reduced by a difference code to a number containing six decimal digits representing the masses of the most intense peaks in each of six consecutive intervals of ten integer masses from forty to ninety. The masses selected are then classed as *unambiguous* or *ambiguous*, the latter being where the relative intensity data of neighbouring peaks might conceivably lead to the selection of a different mass in the same interval when encountering a reasonably distorted version of the spectrum. The *ambiguous* cases are used to generate alternative codes for the same spectrum. A spectrum containing six *unambiguous* mass values will appear as a single entry while n *ambiguous* masses will result in 2^n alternative codes being entered into the library. The abbreviated spectra are assembled in ascending numerical order which can be searched very efficiently using a binary search algorithm. The method has been used to encode 10 000 mass spectra as a dictionary with 60 894 entries. A quarter of test searches

---

[103] J. A. DeHaseth, H. B. Woodruft, S. R. Lowry, and T. L. Isenhour, *Anal. Chim. Acta*, 1978, **103**, 109.
[104] R. G. Dromey, *Anal. Chim. Acta*, 1979, **112**, 133.
[105] R. G. Dromey, *Anal. Chem.*, 1979, **51**, 229.

for particular compounds needed to examine only 0.1% of the total entries for retrieval.

A library of 9628 mass spectra has been converted into a binary code based on 106 mass values which were selected for their information content.[106] The search system used will retrieve the n nearest neighbours of the unknown and, if required, can be operated on a microcomputer system, with a small speed penalty.[107] Account was taken of spectral variations by calculating a 'mis-match probability' based on an examination of 773 cases where two spectra were available for the same compound.[108] The Boolean values in the truth table[107] were replaced with a 'probability of truth'. The retrieval rates were not significantly improved and it was concluded that an improvement in spectrum reproducibility would be the best way for increasing the retrieval rate.

New criteria for selecting significant peaks for coding into a library have been used in a system called SISCOM.[109] Spectra are coded as selected mass numbers and their normal intensity values which exceed a declared threshold and which pass an intensity test, *viz.* the isotope corrected intensity of the peak must exceed the arithmetic mean of its higher and lower homologous neighbours. The coding may exclude quite intense ions and retain low-intensity ions which other systems reject. The matching procedure is based on evaluation of six comparison factors from which the degree of matching of the total or partial spectrum may be deduced. The system has been used to compare mass spectra, from a variety of experimental origins, with three general and one specialized (Steroids) data bases. The SISCOM method is capable of identifying pure compounds and the components in mixtures can often be deduced by intelligent interpretation of the six comparison factors. Furthermore, it is evident that the system will retrieve spectra of compounds which have sub-structural features similar to the unknown even when there is a striking dissimilarity or lack of recognizable characteristic peaks in their full mass spectral patterns.

A system has been described in which binary coded mass spectra on four data bases, operating on a minicomputer system, can be retrieved using forward, reverse, and a combined forward–reverse search method.[110] In the forward binary search, peaks which are present in the unknown but are absent in the library member being considered are registered as differences which contribute to the mis-match score. In the reversed binary search, peaks which are absent in the unknown spectrum but are present in the library spectrum are registered as differences. In the combined forward–reverse search both types of difference are counted, as shown in the Table, and a better matching is achieved.

A computer search system has been used to propose partial structures for unknown steroids.[111] The method uses a method by which key ions are connected with distinct types of sub-structural units.

Previous attempts to identify monoterpenes by library search methods using abbreviated mass spectra have not been entirely satisfactory. A fresh approach

[106] G. Van Marlen, A. Dijkstra, and H. A. Van't Klooster, *Anal. Chim. Acta*, 1979, **112**, 233.
[107] G. Van Marlen and J. H. Van Den Hende, *Anal. Chim. Acta*, 1979, **112**, 143.
[108] G. Van Marlen, A. Dijkstra, and H. A. Van't Klooster, *Anal. Chem.*, 1979, **51**, 420.
[109] H. Damen, D. Henneberg, and B. Weimann, *Anal. Chim. Acta*, 1978, **103**, 289.
[110] J. Kwiatkowski and W. Riepe, *Anal. Chim. Acta*, 1979, **112**, 219.
[111] G. Spiteller, M. Spiteller, M. Ende, and G. A. Hoyer, *Org. Mass Spectrom.*, 1978, **13**, 646.

Table

Spectrum	Binary code (1 = present)			
Unknown	0	1	1	0
Library	0	1	0	1

Search result	Binary code (1 = mis-match)			
Forward search	0	0	1	0
Reverse search	0	0	0	1
Combined forward–reverse search	0	0	1	1

uses all detected peaks between masses 41 and 300. The recorded intensity data are weighted according to the results of an analysis of the variation from multiple scans of pre-recorded standard spectra. This procedure leads to a low weighting for ions which are mostly noise and/or fail to discriminate between structures. The library entries contain mass values together with mean ion abundance and the variation (so-called variance) derived from several standard runs on pure materials. Since the spectra were recorded using on-line gas chromatographic separation, the relative retention times of eluted peaks could be used to restrict the library search routine. The method has been used to identify correctly all the terpenes and phenylpropanoid ethers known to be present in several leaf oils.[112]

A g.c.–m.s. system has been described which will automatically provide quantitative and qualitative analyses for more than 100 components in clinical samples.[113] In a typical application, repetitive mass scans were acquired by a minicomputer system during a gas chromatographic run on an acidified fraction of urine which had been suitably derivatized. Post-run processing involves measurement of g.c. peak areas for quantification and retention indices for tentative identification. These tentative assignments are tested by a reverse library search using a collection of 157 possible components. The system is being used to carry out statistical analyses of urine samples and is being continuously extended to cover a wider range of compounds.

A library of 524 low-resolution mass spectra of steroids has been used to compare different methods of binary coding and matching.[114] The best results were obtained using spectra which were abbreviated to the two most abundant masses in each interval of fourteen and one-bit coding of the intensity data. The results of the library search routines were analysed using an effective recognition rate, $R = 3P/N$, where P is the percentage of correctly identified unknowns and N is the average number of library spectra found as first, second, and third neighbours. An 83% effective recognition rate was achieved when authentic spectra were used as unknowns.

Further examples and some uses for octal-coded mass spectra have appeared.[115]

[112] R. P. Adams, M. Granat, L. R. Hogge, and E. Von Rudloff, *J. Chromatogr. Sci.*, 1979, **17**, 75.
[113] S. C. Gates, M. J. Smisko, C. L. Ashendel, N. D. Young, J. F. Holland, and C. C. Sweeley, *Anal. Chem.*, 1978, **50**, 433.
[114] K. Varmuza, 'Recent Developments in Mass Spectrometry in Biochemistry and Medicine', ed. A. Frigerio, Plenum Press, New York, 1978, Vol. 1, p. 371.
[115] C. Merritt, jun., D. H. Robertson, and R. A. Graham, in ref. 12, p. 1002.

The methods of data acquisition and processing, including library searching routines, which are being developed for the identification of organic pollutants in environmental samples, have been described.[116]

A library of mass spectra of some pesticides has been assembled from g.c.–m.s. data which were encoded by applying some of the principles of information theory.[117] A selective information value, I, was assigned to each peak of mass m and intensity i as in equation (3). N_t is the total number of spectra and N is the number of spectra containing the peak (m, i). This value (I) was then used to select peaks to be encoded for inclusion in the library. Experimental mass spectra of trial mixtures were similarly coded and retrieved from the library by a reverse search algorithm, with encouraging results.

$$I = -\log_2 \frac{N}{N_t} \qquad (3)$$

The efficiency of Modulo N (MODN) data compression, where ion series are extracted by grouping together peaks separated by N a.m.u., have been compared. A MOD14 spectrum is generated by summing the intensities of all peaks in each of the 14 ion series to produce 14 sums which are divided by the sum of all the intensities in the spectrum.[118] The relative efficiency of library searching using MOD14, MOD10, and MOD7 spectra indicates that the greater information in the MOD14 spectra gives better retrieval rates. Very efficient file searching was achieved by grouping spectra according to the intensity of their most abundant ion series.

The above conclusion has been confirmed in an evaluation of mass spectral search algorithms which consistently show the best results with unabbreviated spectra.[7] Two methods of evaluation were used. The first was a conventional trial search to assess ability to identify mass spectra by retrieval from the library. The second method followed the propagation of selected spectra to see how well the algorithm could locate spectra of related compounds. The latter technique offers a new and useful approach to the evaluation of search performance.

Pattern Recognition.—The representation of mass spectra as vectors or points in multidimensional space means that spectra of compounds containing a common feature will tend to cluster in the hyperspace. A linear learning machine (LLM) is a computer program which moves a hyperplane (the multi-dimensional equivalent of a two-dimensional straight line) to a position which divides the data into two sets composing those with and those without a particular training feature.[119] This process, called training, is achieved by moving the hyperplane using an error feedback routine. The result is a weight vector which can be used to classify data which were not used in the training process. It has been pointed out that such methods of developing linear binary pattern clarifiers

---

[116] H. Knoeppel, B. Versino, W. G. Town, G. Schauenburg, A. Piel, J. Poelman, F. Geiss, and I. Norayer, in ref. 12, p. 1013.
[117] E. G. de Jong, J. van Bekkum, H. A. von't Klooster, and J. Freudenthal, in ref. 12, p. 1091.
[118] G. T. Rasmussen, T. L. Isenhour, and J. C. Marshall, *J. Chem. Inf. Comput. Sci.*, 1979, **19**, 98.
[119] P. C. Jurs and T. L. Isenhour, 'Chemical Applications of Pattern Recognition', J. Wiley, New York, 1975.

will fail if the data are linearly inseparable, *i.e.* if a hyperplane does not in fact exist which can divide the data perfectly. Departures from linear separability appear as an inability to produce well trained weight vectors which are then inefficient as binary classifiers. This situation is not uncommon with mass spectral data.[6]

For linearly inseparable data an improved method of training is to use a modified simplex pattern recognition method to produce optimized weight-vectors with binary decision capability. A simplex is a geometrical figure with $(n + 1)$ vertices in an n-dimensional space. A three-dimensional simplex is thus a tetrahedron. Each vertex of the simplex is a weight vector with an associated 'recognition' which is the number of training set patterns which it classifies correctly. The objective of training is to maximize the 'recognition' by moving the simplex. When optimized the simplex will appear to be stranded in a small region of the hyperspace which will define the trained weight vectors. Whilst simplex-developed weight vectors are advantageous when dealing with non-linear data, they may require long computational times to be optimized and they may arrive at local rather than global maxima.

A super-modified simplex method (SMS) has been developed for use with mass spectral data.[120] The new approach is computationally more efficient than earlier examples and shows improvements on them in tests in which weight vectors for the recognition of eleven common functional groups were calculated.

The difficulties encountered in comparing the results from different types of LLM to classify mass spectral features have been recognized and a 'measurement of confidence' in their results has been proposed.[121] Since this confidence measure is dependent on the choice of the sets of test data used in training, it is recommended that, when comparing methods, its minimum value should be used. In general, confidence measures are less optimistic than 'percentage correct predictions'.

The superiority of the K-nearest neighbour (KNN) method of classifying mass spectra has been restated in comparison with other pattern recognition methods.[6] The method has been applied to the sequencing of polypeptides through the identification of pentafluoropropionyl dipeptide methyl esters.[122] Mixtures of dipeptides were formed by enzymic cleavage of the polypeptide and after derivatization these were separated by gas chromatography and their mass spectra recorded using a computerized g.c.-m.s. system. The dipeptide sequences were determined by separate identification of the N— and C— terminal amino-acids using two multicategory KNN analyses. Classification accuracy for 86 dipeptides was 100% in the N— terminal cases falling to 97% for the C— terminal cases.

Linear regression analysis has been used to calculate binary classifiers to recognize steroid substructures from their low-resolution mass spectra.[123] Classifiers for 17 steroidal features were found with an average predictive ability of 84% correct predictions.

[120] S. L. Kaberline and C. L. Wilkins, *Anal. Chim. Acta*, 1978, **103**, 417.
[121] J. A. Richards and A. G. Griffiths, *Anal. Chem.*, 1979, **51**, 1358.
[122] J. N. Ziemer, S. P. Perone, R. M. Caprioli, and W. E. Seifert, *Anal. Chem.*, 1979, **51**, 1732.
[123] H. Rotter and K. Varmuza, *Anal. Chim. Acta*, 1978, **103**, 61.

The construction of decision trees from binary classifiers of mass spectra to build up molecular recognition systems has been discussed in general terms.[124]

Factor analysis is a matrix method for determining the number of independent properties contained in a set of data. Where the data are the mass spectra of a set of related mixtures, the factor analysis will determine the number of components present (see the later section on mixture analysis). Where the data are the mass spectra of related compounds, the factors may correspond to structurally significant fragment ions. This approach has been used to examine the mass spectra of 70 aromatic hydrocarbons.[125] Five significant factors were found which accounted for most of the observed spectral variations, each factor being strongly associated with five particular fragment ions in the mass spectra. By a variation of the presentation of the original data, the factor analysis could be modified to produce a list of dominant fragment ions.

A description of Fisher linear discriminants and their application to the classification of structural features using their mass spectra has appeared.[126] These discriminants provide a useful two-dimensional projection of multi-dimensional patterns which have similar attractive features as other methods for the visual presentation of the data.[127]

A new pattern recognition technique based on training with feedback and using average correlation coefficients for the training and testing of data sets has been developed. Results for twenty binary decision classifiers show good reliability for the weight vectors for limited atomic composition and functional group analysis. These have been combined in one example to produce a total structure determination.[128]

Attention has been paid to the problems associated with comparing the efficiency of binary classifiers. 'Percentage of correctly classified spectra unknown to the classifier' is rejected in favour of a system based on probability and information theory.[129]

Artificial Intelligence.—Interest in deductive methods that use mass spectrometric fragmentation rules to interpret spectra has continued.[15] The well known DENDRAL approach generates all possible structures for a given molecular ion then uses the known rules of mass spectral fragmentation to produce, for each structure, a forecast of its fragmentation pattern which can be compared with the experimental mass spectrum.

A new alternative approach, using high-resolution mass spectral data, is based on a reversal of the logic sequence used in DENDRAL. The computer interpretation of formula and structure (CIFAS)[130] program commences by examining the fragmentation pattern in the experimental mass spectrum using the known set of fragmentation rules. The fragmentation pattern is used to infer the structure without the need for extensive structure generation steps. The system was

[124] W. S. Meisel, M. Jolley, S. R. Heller, and G. W. A. Milne, *Anal. Chim. Acta*, 1979, **112**, 407.
[125] R. W. Rozett and E. M. Petersen, in ref. 12, p. 993.
[126] G. T. Rasmussen, G. L. Ritter, S. R. Lowry, and T. L. Isenhour, *J. Chem. Inf. Comput. Sci.*, 1979, **19**, 255.
[127] J. B. Kruskal, *Psychometrika*, 1964, **29**, 1.
[128] N. H. Mahle and J. W. Ashley, *Comput. Chem.*, 1979, **3**, 19.
[129] K. Varmuza and H. Rotter, in ref. 12, p. 1099.
[130] R. M. Hilmer and J. W. Taylor, *Anal. Chem.*, 1979, **51**, 1361.

developed initially to infer the main functional groups in a molecule and hence put constraints on its structural possibilities. The program does not have the capability of analysing even-electron fragment ions or multi-functional compounds. These restrictions have been removed in a development called FAMOUS[131] (Functional Analysis of Molecules by Optimal Utilization of Spectra). The limited tests reported so far suggest a much improved performance compared with CIFAS for single-function compounds. This is directly attributed to the inclusion of even-electron fragment ions in the system. The extension to include multi-functional compounds is based on the assumption that the total fragmentation of a molecule can be considered to be the sum of the fragmentations induced by the separate functional groups. While this is certainly not universally valid, it has been shown that it will, in some cases, permit accounting for a substantial proportion (*ca.* 90%) of the observed fragment ions. In tests of 25 oxygen-containing multi-functional compounds, which included both aliphatic and aromatic structures, in only four cases were the true structures ranked lower than the third possibility; this result must be viewed as encouraging. It was suggested that this interesting method may have uses in research designed to learn more about the factors which produce the observed fragment ions in a mass spectrum.

A computer program has been described which can be used to select ions in low-resolution mass spectra that contain particular atoms.[132] The method involves a deconvolution of overlapping isotopic patterns and is clearly suited to molecules containing atoms which have high isotopic abundances.

Mixture Analysis.—The resolution of composite low-resolution mass spectra of multicomponent mixtures into concentration and mass spectra of the individual components has been discussed.[133] Three methods, namely a least-squares approximation, a minimum absolute deviation routine, and a Chebyshev approximation, have been discussed in detail. Each method requires reference spectra of those pure compounds which are the dominant components in the mixture. Whichever method is used, the objective is to identify a combination of the references which gives a best fit to the unknown mixture.

Two independent methods of factor analysis, *viz.* matrix rank analysis and principal component analysis, have been combined in a single program.[134] Each technique provides an independent check on the other in determining the number of components in mass spectra of mixtures. In common with other factor analysis methods, it is necessary to have at least as many spectral scans of different mixture compositions as there are components in the mixture. In the present example, these were obtained from repetitive scans of unresolved peaks eluted from a gas chromatograph using a g.c.-m.s. computer system.[134]

A target factor analysis method, previously applied in mass spectral identification,[1] has been adapted to the separation of 'mixed' spectra into pure component spectra and component concentrations.[135] In addition to the usual

---

[131] L. W. McKeen and J. W. Taylor, *Anal. Chem.*, 1979, **51**, 1368.
[132] J. Kwiatowski, *Org. Mass Spectrom.*, 1978, **13**, 513.
[133] D. W. Fausett and J. H. Weber, *Anal. Chem.*, 1978, **50**, 722.
[134] J. M. Halket, *J. Chromatogr.*, 1979, **175**, 229.
[135] F. J. Knorr and J. H. Futtrell, *Anal. Chem.*, 1979, **51**, 1236.

requirement of one mixture composition per component, there is an additional requirement that each component must have spectral purity on at least one of its peaks. This latter constraint may limit the applicability of the method in some cases.

The theory of the errors involved in target factor analysis has been developed and applied to the analysis of mixtures of mass spectra.[136]

An extension of the widely used Probability Based Matching Programme (PBM) employs subtraction of the best-matched components from the mass spectra of mixtures to produce improved identification of minor components of mixtures.[137]

A nearest-neighbour approach to mixture analysis has been described which was based on methods analogous to those used in library searching systems. In a system in which the mass spectra are represented as points in multidimensional space, a library search involves finding those points representing library members which are closest to the point representing the unknown. In the new method, the known is not a single point but a cluster of points representing the mass spectra of a set of related mixtures. These points define a subspace and library members lying nearby are identified as components in the mixture.[138] Generally for an n-component mixture the correct components were found in the n nearest matches in tests on mixtures containing up to five components.

---

[136] E. R. Malinowski, *Anal. Chim. Acta*, 1978, **103**, 339.
[137] B. L. Atwater, R. Venkataraghavan, and F. W. McLafferty, *Anal. Chem.*, 1979, **51**, 1945.
[138] G. T. Rasmussen, B. A. Hohne, R. C. Wieboldt, and T. L. Isenhour, *Anal. Chim. Acta*, 1979, **112**, 151.

8
Gas Chromatography–Mass Spectrometry and High-performance Liquid Chromatography–Mass Spectrometry

BY F. A. MELLON

1 Introduction

The large number of publications describing g.c.–m.s. and its applications (approaching 2000 during the period covered by this report), together with the inclusion of high-performance liquid chromatography–mass spectrometry to the scope of this review, make it unfeasible to refer to every relevant paper. Accordingly, I have tried to emphasize those publications which are genuinely innovatory whilst referring briefly to a number of papers in which g.c.–m.s. or h.p.l.c.–m.s. figures to a large extent. Brief abstracts of all papers in which g.c.–m.s. techniques are described can be found in the excellent *Gas Chromatography–Mass Spectrometry Abstracts*,[1] a boon to all g.c.–m.s. users (as well as beleagured reviewers!).

2 Practical and Instrumental Considerations: G.C.–M.S.

Sensitivity, Interface Techniques, and Chromatographic Performance.—One of the most striking recent improvements in g.c.–m.s. capabilities has been the large increase in sensitivity. Five—ten years ago instrument manufacturers quoted g.c.–m.s. sensitivity as, typically, 10 ng of methyl stearate injected onto the g.c. column yielding a usable mass spectrum. Today instrument manufacturers claim that similar results can be achieved with 100 pg of methyl stearate on most magnetic sector instruments, or with 1 ng of material on a g.c.–quadrupole mass spectrometer system. Mass spectrometer resolution, g.c. peak width, or signal-to-noise ratio have deliberately not been quoted, as mass spectrometer manufacturers generally quote different values for these parameters when specifying sensitivity.

The most noticeable change of emphasis in g.c.–m.s. applications over the past ten years or so has been the increase in the use of capillary columns. This is hardly surprising as capillary columns, as well as their obvious chromatographic advantages, have the additional advantage of being better suited to interfacing with mass spectrometers than are packed g.c. columns, by virtue of their low flow rate requirements and their low phase bleed characteristics. Capillary

---

[1] 'Gas Chromatography–Mass Spectrometry Abstracts', P.R.M. Science and Technology Agency Ltd., London, 1978—1980, Vols. 9—11, currently published bi-monthly.

columns can be directly connected to the ion source of a mass spectrometer without the need for pressure-reducing interfaces, especially if the mass spectrometer ion source is equipped with either high-speed diffusion pumps (*ca.* 1000 l s^{-1} for hydrogen or helium) or with turbomolecular pumps.

New types of capillary column, the flexible fused-silica columns, have been devised.[2] These columns appear to have considerable advantages over conventional capillaries owing to their inertness and mechanical flexibility. The latter property renders them particularly suitable for g.c.–m.s. coupling as they are sufficiently deformable to be fed directly into the ion source of a mass spectrometer. Their major drawback appears to be their poor wetability, limiting coatings to carbowax 20M and SP-2100 silicone phases so far.

Reports which describe improved capillary g.c.–m.s. interfaces have appeared.[3–5] These devices are all essentially modifications of the 'open-split coupling' (OSC) technique.[6] The first of these modified OSC devices[3] was fabricated from glass and allowed a flow of scavenger gas at the interface, obviating 'dead volume tailing' of large peaks at low flow rates and diluting peaks which were too strong at high flow rates. The interface proved effective and also demonstrated the superiority of glass restrictors over platinum capillary restrictors. The problem of absorption of eluting peaks in the mass spectrometer ion source was also investigated. Two other all-glass capillary g.c.–m.s. interfaces were of similar design.[4–5]

The influence of the vacuum on separation efficiency in directly coupled capillary g.c.–m.s. has been assessed.[7,8] With no restriction between the mass spectrometer and the end of the capillary column, the maximum loss in column efficiency was only 25%.[7] Optimum gas velocity was shifted to higher flow rates by the pressure drop, in agreement with the findings of other workers.[8] These investigators also found that at optimum flow velocities reduced pressure operation produced column efficiencies of about 30% lower than for normal g.c. operation. However, at high flow velocities (*i.e.* flow conditions not at an optimum), reduced pressure operation could provide column efficiencies about 100% higher than atmospheric pressure operation.[8]

A method for the preparation of thick film glass capillary columns has been described.[9] Fused, powdered quartz allows the application of a thick film of phase, thus enabling the injection of large samples. The columns are suitable for g.c.–m.s. applications because of their low bleed characteristics. The analysis of materials of low volatility by high-temperature glass capillary g.c.–m.s. using

---

[2] R. Dandeneau and E. H. Zerenner, *J. High Res. Chromatogr. Chromatogr. Commun.*, 1979, **2**, 351.
[3] D. Henneberg, U. Henrichs, H. Husmann, and G. Schomburg, *J. Chromatogr.*, 1978, **167**, 139.
[4] H.-J. Stan and B. Abraham, *Anal. Chem.*, 1978, **50**, 2161.
[5] P. P. Schmid, M. D. Muller, and W. Simon, *J. High Res. Chromatogr. Chromatogr. Commun.*, 1979, **2**, 225.
[6] D. Henneberg, U. Henrichs, and G. Schomburg, *J. Chromatogr.*, 1975, **112**, 343; D. Henneberg, U. Henrichs, and G. Schomburg, *Chromatographia*, 1975, **8**, 449.
[7] F. Vangaever, P. Sandra, and M. Verzele, *Chromatographia*, 1979, **12**, 153.
[8] F. W. Hatch and M. E. Parrish, *Anal. Chem.*, 1978, **50**, 1164.
[9] P. Torline, D. Du Plessis, N. Schnautz, and J. C. Thompson, *J. High Res. Chromatogr. Chromatogr. Commun.*, 1979, **2**, 613.

thermostable columns has been achieved.[10] The columns were coated with OV-1 or SE-54 phases and could be used to chromatograph waxes and triglycerides at temperatures up to 330 °C. The importance of correct injection techniques, with particular emphasis on on-column injection, has been emphasized in investigations of injection techniques onto capillary columns.[11] This work should be of particular interest to g.c.–m.s. users involved in quantitative work with capillary columns. The influence of the syringe needle on the precision and accuracy of vaporizing g.c. injections has also been considered.[12] A number of papers concerned with the preparation and deactivation of capillary columns should be of interest to those involved in g.c.–m.s. These publications include a discussion of surface treatment, deactivation, and coating in glass capillary gas chromatography,[13] a note on the preparation and efficiency of polar SCOT columns,[15] a deactivation procedure for capillaries involving persilylation,[16] and a method for the preparation of apolar glass columns by the barium carbonate procedure.[17]

A chiral stationary phase, Chirasil-Val, has been synthesized and used in the preparation of capillary columns which are able to separate optical antipodes.[18] The excellent thermal stability of this phase renders it especially useful for g.c.–m.s. applications.[19]

Deactivation of an all-glass g.c.–m.s. interface with UCCW 982 has been shown to improve its inertness.[20] Ion source decomposition and/or absorption was also investigated as part of this study. G.c.–m.s. analysis of underivatized chlorophenols at picogram levels has been facilitated by deactivation of the column with 2,4-dichlorophenoxyacetic acid at 200 °C.[21] The deactivation of g.c. columns and mass spectrometer interfaces is a general problem to which no perfect solution has yet been found. Although g.c.–m.s. interfaces generally operate at low pressures, thereby reducing wall collisions and the chances of absorption and/or thermal decomposition, it is still important that they be made as inert as possible. Silanization or coating with Carbowax 20M followed by heat treatment are the most popular deactivation techniques, especially for capillary interfaces. Newer methods which have been applied to the deactivation

---

[10] G. Dielmann, S. Meier, and U. Rapp, *J. High Res. Chromatogr. Chromatogr. Commun.*, 1979, **2**, 343.
[11] K. Grob and K. Grob, jun., *J. Chromatogr.*, 1978, **151**, 311: K. Grob., *J. High Res. Chromatogr. Chromatogr. Commun.*, 1978, **1**, 263.
[12] K. Grob, jun. and H. P. Neukom, *J. High Res. Chromatogr. Chromatogr. Commun.*, 1979, **2**, 15.
[13] P. Sandra and M. Verzele, *Chromatographia*, 1977, **10**, 419.
[14] R. A. Heckman, C. R. Green, and F. W. Best, *Anal. Chem.*, 1978, **50**, 2157.
[15] R. J. Thiecke, J. H. M. Van Den Berg, R. S. Deelder, and J. J. M. Ramaekers, *J. Chromatogr.*, 1978, **160**, 264.
[16] K. Grob, G. Grob, and K. Grob, jun., *J. High Res. Chromatogr. Chromatogr. Commun.*, 1979, **3**, 31.
[17] K. Grob, jun., G. Grob, and K. Grob, *J. High Res. Chromatogr. Chromatogr. Commun.*, 1978, **1**, 149.
[18] H. Frank, G. J. Nicholson, and E. Bayer, *Angew. Chem., Int. Ed. Engl.*, 1978, **17**, 363.
[19] E. Bailey and P. B. Farmer, in 'Advances in Mass Spectrometry', Vol. 8, ed. A. Quayle, Heyden, London, 1980; P. B. Farmer, E. Bailey, J. H. Lamb, and T. A. Connors, *Biomed. Mass Spectrom.*, 1980, **7**, 41.
[20] H. W. Durbeck, I. Buker, and W. Leyman, *Chromatographia*, 1978, **11**, 372.
[21] M. A. White and K. R. Parsley, *Biomed. Mass Spectrom.*, 1979, **6**, 570.

of capillary columns, such as coating with OV-101 followed by heat treatment, may also be applicable to interface deactivation.[22]

Turning now to packed column g.c.–m.s., the new stationary phase NN'-bis(p-phenylbenzylidene)-α,α'-bi-p-toluidine has been shown to be an excellent g.c.–m.s. phase for the analysis of polycyclic aromatic hydrocarbons.[23] Little use has been made of the low-bleed 'non-extractable polymer coated' columns[24] in g.c.–m.s. studies, despite their excellent g.c.–m.s. characteristics of low-bleed and high-temperature operation.

A dense-gas chromatograph–mass spectrometer interface has been designed.[25] Dense gases, *i.e.* those at temperatures greater than their critical temperature and at pressures (20—2000 atm) sufficient to give densities near liquid densities, are often excellent solvents and can thus act as both solvent and carrier (as in liquid chromatography). The special problem of interfacing such a chromatograph with a mass spectrometer was solved by expanding the dense gas in a nozzle–skimmer–collimator system which was evacuated by two high-speed ($2400\,l\,s^{-1}$ and $1200\,l\,s^{-1}$) pumps. Initial results using aromatic hydrocarbon test compounds were promising, although some practical problems, particularly plugging of the nozzle orifice, remain to be solved. Clustering of solute with solvent ions was also observed.

A number of reviews which are either partially or totally concerned with combined g.c.–m.s. techniques have appeared. In particular, combined g.c.–m.s. has been reviewed as a powerful tool in analytical chemistry.[25] The chromatographic–mass spectrometer interface has been discussed (l.c.–m.s. as well as g.c.–m.s.)[27] and techniques of gas chromatography–quadrupole mass spectrometry have been described.[28a] Design considerations of the mass spectrometer ion source for g.c.–m.s. applications have been examined[28b] and a review of the marriage of different types of analytical technique includes descriptions of g.c.–m.s. systems.[29] A survey of new mass spectrometric techniques in organic analysis describes field ionization, chemical ionization, and electron impact studies of drug metabolism by g.c.–m.s. and l.c.–m.s.,[30] and a book entitled 'Practical Mass Spectrometry' contains chapters dealing with g.c.–m.s. applications.[31] A comprehensive review of all aspects of mass spectrometry includes much data on g.c.–m.s. and l.c.–m.s.[32]

Alternative Ionization Techniques.—The use of positive chemical ionization g.c.–m.s. techniques is now so routine that it requires no further comment. However, negative chemical ionization g.c.–m.s. is still fairly unusual and some

[22] G. Schomburg, H. Husmann, and M. Borowitzky, *Chromatographia*, 1979, **12**, 651.
[23] F. Janssen, *Anal. Chem.*, 1979, **51**, 2163.
[24] W. A. Aue, C. R. Hastings, and S. Kapila, *J. Chromatogr.*, 1973, **77**, 299; F. W. Karasek and H. H. Hill, *Res. Dev.*, 1975, **26**, 30; M. M. Daniewski and W. A. Aue, *J. Chromatogr.*, 1978, **147**, 119.
[25] L. G. Randall and A. L. Wahrhaftig, *Anal. Chem.*, 1978, **50**, 1705.
[26] M. C. Ten Noever de Brauw, *J. Chromatogr.*, 1979, **165**, 207.
[27] W. H. McFadden, *J. Chromatogr. Sci.*, 1979, **17**, 2.
[28] (*a*) K. Feser and W. Kogler, *J. Chromatogr. Sci.*, 1979, **17**, 57; (*b*) R. M. Milberg and J. C. Cook, *J. Chromatogr. Sci.*, 1979, **17**, 17.
[29] T. Hirschfeld, *Anal. Chem.*, 1980, **52**, 297A.
[30] D. E. Games, J. L. Gower, M. G. Lee, I. A. S. Lewis, M. E. Pugh, and M. Rossiter, *Proc. Anal. Div. Chem. Soc.*, 1978, **15**, 101.
[31] 'Practical Mass Spectrometry', ed. B. S. Middleditch, Plenum Press, New York, 1979.
[32] A. L. Burlingame, T. A. Baillie, P. J. Derrick, and O. S. Chizhov, *Anal. Chem.*, 1980, **52**, 214R.

interesting techniques and applications have recently been described. In particular, details of an electro-optical ion detector for capillary column g.c.–negative ion mass spectrometry have been published.[33] When using this device the ratio of highest to lowest mass simultaneously detectable is only 1.08; however, the system can be used advantageously for the detection of known compounds. This was exemplified by the detection of hexachlorobenzene at the 10 fg level. An electron capture negative chemical ionization mass spectrometer has successfully detected dopamine, amphetamine, and Δ^2-tetrahydrocannabinol derivatives down to attomole levels.[34] Negative ion counting techniques, involving the use of a multichannel analyser and data acquisition system, have been developed.[35]

Some applications of high-sensitivity field ionization g.c.–m.s. have been described.[36] Improved source design, the 'volcano' ion source, offered considerable improvement in sensitivity. Examples of the application of this system to the analysis of drugs and pesticides were demonstrated.

Gas chromatography–photoionization mass spectrometry has been assessed in the analysis of eighteen compounds.[37] Sensitivity was low compared with other ionization techniques but this was offset to some extent by the production of abundant molecular ions and a reduction in fragmentation. However, it is unlikely that the technique will enjoy widespread application.

An interesting application of the mass spectrometer in a g.c.–m.s. system has been as an element specific detector.[38] Molecules leaving the gas chromatograph were ionized in a microwave induced discharge which was located in the g.c.–m.s. interface. The technique was used to detect F, Cl, Br, I, S, and N in surface water and human adipose tissue samples.

A review of chemical ionization mass spectrometry includes references to g.c.–m.s. and l.c.–m.s. techniques.[39]

Data Processing.—All aspects of data handling in mass spectrometry are reviewed in Chapter 7. Some applications of particular interest to the g.c.–m.s. user include g.c.–m.s. data systems which automatically calculate Kovats retention indices.[40–42] Two of these applications involve identification of lower terpenes[40,41] where the supplementary g.c. data are essential because of similarities between the mass spectra of these compounds. The third investigation describes the application of automatic calculation of Kovats indices in environmental analysis with the aid of a g.c.–m.s. calculator system.[42]

Factor analysis of repetitively scanned spectra in g.c.–m.s. has enabled the detection of hidden impurities.[43] G.c. peak shape need not be strictly maintained

---

[33] B. Hedfjall and R. Ryhage, *Anal. Chem.*, 1979, **51**, 2163.
[34] D. F. Hunt and F. W. Crow, *Anal. Chem.*, 1978, **50**, 1781.
[35] T. Fujii and K. Fuwa, *Anal. Chim. Acta*, 1979, **107**, 335.
[36] R. M. Milberg and J. C. Cook, *J. Chromatogr. Sci.*, 1979, **17**, 43.
[37] N. Washida, H. Akimoto, H. Takagi, and M. Okuda, *Anal. Chem.*, 1978, **50**, 910.
[38] J. Freudenthal, *Int. J. Environ. Anal. Chem.*, 1978, **5**, 311.
[39] R. E. Mather and J. F. J. Todd, *Int. J. Mass Spectrom. Ion Phys.*, 1979, **30**, 1.
[40] T. Saeed, G. Redant, and P. Sandra, *J. High Res. Chromatogr. Chromatogr. Commun.*, 1979, **2**, 75.
[41] R. P. Adams, M. Granat, L. R. Hogge, and E. von Rudloff, *J. Chromatogr. Sci.*, 1979, **17**, 75.
[42] K. R. Betty and F. W. Karasek, *J. Chromatogr.*, 1978, **166**, 111.
[43] J. M. Halket, *J. Chromatogr.*, 1979, **175**, 229; J. M. Halket, in 'Recent Developments in Chromatography and Electrophoresis', ed. A. Frigerio and L. Renoz, Elsevier, Amsterdam, 1979, p. 327.

and no prior knowledge of pure component spectra is required for the successful use of this technique.

Data-blocking cross correlation peak detection has been applied to computerized g.c.–m.s.[44] This resulted in reduction in the rate of data transmission and a reduction in computer time requirements.

The design, implementation, and performance of a high-resolution gas chromatography–high-resolution mass spectrometry–real time computer system for the analysis of complex organic mixtures have been described[45] and details of the principles and applications of a research-orientated g.c.–m.s. data system have appeared.[46] The salient features of a field portable g.c.–m.s.–computer system have been described.[47] The apparatus was used in the analysis of low concentrations of organic compounds in the atmosphere. A data system which accepts g.c.–m.s. information from two mass spectrometers has been implemented.[48] The system also included a second, larger computer for library searching of mass spectra. A g.c.–m.s. data system with a high dynamic range has been constructed.[49]

Computerized pattern recognition has been applied to the g.c.–m.s. identification of pentafluoropropionyl (PFP) dipeptide methyl esters[50] with 100% accurate classification of N-terminal amino-acids and 97% accuracy for C-terminal amino-acids reported for 86 dipeptide esters.

An analogue g.c.–m.s. processor which incorporates a tape recorder, oscilloscope, and appropriate circuitry has been described.[51]

A review of the application of computer techniques to mass spectrometry contains material relevant to g.c.–m.s.[52]

Quantification.—This is discussed in greater detail in Chapter 11. Unusual techniques of particular interest include metastable peak monitoring of a t-butyldimethylsilyl ester of 5α-dihydrotestosterone in the B/E mode down to the 20 pg level[53] and a method of increasing signal-to-noise ratio in g.c.–m.s. by limited mass monitoring.[54]

3 Applications: G.C.–M.S.

Long-chain Compounds.—*Hydrocarbons.* G.c.–m.s. analysis of C_4 to C_{40} alkanes in a single run has been possible when the g.c. column used was filled with PPE 20 phase coated onto graphitized carbon black.[55] The g.c. column was interfaced to a modified mass spectrometer *via* a single-stage jet separator and this system was able to analyse crude oils.

---

[44] W. F. Bryant, M. Trivedi, B. Hinchman, S. Sofranko, and P. Mitachek, *Anal. Chem.*, 1980, **52**, 38.
[45] J. Meili, F. C. Walls, R. McPherron, and A. L. Burlingame, *J. Chromatogr. Sci.*, 1979, **17**, 29.
[46] J. E. Campana, T. H. Risby, and P. C. Jurs, *Anal. Chim. Acta*, 1979, **112**, 321.
[47] R. W. Meier, *Amer. Ind. Hygiene Assoc. J.*, 1978, **39**, 233.
[48] P. D. Bolton and G. W. Trott, *Int. J. Mass Spectrom. Ion Phys.*, 1978, **27**, 419.
[49] E. Muller and U. Krienen, *Z. Anal. Chem.*, 1979, **294**, 241.
[50] J. N. Zeimer, S. P. Perone, R. M. Caprioli, and W. E. Seifert, *Anal. Chem.*, 1979, **51**, 1732.
[51] C. R. Lee, *Biomed. Mass Spectrom.*, 1979, **6**, 165.
[52] F. W. McLafferty and R. Venkataraghavan, *J. Chromatogr. Sci.*, 1979, **17**, 24.
[53] S. J. Gaskell and D. S. Millington, *Biomed. Mass Spectrom.*, 1978, **5**, 557.
[54] R. C. Murphy, S. E. Hattox, and H. Russell Helbig, *Biomed. Mass Spectrom.*, 1978, **5**, 444.
[55] P. Ciccioli, J. M. Hayes, G. Rinaldi, K. B. Denson, and W. G. Meinschein, *Anal. Chem.*, 1979, **51**, 400.

Cuticular wax occurring in the leaves of *Calathea lutea* has been analysed by packed column g.c.–m.s.[56] Odd-numbered n-alkanes in the range C_{25}—C_{37} and even-numbered n-alkanols of C_{20}—C_{32} chain length were found.

The biotransformation of hydrocarbons and related compounds by *Methylcosinus trichosporium* has been examined with the aid of g.c.–m.s.[57] Whole cell suspensions were found to metabolize partially a wide range of compounds, including n-alkanes, alkenes, aromatic, alicyclic, and terpenoid hydrocarbons. In addition alcohols, phenol, pyridine, and ammonia were metabolized.

Fatty Acids, Aldehydes, Ketones, Alcohols, and Esters. A comprehensive study of methods for locating double bonds and cyclopropane rings in fatty acids by g.c.–m.s. has been published.[58] A large number of derivatives were assessed and it was concluded that the trimethylsilyl (TMS) ethers of hydroxymethoxy esters were promising for double-bond location, following preliminary reactions on the double bond(s). The TMS ethers of vicinal diols were suggested as suitable derivatives for the analysis of polyunsaturated fatty acids. G.c.–m.s. analysis of methyl esters of the reaction products arising from the hydroxylation of double bonds and subsequent trimethysilylation has been employed to locate double bonds in fatty acids which contain conjugated double bonds.[59] Ethyl-substituted fatty acids from subcutaneous triacylglycerols in lamb have been analysed by g.c.–m.s. of their methyl esters.[60]

Pyrollidide derivatives of fatty acids have been suggested as suitable derivatives for low-resolution g.c.–m.s. analysis owing to their characteristic fragmentation patterns.[61] However, it has been demonstrated that the effectiveness of these derivatives is limited when the fatty acid contains some oxygen function (ether linkage, epoxide ring, or hydroxy group).[62] This arises because the extra oxygen function enhances cleavage adjacent to itself and suppresses further fragmentation of the chain distal to the pyrollidide group. Complementary derivatization techniques were suggested for epoxy and hydroxy fatty acids.

Benzyl esters have been claimed to be suitable derivatives for capillary g.c. and g.c.–m.s. analysis of volatile fatty acids.[63]

Cyclopentenyl fatty acids have been identified in the seeds of *Hydnocarpus wightiana* by g.c.–m.s.[64] The acids were analysed as their pyrollidide derivatives. Fatty acids in baboon liver lipids have been identified by capillary g.c.–m.s. of their methyl esters.[65] Branched-chain fatty acids from fallow deer perinephric

---

[56] K. E. Malterud, E. Wollenweber, and L. D. Gomez, *Z. Naturforsch., Teil C*, 1979, **34**, 157.
[57] I. J. Higgins, R. C. Hammond, F. S. Sariaslani, D. Best, M. M. Davies, S. E. Tryhorn, and F. Taylor, *Biochem. Biophys. Res. Commun.*, 1979, **89**, 671.
[58] D. E. Minnikin, *Chem. Phys. Lipids*, 1978, **21**, 313.
[59] G. Janssen and G. Parmentier, *Biomed. Mass Spectrom.*, 1978, **5**, 439.
[60] A. Smith and A. G. Calder, *Biomed. Mass Spectrom.*, 1979, **6**, 347.
[61] M. Vetter and W. Walter, *Monatsh. Chem.*, 1975, **106**, 203; B. A. Anderssen and O. Bertelsen, *Chem. Scr.*, 1975, **8**, 91; B. A. Anderssen and R. T. Holman, *Lipids*, 1974, **9**, 185; B. A. Anderssen, W. H. Heimermann, and R. T. Holman, *Lipids*, 1974, **9**, 443; B. A. Anderssen, W. W. Christie, and R. T. Holman, *Lipids*, 1975, **10**, 215.
[62] J. Eagles, G. R. Fenwick, and R. Self, *Biomed. Mass Spectrom.*, 1979, **6**, 462.
[63] R. Liardon and U. Kuhn, *J. High Res. Chromatogr. Chromatogr. Commun.*, 1978, **1**, 47.
[64] V. K. S. Shukla, E. M. Abdel-Moety, E. Larsen, and H. Egsgaard, *Chem. Phys. Lipids*, 1979, **23**, 285.
[65] A. Smith, A. G. Calder, E. R. Morrison, and G. A. Garton, *Biomed. Mass Spectrom.*, 1979, **6**, 345.

triacylglycerols have been characterized by capillary g.c.–m.s.[66] The major fatty acids in flue-cured tobacco have been identified[67] and the metabolism of unsaturated fatty acids by *E. coli*[68] has been studied by g.c.–m.s. The results of the latter investigation showed that the final metabolite of riconoleic acid was 6-hydroxydodec-*cis*-3-enoic acid. Chemical ionization capillary g.c.–m.s. using isobutane as reagent gas has proved to be a suitable technique for the analysis of the TMS derivatives of polyhydroxy fatty acids.[69] The mass spectra of these compounds indicated their molecular weight and the number and position of the hydroxy groups.

Insect juvenile hormones (epoxy methyl farnesoate and homologous derivatives) have been analysed by chemical ionization capillary g.c.–m.s.[70] A variety of c.i. reagent gases was used, of which ammonia appeared to yield the best mass spectra. Despite the use of an SE-30 capillary column (SE-30 is not the best phase for analysis of these unstable, thermally sensitive compounds—Ucon HB-5100 is the phase of choice[71]) it was possible to detect as little as 100 pg of insect juvenile hormone.

The g.c.–m.s. analysis of synthetic aviation oils containing fatty acid tetraesters of pentaerythritol has been described.[72]

Electron impact and chemical ionization g.c.–m.s. have been compared in the analysis of long-chain aldehydes and their *O*-methyl and *O*-t-butyldimethylsilyl (TBDMS) oximes.[73] The results indicated a preference for methane chemical ionization g.c.–m.s. analysis of the underivatized aldehydes.

β-Diketone waxes have been characterized by g.c.–m.s. of their TMS derivatives[74] which exhibit simpler fragmentation patterns than underivatized β-diketones; furthermore, the derivatized hydroxy- and oxo-β-diketones were completely separated on a packed g.c. column, thereby facilitating the analysis.

Alkylaminopyridines and some analogous compounds are suitable derivatives for the characterization of long-chain alcohols by g.c.–m.s.[75] The 2-aminopyridines were found to show the structure of the aliphatic chain most clearly.

Lipids and Sphingolipids. Changes in the acyl lipids and their fatty acids in bacteria grown under photosynthetic and non-photosynthetic conditions have been investigated by g.c.–m.s.[76]

TBDMS derivatization of various glycerols yields compounds which are very suited to g.c.–m.s. analysis.[77] t-Butyldimethylchlorosilane/imidazole was the derivatizing reagent of choice in this study of the molecular species of phos-

---

[66] A. Smith and W. R. H. Duncan, *Lipids*, 1979, **14**, 350.
[67] W. A. Court and J. G. Hendel, *J. Chromatogr. Sci.*, 1978, **16**, 314.
[68] M. Mizugaki, H. Fukuyama, T. Sakamoto, and Y. Yamanaka, *Chem. Pharm. Bull.*, 1978, **26**, 2417.
[69] M.-J. Stan and M. Scheutwinkel-Reich, *Z. Anal. Chem.*, 1979, **296**, 400.
[70] B. Mauchamp, R. Lafont, M. Hardy, and D. Jourdain, *Biomed. Mass Spectrom.*, 1979, **6**, 276.
[71] K. H. Trautmann, A. Schuler, M. Suchy, and H.-K. Wipf, *Z. Naturforsch., Teil C*, 1974, **29**, 161; A. F. Hamnett and G. E. Pratt, *J. Chromatogr.*, 1978, **158**, 387.
[72] A. Zeman, P. Bartl, A. Schaaf, and V. Christ, *Z. Anal. Chem.*, 1978, **290**, 21.
[73] G. Phillipou and A. Poulos, *Chem. Phys. Lipids*, 1978, **22**, 51.
[74] A. P. Tulloch and L. R. Hogge, *J. Chromatogr.*, 1978, **157**, 291.
[75] W. Vetter, W. Meister, and G. Oesterhelt, *Helv. Chim. Acta*, 1978, **61**, 1287.
[76] N. J. Russell and J. L. Harwood, *Biochem. J.*, 1979, **181**, 339.
[77] K. Satouchi and K. Saito, *Biomed. Mass Spectrom.*, 1979, **6**, 396.

pholipids. Characteristic fragments included strong [M-57]⁺ ions formed by cleavage of the t-butyl group, and characteristic fragments of the alk-1'-enyl, alkyl, and acyl groups were also observed. Similar data have been obtained on other diglycerols.[78]

Stereospecific analysis of triacylglycerols by random generation of racemic 1,2-diacylglycerols, synthesis of racemic phosphatidylcholines, and stereospecific release of 1,2-*sn*- and 2,3-*sn*-diacylglycerols by phospholipase C, followed by g.c.–m.s. analysis of the TBDMS derivatives of the diacylglycerols, has been used to determine the fatty acid distribution in lard.[79]

The principal polar lipids of various strains of methanogenic archaebacteria have been determined by g.c.–m.s.[80] Diphytanyl and bidiphytanyl glycerol ethers were identified in these bacteria.

N-Acetyl derivatives of a number of hydroxy sphingamines have been examined by g.c.–m.s. as part of a study of sphingolipid base metabolism.[81] Successful g.c.–m.s. analysis required further derivatization by trimethylsilylation.

The sex-dependent composition of cuticular lipids from the beetle *Rhagonycha fulva* has been determined by g.c.–m.s.[82] Significant differences in the percentages of mono- and di-unsaturated hydrocarbons were found between male and female beetles.

Lipid exchange in plasma lipoproteins following intravenous infusion of an artificial lipid emulsion has been examined with the aid of g.c.–m.s. of the TBDMS derivatives of phosphatidylcholines.[83]

The unsaponified latex from the leaf stalks of *Hoya bella* has been analysed by g.c.–m.s.[84] Surface lipids of trifolium species have also been examined in an attempt to investigate the preferences of some weevils for certain types of clover.[85] Marked differences in alcohol and ester composition were observed.

A survey of lipid derivatives suitable for g.c.–m.s. has appeared[86] and a review of the mass spectrometry of lipids has been published.[87]

Prostaglandins and Related Compounds.—Vapour-phase rearrangement of *N*-acyl prostaglandin carboxamides to nitriles and their TMS esters has been examined with the aid of g.c.–m.s.[88] The ratio of products from PGE_2 carboxamide rearrangement is independent of the *N*-acyl group.

New derivatives of prostaglandins A_1 and E which exhibit very simplified and characteristic mass spectral fragmentation patterns have been reported.[89] The

---

[78] J. J. Myher, A. Kuksis, L. Marai, and S. K. F. Yeung, *Anal. Chem.*, 1978, **50**, 557.
[79] J. J. Myher and A. Kuksis, *Can. J. Biochem.*, 1979, **57**, 117.
[80] T. G. Tornabene and T. A. Langworthy, *Science*, 1979, **203**, 51.
[81] R. J. Kulmacz, A. Kisic, and G. T. Schroepfer, *Chem. Phys. Lipids*, 1979, **23**, 29.
[82] J. Jacob, *Z. Phys. Chem.*, 1978, **359**, 653.
[83] A. Kuksis, W. C. Breckenridge, J. J. Myher, and G. Kakis, *Can. J. Biochem.*, 1978, **56**, 630.
[84] W. J. Baas and C. J. Niemann, *Z. Naturforsch., Teil C*, 1979, **34**, 5.
[85] A. C. Thompson and W. E. Knight, *Phytochem.*, 1978, **17**, 1755.
[86] C. J. W. Brooks, C. G. Edmonds, S. J. Gaskell, and A. G. Smith, *Chem. Phys. Lipids*, 1978, **21**, 403.
[87] R. A. Klein, *Chem. Phys. Lipids*, 1978, **21**, 291.
[88] F. C. Falkner, *Biomed. Mass Spectrom.*, 1979, **6**, 221.
[89] J. Rosello and E. Gelpi, *Biomed. Mass Spectrom.*, 1978, **5**, 531; J. Rosello and E. Gelpi, *J. Chromatogr. Sci.*, 1978, **16**, 177.

prostaglandins were treated with 1:1 mixtures of the silylating reagent BSTFA and a non-aromatic nitrogen heterocycle to yield novel derivatives such as (1). These derivatives were used to detect prostaglandin A and 19-hydroxyprostaglandin A molecules in human seminal plasma.

(1)

Chemical ionization g.c.–m.s. has been utilized in a study of prostaglandins A, B, E, and F.[90] Methane, isobutane, and ammonia c.i. spectra were reported and comparison of the results indicated that ammonia c.i. yielded the best spectra. Differentiation of the α and β isomers of the PGF series and of the *syn* and *anti* isomers of the PGA and PGE series was possible on the basis of their c.i. mass spectra. Further work using ammonia c.i. g.c.–m.s. enabled the identification of the prostaglandins thromboxane B_2 and 6-ketoprostaglandin $F_{1\alpha}$.[91]

The major release product of rabbit peritoneal polymorphonuclear leucocytes has been identified as thromboxane B_2 by g.c.–m.s. of its TMS derivative.[92] Ion cluster techniques, involving the preparation of a synthetic prostaglandin, cloprostenol, labelled with ^{14}C, have facilitated the g.c.–m.s. identification of its metabolites.[93]

G.c.–m.s. has aided studies of the products of arachidonic acid metabolism. A combination of microchemical degradation and g.c.–m.s. has enabled the 'slow reacting substance' (SRS) to be identified as a cysteine-t-hydroxy-7,9,11,14-icosatetranoic acid derivative.[94] The conversion of arachidonate to prostaglandins has been the subject of a number of g.c.–m.s. studies. Microsomal syntheses of prostaglandins in sheep lung preparations have been demonstrated by capillary g.c.–m.s.,[95] and feeding effects on prostaglandin production,[96] as well as other aspects of prostaglandin biosynthesis,[97] have been examined.

Evidence for the existence of 8-isoprostaglandins in human semen has been obtained by g.c.–m.s. studies of their *O*-methyl and *O*-TMS derivatives.[98] 8-Iso-PGE$_1$, 8-iso-PGE$_2$, 8-iso-PGF$_{1\alpha}$, 8-iso-PGF$_{2\alpha}$, and the four corresponding 19-hydroxyprostaglandins were identified. The g.c.–m.s. characteristics of TMS

---

[90] T. Ariga, M. Suzuki, I. Morita, and T. Miyatake, *Anal. Biochem.*, 1978, **90**, 174.
[91] I. Morita, S. Murota, M. Suzuki, T. Ariga, and T. Miyatake, *J. Chromatogr.*, 1978, **154**, 285.
[92] E. M. Davidson, A. W. Ford-Hutchinson, M. J. H. Smith, and J. R. Walker, *Br. J. Pharmacol.*, 1978, **63**, 407p.
[93] G. R. Bourne, S. R. Moss, P. J. Phillips, J. T. A. Webster, and D. F. White, *Biomed. Mass Spectrom.*, 1979, **6**, 359.
[94] R. C. Murphy, S. Hammarstrom, and B. Samuelsson, *Proc. Natl. Acad. Sci. U.S.A.*, 1979, **76**, 4275.
[95] H.-H. Tai, B. Yuan, and A. T. Wu, *Biochem. J.*, 1978, **170**, 441.
[96] H. R. Knapp, O. Oelz, A. R. Whorton, and J. A. Oates, *Lipids*, 1978, **13**, 804.
[97] C. R. Pace-Asciak and G. Ragaraj, *Biochim. Biophys. Acta*, 1978, **529**, 13.
[98] P. L. Taylor, *Prostaglandins*, 1979, **17**, 259.

and TBDMS ethers of prostaglandins have been evaluated in order to assess their relative merits in the analysis of prostaglandin metabolites.[99]

Carbohydrates.—A large number of studies on bacterial carbohydrates has been completed with the aid of g.c.–m.s. For example, a method of 'fingerprinting' the carbohydrates of *Streptococcus mutans* by g.c.–m.s. of their TMS derivatives has been developed.[100] G.c.–m.s., in conjunction with other spectroscopic techniques, has been employed in investigations of the capsular polysaccharides of, amongst others, *Klebsiella* serotypes K23[101], K41,[102] K18,[103] and K31.[104] Similar studies include analyses of isomeric D-ribofuranosylribitol disaccharides from *Haemophilus influenza* type b and *E. coli* K100,[105] and of the capsular antigen from *Streptococcus pneumoniae* type 26.[106] Flavonoid *O*- and *C*-glycosides isolated from the moss *H. ciliata* have been analysed by chemical and degradative studies, including g.c.–m.s.[107]

The first report of the application of c.i. g.c.–m.s. to the analysis of derivatized amino sugars in bacterial lipopolysaccharides contains some interesting data.[108] It was possible to differentiate some 2-amino-2-deoxysugars by both electron impact and methane chemical ionization g.c.–m.s. The *O*-TMS ethers of D-glucose, D-galactose, and D-mannose have been characterized by c.i. g.c.–m.s. using ammonia as the reagent gas.[109] Sufficient fragmentation was observed in the mass spectra to differentiate between diastereoisomers. Chemical ionization of gluconoride derivatives using ammonia or isobutane as reagent gases does not yield sufficient enhancement of molecular ions but introduction of small amounts of pyridine into the ion source yields intense $[M + 80]^+$ ions, greatly simplifying the analysis of these derivatives.[110] Isobutane c.i. capillary g.c.–m.s. of the TMS derivatives of carbohydrates in uremic serum has been reported.[111] TMS derivatives of alditols and aldonic acid yielded quasi-molecular ions, whereas aldoses did not. The g.c.–m.s. and mass spectrometric fragmentation pathways of the *O*-TMS derivatives of hexuronic acids and their lactones have been reported.[112]

The structural analysis of oligosaccharides by tagging the reducing end sugars with 2-aminopyridine and then hydrolysing the molecule, remethylating, and analysing the products by g.c.–m.s. has been demonstrated.[113] A new method for the degradation of the protein part of glycoproteins, leaving the carbohydrate

---

[99] A. R. Brash and T. A. Baillie, *Biomed. Mass Spectrom.*, 1978, **5**, 346.
[100] H. A. S. Aluyi and D. B. Drucker, *J. Chromatogr.*, 1979, **178**, 209.
[101] G. G. S. Dutton, K. L. Mackie, A. V. Savage, and M. D. Stephenson, *Carbohydr. Res.*, 1978, **66**, 125.
[102] J.-P. Joselau, M. Lapeyre, M. Vignon, and G. G. S. Dutton, *Carbohydr. Res.*, 1978, **67**, 197.
[103] G. G. S. Dutton, K. L. Mackie, and M.-T. Yang, *Carbohydr. Res.*, 1978, **65**, 251.
[104] C.-C. Cheng, S.-L. Wong, and Y.-M. Choy, *Carbohydr. Res.*, 1978, **73**, 169.
[105] B. A. Fraser, F.-P. Tsui, and W. Egan, *Carbohydr. Res.*, 1979, **73**, 59.
[106] L. Kenna, B. Lindberg, and J. K. Madden, *Carbohydr. Res.*, 1979, **73**, 175.
[107] B.-G. Osterdahl, *Acta Chem. Scand.*, Ser. B, 1979, **33**, 119.
[108] D. V. Bowser, R. G. Teece, and S. M. Soman, *Biomed. Mass Spectrom.*, 1978, **5**, 627.
[109] T. Murata and S. Takahashi, *Carbohydr. Res.*, 1978, **62**, 1.
[110] L. P. Johnson, S. C. Subba Rao, and C. Fenselau, *Anal. Chem.*, 1978, **50**, 2022.
[111] A. C. Schoots and P. A. Leclercq, *Biomed. Mass Spectrom.*, 1979, **6**, 502.
[112] J. F. Kennedy and S. M. Robertson, *Carbohydr. Res.*, 1978, **67**, 1.
[113] S. Hase, T. Ikenaka, and Y. Matsushima, *Biochem. Biophys. Res. Commun.*, 1978, **85**, 257.

portion intact, has been exemplified using the glycoprotein asialofetnin.[114] Carbohydrate chains were identified by glass capillary g.c.–m.s.

The reactivity of methyl glycosides, oligo-, and poly-saccharides towards trifluoroacetolysis has been studied by g.c.–m.s.[115] The results indicated that trifluoroacetic anhydride/trifluoroacetic acid mixtures might be suitable for the isolation of carbohydrate chains from biological material. Cellulose degradation products formed by thermochemical reactions have been identified with the aid of capillary g.c.–m.s. as a series of aliphatic and alicyclic hydrocarbons, aldehydes, ketones, alcohols, and furans.[116]

The acetylated alditols of 17 methyl esters of L-glycero-D-mannoheptose and D-glycero-D-mannoheptose have been distinguished by capillary g.c.–m.s.[117] The D and L configurations of neutral monosaccharides have been determined by g.c.–m.s. of their TMS (−)-2-butylglycoside derivatives.[118] The g.c.–m.s. of trimethylsilylated 3-ketoses, 2-heptuloses, sedo-heptulose, and coriose has been described[119] and a g.c.–m.s. approach to the identification of aldoses by use of their permethylated aldononitrile derivatives has been devised.[120] A new deacetylation method for 2-acetamido-2-deoxy sugars has been examined by g.c.–m.s.[121] and the distribution of the D-galactosyl side chains in guaran has been determined by capillary g.c.–m.s.[122]

Steroids and Terpenoids.—The value of g.c.–m.s. techniques in steroid research has been stressed in a recent review.[123] A number of steroid derivatives with good g.c.–m.s. properties have been reported. The 20,21 cyclic boronates of aldosterone yield abundant molecular ions and [M–46]$^+$ ions, and also have very simple isobutane c.i. spectra.[124] Reports of the use of TBDMS derivatives of steroids in g.c.–m.s. studies continue to appear. For example, TBDMS-imidazole reacts with hormonal hydroxy steroids and α,β-unsaturated ketosteroids to produce derivatives suitable for g.c.–m.s. quantification.[125] Linked B/E scanning has been used to supplement g.c.–m.s. data obtained on stereoisomeric androstanediol TBDMS ethers,[126] and the stability and g.c.–m.s. of TBDMS ethers of some oestrogens and androstanes have been investigated.[127] New derivatives, dimethylethylsilyl and dimethylisopropylsilyl ethers, have been used successfully in the g.c.–m.s. analysis of bile acids and cholesterol.[128,129] These ethers elute

---

[114] B. Nilsson and S. Svensson, *Carbohydr. Res.*, 1979, **72**, 183.
[115] B. Nilsson and S. Svensson, *Carbohydr. Res.*, 1979, **69**, 292.
[116] P. M. Molton, R. K. Miller, J. M. Donovan, and T. F. Demitt, *Carbohydr. Res.*, 1979, **75**, 199.
[117] J. Radziejewska-Lebrecht, D. H. Shaw, D. Borowiak, I. Fromme, and H. Mayer, *J. Chromatogr.*, 1979, **179**, 113.
[118] G. Gerwig, J. P. Kamerling, and J. F. G. Vliegenhart, *Carbohydr. Res.*, 1978, **62**, 349.
[119] T. Okuda, S. Saito, and M. Hayashi, *Carbohydr. Res.*, 1979, **68**, 1.
[120] F. R. Seymour, E. C. M. Chen, and S. H. Bishop, *Carbohydr. Res.*, 1979, **73**, 19.
[121] B. Nilsson and S. Svensson, *Carbohydr. Res.*, 1978, **62**, 377.
[122] J. Hoffman and S. Svensson, *Carbohydr. Res.*, 1978, **65**, 65.
[123] C. J. W. Brooks, *Philos. Trans. R. Soc. London, Ser. A*, 1979, **293**, 53.
[124] S. J. Gaskell and C. J. W. Brooks, *J. Chromatogr.*, 1978, **158**, 331.
[125] I. A. Blair and G. Phillipou, *J. Chromatogr. Sci.*, 1978, **16**, 201.
[126] S. J. Gaskell, A. W. Pike, and D. S. Millington, *Biomed. Mass Spectrom.*, 1979, **6**, 78.
[127] L. Ballhorn, W. F. Mueller, and F. Korte, *Steroids*, 1979, **33**, 379.
[128] H. Miyazaki, M. Ishibashi, and K. Yamashita, *Biomed. Mass Spectrom.*, 1978, **5**, 469.
[129] H. Miyazaki, M. Ishibashi, and K. Yamashita, *Biomed. Mass Spectrom.*, 1979, **6**, 57.

in regular order, according to the number of hydroxyls originally present, when they are chromatographed on non-polar phases.

The g.c.–m.s. characteristics of TMS ethers of bile alcohols have been examined.[130] Methods for g.c.–m.s. analysis of steroids based on exchange oxime functions have been devised.[131] Conversion of oxime into methoxime functional groups was achieved by acid-catalysed reaction with the appropriate hydroxylamine in pyridine solution and the technique was applied to the characterization of saturated 3-ketosteroids in biological fluid. The influence of the 6-TMS group on the fragmentation of some hydroxy steroids has been investigated as part of a g.c.–m.s. study.[132]

Diethylstilbestrol and related compounds and their acetate, methyl ether, and TMS derivatives have been examined in a g.c.–m.s. study of possible steroid carcinogens.[133] The metabolism of these compounds to catechol has also been studied.[134] A number of g.c.–m.s. studies of marine sterols have been completed. These include black sea sterols,[135] new C_{26} and C_{30} sterols in an oceanic sponge,[136] two new marine sterols from Pacific sponges,[137] and from an Australian sponge,[138] and plant sterols found in aquatic vertebrates.[139] Algal sterols have been analysed by capillary g.c.–m.s. of their TMS and TBDMS ethers[140] and a g.c.–m.s. study of sterols in plants on which livestock graze has demonstrated that no definite correlation between fertilizer application and the sterol content of grasses exists.[141] The biosynthesis of free sterols and sterol esters of *Neurospora crassa* has been investigated by g.c.–m.s.[142] and c.i. g.c.–m.s. has aided the study of the metabolism of sterols by A-ring oxidation.[143] Two new bile alcohols have been isolated from hagfish and identified with the assistance of g.c.–m.s. of their TMS ethers[144] and toad venom sterols have been analysed by capillary g.c.–m.s. of their TMS ethers.[145] The 7-hydroxylation of 3-oxygenated C_{27}—C_{29} sterols in rat liver has been followed by g.c.–m.s.[146] A computerized g.c.–m.s. system

---

[130] G. S. Tint, B. Dayal, A. K. Batta, S. Shefer, F. W. Cheng, G. Salen, and E. H. Mosbach, *J. Lipid Res.*, 1978, **19**, 956.
[131] M. Axelson and J. Sjovall, *J. Steroid Biochem.*, 1978, **9**, 816; M. Axelson, *Anal. Biochem.*, 1978, **86**, 133.
[132] D. J. Harvey and P. Vouros, *Biomed. Mass Spectrom.*, 1979, **6**, 135.
[133] L. L. Engel, P. J. Marshall, J. C. Orr, V. N. Reinhold, and P. Carter, *Biomed. Mass Spectrom.*, 1978, **5**, 582.
[134] J. Weidenfeld, P. Carter, V. N. Reinhold, S. B. Tanner, and L. L. Engel, *Biomed. Mass Spectrom.*, 1978, **5**, 587.
[135] J. J. Boon, W. I. C. Rijpstra, F. De Lange, J. W. Leeuw, M. Yoshioka, and Y. Shimizu, *Nature*, 1979, **277**, 125.
[136] J. A. Ballantine, A. Lavis, and R. J. Morris, *Comp. Biochem. Physiol.*, 1979, **63B**, 119.
[137] C. Delseth, L. Tolelea, P. J. Schever, R. J. Wells, and C. Djerassi, *Helv. Chim. Acta*, 1979, **62**, 101.
[138] W. Horheinz and G. Oesterhelt, *Helv. Chim. Acta*, 1979, **62**, 1307.
[139] T. Takagi, A. Sakai, K. Hayashi, and Y. Itabashi, *Lipids*, 1979, **14**, 5.
[140] B. P. Lisboa and J. M. Halket in 'Recent Developments in Mass Spectrometry in Biochemistry and Medicine', ed. A. Frigerio, Plenum, New York, 1978, Vol. 1, p. 441.
[141] M. Pailer and P. Riedl, *Monatsh. Chem.*, 1978, **109**, 1167.
[142] A. M. Pierce, H. D. Pierce, A. M. Unrau, A. C. Oehlschlager, R. E. Subden, and R. L. Renaud, *Can. J. Biochem.*, 1979, **57**, 112.
[143] G. A. S. Ansari and L. L. Smith, *Chem. Phys. Lipids*, 1978, **22**, 55.
[144] M. Une, K. Kihira, T. Kuramoto, and T. Hoshita, *Tetrahedron Lett.*, 1978, 2527.
[145] P.-Q. Kinh and A. B. Svendsen, *J. Ethnopharmacol.*, 1979, **1**, 197.
[146] L. Aringer, *J. Lipid Res.*, 1978, **19**, 933.

has enabled the characterization of acetylated sterol fractions isolated from latex-bearing plants.[147]

The bile acids from the gall bladder and liver of a 3200 year old Egyptian mummy have been shown to correspond with those of modern man in a g.c.–m.s. study.[148] Bile acid methyl esters have been analysed by c.i. g.c.–m.s. using methane as both carrier and reagent gas.[149] 5β-Cholanoic acids, representing simple chemical and enzymatic degradation products of bile acids, have been investigated and resolved by g.c.–m.s.[150] The free sterols and steroidal acids of tall oil pitch have been characterized by capillary g.c.–m.s.[151]

Many biochemical and biomedical g.c.–m.s. studies of steroids have been completed. For example, the urinary excretion of 5β-pregnane-3α,6α,20α-triol during human gestation has been confirmed by g.c.–m.s. of its TMS ether.[152] A study of steroids and hypertension reports capillary g.c.–m.s. data on TMS derivatives of 3,15- and 3,16-dihydroxy-17-oxo steroids[153] and g.c.–m.s. has aided the identification of the C_{21} steroid 19-*nor*-11-deoxy-corticosterone in the urine of rats with regenerating adrenals.[154] The structure of a previously unidentified component of the urinary steroids of hirsute women has been established as 3α-hydroxy-5β-pregnane-11,20-dione (11-oxopregnanolone) by g.c.–m.s.[155] Computer-assisted capillary g.c.–m.s. has been applied to the investigation of steroids in human seminal fluid; dehydroepiandrosterone and its sulphate conjugate were the most abundant steroids present.[156] Further investigations into the conversion of cholesterol to pregnenolone in the adrenal cortex have shown that one of the intermediates is 5α,6α-epoxycholestan-3β-ol and not the previously assigned 20,22-epoxycholesterol.[157]

A capillary g.c.–m.s. study reports the determination of synthetic anabolic steroids in human urine.[158] The detection of 17α-methyl-17β-hydroxyandrosta-1,4-dien-3-one (2) and two of its major metabolites was possible, following oral administration of (2).

(2)

[147] P. Nielsen, H. Nishimura, Y. Liang, and M. Calvin, *Phytochem.*, 1979, **18**, 103.
[148] A. Kuksis, P. Child, J. J. Myher, L. Marai, I. M. Yousef, and P. K. Lewin, *Can. J. Biochem.*, 1978, **56**, 1141.
[149] G. M. Muschik, L. H. Wright, and J. A. Schroer, *Biomed. Mass Spectrom.*, 1979, **6**, 266.
[150] P. Child, A. Kuksis, and L. Marai, *Can. J. Biochem.*, 1979, **57**, 216; P. Child, A. Kuksis, and J. J. Myher, *Can. J. Biochem.*, 1979, **57**, 639.
[151] B. Holmbom and V. Era, *J. Am. Oil Chem. Soc.*, 1978, **55**, 342.
[152] R.-J. Begue, M. Moriniere, and P. Padieu, *J. Steroid Biochem.*, 1978, **9**, 779.
[153] R.-J. Begue, M. Dumas, M. Moriniere, C. Nivois, and P. Padieu, *Biomed. Mass Spectrom.*, 1979, **6**, 476.
[154] C. E. Gomez-Sanchez, O. B. Holland, B. A. Murry, H. A. Lloyd, and L. Milewich, *Endocrinology*, 1979, **105**, 708.
[155] J. A. Fleetwood, A. L. Latner, P. A. Smith, J. H. Borthwick, C. G. Edmunds, and C. J. W. Brooks, *J. Steroid Biochem.*, 1978, **9**, 1193.
[156] J. McK. Halket, H.-K. Albers, and B. P. Lisboa, *J. Steroid Biochem.*, 1978, **9**, 816.
[157] T. Watabe and T. Sawahata, *Biochem. Biophys. Res. Commun.*, 1978, **83**, 1396.
[158] H. W. Durbeck, I. Buker, B. Scheulen, and B. Telin, *J. Chromatogr.*, 1978, **167**, 117.

G.c.–m.s. studies of biologically important androsterones and androsterols have been completed using the formate derivatives of these compounds.[159] The mass spectra afforded intense [M-HCOOH]$^+$ ions. Electron impact and chemical ionization techniques have been employed in the g.c.–m.s. analysis of steroidal spironolactones.[160] Electron impact spectra exhibited weak molecular ions but CI spectra, particularly when ammonia was used as the reagent gas, yielded molecular weight information readily. A summary of the state of research into the metabolism of cholesterol and steroid hormone production indicates that product identification by g.c.–m.s. of TMS derivatives contributes much information to this type of study.[161]

Naturally occurring stictane triterpenoids have been identified by g.c.–m.s. of their TMS derivatives; ten di- and tri-oxygenated triterpenoids were identified in lichen extracts.[162] Abscissic acid and its metabolites have been identified in seeds by g.c.–m.s.[163] and a g.c.–m.s. study of the metabolism of gibberellin A29 in seeds has revealed the presence of a hitherto untraced metabolite.[164] The early eluting fractions of the alkaline hydrosylate of a leaf extract have been examined by g.c.–m.s. and a number of terpene alcohols were identified.[165]

Amino-acids and Peptides.—The synthesis[18] and application[19] of the g.c.–m.s. analysis of modified amino-acids have already been mentioned. A new, potentially useful series of derivatives of α-amino-acids, 2-trifluoromethyl-3-oxazol-5-ones, have been investigated with special attention to their g.c.–m.s. properties.[166] These derivatives (3), in which R is the side chain of the amino-acid from which they are formed, exhibit intense fragment ions of type (4) and (5) in their mass spectra.

(3)

[R—C=N=CH—CF$_3$]$^+$ [R—C≡NH]$^+$
(4)   (5)

These ions allow the nature of the amino-acid side chain to be deduced directly or indirectly. Another interesting aspect of this work was the evaluation of a number of g.c. phases which resulted in the combination of two phases (SE-30 and OV-225) according to the method of Laub et al.[167] The mixed phase system

---

[159] B. R. Pettit and G. S. King, *Biomed. Mass Spectrom.*, 1979, **6**, 162.
[160] D. R. Boreham, G. C. Ford, N. J. Haskins, C. W. Vose, and R. F. Palmer, *Biomed. Mass Spectrom.*, 1978, **5**, 524.
[161] G. S. Boyd, *Biochem. Soc. Trans.*, 1978, **6**, 893.
[162] P. T. Holland and A. L. Wilkins, *Org. Mass Spectrom.*, 1979, **14**, 160.
[163] A. D. Adesomu, J. I. Okogon, D. E. U. Ekong, and P. Gaskin, *Phytochem.*, 1980, **19**, 223.
[164] V. M. Sponsel and J. McMillan, *Planta*, 1978, **144**, 69.
[165] H. A. Lloyd, S. L. Evans, and H. M. Fales, *Lloydia*, 1978, **41**, 494.
[166] V. Ferrito, R. Borg, J. Eagles, and G. R. Fenwick, *Biomed. Mass Spectrom.*, 1979, **6**, 499.
[167] R. J. Laub, J. H. Purnell, D. M. Summers, and P. S. Williams, *J. Chromatogr.*, 1978, **155**, 1.

yielded good separations of these derivatives. Similar derivatives, the 2-bis-(chlorodifluoromethyl)-4-substituted-1,3-oxazolidin-5-ones, have also shown promise as derivatives for g.c.–m.s. analysis of amino-acids on both packed and capillary columns.[168] Isobutyl ester, $N(O)$-heptafluorobutyrate derivatives have been adopted as suitable derivatives for the quantitative analysis of α-amino-acids using packed and capillary column g.c.–m.s. methods.[169]

C.i. g.c.–m.s. techniques have been used to determine the steric purity of amino-acids and amines.[170] The methyl esters of 2,4,5-trichlorophenoxyacetic acid conjugates of amino-acids have been characterized,[171] and evidence for the presence of these conjugates in soybean callus tissue has been presented.[172]

An extensive study of the g.c.–m.s. characteristics of about 200 TMS derivatized dipeptides, containing the 20 most common amino-acids, has been presented.[173] Chemical ionization and electron impact mass spectra have also been compared in a g.c.–m.s. study of over 40 TMS dipeptides.[174] The c.i. spectra contained more intense mass ions than the e.i. spectra and still exhibited sufficient fragmentation to characterize the peptide sequence, thus c.i. g.c.–m.s. is more sensitive than electron impact for this type of molecule. The g.c.–m.s. properties of a large number of NO-pentafluoropropionyl esters of dipeptides have been determined in both chemical ionization and electron impact modes.[175] An isotope ratio g.c.–m.s. computer technique which is capable of analysing the amino-acid composition of less than 100 μg of protein has been devised.[176] The amino-acids were converted into N-trifluoroacetyl, n-butyl esters, and the entire procedure (excluding hydrolysis) required less than 2.5 hours. A method for calculating the g.c. retention indices of perfluorodideuterioalkylated peptide derivatives has been incorporated into a g.c.–m.s. computer program which is capable of identifying peptides.[177] The synthesis and back-exchange of ^{18}O-labelled amino-acids used as internal standards in g.c.–m.s. have been investigated.[178] It was concluded that back exchange was sufficiently slow to allow the use of these standards in most quantitative g.c.–m.s. studies.

The use of borane as a reducing agent in the sequence analysis of peptides by g.c.–m.s. has been evaluated.[179] Borane reduction was compared with the more usual lithium aluminium deuteride reduction and was found to be superior. Borane reduction followed by derivatization and g.c.–m.s. allowed detection of amounts as low as 0.2 nanomole, an order of magnitude lower than for reduction with lithium aluminium hydride. It was predicted that the application of capillary column techniques with splitless injection to this type of analysis would lower

---

[168] R. Liardon, U. Ott-Kuhn, and P. Husek, *Biomed. Mass Spectrom.*, 1979, **6**, 381.
[169] J. Desgres, D. Boisson, and P. Padieu, *J. Chromatogr.*, 1979, **162**, 133.
[170] C. Wiecek, B. Halpern, A. M. Sargeson, and A. M. Duffield, *Org. Mass Spectrom.*, 1979, **14**, 281.
[171] M. Arjmand, R. H. Hamilton, and R. O. Mumma, *J. Agric. Food Chem.*, 1978, **26**, 898.
[172] M. Arjmand, R. H. Hamilton, and R. O. Mumma, *J. Agric. Food Chem.*, 1978, **26**, 1125.
[173] H. C. Krutzsch and J. J. Pisana, *Biochem.*, 1978, **17**, 2791.
[174] H. C. Krutzsch and T. J. Kindt, *Anal. Biochem.*, 1979, **92**, 525.
[175] W. E. Seifert, R. E. McKee, C. F. Beckner, and R. M. Caprioli, *Anal. Biochem.*, 1978, **88**, 149.
[176] J. J. Rafter, M. Ingelman-Sundberg, and J.-A. Gustafsson, *Biomed. Mass Spectrom.*, 1979, **6**, 317.
[177] B. Van De Graaf, P. J. W. Schuyl, H. C. Beyerman, and A. Knol-Kalkman, *Rec. Trav. Chim. Pays-Bas*, 1978, **97**, 175.
[178] R. C. Murphy and K. L. Clay, *Biomed. Mass Spectrom.*, 1979, **6**, 309.
[179] H. Frank, H. J. C. Das Neves, and E. Bayer, *J. Chromatogr.*, 1978, **152**, 357; H. Frank and D. M. Desiderio, *Anal. Biochem.*, 1978, **90**, 413.

detection limits still further. Urinary peptides have been examined by g.c.–m.s. analysis of their derivatized polyamino-alcohols formed by lithium aluminium deuteride reduction.[180] The structure of the potent toxin cyclochlorotine (6), isolated from *Penicillium islandicum* Sopp, was established by g.c.–m.s. analysis of derivatized hydrolysis products following reduction to polyamino-alcohols.[181]

```
                    Me            OH
                    |             |
                    CH₂           CH₂
                    |             |
         Cl    NH—CH—CO—NH—CH
          \   |                   |
           \  CO                  CO
    Cl—⟨    ⟩                     |
           \ CH                   NH
            \|                    |
             N                    CH—Ph
             |                    |
             CO—CH—NH—CO—CH₂
                 |
                 CH₂
                 |
                 OH
                (6)
```

The mass spectral characteristics of polyamino-alcohol-related derivatives of peptides containing β-phenylalanine, α-aminobutyric acid, and dichloroproline were also determined during the course of this investigation. The complete primary structure of the purple membrane protein bacteriorhodopsin, which contains 248 amino-acid residues, has been established.[182] The sequence was derived by ordering of the CNBr fragments on the basis of methionine containing peptides identified by g.c.–m.s. and by analysis of N-bromosuccinimide fragments which overlapped with the CNBr fragments.

A g.c.–m.s. method for the analysis of ergot-peptide alkaloids has been devised.[183] Instantaneous degradation of the free alkaloids occurs in the injection port of the gas chromatograph at 300 °C. The lactam-type degradation products are then identified by c.i. g.c.–m.s. using methane as both carrier and reagent gas.

Nucleotides, Nucleosides, and Related Compounds.—New derivatives of cytosine-3,N^4-etheno-O-persilyl compounds have proved useful in the g.c.–m.s. analysis of nucleosides and nucleotides.[184] These derivatives, of which (7) is an example, are the first nucleotide derivatives of cytosine to be successfully examined by g.c. Ethenylation of the base blocks further derivatization, thus precluding chromatographic problems associated with trimethylsilylation at the N-4 position in cytosine.

The reaction yielding these derivatives is selective for cytosine and adenine bases and the resulting molecules have good g.c. properties as well as affording more abundant molecular ions than the corresponding TMS derivatives.

---

[180] W. Steiner, A. Otten, Z. Biro, and A. Niederwieser, *Experientia*, 1978, **34**, 918.
[181] R. J. Anderegg, K. Biemann, A. Manmade, and A. C. Ghosh, *Biomed. Mass Spectrom.*, 1979, **6**, 129.
[182] H. G. Khorana, G. E. Gerber, W. C. Herlihy, C. P. Gray, R. J. Anderegg, K. Nihel, and K. Biemann, *Proc. Natl. Acad. Sci. U.S.A.*, 1979, **76**, 5046; *ibid.*, 1979, **76**, 227.
[183] F. J. W. Mansvelt, J. E. Greving, and R. A. De Zeeuw, *J. Chromatogr.*, 1978, **151**, 113.
[184] K. H. Schram, Y. Taniguchi, and J. A. McCloskey, *J. Chromatogr.*, 1978, **155**, 355.

(7)

The g.c.–m.s. of acetyl, methyl, TMS, and homologous alkyl oxime TMS derivatives of cytosine arabinoside and related pyrimidine nucleosides has been studied.[185] The detection limit for methyl acetyl derivatives of cytosine arabinoside was approximately 500 pg but for the methyl trimethylsilyl derivatives it was lower (ca. 50 pg). The oxime derivatives gave less sensitivity.

A study of the mass spectra of suitably derivatized seleno analogues of purine and pyrimidine bases and nucleosides (potential anticancer agents) contains some g.c.–m.s. as well as direct probe data.[186] A number of g.c.–m.s. techniques for the identification of cytokins has been published.[187-189]

Clinical and Metabolic Studies.—Applications of g.c.–m.s. in this area of research continues to expand at a tremendous rate. A number of reviews have been published, including one on the place of g.c.–m.s. in clinical chemistry.[190] Other reviews encompass reports on mass spectrometry and combined techniques in medicine,[191] metabolic profiling,[192] analysis of drugs and metabolites by g.c.–m.s.,[193] drug determination by several techniques including g.c.–m.s.,[194] selected ion monitoring in pharmacology,[195] mass spectrometry in a biomedical environment,[196] volatile constituents of body fluids,[197] and the application of m.s. and g.c.–m.s. to forensic and toxicological problems.[198]

[185] J. Boutagy and D. J. Harvey, *J. Chromatogr.*, 1978, **156**, 153.
[186] J. G. Liehr, C.-L. Weise, P. F. Crain, G. H. Milne, D. S. Wise, L. B. Townsend, and J. A. McCloskey, *J. Heterocyclic Chem.*, 1979, **16**, 1263.
[187] M. Claeys, E. Messens, M. Van Montagu, and J. Schell, *Quant. Mass Spec. Life Sci.*, 1978, **2**, 409; M. Claeys, E. Messens, M. Van Montagu, and J. Schell, *Fresenius' Z. Anal. Chem.*, 1978, **290**, 125.
[188] T. Hashizume, T. Sugiyama, M. Imura, H. T. Cory, M. F. Scott, and J. A. McCloskey, *Anal. Biochem.*, 1979, **92**, 111.
[189] B. Dauphin, G. Teller, and B. Durand, *Planta*, 1979, **144**, 113.
[190] H. M. Liebich, *J. Chromatogr.*, 1978, **146**, 185.
[191] H. M. Liebich, *J. Clin. Chem. Clin. Biochem.*, 1978, **16**, 677.
[192] S. C. Gates and C. C. Sweeley, *Clin. Chem.*, 1978, **24**, 1663.
[193] N. Haskins, *Proc. Anal. Div. Chem. Soc.*, 1978, **15**, 256.
[194] K. U. Bühring, *Arzneim.-Forsch.*, 1978, **28(II)**, 1944.
[195] D. J. Jenden and A. K. Cho, *Biochem. Pharmacol.*, 1979, **28**, 705.
[196] J. D. Baty, *Proc. Anal. Div. Chem. Soc.*, 1979, **16**, 51.
[197] S. D. Sastry, K. T. Bulk, J. Janak, M. Dressler, and G. Preti, in 'Biochemical Applications of Mass Spectrometry', First Supplementary Vol., ed. G. R. Waller and O. C. Dermer, Wiley-Interscience, New York, 1980.
[198] R. M. Smith, *Int. Lab.*, 1978, 115.

A large number of 'metabolic profiling' studies have relied on g.c.–m.s. techniques; for example a detailed description of g.c.–m.s. procedures for the qualitative and quantitative analysis of components in urinary organic acid profiles has been given.[199] The procedure was able to distinguish contributions from at least 150 substances in a single sample and was used to compare three groups of subjects: healthy adults, sick children, and children with neuroblastoma.[200] A rapid and simplified procedure for metabolic profiling of urinary organic acids has been devised.[201] The main improvement comprises a new solid-phase extraction procedure. Several inborn errors of metabolism have been identified by a capillary g.c.–m.s. computer technique.[202] The use of relative retention indices, common to most metabolic profiling methods, was found to yield reproducible values, even between different laboratories. The use of forward library searching techniques and the consequent lack of need for a large data base were emphasized, as was the ability to detect and quantify new components. Capillary g.c.–m.s. techniques under both electron impact and chemical ionization conditions have been used to profile uremic serum,[203] and high-resolution g.c.–high-resolution m.s. has aided identification of organic acids in human urine.[204] The application of pattern recognition and feature extraction to the analysis of volatile constituent metabolic profiles has been described.[205] A normal *versus* pathological prediction rate of 93.75% was achieved by this capillary g.c. technique in which important profile constituents were identified by g.c.–m.s. Approximately two-fifths of 500 detected acidic urine components have been identified by capillary g.c.–m.s.[206] Previously unassigned high molecular weight acids in a metabolic profile have been identified as cinnamoyl glycine and acetyl tributylcitrate (a common plasticizer).[207] Phthalate interferences in the g.c. determination of long-chain fatty acids in plasma have been identified by g.c.–m.s.[208] The organic acid composition of normal human amniotic fluid at different periods of gestation[209] and a study of the volatile components in sera of normal and virus-infected humans[210] have been aided by g.c.–m.s. methods. A number of fatty-acid methyl esters in the kidneys and liver of patients with Reye's syndrome have been identified.[211] Metabolic errors investigated by g.c.–m.s. include propionic acidaemia, in which 3-keto-2-methylvaleric and 3-hydroxy-2-methylvaleric acids were discovered,[212] and aciduria in which 5-hydroxyhexanoic acid was detected.[213] G.c.–m.s. has also shown that alpha

[199] S. C. Gates, N. Dendramis, and C. C. Sweeley, *Clin. Chem.*, 1978, **24**, 1674.
[200] S. C. Gates, C. C. Sweeley, W. Krivit, D. Dewitt, and B. E. Blaisdell, *Clin. Chem.*, 1978, **24**, 1680.
[201] P. J. Anderson, W. L. Fitch, and B. Halpern, *J. Chromatogr.*, 1978, **146**, 481.
[202] W. L. Fitch, P. J. Anderson, and D. H. Smith, *J. Chromatogr.*, 1979, **162**, 249.
[203] A. C. Schoots, F. E. P. Mikkers, C. A. M. G. Cramers, and S. Ringoir, *J. Chromatogr.*, 1979, **164**, 1.
[204] S. Lewis, C. N. Kenyon, J. Meili, and A. L. Burlingame, *Anal. Chem.*, 1979, **51**, 1275.
[205] M. L. McConnell, G. Rhodes, U. Watson, and M. Novotny, *J. Chromatogr.*, 1979, **162**, 495.
[206] M. Spiteller and G. Spiteller, *J. Chromatogr.*, 1979, **164**, 253.
[207] G. K. Brown, O. Stokke, and E. Jellum, *J. Chromatogr.*, 1978, **145**, 177.
[208] C. Geisler, A. R. Swanson, L. Steve, and M. W. Anders, *Clin. Chem.*, 1979, **25**, 308.
[209] T. M. Nicholls, R. Hahnel, and S. P. Wilkinson, *Clin. Chem. Acta*, 1978, **84**, 11.
[210] A. Zlatkis, K. Y. Lee, C. F. Poole, and G. Holzer, *J. Chromatogr.*, 1979, **163**, 125.
[211] M. Sudo, K.-I. Tanioka, T. Momoi, K. Akaishi, and Y. Suzuki, *Clin. Chim. Acta*, 1978, **84**, 179.
[212] R. J. W. Truscott, C. J. Pullin, B. Halpern, J. Hammond, E. Haan, and D. M. Danks, *Biomed. Mass Spectrom.*, 1979, **6**, 294.
[213] R. A. Chalmers and A. M. Lawson, *Biomed. Mass Spectrom.*, 1979, **6**, 445.

keto-acids play an important role in the pathenogenesis of ketotic hypoglycaemia.[214] A g.c.–m.s. study of organic acids in the synovial fluid of arthritic patients has revealed acid concentrations 5—10 times higher than in fluid from normal joints.[215]

Some metabolic errors result in abnormal saccharide excretions and, as with urinary organic acids, many of these compounds have been analysed by g.c.–m.s. Glucose-containing oligosaccharides have been detected in the urine of patients with glycogen storage disease, following partial hydrolysis and analysis as either alditol acetates or permethylation products.[216] Novel saccharides and glycoasparagines have been observed in a g.c.–m.s. investigation of the hydrolysis products of extracts from the urine of patients suffering from fucosidosis.[217] The reduced, acetylated oligosaccharide derivatives isolated from the liver of a patient with Sandhoff–Jatzkewitz disease have been identified by g.c.–m.s.[218] and abnormal deoxyribose metabolites have been detected by g.c.–m.s. of their TMS derivatives in a study of a possible new inborn error of metabolism.[219] A rapid screening for carbohydrate intolerance in children involves g.c.–m.s. analysis of faecal disaccharides[220] and g.c.–m.s. has detected monosaccharide derivatives in human seminal plasma.[221]

A systematic study of drinking water and air samples by g.c.–m.s. was undertaken as part of the search for the cause of Legionnaire's disease before the involvement of micro-organisms was proved.[222] G.c.–m.s. studies have refuted earlier reports of the presence of haemopyrolle and kryptopyrolle in the urine of schizophrenics.[223] A g.c.–m.s. method for the analysis of biologically important β-carbolines has been presented[224] and the presence of 6-hydroxymelatonin in normal human urine has been confirmed.[225] The detection of a fungal metabolite, D-arabinitol, in human serum has been accomplished by g.c.–m.s.[226] and a g.c.–m.s. study of nucleosides in leukaemic patients has been reported.[227] The presence of guanidino compounds in rat and bovine brain has been confirmed by g.c.–m.s. of the dimethylpyrimidyl derivatives of these molecules.[228]

Many studies of drug metabolism have been considerably aided by g.c.–m.s. techniques. Chapter 11 deals with mass spectrometric studies of drug metabolism in detail. Some publications of special note include a capillary g.c.–m.s. analysis

[214] H. J. Sternowsky, Fortschr. Med., 1978, **96**, 1411.
[215] J. Heininger, E. Munthe, J. Pahle, and E. Jellum, J. Chromatogr., 1978, **158**, 297.
[216] A. Lundblad, J. Lundsten, S. Svensson, and A. Hager, Eur. J. Biochem., 1978, **83**, 325.
[217] J. Lundsten, N. E. Norden, S. Sjoblad, S. Svensson, P. A. Ockerman, and M. Gehlhoff, Eur. J. Biochem., 1978, **83**, 513; N. M. K. Ng Ying Kin and L. S. Wolfe, Biochem. Biophys. Res. Commun., 1979, **88**, 696.
[218] N. M. K. Ng Ying Kin and L. S. Wolfe, Carbohydr. Res., 1978, **67**, 522.
[219] R. J. W. Truscott, B. Halpern, J. Hammond, S. Hunt, R. G. H. Cotton, E. A. Haan, and D. D. Danks, Biomed. Mass Spectrom., 1979, **6**, 453.
[220] R. M. Thompson, R. D. Coffin, M. R. Glick, and J. F. Fitzgerald, Clin. Chem. Acta, 1978, **84**, 185.
[221] P. Storstet, O. Stokke, and E. Jellum, J. Chromatogr., 1978, **145**, 351.
[222] I. H. Sufet and P. R. Cairo, Anal. Chem., 1978, **50**, 875A.
[223] P. L. Gender, H. A. Duhan, and H. Rapoport, Clin. Chem., 1978, **24**, 230.
[224] D. W. Showmaker, T. G. Bidder, H. G. Boettger, J. T. Cummins, and M. Evans, J. Chromatogr., 1979, **174**, 159.
[225] T. Miyata, Y. Okano, K. Murao, K. Fukunaga, K. Takahama, and Y. Kase, Life Sci., 1979, **25**, 1731.
[226] T. E. Kiehn, E. M. Bernard, J. W. M. Gold, and D. Armstrong, Science, 1979, **206**, 577.
[227] S. P. Dutta, P. F. Crain, J. A. McCloskey, and G. B. Chheda, Life Sci., 1979, **24**, 1381.
[228] A. Mori, T. Ichimura, and H. Matsumoto, Anal. Chem., 1978, **89**, 393.

of drugs and optically active metabolites on a chiral stationary phase[229] (studies on amino-acids using this phase have already been discussed[18,19]). The phase was used successfully to chromatograph α-amino-alcohols, glycols, aromatic and aliphatic α-hydroxy-carboxylic acids, and amines as well as α-amino-acids. Species differences in the metabolism of an amphetamine metabolite[230] and of isoxepal[231] have been investigated by g.c.–m.s.

The application of g.c.–m.s. to the field of industrial hygiene has been discussed.[232] Examples of its use include the analysis of vinyl chloride during PVC dust inhalation, and the analysis of solvent mixtures. G.c.–m.s. has aided the study of the metabolism of many foreign compounds, for example perchlorethylene and its urinary metabolites have been examined[233] and a study of pentachlorophenol metabolism *in vitro* and *in vivo* has been reported.[234] Hexachlorobenzene toxicology, with particular reference to sulphur-containing metabolites, has been studied.[235] Seven diols and four stereoisomeric 7,8,9,10-tetrols, together with twelve phenols, diol epoxides, and quinones, have been identified in a g.c.–m.s. study of benzo(*a*)pyrene metabolism.[236] The presence of the pesticide endrin and one of its metabolites, 12-ketoendrin, has been confirmed by g.c.–m.s. in a study of foetotoxic effects of endrin in rodents.[237]

Ecological and Environmental Studies.—The sensitivity and specificity of g.c.–m.s. have resulted in an ever increasing expansion of the application of this technique to the detection of pollutants, pesticides, *etc.* in the environment. Review articles relevant to this topic include a perspective of priority pollutants in which g.c.–m.s. analysis of organic pollutants is discussed,[238] surveys of various techniques (including g.c.–m.s.) which have been used to detect organic compounds in waste water[239] and in water and air,[240] and a brief survey of micropollutant analysis by g.c.–m.s.[241] The value of g.c.–m.s. in a broad spectrum approach to environmental analysis has been described,[242] and a review of recent developments in the mass spectrometric determination of trace toxic substances in food has been published.[243] An extensive survey of nitrosamine analysis has emphasized the need for capillary g.c.–high-resolution m.s. studies,[244] and the use of g.c.–m.s. in the analysis of pesticides and pesticide residues has been discussed.[245] A review of xenobiotics in the environment has also been published.[246] A

---

[229] H. Frank, G. J. Nicholson, and E. Bayer, *J. Chromatogr.*, 1978, **146**, 197.
[230] R. T. Coutts, G. R. Jones, and R. E. Townsend, *J. Pharm. Pharmacol.*, 1978, **30**, 415.
[231] H. P. A. Illing and J. M. Fromson, *Drug Metab. Disp.*, 1978, **6**, 310.
[232] G. Choudhury and C. V. Cooper, *Am. Ind. Hygiene Assoc. J.*, 1979, **40**, 39.
[233] W. A. Braun, *J. Chromatogr.*, 1978, **150**, 212.
[234] U. G. Ahlborg, K. Larsson, and T. Thunberg, *Arch. Toxicol.*, 1978, **40**, 45.
[235] G. Koss, W. Koransky, and K. Steinbach, *Arch. Toxicol.*, 1979, **42**, 19.
[236] G. Takahashi, K. Kinoshita, K. Hashimoto, and K. Yasuhira, *Cancer Res.*, 1979, **39**, 1814.
[237] N. Chernoff, R.-J. Kavlock, R. C. Hanisch, and D. A. Whitehouse, *Toxicology*, 1979, **13**, 155.
[238] L. H. Keith and W. A. Telliard, *Environ. Sci. Technol.*, 1979, **13**, 416.
[239] R. A. Hites and V. Lopez, *Anal. Chem.*, 1979, **51**, 1452A.
[240] L. H. Keith, *J. Chromatogr. Sci.*, 1979, **17**, 48.
[241] D. E. Wells, *Proc. Anal. Div. Chem. Soc.*, 1980, **17**, 116.
[242] W. L. Budde and J. W. Eichelberger, *Anal. Chem.*, 1979, **51**, 567A.
[243] R. Self, *Biomed. Mass Spectrom.*, 1979, **6**, 361.
[244] T. A. Gough, *Analyst (London)*, 1978, **103**, 785.
[245] J. A. Sphon and W. C. Brumley in ref. 197, ch. 23.
[246] R. C. Dougherty in ref. 197, ch. 23A.

discussion of the cost-effective analysis of priority pollutants concludes that overall costs are lower for a g.c.–m.s.-based laboratory and, moreover, g.c.–m.s. provides more reliable data than a laboratory equipped with g.c. alone.[247]

Air and Water Pollutants. A procedure for detecting the breakdown products of plastic compounds used in injection moulding operations has been described.[248] Samples were heated in a special apparatus and the trapped vapours were subsequently analysed by g.c. or g.c.–m.s. Trace components in air have been identified by serial mass fragmentographic runs over the entire mass range.[249] Volatiles from airborne particulates have been analysed by g.c.–m.s. computer methods which include g.c. retention time data as well as mass spectrometric data in the identification routine.[250] The isolation and analysis of polar organic fractions from airborne particulate matter have been accomplished by capillary g.c.–m.s. techniques.[251] Twenty compounds were identified and these were mainly polyols, thought to arise by microbial action. A capillary g.c.–m.s. method for the detection and identification of organic pollutants in the atmosphere has been exemplified by the detection of N-nitrosodimethylamine.[252] A multidetection approach to the analysis of organic pollutants in water uses a variety of g.c. methods, including capillary g.c.–m.s., to identify over 100 semi-volatile compounds.[253] The major components were found to be hydrocarbons, chlorinated hydrocarbons, phthalates, fatty acids, and silicones.

A number of g.c.–m.s. studies of industrial wastewater effluents have appeared. One investigation employs g.c.–f.t./i.r. and h.p.l.c.–f.t/i.r. as well as g.c.–m.s. techniques.[254] The spectral data obtained from g.c.–f.t./i.r. complemented the mass spectral data, especially in the ability to distinguish isomers not distinguishable by mass spectrometry alone; the g.c.–m.s. methods were more sensitive. Toxic compounds in the waste water of several textile plants have been studied by g.c.–m.s., both before and after tertiary treatment of the wastewater.[255] Forty-four priority pollutants (ten of which appeared in blank runs) were identified. Halogenated lipophilic compounds in spent bleach liquor from a sulphite pulp mill have been analysed by capillary g.c.–m.s.[256] Volatile organic compounds, including alkanes, alkenes, aldehydes, aromatic hydrocarbons, methyl sulphur, and halogenated compounds, have been identified in coastal seawater with the aid of capillary g.c.–m.s. (in both e.i. and c.i. modes).[257] The contamination of Dutch coastal waters with chlorinated hydrocarbons has

---

[247] D. W. Hoyt and D. E. Smith, *Environ. Sci. Technol.*, 1979, **13**, 534.
[248] A. A. Grote, W. S. Kim, and R. E. Kupel, *Am. Ind. Hygienists' Assoc. J.*, 1978, **39**, 880.
[249] T. Fujii, Y. Yokouchi, and Y. Ambe, *J. Chromatogr.*, 1979, **176**, 165.
[250] F. W. Karasek, D. W. Denney, K. W. Chan, and R. E. Clement, *Anal. Chem.*, 1978, **50**, 82.
[251] B. Wauters, F. Vangaever, P. Sandra, and M. Verzele, *J. Chromatogr.*, 1979, **170**, 133.
[252] J. T. Bursey, D. Smith, R. N. Williams, R. D. Berkley, and E. D. Pellizzari, *Int. Lab.*, 1978, **11**, 20.
[253] R. Ferrand, M. Mazza, P. Payen, and B. P. Cerchar, in 'Aquatic Pollutants: Transformation and Biological Effects', ed. O. Hutzinger, L. H. Van Lelyvand, and B. C. Zoetman, Pergamon, Oxford, 1978, p. 87.
[254] K. H. Shafer, S. V. Lucas, and R. T. Jakobsen, *J. Chromatogr. Sci.*, 1979, **17**, 464.
[255] G. D. Rawlings and M. Samfield, *Environ. Sci. Technol.*, 1979, **13**, 160.
[256] G. Eklund, B. Josefson, and A. Bjorseth, *J. Chromatogr.*, 1978, **150**, 161.
[257] R. P. Schwarzenbach, R. H. Bromund, P. M. Gschwend, and O. C. Zafirou, *Org. Geochem.*, 1978, **1**, 93.

been confirmed by capillary g.c.–m.s. analysis.[258] Organic compounds in air samples collected near a water treatment plant have also been identified by capillary g.c.–m.s.[259] A large number of thiophenes, thiazoles, pyrazines, phenols, alcohols, and unsaturated hydrocarbons, many of them odorous compounds, were found.

An important new application of g.c.–m.s. in environmental analysis is to the 'fingerprinting' of marine pollutant hydrocarbons.[260] Both packed and capillary g.c.–m.s. methods were used to establish characteristic profiles of different petroleum products. Similar analytical techniques have been employed in the determination of petroleum hydrocarbons in water, fish, and sediment following an oil drilling platform accident.[261] The presence of pollutant hydrocarbons in estuarine coastal slime has been confirmed by g.c.–m.s.[262]

Polycyclic Aromatic Hydrocarbons (PAHs). The use of PAHs as qualitative markers in g.c.–m.s. analysis of petroleum in the environment has been suggested.[263] A rapid and simple sample preparation technique, involving ultrasonic extraction, solvent partition, and silica gel chromatography, has been developed as an aid to the g.c.–m.s. analysis of PAHs.[264] Products of the photo-oxidation of phenanthrene under simulated environmental conditions have been identified by capillary g.c.–m.s.[265] and a simple, two-step procedure for the isolation of PAHs from whole smoke condensate prior to g.c.–m.s. analysis has been devised.[266] A computerized capillary g.c.–m.s. system has been used to identify PAHs in particulate matter from industrial environments.[267] More than one hundred PAHs, including heterocyclic molecules, were found. PAHs have been detected in PVC smoke particulates by g.c.–m.s. using a non-extractable polymer-coated g.c. column.[268] The g.c.–m.s. detection of PAHs in smoke condensates from materials containing tobacco substitutes has been described.[269] A modified g.c.–m.s. method was used to detect peaks eluting beyond benzo(a)pyrene. Some potential carcinogens, the K and non-K dihydrodiols of 7,12-dimethylbenz[a]anthracene, have been synthesized and analysed by g.c.–m.s.[270] PAHs mainly derived from anthropogenic combustion sources have been identified and quantified in soils and sediments.[271] Capillary g.c.–m.s. methods have been used to monitor PAH emissions from a wood- and peat-burning boiler, confirming that PAH emissions from peat burning are ten times higher than from wood burning.[272] Coal-tar, which contains some of the most important

[258] J. K. Quirijns, C. G. Van Der Paauw, M. C. Ten Noever De Brauw, and R. H. Devos, *Total Environ.*, 1979, **13**, 225.
[259] M. Hangartner, *Int. J. Environ. Anal. Chem.*, 1979, **6**, 161.
[260] J. Albaiges and P. Albrecht, *Int. J. Environ. Anal. Chem.*, 1979, **6**, 171.
[261] J. R. Law, *Pollution Bull.*, 1978, **9**, 321.
[262] S. Thompson and G. Eglinton, *Estuarine Coastal Mar. Sci.*, 1979, **8**, 75.
[263] H. S. Hertz, W. E. May, S. A. Wise, and F. R. Guenther, *Int. J. Environ. Anal. Chem.*, 1978, **5**, 259.
[264] Y. L. Tan, *J. Chromatogr.*, 1979, **176**, 319.
[265] J. R. Patel, I. R. Politzer, G. W. Griffin, and J. L. Laseter, *Biomed. Mass Spectrom.*, 1978, **5**, 664.
[266] R. J. Levins, *Chromatographia*, 1978, **11**, 736.
[267] A. B. Josreth and G. Eklund, *Anal. Chim. Acta*, 1979, **105**, 119.
[268] J. C. Liao and R. F. Browner, *Anal. Chem.*, 1978, **50**, 1683.
[269] C. D. Briggs and S. J. A. Hawthorne, *Proc. Anal. Div. Chem. Soc.*, 1978, **15**, 181.
[270] L. K. Wong, C.-L. A. Wang, and F. B. Daniel, *Biomed. Mass Spectrom.*, 1979, **6**, 305.
[271] J. G. Windsor and R. A. Hites, *Geochim. Cosmochim. Acta*, 1979, **43**, 27.

carcinogenic PAHs, has been analysed by capillary g.c.–m.s.[273] An homologous series of alkyl-9-fluorenones has been identified in diesel exhaust with the aid of e.i. and c.i. g.c.–m.s.[274] High-resolution mass spectrometry and g.c.–f.t./i.r. studies were also necessary to identify positively these substances.

Organochlorine Residues. G.c.–m.s. analysis of complex mixtures of organochlorine compounds, particularly aromatic molecules, continues to receive much attention. A number of papers concerned with the analysis of chlorinated dibenzodioxins have appeared, provoking some disagreement about the power of the g.c.–m.s. analytical techniques used. For example, a report in which it was stated that low-resolution g.c.–low-resolution m.s. was adequate for the detection of 2,3,7,8-tetrachlorodibenzodioxin (TCDD) at parts per trillion (p.p.t.) levels,[275] contrary to claims by other workers,[276] has prompted a detailed rejoinder.[277] Here, it was stated that l.r.g.c.–l.r.m.s. alone is suitable only if no interferences are found in control samples. In the absence of suitable controls, and when sample clean-up is nonspecific, positive results require confirmation by high-resolution mass spectrometry. The example analysis of beef fat[277] led to three conclusions for that particular analysis: (i) low-resolution g.c.–low-resolution m.s. is sufficient for evaluating dioxin levels >20 p.p.t., (ii) high-resolution mass spectrometry is necessary to check samples yielding positive results for TCDD, (iii) the total analytical system used was not 100% certain of identifying positive samples. Investigations of the negative c.i. g.c.–m.s. detection of polychlorinated dibenzodioxins (PCDDs) have revealed that this technique is 1000-fold more sensitive than methane positive c.i. [278] Possible interferences in the g.c.–m.s. analysis of PCDDs in 2,4,5-trichlorophenol have been identified as chlorinated benzyl phenyl ethers.[279] The synthesis and g.c.–m.s. analysis of the 22 TCDD isomers have been described.[280] The combination of h.p.l.c. separation and g.c.–m.s. analysis allowed assignment of the identity of all 22 isomers.

Papers describing the non-detection of TCDD at levels above 1.2 p.p.t.[281] and the detection of TCDD together with other chlorinated compounds[282] in fly ash have been published. TCDDs formed by photolysis of isomeric hexachlorodibenzo-*p*-dioxins[283] and by combustion of chlorophenols[284] have been analysed by g.c.–m.s. The g.c.–m.s. detection of polychlorinated dibenzofurans (PCDFs), formed by uncontrolled burning of polychlorinated biphenyls (PCBs),

[272] T. Alsberg and U. Stenberg, *Chemosphere*, 1979, **8**, 487.
[273] H. Borwitzky and G. Schomburg, *J. Chromatogr.*, 1979, **170**, 99.
[274] M. D. Erikson, D. L. Newton, E. Pellizzari, K. B. Tomer, and D. Dropkin, *J. Chromatogr. Sci.*, 1979, **17**, 449.
[275] A. di Domenico, F. Merli, L. Boniforti, I. Camoni, A. di Muccio, F. Taggi, L. Vergori, G. Colli, G. Elli, A. Gorni, P. Grassi, G. Invernizzi, A. Jemma, L. Luciani, F. Gattabeni, L. de Angelis, G. Galli, C. Chiabrando, and R. Fanelli, *Anal. Chem.*, 1979, **51**, 735.
[276] R. A. Hummel, *J. Agric. Food Chem.*, 1977, **25**, 1049.
[277] R. A. Hummel and L. A. Shadoff, *Anal. Chem.*, 1980, **52**, 191.
[278] J. R. Hass, M. D. Friesen, D. J. Harvan, and C. E. Parker, *Anal. Chem.*, 1978, **50**, 1474.
[279] L. A. Shadoff, W. W. Blaser, C. W. Kocher, and H. G. Fravel, *Anal. Chem.*, 1978, **50**, 1586.
[280] T. J. Nestrick, L. L. Lamparski, and R. H. Stehl, *Anal. Chem.*, 1979, **51**, 2273.
[281] B. J. Kimble and M. L. Gross, *Science*, 1980, **207**, 59.
[282] G. A. Eiceman, R. E. Clement, and F. W. Karasek, *Anal. Chem.*, 1979, **51**, 2343.
[283] H. R. Buser, *Chemosphere*, 1979, **8**, 251.
[284] B. Jansson, G. Sundstrom, and B. Ahling, *Sci. Total Environ.*, 1978, **10**, 209.

has been reported.[285] The analysis of PCDFs is of particular importance as these compounds have been detected in Yusho oil, a substance associated with rice oil 'Yusho poisoning'.[286—288] The methane/oxygen-enhanced negative c.i. spectra of polychlorinated diphenyl ethers have been determined in a g.c.–m.s. study,[289] and a negative c.i. g.c.–m.s. investigation of octachlorohydroxydiphenyl ethers in technical pentachlorophenol has been reported.[290] G.c.–m.s. techniques have been used to detect PCBs in sediments and oysters,[291] goldfish,[292] household products (including detergents, softeners, disinfectants, soap powder, soap bars, and toilet paper),[293] and in the Severn estuary.[294]

Agrochemicals and Related Compounds. A book devoted to plant growth substances contains information on the analysis of these compounds by techniques which include g.c.–m.s.[295]

A multiresidue analytical method, employing electron capture g.c., t.l.c., and g.c.–m.s. methods, has been applied to the analysis of organochlorine, organophosphorus, dinitrophenyl, and carbamate pesticides in apples.[296] The oxidation products of ten organothiophosphorus pesticides in fruit and vegetables have been determined by g.c.–m.s. and i.r. methods, confirming that the respective P=O oxygen analogues were formed in each case.[297] Important metabolites of organophosphorus pesticides have been identified as their benzyl ester derivatives with the aid of electron impact, charge exchange, and c.i. g.c.–m.s.[298] Electron impact and methane chemical ionization g.c.–m.s. has been applied to the analysis of (alkylamino)-S-triazine herbicides.[299] Soil degradation products of triflaurin ($\alpha\alpha\alpha$-trifluoro-2,6-dinitro-NN-dipropyl-p-toluidine), a selective herbicide, have been investigated comprehensively by direct insertion mass spectrometry and g.c.–m.s.[300] The photolysis products of a synthetic pyrethroid (Pydrin) have been studied by chemical ionization g.c.–m.s. with either methane or isobutane as the reagent gas.[301] The major products were those of photo-induced decarboxylation. Toxic impurities in the herbicide Diuron have been identified by g.c.–m.s.[302]

---

[285] H. R. Buser, H. P. Bosshardt, C. Rappe, and R. Lindahl, *Chemosphere*, 1978, **7**, 419; H. R. Buser and C. Rappe, *ibid.*, 1979, **8**, 157.
[286] H. R. Buser, C. Rappe, and A. Gara, *Chemosphere*, 1978, **7**, 439.
[287] L. R. Kamps, W. J. Trotter, S. J. Young, L. J. Carson, J. A. G. Roach, J. A. Sphon, J. T. Tanner, and B. McMahon, *Bull. Environ. Contam. Toxicol.*, 1978, **20**, 589.
[288] C. Rappe, H. A. Buser, H. Kuroki, and Y. Masuda, *Chemosphere*, 1979, **8**, 259.
[289] K. L. Busch, A. Norstrom, M. M. Bursey, J. R. Hass, and C. A. Nilsson, *Biomed. Mass Spectrom.*, 1979, **6**, 157.
[290] M. Beinzer, D. Griffin, T. Miller, and R. Skinner, *Biomed. Mass Spectrom.*, 1979, **6**, 301.
[291] J. Teichman, A. Bevenue, and J. W. Hylin, *J. Chromatogr.*, 1978, **151**, 155.
[292] E. Herbst, I. Scheuvert, W. Klein, and F. Korte, *Chemosphere*, 1978, **7**, 221.
[293] D. T. Williams and F. M. Benoit, *Bull. Environ. Contam. Toxicol.*, 1979, **21**, 179.
[294] M. Cooke, G. Nickless, A. Povey, and D. J. Roberts, *Sci. Total Environ.*, 1979, **13**, 17.
[295] M. L. Brenner, in 'Plant Growth Substances', ed. N. B. Mandava, American Chemical Society, Washington, D.C., 1979, Chapter 8.
[296] C. E. Johansson, *Pesticide Sci.*, 1978, **9**, 313.
[297] J. Singh and W. P. Cochrane, *J. Assoc. Off. Anal. Chem.*, 1979, **62**, 751.
[298] C. G. Daughton, A. M. Cook, and M. Alexander, *Anal. Chem.*, 1979, **51**, 1949.
[299] P. A. Leclerq and V. Pacakova, *J. Chromatogr.*, 1979, **178**, 193.
[300] T. Golab and J. L. Occolowitz, *Biomed. Mass Spectrom.*, 1979, **6**, 1.
[301] R. L. Holmstead, D. G. Fullmer, and L. O. Ruzo, *J. Agric. Food Chem.*, 1979, **26**, 954.
[302] G. Sundstrom, B. Jansson, and L. Renberg, *Chemosphere*, 1978, **7**, 973.

The analysis of organochlorine residues in the environment, other than pesticides, has already been discussed. Considerable efforts have also been devoted to the g.c.–m.s. analysis of organochlorine pesticides, for example 'carbon skeleton' g.c.–m.s. techniques have been used to analyse DDT and other compounds in soil.[303,304] This method involves the conversion of the organochlorine compounds to hydrocarbons by heterogeneous catalytic hydrodechlorination prior to g.c.–m.s. analysis. The method has also been applied to PCB and polychlorinated alkane analysis.[304] Reductive dechlorination products of the pesticides Aldrin and Dieldrin have also been identified by g.c.–m.s.[305] The absorption, body distribution, metabolism, and excretion of Dieldrin in non-human primates and other laboratory animals have also been studied by g.c.–m.s.[306] The results indicated that non-human primates are a far more reliable model than rats for estimating environmental chemical risks in man.

Organic Geochemistry. Molecular fossils of archaebacteria in kerogen have been identified by g.c.–m.s. as C_{40} hydrocarbons (originally connected to the kerogen polymers by ether linkages).[307] The isolation, g.c.–m.s. identification, and use of paraffinic hydrocarbons (from supercritical-gas extracts of coal) as organic geochemical markers have been discussed.[308]

Sterols and stanols found in the oxic and anoxic waters of the Black Sea have been identified by capillary g.c.–m.s. of their TMS derivatives.[309] Other steroid derivatives detected in sediments by g.c.–m.s. methods include a monoaromatic steroid in marine oil shales and crude oil,[310] sterols in Western North Atlantic coastal areas,[311] and steroid ketones of the South-West African Shelf.[312]

The pyrolysis products of asphaltenes have been shown to be a useful source of geochemical information in a g.c.–m.s. study.[313] Capillary g.c.–m.s. has been applied to the analysis of aliphatic and aromatic hydrocarbons in Antarctic marine sedimentary layers.[314]

Acids and bases in shale oils have been analysed by capillary g.c.–i.r. and g.c.–m.s.[315]

The amino-acids dissolved in interstitial waters of marine sediments have been analysed by capillary g.c.–m.s. of their *N*-heptafluorobutyryl n-butyl esters.[316] Retention times, chemical ionization (methane), and electron impact mass spectra aided the identification of the amino-acids.

---

[303] M. Cooke, K. D. Khallef, G. Nickless, and D. J. Roberts, *J. Chromatogr.*, 1979, **170**, 183.
[304] M. Cooke, G. Nickless, and D. J. Roberts, *J. Chromatogr.*, 1980, **187**, 47.
[305] W. J. Cooper, W. H. Dennis, I. R. De Leon, and J. L. Laseter, *Chemosphere*, 1979, **8**, 191.
[306] W. Muller, G. Nohynek, F. Korte, and F. Coulston, *Z. Naturforsch., Teil C*, 1979, **34**, 340.
[307] W. Michaelis and P. Albrecht, *Naturwissenschaften*, 1979, **66**, 420.
[308] K. D. Bartle, D. W. Jones, H. Pakdel, C. E. Snape, A. Calimli, A. Olcay, and T. Tugrul, *Nature*, 1979, **227**, 284.
[309] R. B. Gagosian and I. Heinzer, *Geochim. Cosmochim. Acta*, 1979, **43**, 471.
[310] J. Schaefle, B. Ludwig, P. Albrecht, and G. Ourisson, *Tetrahedron Lett.*, 1978, 4163.
[311] C. Lee, J. W. Farrington, and R. B. Gagosian, *Geochim. Cosmochim. Acta*, 1979, **43**, 35.
[312] R. B. Gagosian and S. O. Smith, *Nature*, 1979, **277**, 287.
[313] I. Rubinstein, C. Spyckerelle, and O. P. Strausz, *Geochim. Cosmochim. Acta*, 1979, **43**, 1.
[314] H. M. Plall and P. R. Mackie, *Nature*, 1979, **280**, 576.
[315] P. C. Uden, A. P. Carpenter, H. M. Hackett, and D. E. Henderson, *Anal. Chem.*, 1979, **51**, 38.
[316] S. M. Henrichs and J. W. Farrington, *Nature*, 1979, **279**, 319.

Food and Flavour Components.—The appearance of an authoritative review of recent developments in the mass spectrometric estimation of trace toxic substances in food has already been mentioned.[243] The application of g.c.–m.s. to the analysis of food, food additives, and food contaminants has also been reviewed.[317] A review of all the known constituents of tomatoes contains listings of 135 compounds which have been identified by g.c.–m.s. methods.[318]
A simple and efficient external inlet assembly which allows direct g.c.–m.s. analysis of volatile components from raw and processed foods has been described.[319] Phenolic compounds formed by thermal fragmentation during the processing of coffee have been investigated by capillary g.c.–m.s.[320] and another high-resolution g.c.–m.s. study has investigated possible emetics (mainly hydroxylated aromatics) in an acidified roast coffee infusion.[321] A new Amadori compound, 1-deoxy-1-L-theanino-D-fructopyranose, has been identified in green tea by g.c.–m.s. of its TMS derivative.[322] G.c.–m.s. studies of flavour components in beer,[323] gin,[324] and apple brandy,[325] have been completed.

Cigarette smoke flavour components have been analysed in a number of g.c.–m.s. studies.[326-329] The complex constitution of cigarette smoke condensates necessitates the use of glass capillary column chromatography in all of these investigations. Smoke condensate preparations used as flavour additives have also been analysed by capillary g.c.–m.s.[330,331] Over seventy components were identified in one of these studies.[330]

Sixty-five volatile constituents of prickly pears,[332] and an aroma constituent (megastigma-5,8-dien-4-one) common to yellow passion fruit and Virginia tobacco,[333] have been identified. Thirty-two components of cooked artichoke volatiles,[334] components of baked potato volatiles,[335] and of mayonnaise (by direct analysis)[336] have been determined. Volatiles of roast meat,[337] and a meat off-flavour (identified as 2-methyl-6-chlorophenol disinfectant),[338] have been analysed by g.c.–m.s. Flavour qualities in soy protein products[339] and soy sauce[340]

[317] D. Jahr and P. Bennemann, *Z. Anal. Chem.*, 1979, **298**, 337.
[318] K. Herrman, *Z. Lebensm.-Unters. Forsch.*, 1979, **169**, 179.
[319] M. G. Legendre, G. S. Fisher, W. H. Schuller, H. P. Dupuyand, and F. T. Rayner, *J. Am. Oil Chem. Soc.*, 1979, **56**, 552.
[320] W. Rahn, H.-W. Meyer, and W. A. Konig, *Z. Lebensm.-Unters. Forsch.*, 1979, **169**, 346.
[321] W. Rahn and W. A. Konig, *J. High Res. Chromatogr. Chromatogr. Commun.*, 1978, **1**, 69.
[322] T. Anan, *J. Sci. Food Agric.*, 1979, **30**, 906.
[323] T. L. Peppard and J. M. F. Douse, *J. Chromatogr.*, 1979, **176**, 444.
[324] D. W. Clutton and M. Evans, *J. Chromatogr.*, 1978, **167**, 409.
[325] P. Schreier, F. Drawert, and M. Schmid, *J. Sci. Food Agric.*, 1978, **29**, 728.
[326] H. Sakuma, N. Shimojima, and S. Sugawara, *Agric. Biol. Chem.*, 1978, **42**, 359.
[327] S. Ishiguro and S. Sugawara, *Agric. Biol. Chem.*, 1978, **42**, 407.
[328] S. Ishiguro and S. Sugawara, *Agric. Biol. Chem.*, 1978, **42**, 1527.
[329] E. Demole and P. Enggist, *Helv. Chim. Acta*, 1978, **61**, 2318.
[330] W. Baltes and I. Sochtig, *Z. Lebensm.-Unters. Forsch.*, 1979, **169**, 9.
[331] W. Baltes and I. Sochtig, *Z. Lebensm.-Unters. Forsch.*, 1979, **169**, 17.
[332] R. A. Flath and J. M. Takahashi, *J. Agric. Food Chem.*, 1978, **26**, 835.
[333] E. Demole, P. Enggist, M. Winter, A. Furrer, K. H. Schultz-Elte, B. Egger, and G. Ohloff, *Helv. Chim. Acta*, 1979, **62**, 67.
[334] R. G. Buttery, D. G. Guadagni, and L. C. Ling, *J. Agric. Food Chem.*, 1978, **26**, 791.
[335] E. C. Coleman and C.-T. Hao, *J. Agric. Food Chem.*, 1980, **28**, 66.
[336] S. P. Fore, M. G. Legendre, and G. S. Fischer, *J. Am. Oil Chem. Soc.*, 1978, **55**, 482.
[337] P. Dubs and M. Joho, *Helv. Chim. Acta*, 1978, **61**, 2809.
[338] R. L. S. Patterson and C. A. Voyle, *Proc. Anal. Div. Chem. Soc.*, 1979, **16**, 128.
[339] N. Nunomura, M. Sasaki, Y. Asao, and T. Yokotsuka, *Agric. Biol. Chem.*, 1978, **42**, 2123.
[340] E. T. Rayner, M. G. Legendre, and H. P. Dupuy, *J. Am. Oil Chem. Soc.*, 1978, **55**, 454.

have been identified, hexanal being the most important volatile component of the former and thirty-five compounds (of which pyrazines were thought to be the most important) being found in the latter.

Possible corn ear worm attractants in corn husks[341] and volatiles from rice and corn[342,343] have been identified by g.c.-m.s. A total of 146 compounds were analysed in the last of these investigations. The epicuticular wax of rice has also been analysed and found to consist mainly of n-alkanes, esters, aldehydes, and primary alcohols.[344]

The odorous substance of two fungi which have a characteristic, pear-like smell has been identified as methyl cinnamate[345] and the major volatile products of an algal pilot plant have been shown to be a series of *nor*-carotenoids.[346]

A study of heated and irradiated fats, oils, and model substances has led to the identification of twenty hydrocarbons of which hexadecadiene is the most important.[347] A capillary g.c.-m.s. analysis of butter triglycerides has also been completed.[348]

One of the main bitter principles in apricot kernels has been identified, with the aid of electron impact, and ammonia and isobutane chemical ionization g.c.-m.s., as β-D-glucopyranoside.[349]

A g.c.-m.s. method—essentially a profiling technique—for the detection of adulteration in lavender products has been devised.[350]

Compounds associated with the spoilage of meat by bacteria have been identified as the valine catabolites isobutyronitrile and isobutyraldoxime-*O*-methyl ether.[351]

Allelochemicals and Other Secretions.—Allelochemicals are compounds involved in interspecies communication in the animal and plant kingdom. These chemical messengers can be beneficial (*e.g.* aggregation and sex pheromones) or harmful to survival (*e.g.* kairomones). The use of g.c.-m.s. techniques in the analysis of allelochemicals has increased greatly, mainly owing to its sensitivity which allows studies to be performed on very small amounts of material. This is particularly important in allelochemical studies where sample collection may be difficult because of the intractability (or even ferocity!) of the species concerned. Furthermore, in applications such as insect studies, biochemists are highly appreciative of sensitive techniques which allow studies to be performed on tens or hundreds of insects rather than thousands or tens of thousands, thus reducing considerably the amount of labour involved in such investigations.

---

[341] R. G. Buttery, L. C. Ling, and B. G. Chan, *J. Agric. Food Chem.*, 1978, **26**, 366.
[342] M. G. Legendre, H. P. Dupuy, R. L. Ory, and W. O. McIlrath, *J. Agric. Food Chem.*, 1978, **26**, 1035.
[343] T. Tsugita, T. Kurata, and M. Fujimaki, *Agric. Biol. Chem.*, 1978, **42**, 643.
[344] G. Bianchi, E. Lupotto, and S. Russo, *Experientia*, 1979, **35**, 1417.
[345] J. A. Schmitt, *Z. Naturforsch., Teil C*, 1979, **33**, 817.
[346] F. Juttner, *Z. Naturforsch., Teil C*, 1979, **34**, 186.
[347] F. Drawert, B. Beck, and H. Hammerstinol, *Lebensm.-Unters. Forsch.*, 1979, **168**, 99.
[348] P. P. Schmid, M. D. Muller, and W. Simon, *J. High Res. Chromatogr. Chromatogr. Commun.*, 1979, **2**, 675.
[349] A. Kjaer, J. O. Madsen, and M. Sponholtz, *Acta Chem. Scand., Ser. B*, 1978, **32**, 588.
[350] M. J. Prager and M. A. Miskiewicz, *J. Assoc. Off. Anal. Chem.*, 1979, **62**, 1231.
[351] P. A. Gibbs and D. B. Harper, *Biochem. J.*, 1979, **182**, 609.

A survey of the application of mass spectrometry to the study of allelochemicals has appeared recently.[352]

Mammalian Allelochemicals. G.c.–m.s. methods have been employed in the identification of the volatile constituents of human axilla.[353] The major components found included a series of isopropyl esters of fatty acids, several saturated and unsaturated aldehydes and ketones, and some furan derivatives; many of these were of exogenous origin.

Investigations of the chest gland secretions of a species of Bush baby have revealed the presence of benzyl cyanide, *p*-hydroxybenzyl cyanide, and 2-(*p*-hydroxyphenyl) ethanol.[354]

Guinea pig scent glands have been shown to contain a complex mixture of chemicals, mainly fatty acids, alcohols, and ketones.[355]

The major malodorous substance from the anal gland of the stoat has been identified by g.c.–m.s. as the sulphur compound, 2-propyl thietane.[356]

The major component of the sex pheromone of the dog has been found to be methyl *p*-hydroxybenzoate.[357]

Invertebrates. A method for double-bond location in insect-derived alkenes utilizes methoxy mercuration–demercuration techniques followed by g.c.–m.s. analysis of the resulting methoxy derivative.[358] These compounds yield characteristic mass spectrometric cleavage ions on either side of the methoxy group.

A comparative analysis of the mandibular gland secretions in an ant tribe utilized solid-sample g.c.–m.s. methods in which whole mandibular glands were introduced into the g.c. injection port.[359]

Six trinervitene diterpenoids have been identified in the defensive secretions of a termite species[360] and the mono- and sesqui-terpenoid constituents of the defensive secretions of another termite have been analysed.[361] Other defensive secretions identified with the aid of g.c.–m.s. include plagiolactone and epiplagiolactone in chrysomelid larvae,[362] 1,4-napthoquinones in an opilinoid,[363] and *trans–trans* iridodiols in coconut stick insects.[364]

Nerol-derived volatile signals which form the biochemical basis for reproductive isolation between different species have been identified by g.c.–m.s. as compounds of type (8).[365]

---

[352] B. A. Anderson, L. Lundgren, and G. Stenhagen, in ref. 197, ch. 26.
[353] J. Labows, G. Preti, E. Hoelzle, J. Leyden, and A. Kligman, *J. Chromatogr.*, 1979, **163**, 294.
[354] R. M. Crewe, G. V. Burger, M. Le Roux, and Z. Katsir, *J. Chem. Ecol.*, 1979, **5**, 861.
[355] J. L. Wellington, K. J. Byrne, G. Preti, G. K. Beauchamp, and A. B. Smith, *J. Chem. Ecol.*, 1979, **5**, 737.
[356] D. R. Crump, *Tetrahedron Lett.*, 1978, 5233.
[357] M. Goodman, K. M. Gooding, and F. Regner, *Science*, 1979, **203**, 559.
[358] G. J. Blomquist, R. W. Howard, C. A. McDaniel, S. Remaley, L. A. Dwyer, and D. R. Nelson, *J. Chem. Ecol.*, 1980, **6**, 257.
[359] C. Longhurst, R. Baker, and P. E. Howse, *Insect Biochem.*, 1980, **10**, 107.
[360] G. D. Prestwich, *Experientia*, 1978, **34**, 682.
[361] R. Baker, D. A. Evans, and P. G. McDowell, *Tetrahedron Lett.*, 1978, 4073.
[362] F. Sugawara, K. Matsuda, A. Kobayashi, and K. Yamashita, *J. Chem. Ecol.*, 1979, **5**, 929.
[363] D. F. Weimer, J. Mainwald, and T. Eisner, *Experientia*, 1978, **34**, 969.
[364] R. M. Smith, J. J. Brophy, G. W. K. Cavill, and N. W. Davies, *J. Chem. Ecol.*, 1979, **5**, 727.
[365] J. Lofqvist and G. Bergstrom, *Insect Biochem.*, 1980, **10**, 1.

(8)

Aggregation pheromones of ant[366] and bee[367] species have been analysed by g.c.–m.s. The bee pheromone was identified as (Z)-9-hexadecenal and that of the ant consisted of a mixture of linalool, neral, and geranial. A wasp aggregation pheromone has been shown to consist of geranyl acetate and 2-decen-1-ol.[368]

The compound cis-9-pentacosene appears to act as the chemical trigger for egg-warming by some hornets.[369]

Certain species of predatory ant are able to move freely amongst their prey without evoking alarm responses. A g.c.–m.s. study has revealed the reason for this to be the secretion of alcohols (3-octanol, 6-methyl-3-octanol, and undecanol) by these ants rather than the aldehydes and ketones secreted by other predatory species.[370]

The sex pheromone of the female navel orangeworm, a primary pest of almonds, has been assigned the structure (ZZ)-11,13-hexadecadienal with the aid of electron impact and chemical ionization g.c.–m.s.[371] The analysis of the white peach scale sex pheromone, (Z)-3,9-dimethyl-6-isopropenyl-3,9-decadien-1-ol, illustrated a variety of microanalytical techniques, including e.i. and c.i. g.c.–m.s.[372] The sex pheromone of the female corn earworm moth has been analysed as a mixture of hexadecenals with the aid of a capillary g.c.–m.s. computer system.[373] The major constituents of the pheromone glands of Queensland fruit flies have been identified as six aliphatic amides, of which N-3-methylbutylpropanamide was the major component.[374]

Three different g.c.–m.s. systems were used to aid the analysis of wing compounds which maintain reproductive isolation between species of sulphur butterflies.[375] Several aphrodisiac pheromones, including n-hexyl esters of fatty acids, and branched and straight chain hydrocarbons, were found.

Capillary g.c.–m.s. studies with a mixed-phase column have been employed in the identification of volatile compounds emitted by different populations of Douglas fir beetles.[376] Extensive use of limited mass searches aided the computer-assisted identification of esters and terpenes. The metabolism of α-pinene to

[366] G. W. K. Cavill, P. L. Robertson, and N. W. Davies, *Experientia*, 1979, **35**, 989.
[367] A. Hefetz, S. W. T. Batra, and M. S. Blum, *Experientia*, 1979, **35**, 319.
[368] A. Hefetz and S. W. T. Batra, *Experientia*, 1979, **35**, 1138.
[369] H. J. Veith and N. Koeniger, *Naturwissenschaften*, 1978, **65**, 236.
[370] C. Longhurst, R. Baker, and P. E. Howse, *Experientia*, 1979, **35**, 870.
[371] J. A. Coffelt, K. W. Vick, P. E. Sonnet, and R. E. Doolittle, *J. Chem. Ecol.*, 1979, **5**, 955.
[372] R. R. Heath, J. R. McLaughlin, J. H. Tumlinson, T. R. Ashley, and R. E. Doolittle, *J. Chem. Ecol.*, 1979, **5**, 941.
[373] J. A. Klun, J. R. Plimmer, B. A. Bierl-Leonhardt, A. N. Sparks, M. Primiani, O. L. Chapman, G. H. Lee, and G. Lepone, *J. Chem. Ecol.*, 1980, **6**, 165.
[374] T. E. Bellas and B. S. Fletcher, *J. Chem. Ecol.*, 1979, **5**, 795.
[375] J. W. Grula, J. D. McChesney, and O. R. Taylor, *J. Chem. Ecol.*, 1980, **6**, 241.
[376] L. C. Ryker, L. M. Libby, and J. A. Rudinsky, *Environ. Entomol.*, 1979, **8**, 789.

α-pinene oxide by bark beetles and by rat liver microsomal fractions has been compared.[377] Deuterium labelling and c.i. g.c.–m.s. has aided the study of the biosynthetic conversion of the pheromone component ipsdienol to ipsenol in the beetle *Ips paraconfusus*.[378]

A search for possible attractants of the parasite *Heliothis zea* to gravid corn earworms has resulted in the identification of 63 compounds by capillary g.c.–m.s.[379] The sex pheromone of *Heliothis zea* itself consists mainly of the compound (Z)-11-hexadecenal.[380]

A new class of insect pheromones, isopropyl carboxylates, have been found in the hide beetle *Dermestes maculatus* (Deg.).[381] The isopropyl esters of (Z)-5-dodecanoic, palmitoleic, and (Z)-7- and (Z)-5-tetradecanoic acid were amongst those isolated. Isobutane chemical ionization capillary g.c.–m.s. and electron impact g.c.–m.s. have aided the analysis of four sex pheromone compounds isolated from female *Mannestra brassicae* moths.[382]

Mass chromatograms were searched for selected ions of saturated monoenes and dienes in the C_{10}—C_{18} range and these techniques aided the identification of the major biologically active pheromone component as (Z)-11-hexadecenyl acetate.

Certain moths seem to have some means of protection against predation by birds. The possible cause of this in the magpie moth appears to be the presence of pyrrolizidine alkaloids derived from the plant on which it feeds; these alkaloids were identified by g.c.–m.s.[383]

The chemical signals emitted by colonies of a marine bryozo have been found to consist mainly of *cis*- and *trans*-citral, citronellol, nerol, and geraniol.[384]

A new, venomous, constituent, 2-hexyl-5-pentyl-pyrrollidine, has been found to be the main poison gland product of thief ants[385] and an analysis of the venom of the Sydney funnel-web spider by g.c.–m.s. of TMS and PFP derivatives has revealed the presence of γ-aminobutyric acid, spermidine, spermine, tyramine, and octapamine (amongst other compounds).[386]

Pyrolysis G.C.–M.S.—Pyrolytic techniques have been surveyed in a review which includes coverage of pyrolysis g.c.–m.s.[387]

The pyrolysis products of normal human cells from various parts of the body have been identified and include alkanes, alkenes, cyclic compounds, and nitriles.[388]

[377] R. A. White, R. T. Franklin, and M. Algosin, *Pesticide Biochem. Physiol.*, 1979, **10**, 233.
[378] R. H. Fish, L. E. Browne, D. L. Wood, and L. B. Hendry, *Tetrahedron Lett.*, 1979, 1465.
[379] R. A. Flath, R. R. Forrey, J. O. John, and B. G. Chan, *J. Agric. Food Chem.*, 1978, **26**, 1290.
[380] J. A. Klun, J. R. Plimmer, B. A. Bierl-Leonhardt, A. N. Sparks, and O. L. Chapman, *Science*, 1979, **204**, 1328.
[381] W. Francke, A. R. Levinson, T.-L. Jen, and H. Z. Levinson, *Angew. Chem., Int. Ed. Engl.*, 1979, **18**, 796.
[382] D. L. Struble, H. Arn, H. R. Buser, E. Stadler, and J. Freuler, *Z. Naturforsch., Teil C*, 1980, **35**, 45.
[383] M. Benn, J. De Grave, C. Gnanasunderam, and R. Hutchins, *Experientia*, 1979, **35**, 731.
[384] C. Christopherson and J. S. Carle, *Naturwissenschaften*, 1978, **65**, 440.
[385] T. H. Jones, M. S. Blum, and H. M. Fales, *Tetrahedron Lett.*, 1979, 1031.
[386] P. H. Duffield, A. M. Duffield, P. R. Carroll, and D. Morgans, *Biomed. Mass Spectrom.*, 1979, **6**, 105.
[387] W. J. Irwin and J. A. Slack, *Analyst (London)*, 1978, **103**, 673.
[388] E. Reiner, L. E. Abbey, T. F. Moran, P. Papamichalis, and R. W. Schafer, *Biomed. Mass Spectrom.*, 1979, **6**, 491.

Pyrolysis capillary g.c.–m.s. has assisted the analysis of intractable materials.[389] Those materials studied by this method included carbon black filled rubbers, paint resins, textile fibres, polymers, gums, and biological tissue.

Ibuprofein and some analogues, a series of anti-inflammatory propionic acid derivatives, have been characterized in urine samples by pyrolysis g.c.–m.s.;[390] sulphonamides in urine have been analysed similarly.[391]

Other pyrolysis g.c.–m.s. studies include a method for qualitative and quantitative analysis of carboxy-terminated butadiene–nitrile polymers[392] and studies of the thermal degradation of aromatic polymers.[393]

Miscellaneous.—Unknown components in a gas chromatographic analysis of fluoroacyl derivatives of n-alkylamines have been identified by comparison of electron capture and g.c.–m.s. profiles.[394]

Dehydration, reduction, and oxidation reactions in sealed capillaries have been examined by capillary g.c.–m.s.[395] and cis–trans isomeric dimethyl esters produced by photolysis of o-isopropylbenzophenone have been analysed using a g.c.–m.s. calculator system.[396]

Contamination caused by foaming in 'purge and trap' g.c.–m.s. analyses has been suppressed by using silicone surfactants or heat dispersion.[397]

Aza-arenes in complex mixtures have been successfully identified by capillary g.c.–m.s., using carefully deactivated wall-coated capillary columns.[398] Analysis of amines by m.s. and g.c.–m.s. methods has been reviewed.[399]

Aliphatic sulphur compounds have been investigated by methane CI g.c.–m.s.[400] Ion intensities were found to be dependent on the sample pressure in the c.i. source.

The products of homolytic aromatic substitution of coumarin and benzo[b]furan have been identified with the aid of g.c.–m.s. methods.[401]

A number of studies have been devoted to the assessment of derivatives for g.c.–m.s. Included amongst these are PFP derivatives of sulphates,[402] the conversion of aryl sulphate ester salts to alkyl aryl derivatives,[403] and TMS derivatives of sulphonic acids.[404] Boronic derivatives examined include pinacol boronates for the characterization of aromatic boronic acids,[405] boronate esters of 24,25-

---

[389] J. L. Wuepper, *Anal. Chem.*, 1979, **51**, 997.
[390] W. J. Irwin and J. A. Slack, *Biomed. Mass Spectrom.*, 1978, **5**, 654.
[391] W. J. Irwin and J. A. Slack, *J. Chromatogr.*, 1978, **153**, 526.
[392] J. B. Maynard, J. E. Twichell, and J. Q. Walker, *J. Chromatogr. Sci.*, 1979, **17**, 82.
[393] Y. Sugimara and S. Tsuge, *J. Chromatogr. Sci.*, 1979, **17**, 269.
[394] G. S. King, B. R. Pettit, K. Blau, and P. M. Lax, *Biomed. Mass Spectrom.*, 1979, **6**, 482.
[395] G. Stanley, *J. Chromatogr.*, 1979, **178**, 487.
[396] F. W. Karasek, M. F. Tchir, and W. D. Bowers, *J. Chromatogr.*, 1978, **151**, 321.
[397] M. E. Rose and B. N. Colby, *Anal. Chem.*, 1979, **51**, 2176.
[398] M. Novotny, R. Kump, F. Merli, and L. J. Todd, *Anal. Chem.*, 1980, **52**, 401.
[399] D. A. Durden, *Res. Methods Neurochem.*, 1978, **4**, 205.
[400] H. J. Muckel, *Z. Anal. Chem.*, 1979, **295**, 241.
[401] G. Vernin, S. Coen, J. Metzger, and C. Parkanyi, *J. Heterocyclic Chem.*, 1979, **16**, 97.
[402] S. Murray and T. A. Baillie, *Biomed. Mass Spectrom.*, 1979, **6**, 82.
[403] G. Paulson, M. Simpson, J. Giddings, T. Bakke, and G. Stolzenberg, *Biomed. Mass Spectrom.*, 1978, **5**, 413.
[404] O. Stokke and P. Helland, *J. Chromatogr.*, 1978, **146**, 132.
[405] S. Singhawangcha, L.-E. Chen Hu, C. E. Poole, and A. Zlatkis, *J. High Res. Chromatogr. Chromatogr. Commun.*, 1978, **1**, 304.

dihydroxycholecalciferol,[406] and inositol boronic esters.[407] The g.c.–m.s. properties of the ethyl esters of the dithiocarbamic acids of primary and secondary amines have been determined[408] and aminophosphonic acid derivatives have also been examined.[409] The TMS derivatives of arsenic compounds,[410] cephalexin,[411] and iridoids[412] have been investigated, and the g.c.–m.s. use of trialkylsilyl derivatives other than TMS has been examined.[413]

4 High-performance Liquid Chromatography–Mass Spectrometry

The gas chromatograph and mass spectrometer are highly compatible devices but this is not true of the liquid chromatograph and mass spectrometer. The main reasons for this are: firstly, that the eluant used is a liquid in which the solvent concentration is considerably lower than in g.c.; secondly, h.p.l.c. methods are often used because of known thermal instability or involatility of the samples; and finally, h.p.l.c. frequently requires the addition of inorganic salts to act as buffers. Because of these problems, no perfect l.c.–m.s. interface has yet been designed and a number of differing solutions to the problem of l.c.–m.s. coupling have been attempted.

A number of articles reviewing developments in l.c.–m.s. methods have been published.[414—419] One of these review papers contains several brief accounts of l.c.–m.s. applications by different investigators.[418]

L.C.–M.S. Interfaces.—An ideal l.c.–m.s. interface should fulfil the following requirements:

(i) It should not require any modifications to l.c. performance or operating conditions, *i.e.* a wide range of solvents, flow rates, and column types (including reversed phase columns) should be accommodated.

(ii) A high percentage of injected sample should enter the mass spectrometer ion source (high yield) and the ratio of sample to solvent should be substantially increased (good enrichment).

(iii) L.c. resolution should not be degraded.

(iv) It should be possible to operate the mass spectrometer under normal conditions (electron impact or chemical ionization) so that l.c.–m.s. acquired spectra are directly comparable with, for example, direct probe analysis.

---

[406] J. M. Halket, D. G. Lichty, and F. Wightman, *J. Chromatogr.*, 1980, **192**, 435.
[407] J. Wiecko and W. R. Sherman, *J. Am. Chem. Soc.*, 1979, **101**, 979.
[408] G. Blotny, J. Kusmierz, E. Malinski, and J. Szafranek, *J. Chromatogr.*, 1980, **193**, 61.
[409] J. W. Huber, *J. Chromatogr.*, 1978, **152**, 220.
[410] F. T. Henry and T. M. Thorpe, *J. Chromatogr.*, 1978, **166**, 577.
[411] T. Nakagawa, J. Haginaka, M. Masada, and T. Uno, *J. Chromatogr.*, 1978, **154**, 264.
[412] S. Milz and H. Rimpler, *Z. Naturforsch., Teil C*, 1979, **34**, 319.
[413] C. F. Poole and A. Zlatkis, *J. Chromatogr. Sci.*, 1979, **17**, 115.
[414] W. H. McFadden, *J. Chromatogr. Sci.*, 1979, **17**, 2.
[415] P. J. Arpino and G. Guiochon, *Anal. Chem.*, 1979, **51**, 682A.
[416] R. F. Zerilli, *Chromatogr. Symp. Ser.*, 1979, **1**, 59.
[417] B. G. Dawkins and F. W. McLafferty in 'GLC and HPLC Analysis of Drugs', ed. K. Tsiji and W. Morozowich, Marcel Dekker, New York, 1978.
[418] W. H. McFadden, *J. Chromatogr. Sci.*, 1980, **18**, 97.

(v) It should be possible to introduce involatile or thermally labile samples without degrading them.

(vi) The sensitivity of the technique should approach that of g.c.–m.s.

In practice, the problems posed by l.c.–m.s. interfacing are such that these conditions are not realizable simultaneously and all the interfaces so far devised place some sort of restriction on l.c. and/or m.s. operation.

The Atmospheric Pressure Ionization (API) Source. This is very suited to l.c.–m.s. operation as it is capable of accommodating several hundred standard ml per minute of vapour. The source employs a corona discharge or radioactive emitter to ionize the l.c. column effluent which enters the source *via* a heated evaporator tube.[420] Products of ion–molecule reactions in the discharge are sampled through a micropore aperture and the ions are analysed by a quadrupole mass spectrometer computer system. This technique is capable of analysing sample components down to nanogram levels but suffers from a number of severe restrictions. Firstly, solvents are limited to those with low proton or electron affinity, and sample types must have correspondingly high affinities; secondly there are problems with solvent impurities, which tend to reduce sensitivity (femtogram sensitivities are sometimes obtainable when the API source is operated without l.c. coupling); thirdly, nonvolatile samples cannot be handled; and finally, the use of buffered solutions is prohibited.

Direct Introduction of L.C. Effluent. Historically, the first attempt to introduce l.c. effluent directly into the ion source was made by Tal'Rose and co-workers, using a conventional electron impact mass spectrometer.[421] A small solvent flow, of the order of nl min^{-1} was introduced into the ion source. This technique was impractical for l.c.–electron impact m.s. because of the high split ($>10^4$) for the l.c. effluent. Subsequent developments in direct introduction have allowed much more favourable split ratios by using l.c. c.i. m.s. with the l.c. solvent providing the chemical ionization plasma.[422,423] Many h.p.l.c. solvents are suited to this technique, although the complexity of ion–molecule reactions can be confusing

---

[419] D. E. Games, *Proc. Anal. Div. Chem. Soc.*, 1980, **17**, 110.

[420] (*a*) E. C. Horning, D. I. Carroll, I. Dzidic, K. D. Haegele, M. G. Horning, and R. N. Stillwell, *J. Chromatogr.*, 1974, **99**, 13; (*b*) E. C. Horning, D. I. Carroll, I. Dzidic, K. D. Haegele, M. G. Horning, and R. N. Stillwell, *J. Chromatogr. Sci.*, 1974, **12**, 725; (*c*) E. C. Horning, D. I. Carroll, I. Dzidic, K. D. Haegele, M. G. Horning, and R. Stillwell, *Kagaka No Ryoiki Zokan*, 1976, **109**, 85; (*d*) E. C. Horning, D. I. Carroll, I. Dzidic, K. D. Haegele, S.-W. Lun, C. U. Oertli, and R. N. Stillwell, *Clin. Chem.*, 1977, **23**, 13; (*e*) E. C. Horning, D. I. Carroll, I. Dzidic, and R. N. Stillwell, *Pure Appl. Chem.*, 1978, **50**, 113; (*f*) D. I. Carroll, I. Dzidic, R. N. Stillwell, K. D. Haegele, and E. C. Horning, *Anal. Chem.*, 1975, **47**, 2369.

[421] V. L. Tal'Rose, G. V. Karpov, I. G. Gordoetshii, and V. E. Skurat, *Russ. J. Phys. Chem.*, 1968, **42**, 1658; V. L. Tal'Rose, V. D. Grishin, V. E. Skurat, and G. D. Tantsyrev in 'Recent Developments in Mass Spectrometry', ed. K. Ogta and T. Hagakawa, University Park Press, Baltimore, Maryland, U.S.A., 1970, p. 1218.

[422] (*a*) M. A. Baldwin and F. W. McLafferty, *Org. Mass Spectrom.*, 1973, **7**, 1111; (*b*) P. J. Arpino, B. G. Dawkins, and F. W. McLafferty, *J. Chromatogr. Sci.*, 1974, **12**, 574; (*c*) P. J. Arpino, M. A. Baldwin, and F. W. McLafferty, *Biomed. Mass Spectrom.*, 1974, **1**, 80; (*d*) F. W. McLafferty, R. Knutti, R. Venkataraghavan, P. J. Arpino, and B. G. Dawkins, *Anal. Chem.*, 1975, **47**, 1503; (*e*) F. W. McLafferty and B. G. Dawkins, *Biochem. Soc. Trans.*, 1975, **3**, 856; (*f*) B. G. Dawkins, P. J. Arpino, and F. W. McLafferty, *Biomed. Mass Spectrom.*, 1978, **5**, 1; (*g*) P. J. Arpino, G. Guiochon, P. Krien, and G. Devant, *J. Chromatogr.*, 1979, **185**, 529.

[423] J. D. Henion, *Anal. Chem.*, 1978, **50**, 1687; J. D. Henion, *Adv. Mass Spectrom.*, 1978, **7**, 865.

in some cases. The highest rate of liquid introduction achievable by direct coupling currently appears to be about 100 μl min^{-1}. Buffered solvents are incompatible with direct introduction but a continuous flow extraction technique in which apolar solvents are used to extract sample from buffered solvents prior to introduction to the mass spectrometer has been developed.[424]

Commercially developed l.c.–m.s. interfaces of the direct introduction type are currently available from Ribermag and Hewlett–Packard. These manufacturers also employ cryocooling to aid solvent pumping, an improvement first described by Arpino et al.[422g]

Jet-type L.C.–M.S. Interfaces. Several workers have attempted to improve l.c.–m.s. couplings by enrichment using jet-type interfaces.[425—428] A variety of techniques has been adopted, ranging from a simple modified jet separator,[425] through a vacuum nebulizing interface in which coaxially introduced helium provides the nebulizing energy,[426] to laser[427] and ultrasonic[428] vaporization at the expanding jet. Whilst all of these techniques showed some promise, and indeed permitted mass spectrometer operation in electron impact as well as chemical ionization mode, sensitivity and enrichment levels have been somewhat disappointing.

Membrane Interfaces. Semipermeable membrane interfaces have a long history of application to g.c.–m.s. coupling, so it is not surprising that an attempt has been made to apply modified versions of this device to l.c.–m.s. coupling.[429] The device appears to be of limited applicability as it is restricted to certain solvent–solute systems. Furthermore, significant peak broadening was observed with the test mixture of aromatic hydrocarbons.

Mechanical Transfer. The first mechanical transfer device to be used with any degree of success in l.c.–m.s. interfacing was a moving wire system which passed directly into the ion source of a quadrupole mass spectrometer.[430] This system showed promise but was limited by the inability of the wire to take up large quantities of l.c. effluent, and therefore required a 100:1 split. The problem of increasing the sample capacity was eventually solved by other workers who used a moving belt, instead of a wire, to transfer the sample.[431] Solvent flows as high as 1—5 standard ml min^{-1} of volatile solvent can be accommodated by this type of interface, although this is reduced considerably for more polar, less volatile solvents. The belt system appears to be the closest to a universal l.c.–m.s. interface so far devised and even buffered solvents can be used, provided the buffer is removed from the belt after use. The belt material used was either stainless steel or polyimide.

---

[424] B. L. Karger, D. P. Kirby, P. Vouros, R. L. Foltz, and B. Hidy, *Anal. Chem.*, 1979, **51**, 2324.
[425] T. Takeuchi, Y. Hirata, and Y. Ohumura, *Anal. Chem.*, 1978, **50**, 660.
[426] (a) S. Tsuge, Y. Hirata, and T. Takeuchi, *Anal. Chem.*, 1979, **51**, 166; (b) Y. Hirata, T. Takeuchi, S. Tsuge, and Y. Yoshida, *Org. Mass Spectrom.*, 1979, **14**, 126.
[427] C. R. Blakley, M. J. McAdams, and M. L. Vestal, *J. Chromatogr.*, 1978, **158**, 261.
[428] H. R. Useth, R. G. Gorth, and J. H. Futrell, 'Proceedings of the 26th Annual Conference of Mass Spectrometry and Allied Topics', St. Louis, Missouri, 1978, p. 659.
[429] P. R. Jones and S. K. Yang, *Anal. Chem.*, 1975, **47**, 1000.
[430] R. P. W. Scott, G. C. Scott, M. Munroe, and J. Hess, *J. Chromatogr.*, 1974, **99**, 395.
[431] W. H. McFadden, H. L. Schwartz, and S. Evans, *J. Chromatogr.*, 1976, **122**, 389.

A modified version of the belt transfer device removes solvent and then reduces sample molecules to the parent hydrocarbons which are swept into the ion source of a mass spectrometer *via* a conventional g.c.–m.s. interface.[432]

Belt interfaces are commercially available and are currently manufactured by Finnigan instruments and V.G. Micromass.

Applications.—Varying degrees of success have been achieved, but l.c.–m.s. data have been obtained on a wide variety of molecules. These are summarized in the Table, together with the type of l.c.–m.s. coupling used.

Table Applications of l.c.–m.s. coupling methods to different sample types

Interface and ionization methods	Applications
Direct Introduction (API)	alkaloids,[420a,c] aromatic compounds,[420b] drugs,[420c–e] insecticides,[420e] PAHs,[420e,f] steroids,[420b,c] urinary bases[420e]
Direct Introduction (c.i.)	acids,[433] alkaloids,[441,442] amino-acids and peptides,[422a,e,f] aromatic compounds,[424] cannabinoids,[443] drugs,[423,441] lipids and other long chain molecules,[422b,d,433] pesticides,[424] steroids[422b,c]
Jet Enrichment (e.i. and/or c.i.)	amino-acids,[426b] aromatic amines,[426a] aromatic hydrocarbons,[425,427,428] drugs,[426a,b] fatty acid esters,[425] nucleosides,[427] polymers,[426b] priority pollutants[448]
Membrane (e.i.)	aromatic hydrocarbons[429]
Mechanical Transfer (e.i. and/or c.i.)	amino-acids and peptides,[430,437] heteroaromatics,[435] drugs,[437,445] herbicides and their photolysis products,[447] lipids,[432,437,440] liquid crystals,[444] natural products,[430,436,437,439] organic acids and bases,[449] PAHs,[434] PCBs,[446] pesticides,[431,449] porphyrins,[440] saccharides,[436,438] steroids[431,436]

General Comments and Future Developments.—As can be seen from the Table, a wide (and ever increasing) range of sample types has been studied by the different l.c.–m.s. techniques. The technique is still in its infancy and no ideal interface has yet been developed; some applications are best suited to a specific type of interface. It does appear that the belt system has been used to analyse a wide variety of molecules so that one can perhaps look for possible improvements to the belt interface in order to increase the power of this system still further. For example, new belt materials might increase ease of desorption from the belt surface (one of the original materials, kapton, already has advantages over probe introduction of samples as it permits volatilization of labile samples at lower temperatures) or coating of the belt with deactivating materials may be possible (some surface deactivation techniques which have been applied to

---

[432] W. L. Erdahl and O. S. Privett, *Lipids*, 1977, **2**, 797; W. L. Erdahl and O. S. Privett, *Chem. Phys. Lipids*, 1978, **21**, 361.
[448] M. J. McAdams and M. L. Vestal in ref. 418.
[449] B. L. Karger, D. P. Kirby, and P. Vouros in ref. 418.

g.c. could be tried). Direct ionization from the belt surface may be possible, either by passing it directly into the ion plasma of a c.i. source or by ion bombardment.[450] Ion bombardment techniques might also be applicable to the jet enrichment methods.

Several workers have already realized the advantages of micro l.c. methods in l.c.–m.s. coupling. The low flow rates used in micro l.c. (generally in the range 1—20 μl min^{-1}) are more compatible than normal l.c. flows with mass spectrometer pumping systems. Micro l.c. therefore holds promise for all types of l.c. coupling except, possibly, the membrane type. It appears to be especially compatible with direct introduction and jet enrichment methods but could also be of advantage in belt transfer systems, allowing the use of a narrower belt and thus decreasing the surface area from which the sample is evaporated.

---

[450] A. Benninghoven and W. K. Sichtermann, *Anal. Chem.*, 1978, **50**, 1180.

9
Reactions of Negative Ions in the Gas Phase

BY J. H. BOWIE, V. C. TRENERRY, AND G. KLASS

1 Introduction

This report is concerned with the formation and reactions of organic negative ions in the gas phase. It does not cover the ion chemistry of atomic negative ions, nor of negative ions derived from inorganic molecules. The policy of earlier reviews[1] has been retained, *i.e. selective* references are listed in each section, and only those of *particular* interest in the context of this chapter are considered in any detail.

During the review period, research has been concentrated in a number of major areas including (i) electron capture by organic and organometallic compounds, and fragmentation mechanisms of the resulting molecular anions, (ii) negative ion chemical ionization mass spectrometry, (iii) negative ion-molecule reactions, and (iv) charge inversion reactions of negative ions. These topics will be covered below.

The observation[2] that negative ions may be produced using a commercial field ionization source is of particular interest. This allows unimolecular decompositions of negative ions (formed in the FI Source) to be studied in the time range 10^{-5}—10^{-11} s by a technique developed for positive ions.[3]

A number of books[4-9] and reviews,[10-15] concerned with aspects of negative ion chemistry, have been published during the past two years. The texts edited

---

[1] J. H. Bowie, in this series, 1979, Vol. 5, p. 262.
[2] J. Vander Greef and N. M. M. Nibbering, *Int. J. Mass Spectrom. Ion Phys.*, 1979, **31**, 71.
[3] H. D. Beckey, M. D. Migahed, and F. W. Röllgen, *Int. J. Mass Spectrom. Ion Phys.*, 1973, **10**, 471.
[4] R. G. Cooks, 'Collision Spectroscopy', Plenum Press, New York, 1978.
[5] 'High Performance Mass Spectrometry, Chemical Applications' (American Chemical Society Symposium Series, Series 70), Americal Chemical Society, Washington, D.C. 1978.
[6] 'Recent Developments in Mass Spectrometry in Biochemistry and Medicine', Plenum Press, New York, 1978.
[7] 'Ion Molecule Reactions', Part I and II, ed. J. L. Franklin (Benchmark Papers in Physical Chemistry and Chemical Physics, Vol. 3), Dowden, Hutchinson, and Ross, 1979.
[8] 'Gas Phase Ion Chemistry', Part I and II, ed. M. T. Bowers, Academic Press, New York, 1979.
[9] A. Giraud and M. Petit, 'Ionospheric Techniques and Phenomena', Reidel Publ. Co., London, 1978.
[10] L. G. Christophorou, in 'Advances in Electronics and Electron Physics', Academic Press, New York, 1978, Vol. 46, pp. 55–126.
[11] W. J. Richter and H. Schwarz, *Angew. Chem., Int. Ed. Engl.*, 1978, **17**, 424.
[12] R. A. Khmel'nitskii and R. B. Terent'ev, *Russ. Chem. Rev.*, 1979, **48**, 463.
[13] H. Budzikiewicz, in 'Mass Spectrometry', Part A. Practical spectroscopy series, Marcel Dekker Inc., New York, Basel, 1979, Vol. 3, Ch. 1.
[14] K. D. Jordan, *Acc. Chem. Res.*, 1979, **12**, 36.
[15] J. H. Bowie, *Acc. Chem. Res.*, 1980, **13**, 76.

by Bowers[8] and the review written by Christophorou[10] are particularly recommended.

2 Fragmentation Mechanisms of Negative Ions Formed by Electron Capture (or Dissociative Electron Capture)

Organic Negative Ions.—Studies of the energetics of formation of negative ions from fluoroalkanes continue.[16–18] In an elegant experiment, potassium atom beams (1—30 eV) have been used to induce electron transfer with a variety of fluoro- and chloro-methanes.[19] This technique has allowed the first observations of parent radical anions for CCl_4, $CFCl_3$, and CF_2Cl_2.

Theoretical calculations and electron transmission spectroscopy have been used to estimate the energies of the temporary negative ion states associated with the three low-lying π^* orbitals of benzene and fluorobenzenes,[20–22] cf. ref. 10. Electron affinities have been measured (or calculated) by various techniques for halomethanes,[19] haloanils,[23] benzoquinone and perfluorobenzoquinone,[24] and for the $[M - H^+]^-$ ions derived from p-substituted acetophenones.[25]

Negative ion spectra have been measured for the following systems: methyl esters of fluorine-substituted benzoic acids,[26] polyunsaturated esters,[27] alkoxycarbonyl compounds,[28] p-nitrobenzyl esters of fatty acids,[29,30] thioamides,[31,32] aryl-*ONN*-azoxy compounds,[33] and cyanopyridines.[34] An argon discharge source[35] has been used to measure the negative ion spectra of products obtained

---

[16] D. L. McCorkle, A. A. Christodoulides, and L. G. Christophorou, *Bull. Am. Phys. Soc.*, 1979, **24**, 134.
[17] A. A. Christodoulides, L. G. Christophorou, R. Y. Pai, and C. M. Tung, *J. Chem. Phys.*, 1979, **70**, 1156, 1169.
[18] E. Illenberger, H.-U. Sheunemann, and H. Baumgärtel, *Chem. Phys.*, 1979, **37**, 21.
[19] H. Dispert and K. Lacmann, *Int. J. Mass Spectrom. Ion Phys.*, 1978, **28**, 49.
[20] J. R. Frazier, L. G. Christophorou, J. G. Carter, and H. C. Schweinler, *J. Chem. Phys.*, 1978, **69**, 3807.
[21] K. D. Jordan and P. D. Burrow, *J. Chem. Phys.*, 1979, **71**, 5384.
[22] L. G. Christophorou and H. C. Schweinler, *J. Chem. Phys.*, 1979, **71**, 5385.
[23] C. D. Cooper, W. F. Frey, and R. N. Compton, *J. Chem. Phys.*, 1978, **69**, 2367.
[24] J. E. Bloor, R. A. Paysen, and R. E. Sherrod, *Chem. Phys. Lett.*, 1979, **60**, 476.
[25] R. L. Jackson, A. H. Zimmerman, and J. I. Brauman, *J. Chem. Phys.*, 1979, **71**, 2088.
[26] V. I. Khvostenko, V. V. Takhistov, V. Fal'ko, and O. S. Sokolova, *Bull. Acad. Sci. USSR, Div. Chem. Sci.*, 1978, **27**, 1352.
[27] A.Sh. Sultanov, U. M. Dzhemilev, V. I. Khvostenko, G. A. Tolstikov, R. I. Khusnutdinov, and S. R. Rafikov, *Dokl. Chem. (Engl. Transl.)*, 1978, **238**, 32.
[28] V. I. Khvostenko, V. S. Fal'ko, G. V. Leplyanin, V. V. Takhistov, and O. S. Sokolova, *Bull. Acad. Sci. USSR, Div. Chem. Sci.*, 1978, **27**, 510.
[29] Y. Hirata, T. Takeuchi, and K. Matsumoto, *Anal. Chem.*, 1978, **50**, 1943.
[30] C. L. Chan and C. L. Brown, *Biomed. Mass Spectrom.*, 1978, **5**, 380.
[31] K. Clausen, B. S. Pedersen, S. Scheibye, S.-O. Lawesson, and J. H. Bowie, *Org. Mass Spectrom.* 1979, **14**, 101.
[32] K. Clausen, B. S. Pedersen, S. Scheibye, S.-O. Lawesson, and J. H. Bowie, *Int. J. Mass Spectrom. Ion Phys.*, 1979, **29**, 223.
[33] G. A. Vagilo, R. P. Ferrari, V. Mortarini, A. Gasco, and R. Calvino, *Org. Mass Spectrom.*, 1979, **14**, 668.
[34] G. Holzmann and H. W. Rothkopf, *Org. Mass Spectrom.*, 1978, **13**, 636.
[35] M. von Ardenne, *Z. Angew. Phys.*, 1959, **11**, 121; M. von Ardenne, K. Steinfelder, and R. Tümmler, 'Electronenanlagerungs Massenspektrographic organischer Substanzen,' Springer Verlag, Berlin, 1971.

from the hydrolysis of N-(2-cyanoethylene)urea,[36] and the spectra of α-amino-acids[37] and 2,3,4,6-tetra-O-acetyl-β-D-glucopyranose esters.[38]

p-Nitrobenzyl esters may be used for the analysis of fatty-acid mixtures since the base peak corresponds to the carboxylate anion shown in (1).[29,30] Molecular anions are small in these spectra; if molecular anions need[15] to be monitored, the p-nitrobenzoate is the better derivative.[39] The basic negative ion cleavages of thioamides are shown in (2).[31,32] Molecular anions of o-nitrothioacetanilides[31]

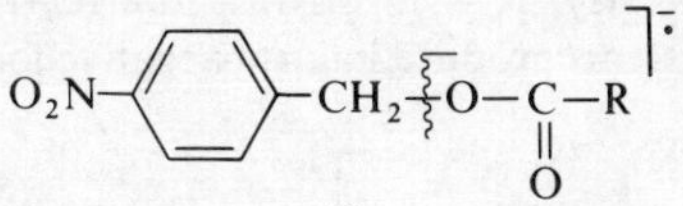

$$O_2N--CH_2-O-C-R \quad (1) \qquad R^1-C-N\genfrac{}{}{0pt}{}{R^2}{R^3} \quad (2)$$

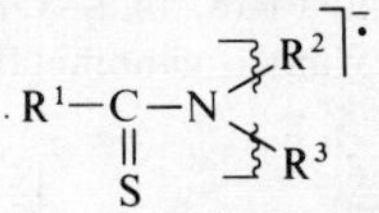

and o-nitrothiobenzamides[32] undergo pronounced rearrangement reactions. For example, the major peaks in the spectrum of o-nitro-N-methylthioacetanilide are due to the ions $MeCO_2^-$ and $MeCOS^-$ which may arise as shown in (3) and (4).[31] Aryl azoxy ester parent anions fragment as shown in (5),[33] while α-amino-acids yield $M^{-\bullet}$, $[M - H^\bullet]^-$, and $[M - H_2O]^{-\bullet}$ ions as well as fragmenting as shown in (6).[37]

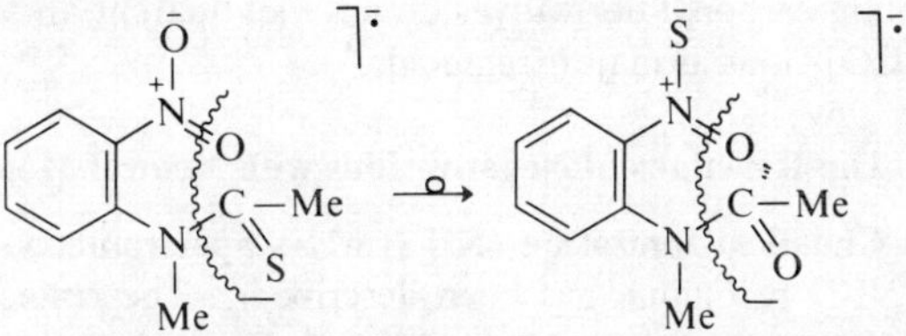

(3) → (4)

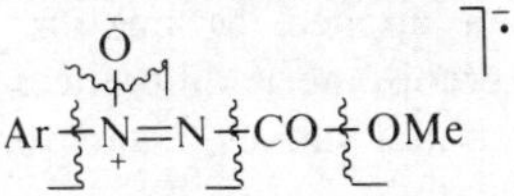

$$Ar-N=N-CO-OMe \quad (5) \qquad R-CH-CO_2H,\ NH_2 \quad (6)$$

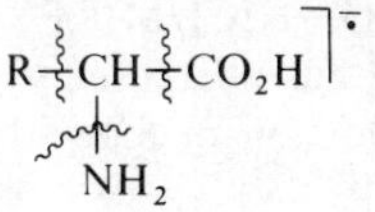

Organometallic Negative Ions.—The application of negative ion mass spectrometry to organometallic systems continues, and the following classes of compound have been studied: organophosphorus compounds,[40] selenium analogues of 2,1,3-benzoxodiazole,[41] alkyl and aryl siloxanes,[42,43] oxyacetates

---

[36] M. von Ardenne, H. Berg, H. Fritzche, P. G. Reitnauer, and R. Tümmler, *J. Prakt. Chem.*, 1977, **319**, 848.
[37] D. Voigt and J. Schmidt, *Biomed. Mass Spectrom.*, 1978, **5**, 44.
[38] D. Voigt and H. Lehmann, *J. Prakt. Chem.*, 1979, **321**, 243.
[39] J. H. Bowie and B. J. Stapleton, *Aust. J. Chem.* 1975, **28**, 1011.
[40] H. J. Meyer, F. C. V. Larsson, S.-O. Lawesson, and J. H. Bowie, *Bull. Soc. Chim. Belg.*, 1978, **87**, 517.
[41] M. R. Arshadi, *Org. Mass Spectrom.* 1978, **13**, 379.
[42] V. N. Bochkarev, A. N. Polivanov, V. S. Fal'ko, L. M. Blekh, V. I. Khvostenko, and E. A. Cherlnyshev, *J. Gen. Chem. USSR (Engl. Transl.)*, 1978, **48**, 783.
[43] A. N. Polivanov, A. A. Bernadskii, V. S. Fal'ko, N. P. Telegina, V. I. Khvostenko, and V. N. Bochkarev, *J. Gen. Chem. USSR (Engl. Transl.)*, 1978, **48**, 359.

of beryllium,[44] organomercury compounds,[45,46] osmium carbonyls,[47] iron, chromium, molybdenum, and tungsten tricarbonyls,[48] salicylaldehydato and 2-hydroxy-1-naphthaldehydato complexes of Co^{II}, Ni^{II}, and Cu^{II},[49] tris- and bis-(dipivaloylmethanato) metal chelates,[50] tris(hexafluoroacetylacetonato) metal chelates,[51] and nickel(II) β-diketonate complexes.[52]

Organophosphorus compounds of formula (7) (X = O or S) fragment by simple cleavage as indicated,[40] but the basic cleavage of molecular anions corresponding to (8) (Y = O, S, or NH; X = O or S) is the rearrangement (8) → (9).[40] Many organometallic systems produce intense negative ion currents

$$
\begin{array}{ccc}
R^1\!\!-\!\!O\!\!\diagdown\!\!\underset{\displaystyle P}{}\!\!\!=\!\!X \\ R^1\!\!-\!\!O\!\!\diagup\!\!\!\!\overbrace{X\!\!\}\!\!R^2} \end{array}\bigg]^{\bar{\cdot}}
\qquad
\begin{array}{c} R^1\diagdown\underset{\displaystyle}{\overset{S}{\underset{\|}{P}}}\!\!\!X \\ R^1\diagup\overset{\displaystyle|}{Y}\!\!-\!\!\overset{\displaystyle\|}{C}\!\!-\!\!R^2 \end{array}\bigg]^{\bar{\cdot}}
\xrightarrow{}
\begin{array}{c} R^1\diagdown\diagup S^- \\ P \\ R^1\diagup\diagdown X \end{array}
$$

(7)   (8)   (9)

due mainly to molecular and ligand ions. This is particularly significant in the cases of cycloheptatriene derivatives of zero-valent Fe, Cr, Mo, and W tricarbonyls, which form intense molecular anions in apparent violation of the rare-gas rule.[48] In contrast, osmium carbonyl derivatives do not yield parent anions, instead they form $[M - nCO]^-$ ions as major fragments.[47]

3 The Reactions of Negative Ions with Neutral Molecules

Negative Ion Chemical Ionization (NICI) Mass Spectrometry.—The development of the NICI technique has been described.[53] The collective term NICI arguably includes some of the most sensitive of all mass spectrometric analytical techniques. Various techniques have been applied to problems during the review period. HO^-/NICI/MIKES has been applied to mixture analysis,[54] and HO^-/NICI has been used to detect stereoisomeric differences in cyclic diols,[55] for the detection of methadone and its derivatives,[56] and for the analysis of steroids.[57]

---

[44] A.Sh. Sultanov, A. I. Grigor'ev, V. I. Khvostenko, V. K. Mavrodiev, and L. N. Reshetova, *Dokl. Phys. Chem.* (*Engl. Transl.*), 1978, **237**, 1128.
[45] A. N. Nesemyanov, V. I. Khvostenko, Yu. S. Nekrasov, I. I. Kritskaya, O. G. Khvostenko, and G. A. Tolstikov, *Dokl. Phys. Chem.* (*Engl. Transl.*), 1978, **241**, 699.
[46] A. N. Nesmeyanov, V. T. Aleksanyan, L. I. Denisovich, Yu. S. Nekrasov, E. I. Fedin, V. I. Khvostenko, and I. I. Kritskaya, *J. Organomet. Chem.*, 1979, **172**, 133.
[47] G. A. Vaglio, *J. Organomet. Chem.*, 1979, **169**, 83.
[48] M. R. Blake, J. L. Garnett, I. K. Gregor, and S. B. Wild, *J. Organomet. Chem.*, 1979, **178**, C37.
[49] Y. Hirata, K. Matsumoto, and T. Takeuchi, *Org. Mass Spectrom.*, 1978, **13**, 264.
[50] J. L. Garnett, I. K. Gregor, and M. Guilhaus, *Org. Mass Spectrom.*, 1978, **13**, 591.
[51] D. R. Dakternieks, I. W. Fraser, J. L. Garnett, and I. K. Gregor, *Org. Mass Spectrom.*, 1979, **14**, 330.
[52] D. R. Dakternieks, I. W. Fraser, J. L. Garnett, and I. K. Gregor, *Org. Mass Spectrom.*, 1979, **14**, 676.
[53] K. R. Jennings, in this series, 1977, Vol. 4, p. 203.
[54] G. A. McClusky, R. W. Kondrat, and R. G. Cooks, *Anal. Chem.*, 1978, **50**, 1222; *J. Am. Chem. Soc.*, 1978, **100**, 6045.
[55] F. J. Winkler and D. Stahl, *J. Am. Chem. Soc.*, 1978, **100**, 6779.
[56] A. L. C. Smit and F. H. Field, *Biomed. Mass Spectrom.*, 1978, **5**, 572.
[57] T. A. Roy, F. H. Field, Y. Y. Lin, and L. L. Smith, *Anal. Chem.*, 1979, **51**, 272.

Methane/NICI has been used to detect melatonin[58] and clonazepam[59] in human plasma, and also impurities in commercial pentachlorophenol.[60] Addition of oxygen to the methane reactant gas has been used for enhanced detection of polychloroanisoles[61] and chlorinated diphenyl ethers,[62] while CH_4/O_2, NO, CO, and PF_3 mixtures have been used for the analysis of metal Schiff-base complexes.[63] Ammonia, CH_4, or N_2O may be used as a reagent gas for NICI of phenothiazines.[64] Various chloromethanes can be used for Cl^-/NICI biologically important materials.[65,66] For example, CF_2Cl_2 (Cl^-)/NICI of olgose produces an $[M + Cl^-]^-$ peak which undergoes the fragmentations shown in (10), thus enabling the carbohydrate sequence of the oligosaccharide to be determined.[66] The reagent gases CF_4, SO_2F_4, and SF_6 have been tested for certain organometallic molecules.[67]

(10)

The first radical addition reactions have been reported under chemical ionization conditions.[68] 7,7,8,8-Tetracyanoquinodimethane (TCNQ) under CH_4/NICI conditions yields $[M + H]^-$, $[M + Me]^-$, and $[M + Et]^-$ ions. These must be formed by radical reactions, since H^-, Me^-, and Et^- are not present in the methane plasma.

Ion Cyclotron Resonance and Flowing Afterglow Studies.—In this section coverage is restricted to the reactivity of organic anions, and the mechanisms of bimolecular organic reactions.

A most important application of ICR spectrometry in the past decade has been its use for the determination of the gas-phase acidities and basicities of organic molecules with respect to the proton. These topics have been described in two excellent reviews.[69,70] More recently, an important paper[71] has reported

---

[58] A. J. Lewy and S. P. Markey, *Science*, 1978, **201**, 741.
[59] W. A. Garland and B. H. Min, *J. Chromatogr.*, 1979, **172**, 279.
[60] M. Deinzer, D. Griffin, and T. Miller, *Biomed. Mass Spectrom.*, 1979, **6**, 301.
[61] K. L. Busch, J. R. Hass, and M. M. Bursey, *Org. Mass Spectrom.*, 1978, **13**, 604.
[62] K. L. Busch, A. Norstrom, M. M. Bursey, J. R. Hass, and C.-A. Nilsson, *Biomed. Mass Spectrom.*, 1979, **6**, 157.
[63] E. Baumgartner and J. G. Dillard, *Inorg. Chim. Acta*, 1979, **32**, 11.
[64] R. Ryhage and H. Brandenberger, *Biomed. Mass Spectrom.*, 1978, **5**, 615.
[65] A. K. Bose, H. Fujiwara, and B. N. Pramanik, *Tetrahedron Lett*, 1979, 4017.
[66] A. K. Ganguly, N. F. Cappuccino, H. Fujiwara, and A. K. Bose, *J. Chem. Soc., Chem. Commun.*, 1979, 148.
[67] E. Baumgartner, and J. G. Dillard, *Org. Mass Spectrom.*, 1979, **14**, 360.
[68] C. N. McEwen and M. A. Rudat, *J. Am. Chem. Soc.*, 1979, **101**, 6470.
[69] D. H. Aue and M. T. Bowers, in 'Gas-Phase Ion Chemistry', ed. M. T. Bowers, Academic Press, London, 1979, Vol. 2, Ch. 9.
[70] J. E. Bartmess and R. T. McIver, in ref. 69, Ch. 11.
[71] J. E. Bartmess, J. A. Scott, and R. T. McIver, *J. Am. Chem. Soc.*, 1979, **101**, 6046.

the gas-phase acidities of eighty organic molecules. In this work, the pulsed ICR technique was used to determine thermodynamic parameters for a series of proton transfer reactions of the type $A^- + BH \rightleftarrows AH + B^-$

A number of nucleophilic reactions in the gas phase have been reported. The reaction (flowing afterglow) between HO^- and propyne results in the removal of the acetylenic proton and a methyl proton in the approximate ratio (2 : 3).[72] Reactions of various carbanions with both triplet and singlet oxygen are complex, and may yield a range of products.[73] For example, although the $[M - H]^-$ anions of toluene and m- and p-xylene react slowly with $^3O_2$ to yield HO^- almost exclusively, the o-xylene anion shows three major pathways (Scheme 1).[73]

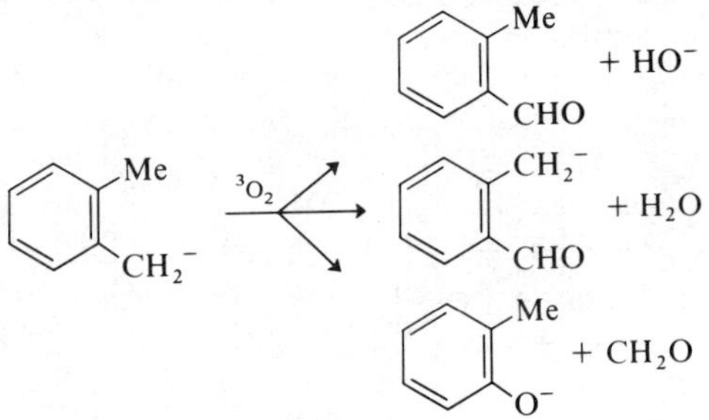

$$o\text{-}C_6H_4(Me)CH_2^- \xrightarrow{^3O_2} \begin{cases} o\text{-}C_6H_4(Me)CHO + HO^- \\ o\text{-}C_6H_4(CH_2^-)CHO + H_2O \\ o\text{-}C_6H_4(Me)O^- + CH_2O \end{cases}$$

Scheme 1

Deuterium exchange of hydrogens of various carbanions RCH_2^- can be effected using D_2O, MeOD, or CF_3CH_2OD in flowing afterglow experiments.[74] The reaction (ICR) between $CH_2=C^{-\bullet}$ and N_2O yields $CH_2=C=N^{-\bullet}$,[75] whereas that between $O^{-\bullet}$ and propene produces a long-lived allyl anion.[76]

Reactions of nucleophiles at carbonyl centres continue to excite interest; previous work in this area has been reviewed.[15] Perhaps the most interesting aspect concerns the nature of the transition state(s) [and/or reactive intermediate(s)] in the reaction sequence; viz, do they correspond to H-bonded complexes, tetrahedral structures, or association complexes which look like reactants or products? Examples of all cases have been cited previously.[15] The problem has been approached[77] by measuring the chlorine isotope effects for losses of HCl from adducts formed with carbonyl-containing compounds and Cl^- (i.e. from Cl^-/NICI spectrometry). The $[M + Cl]^-$ adducts from benzoic acid and acetophenone eliminate HCl with $k^{35}{}_{Cl}/k^{37}{}_{Cl} = 1.19$ and 1.04 respectively and this has been interpreted in terms of an H-bonded complex (11) in the former case and a tetrahedral species (12) in the latter. Kinetic evidence has been reported[78]

---

[72] C. H. DePuy, V. M. Bierbaum, L. A. Flippin, J. J. Grabowski, G. K. King, and R. J. Schmitt, *J. Am. Chem. Soc.*, 1979, **101**, 6443.
[73] R. J. Schmitt, V. M. Bierbaum, and C. H. DePuy, *J. Am. Chem. Soc.*, 1979, **101**, 6443.
[74] C. H. DePuy, V. M. Bierbaum, G. K. King, and R. H. Shapiro, *J. Am. Chem. Soc.*, 1978, **100**, 2921.
[75] J. H. J. Dawson and N. M. M. Nibbering, *J. Am. Chem. Soc.*, 1978, **100**, 1928.
[76] J. H. J. Dawson, A. J. Noest, and N. M. M. Nibbering, *Int. J. Mass Spectrom. Ion Phys.*, 1979, **29**, 205.
[77] D. Zakett, R. G. A. Flynn, and R. G. Cooks, *J. Phys. Chem.*, 1978, **82**, 2359.
[78] O. I. Asubiojo and J. I. Brauman, *J. Am. Chem. Soc.*, 1979, **101**, 3715.

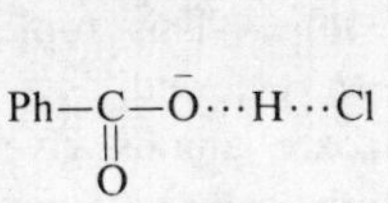

Ph—C—Ō···H···Cl   Ph—C—OH
   ‖ |
   O   Cl
   (11)   (12)

with Ō on top of the C in (12)

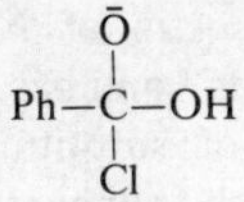

which suggests that the reaction (ICR) between a nucleophile and an acid halide proceeds through an association complex which looks like reactants, then a tetrahedral transition state, followed by an association complex that looks like products. Reactions between MeO⁻ and formaldehyde (flowing afterglow)[79] and MeO⁻ and acetone (ICR)[80] produce detectable 1 : 1 adducts. Whether the adduct in the latter case corresponds to the less thermodynamically stable complex (13), or the more stable tetrahedral structure (14), is a matter of conjecture.

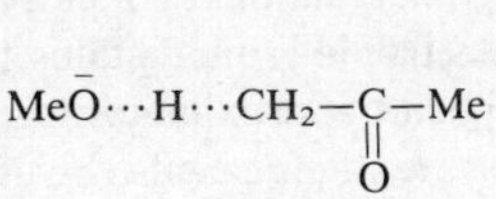

MeŌ···H···CH₂—C—Me   Me—C—Me
              ‖ |
              O   OMe
    (13)   (14)

with Ō on top of the C in (14)

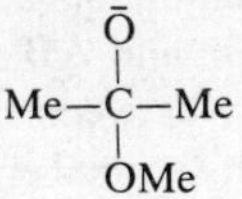

The following reactions have all been studied using ICR techniques. The reactions between MeO⁻ (or HO⁻) with ketones[80] and alkyl nitrites[81] produce ambident anions, e.g. MeCOCH₂⁻ from acetone,[80] and ⁻CH₂CN from acetonitrile.[81] Both types of ion react with methyl nitrite through the 'soft' carbanion centre to form (15; Y = CN or COR), which decomposes to yield (16). The M − 1 ions are also formed by the reactions of MeO⁻ with αβ-unsaturated ketones and nitriles.[82] Treatment of carboxylic acids with CF₃O⁻ yields (17) and CF₂O.[83,84] The species (17) undergoes a number of interesting reactions including transfer of F to other carboxylic acids, and the formation of an ion (18). The H-bound dimer (18) is analogous to similar species formed between RO⁻ and ROH.[85] The methoxide anion reacts both with 2-fluoroethanol to form ⁻CH₂CHO,[86] and with trimethylphosphate to form (MeO)₂PO₂⁻.[87]

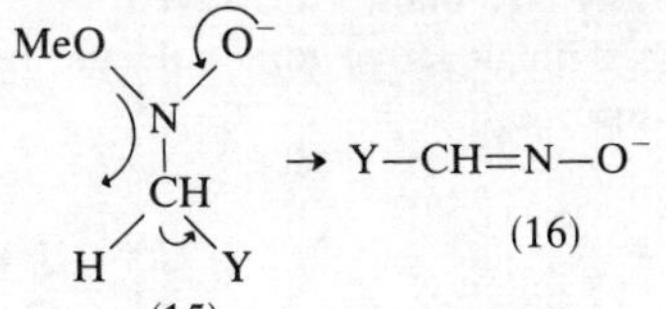

         MeO   O⁻
           \ /
            N
            |
            CH   →   Y—CH=N—O⁻
           / \   (16) 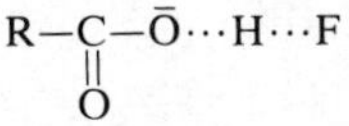
          H   Y
           (15)

R—C—Ō···H···F   R—C—Ō···H—O—C—R
  ‖   ‖   ‖
  O   O   O
  (17) 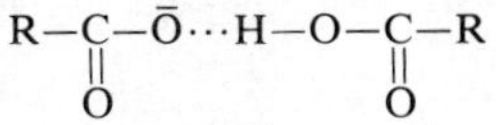   (18)

[79] D. K. Bohme, G. I. Mackay, and S. D. Tanner, *J. Am. Chem. Soc.*, 1980, **102**, 407.
[80] G. Klass and J. H. Bowie, *Aust. J. Chem.*, 1980, **33**, 2271.
[81] J. H. J. Dawson, and N. M. M. Nibbering, *Int. J. Mass Spectrom. Ion Phys.*, 1980, **33**, 3.
[82] J. E. Bartmess, *J. Am. Chem. Soc.*, 1980, **102**, 2483.
[83] R. L. Clair and T. B. McMahon, *Can. J. Chem.*, 1979, **57**, 473.
[84] R. L. Clair and T. B. McMahon, *Can. J. Chem.*, 1980, **58**, 307.
[85] L. K. Blair, P. C. Isolandi, and J. M. Riveros, *J. Am. Chem. Soc.*, 1973, **95**, 1057.
[86] R. L. Clair and T. B. McMahon, *Int. J. Mass Spectrom. Ion Phys.*, 1980, **33**, 21.
[87] R. V. Hodges, S. A. Sullivan, and J. L. Beauchamp, *J. Am. Chem. Soc.*, 1980, **102**, 935.

Charge Inversion Spectra of Negative Ions.—This topic was discussed in Volume 5 of this series[88] and has since been reviewed.[15]

Charge inversion of substituted benzoate anions gives rise to unstable benzoate cations which fragment characteristically by competitive losses of O, CO_2, and $(CO_2 + C_2H_2)$.[89] The positive ion spectra produced by charge inversion of the formate and acetate anions indicate that rearrangement reactions do not accompany or precede the positive ion fragmentations.[90] A consideration of the CID decompositions of $C_5H_5^+$ ions produced by EI from various C_5H_6 isomers, together with the charge inversion spectra of $C_5H_5^-$ ions (produced from the same neutrals), suggests that $C_5H_5^+$ ions formed by EI do not correspond to the cyclopentadienyl cation.[91]

Translational energy measurements (carried out with a V.G. M.M. ZAB 2F instrument) for a number of charge inversion reactions $AB^- + N \rightarrow AB^+ + N + 2e$ show that if AB^- and AB^+ are in electronic ground states the difference in translational energy between the two species approximates to the sum of IE_{AB} and $EA_{(AB^-)}$.[92] Translational energy measurements can also be used to determine energy requirements for dissociative charge inversion reactions, *i.e.* $AB^- \rightarrow [AB^+]^* \rightarrow A^+ + B$. For example, the charge inversion spectrum of the nitrobenzene molecular anion shows a process corresponding to either $PhNO_2^{+\cdot} \rightarrow (PhNO_2^{+\cdot} - NO_2^{\cdot})^+$ or $PhNO_2^{+\cdot} \rightarrow [PhNO_2^{+\cdot} - (NO^{\cdot} + O^{\cdot})]^+$. Translational energy measurements demonstrate the operation of the first of these processes, *i.e.* the loss of an intact NO_2 group.[93]

4 Concluding Remarks

In the mid 1970s, no more than a handful of research groups were working actively on the gas-phase chemistry of organic negative ions. The foregoing review shows how that situation is changing. Perhaps the increasing interest is due, in part, to the realization of many research groups that much of the truly innovative work has now been done for positive ions, and that it is time to move to new fields. It is certainly also due to the advent of NICI mass spectrometry, and to a renewed interest in negative ion–molecule reactions. We hope that the present trend will continue.

---

[88] J. H. Bowie, this series, 1979, Vol. 5, p. 281.
[89] J. H. Bowie and P. Y. White, *Austr. J. Chem.*, 1978, **31**, 1511.
[90] M. M. Bursey, D. J. Harvan, C. E. Parker, L. G. Pederson, and J. R. Hass, *J. Am. Chem. Soc.*, 1979, **101**, 5489.
[91] M. M. Bursey, J. R. Hass, and D. J. Harvan, *Tetrahedron Lett.*, 1979, 1725.
[92] I. Howe, J. H. Bowie, J. E. Szulejko, and J. H. Beynon, *J. Chem. Soc., Chem. Commun.*, 1979, 983.
[93] J. E. Szulejko, J. H. Bowie, I. Howe, and J. H. Beynon, *Internat. J. Mass Spectrom. Ion Phys.*, 1980, **34**, 99.

10
Natural Products

BY D. E. GAMES

1 Introduction

A major event has been the publication of the First Supplementary Volume of 'Biochemical Applications of Mass Spectrometry'.[1] Of particular relevance to this report are chapters on fatty acids,[2a] lipids,[2b] steroids,[2c] bile acids,[2d] carbohydrates,[2e] terpenoids and carotenoids,[2f] amino-acids,[2g] peptide sequencing,[2h] nucleic acids,[2i] antibiotics,[2j] hormones,[2k] tetrapyrroles,[2l] alkaloids,[2m,n] pheromones,[2o] and flavonoids.[2p] The recent biannual review on mass spectrometry in *Analytical Chemistry*[3] continues to maintain its high critical standard and is another indispensible source of information.

In the previous report,[4] the relative merits of the wide range of new mass spectral techniques available for structural studies of natural products were discussed. Life has not been simplified in the intervening period owing to the presentation of often exaggerated claims for the solution of problems by one particular mass spectral technique. In the author's experience, problems are normally solved by an accumulation of data obtained from a variety of mass spectral and other techniques. The introductory portion of this present report will concentrate on recent technical innovations which should prove of particular assistance to the natural product chemist.

The main thrust of current improvements in mass spectral techniques has been in three areas.

(i) Development of improved methods for obtaining mass spectral data from thermally labile and/or compounds of low volatility.

(ii) Development of techniques for handling high-mass compounds. Although this is separately categorized, techniques developed under (i) are obviously necessary.

(iii) Improved methods for handling complex mixtures.

[1] 'Biochemical Applications of Mass Spectrometry', First Supplementary Volume, ed. G. R. Waller and O. C. Dermer, Wiley-Interscience, New York, 1980.
[2] (*a*) G. Odham, in ref. 1, p. 153; (*b*) G. W. Wood, *ibid.*, p. 173; (*c*) H. Budzikiewicz, *ibid.*, p. 211; (*d*) W. H. Elliot, *ibid.*, p. 229; (*e*) T. Radford and D. C. DeJongh, *ibid.*, p. 255; (*f*) C. R. Enzell and I. Wahlberg, *ibid.*, p. 311; (*g*) W. Vetter, *ibid.*, p. 439; (*h*) K. Biemann, *ibid.*, p. 469; (*i*) C. Hignite, *ibid.*, p. 527; (*j*) D. B. Borders and R. T. Hargreaves, *ibid.*, p. 567; (*k*) C. J. W. Brooks and S. J. Gaskell, *ibid.*, p. 611; (*l*) R. C. Dougherty, *ibid.*, p. 693; (*m*) S. D. Sastry and K. M. Madyastha, *ibid.*, p. 751; (*n*) M. Hesse, *ibid.*, p. 797; (*o*) B. A. Andersson, L. Lundgren, and G. Stenhagen, *ibid.*, p. 855; (*p*) T. J. Mabry and A. Ulubelen, *ibid.*, p. 1131.
[3] A. L. Burlingame, T. A. Baillie, P. J. Derrick, and O. S. Chizhov *Anal. Chem.*, 1980, **52**, 214R.
[4] D. E. Games, in 'Mass Spectrometry', ed. R. A. W. Johnstone (Specialist Periodical Reports), The Chemical Society, London, 1979, Vol. 5, p. 285.

Mass spectral techniques for obtaining spectral data from involatile and thermally unstable molecules have been reviewed.[5]

Laser desorption mass spectrometry has been shown[6] to be useful for the characterization of polar involatile bio-organic molecules. Using a sub-microsecond laser pulse, cationized species of intact molecules and some fragment ions were obtained from adenosine, adenosine-5'-monophosphate, an octapeptide, and mono-, di-, tri-, and tetra-saccharides. In the saccharide case sequence ions were also present. The power of the method was exemplified in the spectrum of digitonin, a pentaglycoside obtained from digitalis. A disadvantage in this approach is that photoplate or simultaneous electro-optical ion detection have to be used for spectral recording. Use of a continuous wave CO_2 laser of low intensity ≥ 20 W cm^2 gives longer desorption times, enabling spectra to be obtained with a quadrupole instrument.[7] Using this approach, saccharides yielded similar data to those obtained with the pulsed laser.

A comprehensive review of the use of field desorption (FD) mass spectrometry has appeared[8] and laser-supported FD has proved useful in providing spectra from more difficult molecules.[9,10] Desorption chemical ionization (DCI)[11] is becoming an increasingly important technique and may well soon overtake the use of FD in many areas. The major advantages of DCI are that non-activated rhenium[12,13] or platinum[14] wires can be used, more fragmentation is induced than in FD spectra, and the technique often provides spectra from samples obtained from biological sources, without the extensive clean-up procedures usually necessary to obtain good FD spectra. Despite impressive data recently reported using negative DCI,[14] it is clear that the technique will not provide molecular weight information in all cases and FD will remain necessary for the solution of many problems.[12]

The most interesting development in FD has been the increased use of collision-induced dissociation (CID) with B/E linked scanning. Combination of the techniques enables more structural features to be obtained. Examples of substances to which this combination has been applied include suicidal covalent adducts of cytochrome P-450,[15] benzo[a]pyrene-bound dinucleotides,[16] the linkage isomers of disaccharides,[17] and phaeophytin a.[18]

---

[5] G. D. Daves, jun., *Acc. Chem. Res.*, 1979, **12**, 359.
[6] M. A. Posthumus, P. G. Kistemaker, H. L. C. Meuzelaar, and M. C. Ten Noever de Brauw, *Anal. Chem.*, 1978, **50**, 985.
[7] R. Stoll and F. W. Röllgen, *Org. Mass Spectrom.*, 1979, **14**, 642.
[8] H.-R. Schulten, *J. Mass Spectrom. Ion. Phys.*, 1979, **32**, 97.
[9] H.-R. Schulten, T. Komori, T. Hohara, R. Higuchi, and T. Kawasaki, *Tetrahedron* 1978, **34**, 1003.
[10] H.-R. Schulten and H. M. Schiebel, *Naturwissenschaften* 1978, **65**, 223.
[11] D. F. Hunt, J. Shabanowitz, F. K. Botz, and D. A. Brent, *Anal. Chem.*, 1977, **49**, 1160.
[12] U. Rapp, G. Dielmann, D. E. Games, J. L. Gower, and E. Lewis, in 'Advances in Mass Spectrometry', ed. A. Quayle, Heyden, London, 1980, Vol. 8, p. 1660.
[13] P. J. Arpino and G. Devant, *Analusis*, 1979, **7**, 348.
[14] A. P. Bruins, *Anal. Chem.*, 1980, **52**, 605; in ref. 12, p. 246.
[15] P. Ortiz de Montellano, G. S. Yost, B. A. Mico, S. E. Dinizo, M. A. Correia, and H. Kambara, *Arch. Biochem. Biophys.* 1979, **197**, 524.
[16] K. M. Straub and A. L. Burlingame, in ref. 12, p. 1127.
[17] H. Kambara and A. L. Burlingame, *27th Annual Conference on Mass Spectrometry and Allied Topics*, Seattle, 1979, 697.
[18] A. H. Jackson, *Philos. Trans. R. Soc. London, Ser. A*, 1979, **293**, 21.

More detailed reports of organic secondary ion mass spectrometry have appeared[19-21] and the area has been reviewed.[21] Whereas it is apparent that the technique is better than conventional EI in handling compounds of low volatility, at present no major advantages over other reported soft ionization methods are apparent. A similar comment is applicable to electrohydrodynamic ionization mass spectrometry which has been used to study pyrimidines, purines, and nucleotides.[22]

Californium-252 plasma desorption mass spectrometry has been reviewed[23] and continues to produce spectacular results. A major advantage of the technique is that only a relatively small portion of the sample is used to make the measurement and hence the technique is virtually non-destructive and samples can be recovered to make other measurements. The mass range and use of the technique are well illustrated by recent studies of protected tetra-, penta-, hexa-, and deca-nucleotides[24] where molecular weight data were obtained from the positive ion spectra whilst the negative ion spectra enabled the sequences to be determined. The value of combining soft ionization techniques with microscale chemical reactions is well illustrated[25] by Californium-252 plasma desorption data obtained from palytoxin, a powerful marine toxin, and its N-acetyl and N-acetylperhydro derivatives.

In addition to the DCI studies discussed earlier, a number of other improved sample introduction techniques based on in-beam sample introduction for obtaining CI data[26] have been reported. Teflon[27] and Vespel tips[28,29] and glass and copper rod extensions coated with a film of SE 30[30,31] all provide considerably improved CI spectra from thermally labile compounds of low volatility.

The availability of high-mass calibrants is vital for studies at high mass. Details of the use of Fomblin, the polyperfluoropropylene oxide mixture, as an EI calibrant have appeared.[32,33] In the FD mode, polystyrene and polypropylene glycol appear to have utility as mass reference compounds up to MW 10 000 at least.[34] A high-resolution Tandem mass spectrometer of increased sensitivity and mass range has been developed[35] and should have considerable use in handling higher-mass compounds of biological origin.

[19] A. Benninghoven and W. K. Sichtermann, *Anal. Chem.*, 1978, **50**, 1180.
[20] A. Benninghoven, in 'Trace Organic Analyses: A New Frontier in Analytical Chemistry', ed. H. S. Hertz and S. N. Chesler, N.B.S., Washington D.C., 1979, p. 627.
[21] R. J. Day, S. E. Unger, and R. G. Cooks, *Anal. Chem.*, 1980, **52**, 557A.
[22] S.-T. F. Lai and C. A. Evans, jun., *Biomed. Mass Spectrom.*, 1979, **6**, 10.
[23] R. D. Macfarlane, in ref. 1, p. 1209.
[24] C. J. McNeal, S. A. Narang, R. D. Macfarlane, H. H. Hsiung, and R. Brousseau. *Proc. Natl. Acad. Sci. USA*, 1980, **77**, 735.
[25] R. D. Macfarlane, D. Vernura, K. Ueda, and Y. Hirata, *J. Am. Chem. Soc.*, 1980, **102**, 875.
[26] M. A. Baldwin and F. W. McLafferty, *Org. Mass Spectrom.*, 1973, **7**, 1353.
[27] G. Hansen and B. Munson, *Anal. Chem.*, 1978, **50**, 1130; *ibid.*, 1980, **42**, 245.
[28] R. J. Cotter, *Anal. Chem.*, 1979, **51**, 317.
[29] R. J. Cotter and C. Fenselau, *Biomed. Mass Spectrom.*, 1979, **6**, 287.
[30] J. P. Thenot, J. G. Nowlin, D. I. Carroll, F. E. Montgomery, and E. C. Horning, *Anal. Chem.*, 1979, **51**, 1101.
[31] D. I. Carroll, I. Dzidic, M. G. Horning, F. E. Montgomery, J. G. Nowlin, R. N. Stillwell, J. P. Thenot, and E. C. Horning, *Anal. Chem.*, 1979, **51**, 1858.
[32] W. V. Ligon, jun., *Anal. Chem.*, 1978, **50**, 1228.
[33] G. Warburton, M. L. Aspinal, K. T. Taylor, and D. Hazelby, in ref. 12, p. 1953.
[34] T. Matsuo, H. Matsuda, and I. Katakuse, *Anal. Chem.*, 1979, **51**, 1329.
[35] F. W. McLafferty, P. J. Todd, D. C. McGilvery, and M. A. Baldwin, *J. Am. Chem. Soc.*, 1980, **102**, 3360.

Mass spectrometry–mass spectrometry (m.s.–m.s.), *i.e.* experiments in which ions in a mass-selected beam are fragmented by collision and the ionic fragments so formed are mass analysed, is being strongly advocated as a preferable alternative to g.c.–m.s. and l.c.–m.s. for the analysis of mixtures, mainly in terms of the speed of analysis.[36,37] This, and the utility of other metastable techniques in this context, has been reviewed.[38,39] Whereas most studies have employed reversed-geometry double-focusing instruments, the use of three quadrupoles in line has been advocated in this area in terms of the enhanced efficiency of CID and overall sensitivity obtained.[40,41] The approach has been applied, using negative CI, to the analysis of priority pollutants in industrial sludge[42] and polypeptide sequencing.[43] Development of a computerized mass spectrometer linked-scan system for recording metastable ions from a conventional geometry double-focusing instrument has been reported[44] and applied to the identification of *O*-methylabscisic acid in a sugar cane extract. The use of selected fragment scans for loss of neutral fragments[45] could be particularly useful in m.s.–m.s. in the screening of plant extracts for natural products where closely related series of compounds are often present. Improved data should be obtainable from the recently developed high-resolution tandem mass spectrometer[35] which allows high resolution to be used in the first m.s. region before CID is performed. Another advantage is the enhanced mass range for ions from which CID spectra can be obtained.

The technique of m.s.–m.s. has been used to identify coniine in hemlock,[46] constituents of mushrooms,[47,48] and cocaine and cinnomoylcocaine in cocoa leaf.[49] One of the dangers of this simplistic approach to compound identification is illustrated by the recent finding that the m.s.–m.s. identification of l-methyladenine in the pyrolysate of salmon sperm DNA[50] is an artifact of the pyrolysis.[51] Other dangers can be encountered in direct mixture analysis from the presence of artifact peaks, and a methodology for delineating these peaks has been described.[52]

Although it is apparent that m.s.–m.s. techniques are useful for the rapid screening of samples for the possible presence of known compounds, on present

---

[36] R. W. Kondrat and R. G. Cooks, *Anal. Chem.*, 1978, **50**, 81A.
[37] R. G. Cooks, in ref. 20, p. 609; F. W. McLafferty and F. M. Bockhoff, *Anal. Chem.*, 1978, **50**, 69.
[38] F. W. McLafferty, *Acc. Chem. Res.*, 1980, **13**, 33.
[39] J. H. Beynon and R. M. Capriolo, in ref. 1, p. 89.
[40] R. A. Yost and C. G. Enke, *J. Am. Chem. Soc.*, 1978, **100**, 2274; *Anal. Chem.*, 1979, **51**, 1251A.
[41] R. A. Yost, C. G. Enke, D. C. McGilvery, D. Smith, and J. D. Morrison, *Int. J. Mass Spectrom. Ion Phys.*, 1979, **30**, 127.
[42] D. F. Hunt, J. Shabanowitz, and A. B. Giordani, *Anal. Chem.*, 1980, **5**, 386.
[43] D. F. Hunt, A. M. Buko, J. Ballard, and J. Shabanowitz, in ref. 17, p. 680
[44] W. F. Haddon, *Anal. Chem.*, 1979, **51**, 983.
[45] D. Zakelt, A. E. Schoen, R. W. Kondrat, and R. G. Cooks, *J. Am. Chem. Soc.*, 1979, **101**, 6781.
[46] R. W. Kondrat and R. G. Cooks, *Science*, 1978, **199**, 978.
[47] G. A. McClusky, R. G. Cooks, and A. M. Knevel, *Tetrahedron Lett.*, 1978, 4471.
[48] S. E. Unger and R. G. Cooks, *Anal. Lett.*, 1979, **12**, 1157.
[49] R. W. Kondrat, G. A. McClusky, and R. G. Cooks, *Anal. Chem.*, 1978, **50**, 2017.
[50] A. E. Schoen, R. G. Cooks, and J. L. Wiebers, *Science*, 1979, **203**, 1249.
[51] D. L. Smith, P. F. Crain, and J. A. McCloskey, *Science*, 1980, in the press.
[52] M. H. Bozorgzadeh, A. G. Brenton, J. L. Wiebers, and J. H. Beynon, *Biomed. Mass Spectrom.*, 1979, **6**, 340.

evidence, considerable care should be exercised in the interpretation of the results, particularly if there is the possibility of isomeric species being present. G.c.–m.s. and l.c.–m.s. will have to be used to confirm findings and thus remain the methods of choice for the examination of natural materials for new natural products unless the chromatographic technique is used in combination with m.s.–m.s.

Considerable improvements have been effected in l.c.–m.s. over the past two years and the reader is referred to recent reviews[53—57] of the area and to Chapter 8 in this volume. The applicability of l.c.–m.s. in chemotaxonomic studies and in the search for new natural products is shown by papers on cannabinoids,[58] nucleosides,[59—61] peptides,[62,63] sugars,[61] glycosides,[61] alkaloids,[63—65] oxygen heterocyclics,[57,63,65,66] lipids,[67,68] and petroporphyrins[68] and a rapid growth in the use of the technique is envisaged as it becomes more widely available.

A technique utilizing ammonia and deuterioammonia as CI reagent gases has been reported for the estimation of active hydrogens.[69] Negative CI has proved to be useful in obtaining data from a wide range of natural products[70] and using a freon as a reagent gas has enabled molecular weight data to be obtained from a pentasaccharide.[71]

A general method for establishing whether retention or inversion occurs at phosphorus in phosphoryl transfer reactions illustrates a novel use of mass spectral methods.[72] $1(R)$-$^{16}O,^{17}O,^{18}O$-Phospho-(5)-propane-1,2-diol was prepared. The configuration at P was determined by ring closure to a cyclic diester, followed by methylation with diazomethane. Separation of the *syn*- and *anti*-isomers was then effected. Treatment with methanol was followed by linked-scan metastable ion analysis which enabled the configuration to be established.

---

[53] P. J. Arpino and G. Guiochon, *Anal. Chem.*, 1979, **51**, 682A.
[54] W. H. McFadden, *J. Chromatogr. Sci.*, 1979, **17**, 2.
[55] W. H. McFadden. *J. Chromatogr. Sci.*, 1980, **18**, 97.
[56] D. E. Games, *Anal. Proc.*, 1980, 110.
[57] D. E. Games, *Anal. Proc.*, 1980, 322.
[58] P. J. Arpino and P. Krien, *J. Chromatogr. Sci.*, 1980, **18**, 104.
[59] C. R. Blackley, M. J. McAdams, and M. L. Vestal, *J. Chromatogr.*, 1978, **158**, 261.
[60] A. Melera, in ref. 17, p. 363.
[61] D. E. Games and E. Lewis, *Biomed. Mass Spectrom.*, 1980, **7**, 433.
[62] B. G. Dawkins, P. J. Arpino, and F. W. McLafferty, *Biomed. Mass Spectrom.*, 1978, **5**, 1.
[63] D. E. Games, C. Eckers, J. L. Gower, P. Hirter, M. E. Knight, E. Lewis, K. R. N. Rao, and N. C. Weerasinghe, in 'Current Developments in the Clinical Applications of HPLC, GC and MS', ed. A. Lawson, Academic Press, New York, 1980, p. 97.
[64] J. D. Henion, *Anal. Chem.*, 1978, **50**, 1687.
[65] C. Eckers, D. E. Games, E. Lewis, K. R. N. Rao, M. Rossiter, and N. C. A. Weerasinghe, in ref. 12, p. 1396.
[66] D. E. Games, J. L. Gower, M. G. Lee, I. A. S. Lewis, M. E. Pugh, and M. Rossiter, in 'Blood Drugs and other Analytical Challenges', ed. E. Reid, Ellis Horwood, Chichester, 1978, p. 185.
[67] O. S. Privett and W. L. Erdahl, *Chem. Phys. Lipids*, 1978, **21**, 361.
[68] W. H. McFadden, D. C. Bradford, G. Eglington, S. K. Hajlbrahim, and N. Nicolaides, *J. Chromatogr. Sci.*, 1978, **17**, 518.
[69] Y. Y. Lin and L. L. Smith, *Biomed. Mass Spectrom.*, 1979, **6**, 15.
[70] A. K. Bose, H. Fujiwara, and P. N. Pramanik, *Tetrahedron Lett.*, 1979, 4017.
[71] A. K. Ganguly, N. F. Cappuccino, H. Fujwara, and A. K. Bose, *J. Chem. Soc., Chem. Commun.*, 1979, 148.
[72] S. J. Abbott, S. R. Jones, S. A. Weinmann, and J. R. Knowles, *J. Am. Chem. Soc.*, 1978, **100**, 2558.

2 Alkaloids

The area has been reviewed.[2m,n] Detailed EI mass spectral studies have been reported on lysergic acid and 9,10-dihydrolysergic acid and the data compared with electron-attachment negative ion mass spectra.[73] Similar studies have been performed with the sugar esters of these compounds.[74] EI, CI, and FD mass spectral studies on Amaryllidaceae alkaloids have been used together with g.c.–m.s. of trimethylsilylated derivatives, high-performance liquid chromatography, and ^{1}H n.m.r. to identify tentatively three new alkaloids from *Crinum ornatum* and one from *C. natans*.[75] Combined l.c.–m.s. studies of extracts of *Crinum glaucum* and Chinchona alkaloids show the potential of the technique in assisting in the discovery of new alkaloids.[65]

The role of mass spectrometry in determining the structures of some new diterpenoid alkaloids from *Delphinium ajacis* has been described.[76] Mass spectral techniques have assisted in structural assignments to many new alkaloids, as exemplified by aphelandrin (1), a new spermine alkaloid,[77] and the new alkaloids from *Catha edulis*.[78] MIKES has been applied in the structural assignment of the new peptide alkaloid scutianine H (2).[79] MIKES of the base peak (m/z 114)

(2) R^1 = dimethylisoleucine
R^2 = CH(OH)Ph

in its spectrum was compared with similar spectra obtained from alkaloids of known structure and enabled the group R^1 to be identified as an *NN*-dimethylisoleucine moiety rather than *NN*-dimethyl-leucine.

Studies of the biosynthesis of slaframine (3) utilized mass spectral analysis of [^{2}H]-enriched slaframine isolated from feeding experiments with 2-[^{2}H]-acetic

---

[73] J. Schmidt, R. Kraft, and D. Voigt, *Biomed. Mass Spectrom.*, 1978, **12**, 674.
[74] J. Schmidt, K. Seifert, and S. Johne, *Pharmazie*, 1979, **34**, 77.
[75] O. S. Onyiruka and A. H. Jackson, *Isr. J. Chem.*, 1978, **17**, 185.
[76] G. R. Waller and R. H. J. Lawrence, in 'Recent Developments in Mass Spectrometry in Biochemistry and Medicine', ed. A. Frigerio, Plenum Press, New York, 1978, Vol. 1, p. 429.
[77] P. Datwyler, H. Bosshardt, H. O. Bernhard, M. Hesse, and S. Johne, *Helv. Chim. Acta*, 1978, **61**, 2646.
[78] R. L. Baxter, L. Crombie, D. J. Simmonds, D. A. Whitting, O. J. Braenden, and K. Szendrei, *J. Chem. Soc., Perkin Trans. 1*, 1979, 2965; R. L. Baxter, L. Crombie, D. J. Simmonds, and D. A. Whiting, *ibid.*, p. 2972; L. Crombie, W. M. L. Crombie, D. A. Whiting, and K. Szendri, *ibid.*, p. 2976; R. L. Baxter, W. M. L. Crombie, L. Crombie, D. J. Simmonds, D. A. Whiting, and K. Szendri, *ibid.*, p. 2982.
[79] A. F. Morel, R. V. F. Bravo, F. De. A. M. Reis, and E. A. Ruveda, *Phytochemistry*, 1979, **18**, 473.

acid and *cis*- and *trans*-1,3,3-[²H]-1-hydroxyoctahydroindotizine together with more traditional ³H- and ¹⁴C-feeding experiments.⁸⁰

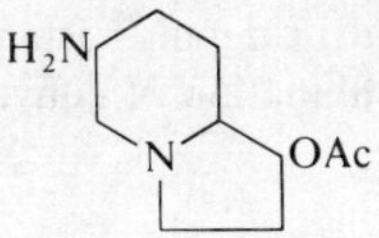

[Structure (3): piperidine ring with H₂N substituent and OAc group]

(3)

3 Aromatic Compounds and Oxygen Heterocycles

Off-line l.c.–m.s. has been developed for the analysis of capsaicinoids.[81] Whilst this technique is suitable for the analysis of well-resolved l.c. components, combined l.c.–m.s. is preferable when complex plant extracts are studied as is illustrated by recent l.c.–m.s. studies of coumarins in *Imperatoria ostruthium*,[57] *Mammea americana*,[65] and *Calophyllum inophyllum*.[66]

Based on studies of the mass spectral fragmentations of a series of flexirubin pigments of known structure and model compounds, stemphol, a product of *Stemphylium majusculum*, has been assigned structure (4).[82] CI studies of

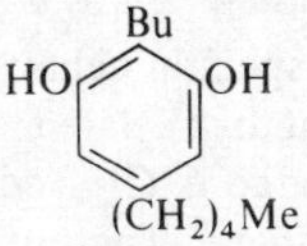

[Structure (4): benzene ring with Bu, two OH groups, and (CH₂)₄Me substituent]

(4)

phloroglucinol derivatives from *Hagenia abyssinica*[83] and ferns of the genus *Dryopteris*[84] have been reported. In the latter case, the spectra were complicated by thermal degradation prior to protonation. A detailed comparison of the EI, CI, and FD spectra of natural anthrones and anthraquinones is reported. FD in combination with EI is particularly useful for characterizing *O*-glycosides of this type.[85]

A number of studies of humic and fulvic acids by pyrolysis–mass spectrometry have appeared[85–88] and isobutane CI has assisted in structural studies of lignins.[89]

Using glass capillary g.c.–m.s. a number of new cannabinoids were shown to be present in extracts of Hashish.[90] The structure (5), a new cannabicoumaronon,

---

[80] E. C. Clevenstine, H. P. Broquist, and T. M. Harris, *Biochemistry*, 1979, **18**, 3658; E. C. Clevenstine, P. Walter, T. M. Harris, and H. P. Broquist, *ibid.*, p. 3663.
[81] F. Heresh and J. Jurenitsch, *Chromatographia*, 1979, **12**, 647.
[82] H. Achenbach and W. Kohl, *Chem. Ber.*, 1979, **112**, 209.
[83] M. Lounasmaa and P. Varenne, *Planta Med.*, 1978, **34**, 153.
[84] M. Lounasmaa, *Planta Med.*, 1979, **37**, 151.
[85] F. J. Evans, M. G. Lee, and D. E. Games, *Biomed. Mass Spectrom.*, 1979, **6**, 374.
[86] T. V. Nazarova, I. M. Lukashenko, R. A. Kheml'nitskii, and V. A. Chernikov, *Izv. Timiryazevsk. Skh. Akad.*, 1978, 176.
[87] K. Haider and H. L. C. Meuzelaar, *Agrochimica*, 1978, **22**, 353.
[88] G. Minderman, *Neth. J. Agric. Sci.*, 1979, **27**, 277.
[89] J. Metzger, *Fresenius Z. Anal. Chem.*, 1979, **295**, 45.
[90] H. Grote and G. Spiteller, *J. Chromatogr.*, 1978, **154**, 13; *Tetrahedron*, 1978, **34**, 3207.

was assigned to one of them on the basis of its mass spectrum and chemical interconversions.

The mechanism of elimination of water under EI conditions in desmethylencecalin (6) has been studied in detail.[91] Isomeric methoxyfuranocoumarins can be differentiated by a combination of conventional EI mass spectrometry

(5)   (6)

with CID of molecular ions and selected fragment ions.[92] Deuterium labelling and high-resolution studies[93] of the EI fragmentation pathways of 7-flavan-4-ols are consistent with previously proposed ion structures but indicate that the mechanisms for certain H-rearrangements require revision. On the basis of the relative abundance of $M^{+\cdot}$ and $[M-15]^+$ peaks, 5-hydroxy-6,7-dimethoxy- and 5,7,8-trimethoxy-flavones can be differentiated from 5-hydroxy-7,8-dimethoxy- and 5,6,7-trimethoxy-flavones, respectively.[94] Mass spectral studies are also reported for a series of hydroxy-flavanones, flavones, and flavonols and their permethylated derivatives.[95]

Detailed studies of the mass spectral behaviour of permethylated, isopropylidine permethylated and trimethylsilylated flavonoid glycosides have been described. Isomeric permethylated 6-C-glycosylflavones O-glycosylated on a phenolic group are distinguishable.[96] The position of the O-glycosidic bond is assignable on the basis of mass spectral study of the permethylated acid hydrolysis product of the O-glycosyl-6-C-glycosylflavone;[97] the methodology has been applied to study the O-glycosylation of isovitexin in petals of *Melandrium album*.[98] Studies of the EI mass spectrometry of twenty mono-, di-, and tri-flavanoid glycosides in the form of isopropylidene, permethyl derivatives showed that their mass spectra could differentiate between stereoisomeric sugar moieties, determine the interglycosidic linkages in di- and tri-glycosides, and differentiate between furanoid and pyranoid ring forms in the sugars.[99] A similar study of

---

[91] M. Nakayama, S. Eguchi, S. Hayashi, M. Tsukayama, and H. Nakata, *Org. Mass Spectrom.*, 1979, **14**, 672.
[92] S.-Y. Tang, J. C. McCowan, M. Singh, P. Galatsis, B. E. Ellis, R. K. Boyd, and S. A. Brown, *Can. J. Chem.*, 1979, **57**, 1995.
[93] T. Radford, J. C. Sweeny, A. C. Iacobucci, J. D. Hribar, and L. A. Weibel, *Org. Mass Spectrom.*, 1979, **14**, 299.
[94] M. Goudard, J. Favre-Bouvin, A. Strelisky, M. Nogradi, and J. Chopin, *Phytochemistry*, 1979, **18**, 186.
[95] G. P. Kononenko, S. A. Popravko, B. V. Rozynov, V. G. Zaikin, and N. S. Wulfson, *Bioorg. Khim.* 1980, 267.
[96] M. L. Bouillant, A. Besset, J. Favrebonvin, and J. Chopin, *Phytochemistry*, 1978, **17**, 527.
[97] M. L. Bouillant, A. Besset, J. Favrebonvin, and J. Chopin, *Phytochemistry*, 1979, **18**, 690.
[98] E. Besson, A. Besset, M. L. Bouillant, J. Chopin, J. van Brederode, and G. van Nigtevecht, *Phytochemistry*, 1979, **18**, 657.
[99] O. Seligmann and H. Wagner, *Tetrahedron*, 1978, **34**, 3299.

trimethylsilylated mono- and di-glycosides has been conducted.[100] An alternative strategy in structural study is utilization of EI and FD spectra of underivatized glycosides.[101]

4 Isoprenoids

The area has been reviewed.[2f] A detailed study of the origin of the base peak in the EI spectrum of limonene indicates that its most probable mechanism of formation involved a retro Diels–Alder reaction of the structurally intact molecular ion.[102] OH⁻ negative CI has been shown to be useful in providing molecular weight information and in characterizing the acid portion of esters in components of essential oils.[103] The technique has been applied in combination with EI in studies of the essential oils of *Liquidambar styraciflua*[104] and *Valeriana officinalis*.[105] Field ionization (FI) in combination with CI and EI offers an alternative approach but does not have the sensitivity of CI and EI.[106]

Ester iridoids of the valepotriate type[107] have been studied by EI and the EI and FD spectra of oleuropein and ligustroside acetyl derivatives have been discussed.[108] Studies of the mass spectral behaviour of germacronolide and eudesmane sesquiterpenes are reported.[109–111] Pairs of gibberellins, differing from each other only by the presence or absence of a double bond, were separated in the form of their *p*-nitrobenzylesters using high-performance liquid chromatography with a silver nitrate-impregnated silica column; mass spectrometry was used to identify them.[112] High-resolution and metastable-ion studies have assisted in characterizing the EI behaviour of GA_3-type gibberellins.[113] Studies of the mass spectral fragmentations of triterpenoids of the dammarane,[114] stictane,[115] meliacin,[116] hopane,[117] and allobetulane[118] types and of sesquiterpenes of the taraxane type[119] have appeared.

Using hydrogen and isobutane as CI reagent gases, a detailed study of carotenoids with symmetrical and unsymmetrical end groups has been reported. Both reagent gases gave similar results.[120] CI was preferred to EI as a method

[100] H. Schels, H. D. Zinsmeister, and K. Pfleger, *Phytochemistry*, 1978, **17**, 523.
[101] K. M. Biswas, M. E. Ali, A. H. Jackson, and D. E. Games, *J. Indian Chem. Soc.*, 1978, **55**, 1240.
[102] D. Harris, S. McKinnon, and R. K. Boyd, *Org. Mass Spectrom.*, 1979 **14**, 265.
[103] A. P. Bruins, *Anal. Chem.*, 1979, **51**, 967.
[104] D. H. E. Tattje, R. Bos, and A. P. Bruins, *Planta Med.*, 1980, **38**, 79.
[105] H. Hendriks and A. P. Bruins, *J. Chromatogr*, 1980, **190**, 321.
[106] D. W. Knight, M. E. Knight, M. Rossiter, A. Jones, and D. E. Games, in ref. 12, p. 1116.
[107] S. S. Popov and N. V. Handjieva, *Biomed. Mass Spectrom.*, 1979, **6**, 124.
[108] A. Selva, V. Vettori, S. Popov, and N. L. Marekov, *Boll. Chim. Farm.*, 1978, **117**, 77.
[109] D. L. Perry, D. Desiderio, and N. H. Fischer, *Org. Mass Spectrom.*, 1978, **13**, 325.
[110] P. I. Zakharov, P. B. Terent'ev, O. A. Konovalova, and K. S. Rybalko, *Khim. Prir. Soedin.*, 1978, 337.
[111] U. A. Abdullaev, Ya. V. Rashkes, V. A. Tarasov, and Sh. Z. Kasymov, *Khim. Prir. Soedin.*, 1979, 34.
[112] E. Heftmann, G. A. Saunders, and W. F. Haddon, *J. Chromatogr.*, 1978, **156**, 71.
[113] D. Voigt, G. Adam, J. Schmidt, and P. Franke, *Org. Mass Spectrom.*, 1978, **13**, 599.
[114] B. H. Han and J. H. Kim, *Soul Taehakkyo Saengyak Yonguso Opjukjip*, 1978, **17**, 12.
[115] P. T. Holland and A. L. Williams, *Org. Mass Spectrom.*, 1979, **14**, 160.
[116] G. B. Marcelle and B. S. Mootoo, *Rev. Latinoam. Quim.*, 1979, **10**, 71.
[117] J. Schmidt and S. Huneck, *Org. Mass Spectrom.*, 1979, **14**, 656.
[118] J. Schmidt and S. Huneck, *Org. Mass Spectrom.*, 1979, **14**, 646.
[119] J. S. Pyrek and E. Barandwska, *Pol. J. Chem.*, 1978, **52**, 97; J. S. Pyrek, *ibid.*, 1979, **53**, 1795.
[120] J. Carnevale, E. R. Cole, D. Nelson, and J. S. Shannon, *Biomed. Mass Spectrom.*, 1978, **5**, 641.

for characterizing carotenoids on the basis of its better sensitivity, improvement of the relative abundances of higher mass ions, and the fact that the diagnostic features of the $[M-92]^+/[M-106]^+$ ratio were retained. CI has also been advocated as a useful method for studying marine sterols.[121] Oxidation of the sterols with RuO_4 followed by esterification of the crude oxidation product with diazomethane and subsequent CI study with ammonia as reagent gas enables double bonds in the sterol side-chain to be located. A detailed study of the EI fragmentation behaviour of 3β-hydroxy-Δ^5 sterols with a C-24 or C-24(8) unsaturated side-chain has been undertaken.[122] This latter approach together with 360 MHz ^1H n.m.r. has been successful in identifying a range of new marine sterols.[123]

Using CI, high-resolution FD and EI with X-ray crystallography and n.m.r. the plant growth promoting factor in Brassica napus pollen, brassinolide, has been identified as a C_{28} sterol with a novel ring B lactone structure (7).[124]

(7)

Cardiac glycosides, particularly when large numbers of sugars are present, represent a challenge for mass spectral techniques both in terms of molecular weight data and sequencing of the sugar units. Pulsed laser desorption as discussed earlier has been successful with compounds of this type.[6] Positive DCI[12] is also useful, but negative DCI with OH^- as reagent ion appears to be superior,[14] providing abundant $[M-H]^-$ ions and sequence ions for the sugar moieties. For simpler glycosides, i.e. one or two sugar units, direct-probe ammonia CI enables the molecular weight to be ascertained and provides abundant fragment ions by cleavage of the glycosidic linkages.[125] FD has previously been the technique of choice in this area and has been applied in the identification of the cardiac glycosides present in the arrow poison from Lophopetalum toxicum.[126] In studies of series of physiologically active glycosidic

---

[121] A. K. Bose, P. N. Pramanik, B. G. Pujar, and H. Fujiwara, in 'Drugs and Food from the Sea', ed. P. N. Kaul and C. J. Sindermann, University of Oklahoma Press, Norman, Oklahoma, 1978, p. 181.
[122] I. J. Massey and C. Djerassi, J. Org. Chem., 1979, 44, 2448.
[123] B. N. Ravi, W. C. M. C. Kokke, C. Delseth, and C. Djerassi, Tetrahedron Lett., 1978, 4379; C. Delseth, L. Tolela, P. J. Scheuer, R. J. Wells, and C. Djerassi, Helv. Chim. Acta., 1979, 62, 101; W. C. M. C. Kokke, C. S. Pak, W. Fenical, and C. Djerassi, ibid., p. 1310; C. Tarchini, M. Rohmer, and C. Djerassi, ibid., p. 1210.
[124] M. D. Grove, G. F. Spencer, W. K. Rohwedder, N. Mandava, J. F. Worley, J. D. Warthen, jun., G. L. Steffens, J. L. Flippen-Anderson, and J. C. Cook, jun., Nature, 1979, 281, 216.
[125] J. Vine, L. Brown, J. Boutagy, R. Thomas, and D. Nelson, Biomed. Mass Spectrom., 1979, 6, 415.
[126] H. Wagner, H. Habermeier, A. Liptak, and H.-R. Schulten, Planta Med., 1979, 37, 381.

saponins, laser-supported FD has proved useful in providing molecular weight and sequence data. FD cleavages similar to those occurring in acidic solvolysis were observed.[127]

5 Steroids

Detailed reviews of the area have appeared.[2c,128] A systematic study[129] of the behaviour of thirty-four C_{27} steroids under CI conditions with methane, isobutane, and ammonia has resulted in the observation of characteristic behaviour which can assist in functional group identification. MIKES has been shown to be useful in the differentiation of steroids with similar tetracyclic carbon skeletons but possessing diverse side-chains.[52] The authors were particularly careful to articulate on the presence of extraneous peaks which can cause problems in interpretation of MIKES data. Linked-scan (B/E) studies of isomeric androstanediol t-butyldimethylsilyl ethers showed the technique to be capable of differentiating isomeric species.[130]

Studies of the EI mass spectral fragmentations of 3,6,20-trihydroxypregnanes,[131] $\alpha\beta$-unsaturated 3-keto steroids,[132] steroidal diketones with 'unnatural' stereochemistry at positions 8, 9, and 14,[133] and 5-methyl-19-nor-5β-cholest-9-enes[134] have been reported. Identification of partial structure of unknown steroids using a key-ion index and computer-aided retrieval system has been described.[135] The EI and CI behaviour of steroidal spirolactones has been studied,[136] as has the hydroxy ion negative CI behaviour of a series of steroids.[137] The influence of the 6-trimethylsilyl group on the EI fragmentation of the trimethylsilyl derivatives of some 6-hydroxy- and 3,6-dihydroxy-steroids has been investigated[138] and the EI mass spectra of formate derivatives of androsterones and androsterols have been reported.[139] Bile acid esters have been the subject of both EI[140] and CI[141] studies.

The stereochemistry of vitamin D_3 and its analogues has been extensively studied.[142] Methods for determining the configuration of 5,6-double bonds,[143] epimeric diols,[144] and the relative stabilities of stereoisomers[145] have all been

[127] H.-R. Schulten, T. Komori, T. Hohara, R. Higuchi, and T. Kawasaki, *Tetrahedron*, 1978, **34**, 1003; T. Komori, M. Kawamura, K. Miyahara, T. Kawasaki, O. Tanaka, S. Yahara, and H.-R. Schulten, *Z. Naturforsch., Teil C*, 1979, **34**, 1094.
[128] C. Djerassi, *Pure Appl. Chem.*, 1978, **50**, 171.
[129] Y. Y. Lin and L. L. Smith, *Biomed. Mass Spectrom.*, 1978, **5**, 604.
[130] S. J. Gaskell, A. W. Pike, and D. S. Millington, *Biomed. Mass Spectrom.*, 1979, **6**, 78.
[131] A. Grupe and G. Spiteller, *Org. Mass Spectrom.*, 1978, **13**, 448.
[132] F. J. Brown and C. Djerassi, *J. Am. Chem. Soc.*, 1980, **102**, 807.
[133] D. G. Patterson, A. Lavanchy, and C. Djerassi, *Org. Mass Spectrom.*, 1980, **15**, 41.
[134] F. Turecek and P. Kocovsky, *Coll. Czech. Chem. Commun.*, 1979, **44**, 429.
[135] G. Spiteller, M. Spiteller, M. Ende, and G. A. Hoyer, *Org. Mass Spectrom.*, 1978, **13**, 646.
[136] D. R. Boreham, G. C. Ford, N. J. Haskins, C. W. Vose, and R. F. Palmer, *Biomed. Mass Spectrom.*, 1978, **5**, 524.
[137] T. A. Roy, F. H. Field, Y. Y. Lin, and L. L. Smith, *Anal. Chem.*, 1979, **51**, 272.
[138] D. J. Harvey and P. Vouros, *Biomed. Mass Spectrom.*, 1979, **6**, 135.
[139] B. R. Pettit and G. S. King, *Biomed. Mass Spectrom.*, 1979, **6**, 162.
[140] J. R. Dias and B. Nassim, *J. Org. Chem.*, 1980, **45**, 337.
[141] G. M. Muschik, L. H. Wright, and J. A. Schroer, *Biomed. Mass Spectrom.*, 1979, **6**, 266.
[142] Z. V. I. Zaretskii, *Isr. J. Chem.*, 1978, **17**, 198.
[143] Z. V. I. Zaretskii, *Biomed. Mass Spectrom.*, 1978, **5**, 576.
[144] Z. V. I. Zaretskii, *Steroids*, 1979, **33**, 595.
[145] Z. V. I. Zaretskii, *Nouv. J. Chim.*, 1978, **2**, 531.

developed. A combination of ^2H n.m.r. and mass spectrometry was utilized in studies of the vitamin D_3–previtamin D_3 equilibrium.[146]

6 Antibiotics

Underivatized aminoglycoside antibiotics when studied by in-beam EI techniques yield spectra which are a mixture resulting from the ions obtained by conventional EI and ions derived from $[M + 1]^+$ ions.[147] $[M + 1]^+$ Ions were obtained from mono-, di-, and tri-aminoglycosides but were absent in the tetra-aminoglycoside lividomycin. The use of CI in structural studies of underivatized and N-acetylated-NO-permethylated aminoglycoside antibiotics has been investigated.[148] EI studies of NO-permethylated and N-trideuterioacetylated, NO-permethylated derivatives have been used to characterize partially N-acetylated derivatives of Kanamycin A.[149]

Flambamycin is a member of the orthosomycin class of natural product. The role of mass spectral evidence in its structural assignment has been discussed in considerable detail.[150] Iturine C (8; n = 1 or 0), an antibiotic from

```
┌──────────────── L-Asp-D-Tyr-D-Asn-L-Gln ────────────────┐
│                                                         │
└──────── CO—CH₂—CHNH-L-Ser-D-Asn-L-Pro ──────────────────┘
                  │
              (CH₂)₈CHMe(CH₂)ₙMe
```

$$\text{CO--CH}_2\text{--CHNH-L-Ser-D-Asn-L-Pro}$$
$$|$$
$$(CH_2)_8 CHMe(CH_2)_n Me$$

(8)

Bacillus subtilis, has been shown to be a 54:46 mixture of cyclic peptides. Its amino-acid sequence was determined by comparison of the mass spectrum of its permethylated derivative with that obtained from permethylated iturine A.[151] CI and FD mass spectrometry together with X-ray crystallography were used in the structural elucidation of 23672 RP, a new macrolide antibiotic from *Streptomyces chryseus*.[152] In contrast, a combination of EI, FD, and n.m.r. data were used in the structural study of the polyether antibiotic K-41.[153]

A major problem in this area is early recognition of previously identified compounds. FD mass spectrometry has been shown to be useful in this context. FD studies of an intact and derivatized pentaene macrolide antibiotic and some of its degradation products enabled its early identification as fungichromin.[154] Combined l.c.–m.s. should also be useful in this context and some cephalosporins[63] and penicillins[65] have been shown to be amenable to study by the technique.

---

[146] M. Sheves, E. Berman, Y. Mazur, and Z. V. I. Zaretskii, *J. Am. Chem. Soc.*, 1979, **101**, 1882.
[147] M. Ohashi, S. Yamada, H. Kudo, and N. Nakayama, *Biomed. Mass Spectrom.*, 1978, **10**, 578.
[148] A. M. Sepulchre, P. Varenne, B. C. Das, S. D. Gero, and H. E. Audier, *Nouv. J. Chim.*, 1978, **2**, 405.
[149] D. C. DeJongh, M. S. B. Mayar, G. Patil, and S. Hanessian, *Tetrahedron*, 1978, **34**, 189.
[150] W. D. Ollis, S. Jones, C. Smith, and D. E. Wright, *Tetrahedron*, 1979, **35**, 1003.
[151] F. Peypoux, F. Besson, G. Michel, L. Delcambe, and B. C. Das, *Tetrahedron*, 1978, **34**, 1147.
[152] B. Arnoux, C. Pascard, L. Raynaud, and J. Lunel, *J. Am. Chem. Soc.*, 1980, **102**, 3605.
[153] J. L. Occolowitz, D. E. Dorman, and R. L. Hamill, *J. Chem. Soc., Chem. Commun.*, 1978, 683.
[154] R. C. Pandey, C. C. Kalita, A. A. Aszalos, R. Geoghegan, jun., A. L. Garretson, J. C. Cook, jun., and K. L. Rinehart, jun., *Biomed. Mass Spectrom.*, 1980, **7**, 93.

FD studies of nucleoside antibiotics[155] and the EI mass spectra of rimocidin and various of its derivatives[156] are described. In direct probe studies, many penicillins and cephalosporins undergo thermal degradation. Pyrolysis mass spectrometry has been studied as a method for characterization of members of these classes of antibiotics.[157]

7 Nucleic Acid Components and Cytokinins

The EI spectra of $O^6,5^1$- and $O^2,2'$-anhydrouridine have been studied[158] as models for the fragmentation behaviour of pyrimidine anhydronucleosides. Methyl-substituted uracils have been shown to undergo retro Diels–Alder reactions in the mass spectrometer with elimination of HNCO or MeNCO, depending on the substitution at N-3.[159] This is followed by sequential loss of CO and H. The extent of other fragmentation was found to be dependent on the number and location of the methyl groups. Studies of the EI mass spectrometry of trimethylsilyl derivatives of 3'- and 5'-monophosphoric acids of guanosine, uridine, cytidine, and adenosine show that this method can be used to differentiate between the two types of isomer.[160] Sterically crowded trialkylsilyl derivatives of ribo- and 2'-deoxyribo-nucleosides have been subjected to EI study,[161] and 3,N^4-etheno-O-trimethylsilyl derivatives of nucleosides and nucleotides of cytosine have been advocated in preference to trimethylsilyl derivatives because of the enhanced abundance of their molecular ions.[162] The low- and high-resolution EI spectra of mono-benzylated nucleosides have been studied[163] with a view to utilizing these derivatives in the characterization of blocked nucleosides.

The electrohydrodynamic ionization mass spectra of pyrimidines, purines, nucleosides, and nucleotides have been reported.[22] Utilization of Californium-252 plasma desorption for the characterization of protected oligodeoxyribonucleotides[24] has already been discussed in this report. Although FD mass spectrometry works well with synthetic nucleosides and nucleotides problems are encountered with samples of biological origin. Extensive clean-up procedures have to be resorted to if adequate spectral data are to be obtained. An alternative strategy involves derivatization and is illustrated by FD studies of methylation products of dinucleoside phosphates.[164] The use of FD in combination with CID and B/E linked scans has been shown to have considerable potential in structural studies of polynucleotides.[165]

The use of mass spectral methods in studies of nucleic acid complexes with 7,12-dimethylbenz[a]anthracene and trigonelline have been reviewed.[166] The

[155] K. Fukushima and T. Arai, *J. Antibiot.*, 1978, **31**, 377.
[156] F. Falkowski, J. Zielinski, J. Golik, E. Bylec, and E. Borowski, *J. Antibiot.*, 1978, **31**, 742.
[157] M. D. Mueller, J. Seibl, and W. Simon, *Anal. Chim. Acta*, 1978, **100**, 263.
[158] G. Puzo, K. H. Schram, J. G. Liehr, and J. A. McCloskey, *J. Org. Chem.*, 1978, **43**, 767.
[159] S. K. Saha and W. Pfleiderer, *Indian J. Chem.*, Sect. B., 1978, **16**, 246.
[160] H. Budzikiewicz and G. Feistner, *Biomed. Mass Spectrom.*, 1978, **5**, 512.
[161] M. A. Quilliam, K. K. Ogilvie, K. L. Sadana, and J. B. Westmore, *Org. Mass Spectrom.*, 1980, **15**, 207.
[162] K. H. Schram, Y. Taniguchi, and J. A. McCloskey, *J. Chromatogr.*, 1978, **155**, 355.
[163] H. T. Cory, K. Yamaizumi, D. L. Smith, D. R. Knowles, A. D. Broom, and J. A. McCloskey, *J. Heterocyclic Chem.*, 1979, **16**, 585.
[164] M. Linscheid, G. Feistner, and H. Budzikiewicz, *Isr. J. Chem.*, 1978, **17**, 163.
[165] K. M. Straub and A. L. Burlingame, in ref. 12, p. 1127.
[166] K. Nakanishi and J. L. Occolowitz, *Philos. Trans. R. Soc. London, Ser A.* 1979, **293**, 3.

structure of a highly modified nucleoside (9) from mammalian transfer RNA has been elucidated[167] on the basis of high-resolution mass spectral data of a series of methylated, deuteriomethylated, trimethylsilylated, and deuteriotrimethylsilylated derivatives. Using a combination of EI and CI studies of trimethylsilyl and acetyl derivatives and underivatized material, a number of zeatin glycosides and other metabolites were identified in sweet-corn kernels.[168]

(9)

Analysis of DNA and RNA by pyrolysis m.s. has been reviewed.[169] Earlier studies in this area[170] have been supplemented by a high-resolution m.s. study of the ions used to characterize bases,[171] by MIKES identification of modified bases in DNA,[50] and by an investigation of the probe pyrolysis of oligoribonucleotides.[172] Although adenine and uracil could be readily identified in this study, evidence was difficult to obtain for guanine and cytosine. However, recent studies[51] indicate that caution should be exercised in this area with identification based on pyrolysis studies. It would appear that combined l.c.–m.s. or derivatization methods in combination with g.c.–m.s. may produce more reliable data.

8 Pyrrole Pigments

The use of mass spectral techniques in this area has been reviewed.[2,18,173] Studies of phycoerythrobilin and phycocyanobilin, using high-performance liquid chromatography and rapid-heating techniques[174] in combination with ammonia

[167] Z. Yamaizumi, S. Nishimura, K. Limburg, M. Raba, H. J. Gross, P. F. Crain, and J. A. McCloskey, *J. Am. Chem. Soc.*, 1979, **101**, 2224.
[168] R. E. Summons, B. Entsch, D. S. Letham, B. I. Gollnow, and J. K. MacLeod, *Planta*, 1980, **147**, 422.
[169] J. L. Wiebers, in 'High Performance Mass Spectrometry, Chemical Applications' (American Chemical Society Symposium Series, Series 70), ed. M. L. Gross, American Chemical Society, Washington, D.C. 1978, p. 248.
[170] J. L. Wiebers, *Nucleic Acid Res.*, 1976, **3**, 2959; J. L. Wiebers and J. A. Shapiro, *Biochemistry*, 1977, **16**, 1044.
[171] D. Gaudin and K. Jankowski, *Org. Mass Spectrom.*, 1980, **15**, 78.
[172] M. L. Gross, P. A. Lyon, A. Dasgupta, and N. K. Gupta, *Nucleic Acid Res.*, 1978, **5**, 2695.
[173] H. Budzikiewicz, in 'Porphyrins', 3 (Part A), ed. D. Dolphin, Academic Press, New York, 1978, p. 395.
[174] R. J. Beuhler, R. C. Pierce, L. Friedman, and H. W. Siegelman, *J. Biol. Chem.*, 1976, **251**, 2405.

CI have established that each pigment exists in two principal isomeric forms.[175] The molecular weights of the pigment-free acids were established as 586 in agreement with EI studies of their methyl esters[176] thereby clarifying previous results of underivatized material.[177]

Methane CI has been shown to be an easier method for the analysis of petroporphyrin mixtures than low-eV EI.[178] Even more information in this area is obtainable by l.c.–m.s.[68] The mass spectra of *meso*-substituted porphyrins have been described.[179] FD mass spectrometry, when used in the laser supported mode, has been used to study vitamin B_{12} and various corrins.[180,181] The hydration behaviour of chlorophyll *a* has been studied by FD[182] and the technique has assisted in the characterization in bacteriochlorophylls *c* from *Chloropseudomonas ethylicum*.[183]

9 Carbohydrates

Two reviews of the use of mass spectral techniques in this area have appeared.[2e,184] Considerable interest continues in the mass spectral behaviour of derivatized carbohydrates. High-performance liquid chromatography is becoming increasingly important in studies of these compounds and the mass spectral behaviour of some u.v.-absorbing derivatives of mono- and di-saccharides[185] and fluorescent 2-aminopyridine derivatives of oligosaccharides[186] have been described. The advent of l.c.–m.s. makes the necessity to have u.v. or fluorescent derivatives less important and l.c.-m.s. of both underivatized[61] and permethylated saccharides[187] have been described. Studies[188] of the EI mass spectrometry behaviour of selectively deuteriated permethyl ethers of arabino-, gluco-, and galacto-pyranose have resulted in partial revision of proposed fragmentation schemes. The mass spectral behaviour of inosital cyclic boronic esters,[189] methyl-*o*-methylglucopyranosiduronamides,[190] partially methylated methylpentopyranosides,[191] and alditolacetates[192] have been described. A detailed study using high-resolution deuterium labelling and linked-scan (B/E)

---

[175] E. Fu, L. Friedman, and H. W. Siegelman, *Biochem. J.*, 1979, **179**, 1.
[176] D. J. Chapman, W. J. Cole, and H. W. Siegelman, *J. Am. Chem. Soc.*, 1967, **89**, 5976; W. J. Cole, D. J. Chapman, and H. W. Siegelman, *Biochemistry*, 1968, **7**, 2929.
[177] H. L. Crespi, L. J. Boucher, G. O. Norman, J. J. Katz, and R. C. Dougherty, *J. Am. Chem. Soc.*, 1967, **89**, 3642; B. L. Schram and H. H. Kroes, *Eur. J. Biochem.*, 1971, **19**, 581.
[178] G. J. Shaw, J. M. E. Quirke, and G. Eglington, *J. Chem. Soc., Perkin Trans. 1*, 1978, 1655.
[179] G. V. Ponomarev, V. P. Suboch, and A. N. Lyashko, *Khim. Geterotsikl. Soedin.*, 1978, 773.
[180] H.-R. Schulten and H. M. Schiebel, *Naturwissenschaften*, 1978, **65**, 223.
[181] H. M. Schiebel and H.-R. Schulten, *Tetrahedron*, 1979, **35**, 1191.
[182] R. C. Dougherty, P. A. Dreifuss, J. Sphon, and J. J. Katz, *J. Am. Chem. Soc.*, 1980, **102**, 416.
[183] K. M. Smith, M. J. Bushell, J. Rimmer, and J. F. Unsworth, *J. Am. Chem. Soc.*, 1980, **102**, 2437.
[184] O. S. Chizhov and A. Y. Ott, *Usp. Biol. Khim.*, 1978, **19**, 151.
[185] R. M. Thompson and D. A. Cory, *Biomed. Mass Spectrom.*, 1979, **6**, 117.
[186] S. Hase, T. Ikenaka, and Y. Matsushima, *Biochem. Biophys. Res. Commun.*, 1978, **85**, 257.
[187] C. N. Kenyon, M. McNeil, and A. Darvill, *28th Annual Conference on Mass Spectrometry and Allied Topics*, New York, 1980, paper RPMP11.
[188] A. Y. Ott, B. M. Zolotarev, and O. S. Chizhov, *Izv. Akad. Nauk SSR, Ser. Khim.*, 1978, 377; *ibid.*, 1979, 191.
[189] J. Wiecko and W. R. Sherman, *J. Am. Chem. Soc.*, 1979, **101**, 979.
[190] V. Mihalov, V. Kovacik, and P. Kovac, *Carbohydr. Res.*, 1979 **70**, 239.
[191] V. Mihalov, V. Kovacik, and P. Kovac, *Carbohydr. Res.*, 1979, **73**, 267.
[192] A. R. Rao and N. Roy, *Carbohydr. Res.*, 1979, **74**, 349.

techniques of the EI-induced fragmentations of methyl- and phenyl- 4,6-O-benzylidene-2,3-di-O-methyl-α-D-glucopyranoside has been reported.[193] Mass spectral analysis of alditol derivatives has proved useful in the elucidation of the structure of a new monosaccharide in the *Vibrio cholerae* O-antigen[194] and the identification of N-glycolylneuraminyl-(2 → 8)-N-glycolylneuraminyl groups in a trout egg glycoprotein.[195] Peracetates of aldononitriles were used for the mass spectral identification of methylated monosaccharides obtained from some unusual dextrans.[196] A comparative study of the anionic and cationic mass spectra of 2,3,4,6-tetra-O-acetyl-β-D-glucopyranosyl esters has been conducted;[197] the CI mass spectra of derivatized aldoses have proved useful in their characterization.[198]

A systematic study of mass spectral methods for the characterization of permethylated disaccharides has been undertaken. EI studies,[199] which extend an earlier report,[200] have resulted in the formulation of a scheme which enables discrimination between all the various types of glycosidic linkages. The structure of the m/z 101 ion in the spectra of these derivatives has been formulated as (10) on the basis of ^{13}C-labelling, metastable ion, and collisional activation

$$\text{MeO}-\underset{|}{\text{C}}=\underset{|}{\text{C}}-\underset{|}{\text{C}}=\overset{+}{\text{O}}\text{Me}$$
$$\text{H} \quad \text{H} \quad \text{H}$$
(10)

spectra.[201] CI spectra show more pronounced differences for isomeric species than EI and provide a more reliable differentiation between (1→ 2) and (1→ 4) linked disaccharides as well as better characterization of stereoisomers in the case of (1→ 4) and (1→ 6) linked disaccharides.[202] Of the reagent gases studied, ammonia was preferred for assignment of molecular weight. Isobutane, although preferable to methane and ammonia in differentiating glycosidic links, was inferior to methane in the differentiation of stereoisomers. CID spectra obtained from the CI-generated quasimolecular ions were found to be effective also for the differentiation of stereoisomers, and best results were obtained when ammonia or trimethylamine with a trace of water were used as reagent gases.[203] Studies of the CID spectra of the alkali attachment ions formed by the FD of monosaccharides indicate that the technique may be useful for differentiation of isomeric species of this type.[204]

---

[193] N. Kyriakidis, T. C. Smale, and E. S. Waight, *Org. Mass Spectrom.*, 1979, **14**, 532.
[194] L. Kenne, N. Linderberg, P. Unger, T. Holme, and J. Holmgren. *Carbohydr. Res.*, 1979, **68**, C14.
[195] S. Inoue and G. Matsumura, *Carbohydr. Res.*, 1979, **74**, 361.
[196] F. R. Seymour, E. C. M. Chen, and S. H. Bishop, *Carbohydr. Res.*, 1979, **68**, 113.
[197] D. Voigt and H. Lehmann, *J. Prakt. Chem.*, 1979, **321**, 243.
[198] F. R. Seymour, E. C. M. Chen, and S. H. Bishop, *Carbohydr. Res.*, 1979, **73**, 19.
[199] E. G. deJong, W. Heerma, J. Haverkamp, and J. P. Kamerling, *Biomed. Mass Spectrom.*, 1979, **6**, 72.
[200] J. Moor and E. S. Waight, *Biomed. Mass Spectrom.*, 1975, **2**, 36.
[201] E. G. deJong, W. Heerma, C. A. X. C. F. Sicherer, and G. Dijkstra, *Biomed. Mass Spectrom.*, 1980, **7**, 122.
[202] E. G. deJong, W. Heerma, and C. A. X. C. F. Sicherer, *Biomed. Mass Spectrom.*, 1979, **6**, 242.
[203] E. G. deJong, W. Heerma, and G. Dijkstra, *Biomed. Mass Spectrom.*, 1980, **3**, 127.
[204] F. W. Rollgen, U. Giessmann, F. Borchers, and K. Levsen, *Org. Mass Spectrom.*, 1978, **13**, 459.

Mass spectral study of permethylated derivatives of isomeric D-ribofuranosyl-ribitol disaccharides from the capsular polysaccharides of *Haemophilus influenzae* type b and *Escherichia coli* k enabled their characterization.[205] A more complex study[206] was necessary to sequence the oligosaccharide portion of the glycopeptide from human transferrin. Mass spectral examination of the *N*-acetylated glycopeptide, in methylated, methylated/reduced, and methylated/reduced/silylated forms, was necessary.

Examples of the use of FD in structural studies of complex carbohydrate components[207] and further details of the characterization of oligosaccharides and glycolipids by the cationization method[208] have appeared. The use of other ionization methods in this area has been discussed in the introduction.

10 Amino-acids and Peptides

Amino-acids.—The mass spectra of amino-acids have been reviewed.[2g] Studies of the EI mass spectral behaviour of trimethylsilyl amino-acids,[209] trimethylsilyl glycine conjugated,[210] and oxazolidinone derivatives[211] of amino-acids have been reported. EI and FI mass spectrometry have been compared as methods for the analysis of *N*-dithiocarbamic derivatives of amino-acids.[212] The proton affinities of amino-acids in CI have been studied[213] and the CI mass spectra of trimethylsilyl[214] and thiohydantoin[215] derivatives reported. FD mass spectrometry has been shown to be useful in the identification of fluorescamine derivatives of amino-acids[216] and a detailed study of the FD fragmentation of methionine has appeared.[217]

Peptides.—Peptide sequencing by mass spectral methods has been extensively covered in a number of review articles.[2h,218] EI studies of a variety of dipeptide decanoyl methyl derivatives,[219] Schiff base peptide esters,[220] and acetylated, permethylated, and reduced oligopeptides[221] are reported. Study of the EI spectra of peptides with trifluoroacetyl perdeuterioleucine as the *N*-terminal

---

[205] B. A. Fraser, F.-P. Tsui, and W. Egan, *Carbohydr. Res.*, 1979, **73**, 59.
[206] K. A. Karlsson, I. Pascher, B. E. Samuelsson, J. Finnie, T. Krusius, and H. Rauvala, *FEBS Lett.*, 1978, **94**, 413.
[207] V. H. Reinhold, in 'Glyco conjugate Research,' ed. J. D. Gregory and R. W. Jeanloz, Academic Press, New York, 1979, p. 265.
[208] J. C. Prome and G. Puzo, *Isr. J. Chem.*, 1978, **17**, 172.
[209] H. Iwase, Y. Takeuchi, and A. Murai, *Chem. Pharm. Bull.*, 1979, **27**, 1307.
[210] P. V. Fennessey and S. S. Tjoa, *Org. Mass Spectrom.*, 1980, **15**, 202.
[211] R. Liardon, U. Ott-Kuhn, and P. Husek, *Biomed. Mass Spectrom.*, 1979, **6**, 381.
[212] J. Szafranek, G. Blotny, and P. Vouros, *Tetrahedron*, 1978, **34**, 2763.
[213] M. Meot-Ner, E. P. Hunter, and F. H. Field, *J. Am. Chem. Soc.*, 1979, **101**, 686.
[214] H. Budzikiewicz and G. Meissner, *Org. Mass Spectrom.*, 1978, **13**, 608.
[215] K. Okada and A. Sakuno, *Org. Mass Spectrom.*, 1978, **13**, 535.
[216] K. E. Murray and D. L. Ingles, *Chem. Ind.*, 1979, 476.
[217] J. van der Greef, N. M. M. Nibbering, H.-R. Schulten, and W. D. Lehmann, *Z. Naturforsch.*, Teil B, 1978, **33**, 770.
[218] J. D. Priddle, *Int. Rev. Biochem.*, 1979, **24**, 1; R. E. Lovins, *Pract. Spectrosc.*, 1979, **3**, 19; K. Biemann, *Pure Appl. Chem.*, 1978, **50**, 149; D. H. Williams, *Pure Appl. Chem.*, 1978, **50**, 219; H. R. Morris, *Philos. Trans. R. Soc. London, Ser. A*, 1979, 293, 39.
[219] A. A. Akhren, V. D. Matveentsev, V. L. Chashchin, O. A. Strel'chenok, V. P. Suboch, and V. I. Vil'chinskaya, *Zh. Org. Khim.*, 1978, **14**, 1160, 1374, 1378.
[220] K. Jayasimhulu and R. A. Day, *Org. Mass Spectrom.*, 1978, **13**, 540.
[221] V. K. Mahajan and D. M. Desiderio, *Biochem. Biophys. Res. Commun.*, 1978, **82**, 1104.

blocking group enables distinction between leucine and isoleucine in the same position of the peptide, provided both samples are available.[222]

The isobutane CI spectra of trimethylsilyl derivatives of di-[214,223] and tripeptides[214] have been studied. CI has been shown to enable identification of molecular weight at lower levels than EI.[223] CI was also found to be preferable to EI in providing sequence information for depsipeptides and their permethyl derivatives.[224] FD studies of peptides continue to appear.[225–229] The technique is best used in combination with EI mass spectrometry if sequence information is required.[225,226]

Detailed studies of peptide sequencing using EI and MIKES of derivatized peptides are described[230] and the ratio of metastable to daughter ion abundance in the first field region has been shown to be influenced by the amino-acid sequence in derivatized peptides.[231] Negative CI, in combination with CID–MIKES, has been shown to provide useful sequence information from underivatized peptides and their mixtures.[232] Using suitably derivatized peptide mixtures, advantage has been taken of the enhanced sensitivity obtainable in negative CI and sequence data have been obtained from these mixtures using a triple quadrupole system.[43]

An ability to obtain sequence information from very small samples of peptide is important. A useful technique for assisting in this context has been reported involving the use of a carrier peptide of similar volatility to the unknown to enhance sensitivity.[233] Other reports on approaches to peptide sequencing include CI methods[234] and a combination of FD and Edman degradation.[235]

Mass spectral analysis of variant forms of a peptide obtained from bee venom showed that N-formyl lysine residues were present.[236] Application of mass spectral sequencing methods are illustrated in determination of the N-terminal sequences of blood coagulation factor X_1 and X_2 light chains,[237] the structure of the toxin cyclochlorotine isolated from *Penicillium islandicum*,[238] the partial

---

[222] R. Waern and H. Falter, *Biochem. Biophys. Res. Commun.*, 1978, **81**, 448.
[223] H. C. Krutzsch and T. J. Kindt, *Anal. Biochem.*, 1979, **92**, 525.
[224] B. C. Das, P. Varenne, and A. Taylor, *J. Antibiot.*, 1979, **32**, 569.
[225] I. Luederwald, M. Przybylski, H. Ringsdorf, and D. Silberhorn, *Z. Naturforsch., Teil B*, 1978, **33**, 805.
[226] I. Luederwald, M. Przybylski, H. Ringsdorf, D. Silberhorn, H. Halbacher, and W. Voelter, *Z. Naturforsch., Teil B*, 1978, **33**, 809.
[227] M. Przybylski, I. Luederwald, E. Krass, W. Voelter, and S. D. Nelson, *Z. Naturforsch., Teil B*, 1979, **34**, 736.
[228] W. Frick, E. Barofsky, C. D. Daves, jun., D. F. Barofsky, D. Chang, and K. Folkers, *J. Am. Chem. Soc.*, 1978, **100**, 6221.
[229] I. Katakuse, T. Matsuo, H. Wollnik, and H. Matsuda, *Org. Mass Spectrom.*, 1979, **14**, 457.
[230] U. P. Schlunegger and P. Hirter, *Isr. J. Chem.*, 1978, **17**, 168.
[231] Z. V. I. Zaretskii and P. Dan, *Biomed. Mass Spectrom.*, 1979, **6**, 45.
[232] C. V. Bradley, I. Howe, and J. H. Beynon, *Chem. Commun.*, 1980, **12**, 562.
[233] A. D. Auffret and D. H. Williams, *J. Chem. Soc., Chem. Commun.*, 1979, 692.
[234] Y. Yasuhiko, H. Nomura, M. Sano, Y. Kudo, and S. Sasaki, *Bull. Chem. Soc. Jpn.*, 1978, **51**, 1909; Y. Yotsui, H. Nomura, M. Sano, Y. Kudo, and S. Sasaki, *Bunseki Kagaku*, 1979, **28**, 721.
[235] Y. Shimonishi, Y.-M. Hong, T. Matsuo, I. Katakuse, and H. Matsuda, *Chem. Lett.*, 1979, 1369.
[236] S. Doonan, A. J. Garman, J. M. Hanson, A. G. Loudon, and C. A. Vernon, *J. Chem. Soc., Perkin Trans. 1*, 1978, 1157.
[237] H. C. Thogersen, T. E. Petersen, L. Sottrup-Jensen, S. Magnusson, and H. R. Morris, *Biochem. J.*, 1978, **175**, 613.
[238] R. J. Anderegg, K. Biemann, A. Manmade, and A. C. Ghosh, *Biomed. Mass Spectrom.*, 1979, **6**, 129.

primary structure of bacteriorhodopsin,[239] and the amino-acid sequences of the tryptic peptides of cowpea chlorotic mottle virus protein.[240] In the latter case sequences were determined by a combination of enzyme hydrolysis, mass spectrometry, and Edman degradation. In spite of the availability of automated Edman degradation, in two cases mass spectral methods succeeded where Edman degradation failed.

11 Fatty Acids and Lipids

Reviews have appeared on mass spectral methods for studies of fatty acids[2a] and the location of double bonds and cyclopropane rings in their structure.[241] General reviews of the mass spectrometry of lipids[2b,242] have appeared and the use of l.c.–m.s.[243] and soft ionization methods[244] in their analysis has been discussed. Structural studies of glycolipids have also been reviewed.[245,246]

The mass spectral fragmentation patterns of long-chain polyunsaturated fatty-acid methyl esters and their thiol esters have been discussed.[247] Fingerprinting of mixtures of fatty acids can be effected by negative EI mass spectrometry of their p-nitrobenzoates[248] since the spectra are dominated by carboxylate anions. FD m.s. has also been examined in this context.[249,250] Interpretation of the spectra of free acids is complicated by the presence at high mass of 2 molar ions and by anhydride and polyester (from hydroxy acids) formation.[249] Studies of sodium carboxylates showed the approach to be even less useful for mixture analysis because of the complexity of the spectra produced.[250]

Location of the site of unsaturation in fatty acids has received further considerable attention. Oxidation with osmium tetroxide, followed by trimethylsilylation and CI studies of the products,[251] and ozonolysis followed by esterification and g.c.–m.s. of the products[252] have both been used. Pyrrolidine derivatives have been used for the location of double bonds in C_{20} fatty acids from bovine lens.[253] The limitations of this approach in studies of oxygen-containing fatty acids have been delineated.[254] The applicability of pyrrolidine derivatives has been extended to the location of triple bonds in isomeric octadecynoic acids.[255]

---

[239] G. E. Gerber, R. L. Anderegg, W. C. Herlihy, C. P. Gray, K. Biemann, and H. G. Khorana, *Natl. Acad. Sci. USA*, 1979, **76**, 227.
[240] M. W. Rees, R. Casey, M. N. Short, J. J. Sexton, J. F. March, J. Eagles, K. R. Parsley, and R. Self, *Biomed. Mass Spectrom.*, 1980, **7**, 132.
[241] D. E. Minnikin, *Chem. Phys. Lipids*, 1978, **21**, 313.
[242] R. A. Klein, *Chem. Phys. Lipids*, 1978, **21**, 291.
[243] O. S. Privett and W. L. Erdahl, *Chem. Phys. Lipids*, 1978, **21**, 361.
[244] D. E. Games, *Chem. Phys. Lipids*, 1978, **21**, 389.
[245] M. Breimer, G. C. Hansson, A. L. Karlsson, G. Larson, H. Leffler, I. Pascher, W. Pimlott, and B. E. Samuelsson, *Dev. Plant Biol.*, 3 (*Adv. Biochem. Physiol. Plant Lipids*), 1979, 281; K. A. Karlsson, *Prog. Chem. Fats Other Lipids*, 1978, **16**, 207.
[246] H. Egge, *Chem. Phys. Lipids*, 1978, **21**, 349.
[247] M. G. Hussain and F. D. Gunstone, *Bangladesh J. Sci. Ind. Res.*, 1979, **14**, 105.
[248] Y. Hirata, T. Takeuchi, and K. Matsumoto, *Anal. Chem.*, 1978, **50**, 1943.
[249] G. W. Wood, E. J. Oldenberg, P.-Y. Lau, and D. L. Wade, *Can. J. Chem.*, 1978, **56**, 1372.
[250] G. W. Wood, E. J. Oldenberg, and P.-Y. Lau, *Can. J. Chem.*, 1978, **56**, 2750.
[251] T. Ariga, E. Araki, and T. Murata, *Igakuto Seibutsugaku*, 1977, **91**, 69.
[252] T. Takayama, N. Qureshi, and H. K. Schnoes, *Lipids*, 1978, **13**, 575.
[253] A. J. Vallicenti, C. J. Chapman, and R. T. Holman, *Lipids*, 1978, **13**, 190.
[254] J. Eagles, G. R. Fenwick, and R. Self, *Biomed. Mass Spectrom.*, 1979, **6**, 462.
[255] A. J. Vallicenti, W. H. Heimermann, and R. T. Holman, *J. Org. Chem.*, 1979, **44**, 1068.

Mass spectral methods have been used in the identification of mycolates from mycobacteria[256] and soil.[257] Combined l.c.–m.s. should prove extremely useful in this area and the effectiveness of this approach is exemplified in studies of meibomian gland waxes.[68]

The mass spectral fragmentation behaviour of deuteriated analogues of prostaglandin $F_2\alpha$ has been studied[258] and ammonia has been advocated as CI reagent gas for the identification of trimethylsilylated prostaglandins of the A, B, and E series.[259] The EI mass spectral behaviour of derivatized prostaglandins of the A, B, E, and F series has been reviewed.[260] The mass spectral behaviour of new trialkylsilyl derivatives of prostaglandins has been described.[261]

Considerable effort has been expended over many years in elucidating the structure of Slow-Reacting Substances (SRS). A mass spectrum of the trimethylsilyl ether of the N-acetyl methyl ester of SRS-A isolated from guinea-pig lung[262] and from rat basophil leukaemia cells[263] together with analytical protein chemical techniques resulted in its structural assignment (11). Another substance

(11)

isolated from murine mastocytoma cells (leukotriene C) differs from SRS-A in the presence of a glutamic acid residue.[264] The structure determination of this latter compound was assisted by microscale chemical degradation, and g.c.–m.s. Stereospecific synthesis leucotriene C has established the configuration of the compound and confirmed the structural assignments.[265]

Metastable defocusing and MIKES have been used to characterize the molecular species in mixtures of triglycerides, glycerophospholipids, and ornitholipids.[266] In contrast, the structure and molecular species of cerebrosides

---

[256] T. Kusaka and N. Akimori, *Koenshu-Iyo Masu Kenkyukai*, 1979, **4**, 339.
[257] I. M. Lukashenko, G. G. Gutiev, E. S. Brodskii, and R. A. Khmel'nitskii, *Izv. Timiryazevsk. Skh. Akad.*, 1979, 191.
[258] G. Horvath and G. Ambrus, *Biomed. Mass Spectrom.*, 1978, **5**, 544.
[259] T. Ariga, M. Suzuki, I. Morita, S. I. Murota, and T. Miyakake, *Anal. Biochem.* 1978, **90**, 174.
[260] D. M. Desiderio, *GC–MS News*, 1979, **7**, 32.
[261] H. Miyazaki, M. Ishibashi, K. Yamashita, and M. Katori, *J. Chromatogr.*, 1978, **153**, 83; A. R. Brash and T. A. Baillie, *Biomed. Mass Spectrom.*, 1978, **5**, 346.
[262] H. R. Morris, G. W. Taylor, P. J. Piper, and J. R. Tippins, *Nature*, 1980, **285**, 104.
[263] H. R. Morris, G. W. Taylor, P. J. Piper, M. N. Samhoun, and J. R. Tippins, *Prostaglandins*, 1980, **19**, 185.
[264] R. C. Murphy, S. Hammarstrom, and B. Samuelsson, *Proc. Natl. Acad. Sci. USA*, 1979, **76**, 4275.
[265] E. J. Corey, D. A. Clark, G. Goto, A. Marfat, C. Mioskowski, B. Samuelsson, and S. Hammarstrom, *J. Am. Chem. Soc.*, 1980, **102**, 1436.
[266] S. G. Batrokov, V. L. Sadovskaya, V. N. Galyashin, B. V. Rosynov, and L. D. Bergel'son, *Bioorg. Khim.*, 1978, **4**, 1220, 1398; S. G. Bartrakov, V. L. Sadovskaya, B. V. Rozynov, and L. D. Bergel'son, *Bioorg. Khim.*, 1978, **4**, 1398.

have been assigned on the basis of isobutane CI of their trimethylsilyl derivatives.[267] CI has also been used to study the thermolysis of an underivatized ornithine-containing lipid from *Thiobacillus thiooxidans*.[268] FD mass spectrometry of phosphatidylsulfacholines has been reported and the technique is useful in assignment of molecular weight and assisting in structural assignments.[269]

Several blood group type glycolipids from the small intestine of rabbit have been identified by selected ion monitoring of key fragment ions obtained from slow evaporation of permethylated and reduced derivatives into the mass spectrometer ion source.[270] A combination of mass spectral and n.m.r. methods was used to assign structures to two gangliosides from minipig posterior root ganglion.[271] The structure of the 'cord factors' of *Corynebacterium diphtheriae*[272] and *Mycobacterium paraffinicum*[273] have been determined. In the former case, a combination of FD of underivatized and EI mass spectrometry of trimethylsilylated materials was used whilst the latter study used EI and FI.

---

[267] T. Murata, T. Ariga, M. Oshima, and T. Miyatake, *J. Lipid Res.*, 1978, **19**, 370.
[268] D. R. Hilker, H. W. Knoche, and M. L. Gross, *Biomed. Mass Spectrom.*, 1979, **6**, 356.
[269] G. W. Wood, P.-A. Tremblay, and M. Kates, *Biomed. Mass Spectrom.*, 1980, **7**, 11.
[270] M. E. Breimer, G. C. Hansson, K.-A. Karlsson, H. Leffler, W. Pimlott, and B. E. Samuelsson, *Biomed. Mass Spectrom.*, 1979, **6**, 231.
[271] B. E. Samuelsson, *Adv. Exp. Med. Biol.*, 1980, **125**, 63.
[272] G. Puzo, G. Tissie, C. Lacave, H. Aurelle, and J. C. Prome, *Biomed. Mass Spectrom.*, 1978, **5**, 699.
[273] S. G. Batrakov, V. L. Sadovskaya, B. V. Bozynov, T. V. Koronelli, and L. D. Bergelison, *Bioorg. Khim.*, 1978, **4**, 667.

11
The Use of Mass Spectrometry in Pharmacokinetic and Drug Metabolism Studies

BY L. E. MARTIN

1 Introduction

The generalized literature on g.c.–m.s. and h.p.l.c.–m.s. is reviewed in Chapter 8 and this chapter only deals with the use of mass spectrometry in relation to the quantitative determination and qualitative identification of drugs and metabolites.

References in the literature relevant to the use of mass spectrometry in pharmacokinetic and metabolic studies on drugs over the period from May 1978 to April 1980 have been included. In the case of papers published in journals not directly read by the Reporter the sources for references have been the Mass Spectrometry Bulletin, U.K.C.I.S., and CA Selects, Mass Spectrometry.

Recent books of interest to workers in the biomedical field include a further three volumes in the series on the analysis of drugs by g.c.–m.s.[1-3] Symposia published during the period under review are those from the Second International Symposium on Quantitative Mass Spectrometry held in Ghent in 1978,[4] the Fifth International Symposium on Biochemistry and Medicine held in Rimini in 1977,[5] and the Symposium on the Stable Isotopes held in London in 1977.[6]

Relevant reviews include those on the application of quantitative mass spectrometry to pharmacokinetic studies on drugs and metabolites,[7-9] the use of g.c. pyrolysis mass spectrometry in biomedical studies,[10] negative ion mass spectrometry,[11] and h.p.l.c.–m.s.[12]

---

[1] B. J. Gudzinowicz, and M. J. Gudzinowicz, 'Central Nervous Stimulants' (Analysis of Drugs and Metabolites by G.C.–M.S.), Marcel Dekker, New York, 1978, Vol. 4.
[2] 'Analgesics, Local Anaesthetics, and Antibiotics', ref. 1, Vol. 5.
[3] 'Cardiovascular, Antihypersensitive, Hypoglycemic and Thyroid-Related Agents', ref. 1, Vol. 6.
[4] 'Quantitative Mass Spectrometry in Life Science II', ed. A. P. de Leenheer, R. Ronucci, and C. van Petegrem, Elsevier, Amsterdam, 1978.
[5] 'Recent Developments in Mass Spectrometry in Biochemistry and Medicine', ed. A. Frigerio, Plenum Press, New York and London, 1978, Vol. 2.
[6] 'Stable Isotopes', ed. T. A. Baillie, MacMillan, London, 1978.
[7] W. D. Lehmann and H.-R. Schulten, *Angew. Chem., Int. Ed. Engl.*, 1978, **17**, 221.
[8] G. H. Draffan, J. D. Gilbert, and H. T. Gilbert, in 'Drug Metabolism in Man', ed. J. W. Gorrod and A. H. Beckett, Taylor and Francis, London, 1978, 193.
[9] E. L. Ghisalberti, 'Drug Fate and Metabolism', ed. E. R. Garrett and J. L. Hirtz, Marcel Dekker, New York, 1979, Vol. 3, p. 1.
[10] W. J. Irwin and J. A. Slack, *Analyst (London)*, 1978, **103**, 673.
[11] H. Brandenberger and R. Rhyage, in 'Blood Drugs and Other Analytical Challenges', ed. E. Reid, Ellis Horwood, Chichester, 1978, Vol. 7, p. 173.
[12] G. Guiochen and P. J. Arpino, *Anal. Chem.* 1979, **51**, 683.

2 Quantitative Mass Spectrometry

The increase in potency of many drugs has made increasing demands for the bioanalyst to develop highly sensitive methods for determining drugs and metabolites in biological fluids. Radioimmunoassay, fluorimetric, and g.c. electron capture methods for determining drugs can offer the same sensitivity as mass spectrometry, but none of these methods offers the combination of sensitivity and specificity that is unique to mass spectrometry.

During the past two years the number of papers has doubled in which mass spectrometry has been used for the quantitative determination of drugs and their metabolites in biological samples. Every step in a drug metabolism study should be studied in detail to avoid artifacts which can be the result of being unaware of the pitfalls which can occur during the collection, preparation, and analysis of biological samples.[13,14]

Internal Standards Used for Analysis by Mass Spectrometry.—It is pleasing to record that most investigators are now adding internal standards to the sample prior to analysis. There are still some instances where the internal standard is added to a sample obtained after prior extraction or chromatographic separation of the drug from plasma or urine. In these instances no correction is made for loss during the initial analytical procedures.

The sources of error and criteria for selection of internal standards for quantitative mass spectrometry have been reviewed in two papers.[15,16] Also in these reviews the errors due to sampling, solvent extraction, derivatization, adsorption of compounds from dilute solution onto glassware, and decomposition in the g.c. are discussed. It was concluded that the use of a homologue internal standard was more than adequate and it offered the advantage of being more readily available and cheaper than a stable isotope analogue. Although this conclusion may be true in many cases, in the Reporter's laboratory it has been necessary for some analyses to use a stable isotope analogue as a carrier to reduce losses during analysis.

Extra sensitivity is achieved by using both a stable isotope-labelled analogue of the drug as a 'carrier' for the drug plus an internal standard with a different retention time from the drug but which yields an ion of the same m/z value as the drug. The mass spectrometer can then focus on the one ion and any instability due to voltage switching is eliminated and the signal from the electron multiplier is integrated for a longer period.

Direct Probe Methods.—The direct probe integrated ion current method for quantification of drugs and metabolites in solvent extracts of biological samples offers a simple and rapid means of analysing large numbers of samples. Interference from endogenous materials, especially cholesterol and lipids, in the sample extract can cause difficulties at low resolution so that, in much of the earlier work, high-resolution mass spectrometry was used.

---

[13] R. C. Veith, V. A. Raisys, and C. Perera, *Comm. Psychopharmacol.* 1978, **2**, 491.
[14] A. H. Beckett and D. A. Cowan, in ref. 8, p. 257.
[15] B. J. Millard, *GC-MS News*, 1978, **6**, 64.
[16] B. J. Millard, *GC-MS News*, 1978, **6**, 80.

Stable isotope-labelled analogues added to the sample prior to solvent extraction are most commonly used as internal standards and this procedure serves to compensate for variation in efficiency of extraction and sample loss during transfer to the probe and into the source of the mass spectrometer. Almost any compound which does not give an ion of m/z value characteristic of the compound being determined can be used for measuring losses occurring during transfer to the probe and introduction into the source. Methyl stearate is an easily available independent internal standard which has been used for this purpose.[17]

Low-resolution chemical ionization mass spectrometry using a deuterium-labelled analogue has been used for the determination of 1,3-bis(2-chloroethyl)-1-nitrosourea (BCNU), an anti-tumour drug, in biological fluids. Solvent extracts of the sample were evaporated directly onto the probe and protonated molecular ion intensity ratios of the ions m/z 214 and 216 from the drug and m/z 222 and 224 from the deuteriated standard were measured.[18]

Diphenylhydantoin (DPH), phenobarbital (PB), and pyrimidone (P) have been simultaneously determined in serum using methane CI mass spectrometry; 5-(-p-methylphenyl)-5-phenylhydantoin and 5-ethyl-5-p-tolylbarbituric acid were used as internal standards.[19] The probe was heated and the compounds selectively volatized into the source. The abundances of the protonated molecular ions of drugs and standards were measured using a multichannel analyser. At a signal-to-noise ratio of three, the sensitivity was 1×10^{-10} g for P and 1×10^{-9} g for DPH and PB.

Repetitive scanning, under low resolution CI conditions using isobutane gas, was used for the determination of DPH, mephobarbital, carbamazepine, PB, and P in serum.[20] Palmitic acid present in the extract and deuteriated DPH both gave ions m/z 256 but, as palmitic acid volatilized from the probe before the deuteriated DPH, it did not interfere in the assay. The coefficient of variation was 2—3% for drugs in the therapeutic concentration range. Low-resolution repetitive CI scanning has also been used to monitor solvent extracts of urine containing narcotics, tricyclic antidepressants, amphetamines, and pentazocine. The detection limits were in the sub-microgram per cm^3 range.[21]

Repetitive scanning, low-resolution mass spectrometry using deuteriated analogue standards has been used to determine the anti-inflammatory drug Sulindac (1) and two of its metabolites (2) and (3) in benzene extracts from serum.[22] The [MH]$^+$ ions were used to measure the drug and metabolites in plasma. The M-16 ion from the drug (1) yields an ion of the same mass as the molecular ion of metabolite (2) but the vapour pressure of the latter was greater than that of (1) or metabolite (3) and by careful heating of the direct probe it was possible to separate compounds (1) and (2).

---

[17] F. Heresch, E. R. Schmid, and A. Weiszbart, *Biomed. Mass Spectrom.*, 1979, **6**, 566.
[18] R. J. Weinkam, J. H. Wen, D. E. Furst, and V. A. Levin, *Clin. Chem.* 1978, **24**, 45.
[19] M. Lehrer and A. Karmen, *Isr. J. Chem.*, 1978, **17**, 206.
[20] R. J. W. Truscott, D. G. Burke, J. Korth, B. Halpern, and R. Summons, *Biomed. Mass Spectrom.*, 1978, **5**, 477.
[21] R. Saferstein, J. J. Manura, T. A. Brettell, and K. De Paritosh, *J. Anal. Toxicol.*, 1978, **2**, 245.
[22] R. W. Walker, V. F. Gruber, A. Rosenberg, F. J. Wolf, and W. J. A. Vanden-Heuvel, *Anal. Biochem.*, 1979, **95**, 579.

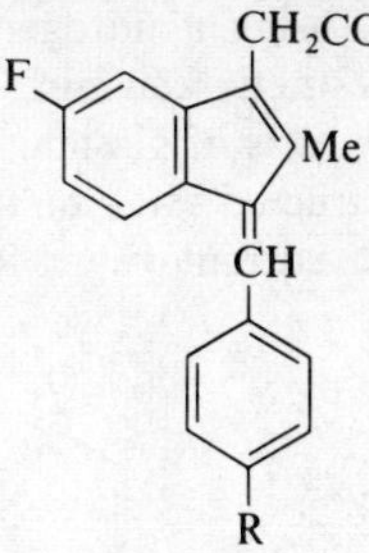

(1) R = SOMe
(2) R = SMe
(3) R = SO₂Me

Selected Ion Recording *via* **G.C.–M.S.**—*A Comparison Between G.C.–M.S. and Other Analytical Techniques.* The use of g.c.–m.s. as an analytical procedure for the monitoring of drugs continues to increase. The technique is expensive and in some cases less expensive techniques, *e.g.* fluorimetry, radioimmunoassay, radio-derivative analysis and g.c. analysis using electron capture or nitrogen specific detectors, can be used instead of g.c.–m.s.

The determination of nadolol (4), a β-adrenergic receptor blocking agent, has been compared using spectrophotofluorimetry and g.c.–m.s. *N*-Methyl-nadolol (5) was used as the internal standard.[23] When the data from two assays were correlated by linear regression analysis the correlation coefficient was 0.9. It was concluded that for processing large numbers of samples the fluorimetric method was the more rapid but that the g.c.–m.s. method was more specific at low serum nadolol concentrations. The detection limit was about 7.0 ng ml⁻¹ serum.

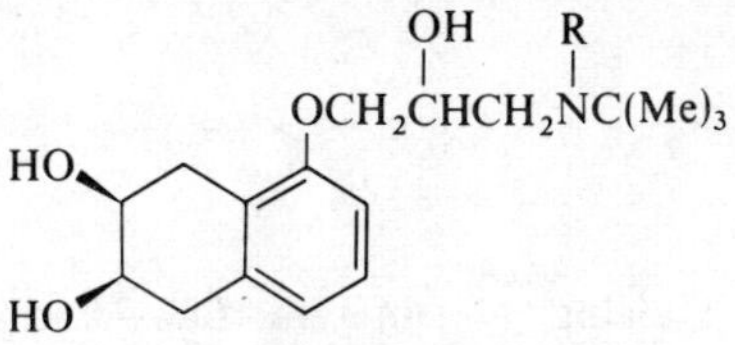

(4) R = H
(5) R = Me

A radioenzymic assay using chloramphenicol acetyl transferase, g.c. with electron capture detection, and a g.c.–m.s. method have been compared for determining chloramphenicol in serum or urine.[24] The internal standard used for the latter two assays was D(−)-*threo*-*N*-[β-hydroxy-α-(hydroxymethyl)-*p*-nitrophenethyl] acetamide. All three assays were found to correlate well and the g.c. electron capture method was recommended for routine use.

[23] P. T. Funke, M. F. Malley, E. Ivashkiv, and A. I. Cohen, *J. Pharm. Sci.* 1978, **67**, 653.
[24] L. K. Pickering, J. L. Hoecker, W. G. Kramer, J. G. Liehr, and R. M. Caprioli, *Clin. Chem.* 1979, **25**, 300.

The use of g.c.-m.s. and g.c. using a nitrogen-specific detector has been compared to determine the anaesthetic ketamine (6) and two metabolites (7) and (8) in methylene chloride extracts of plasma.[25] Diphenydramine was used as the internal standard. Reconstructed total ion current chromatograms were plotted and integrated total ion current values corresponding to the peaks

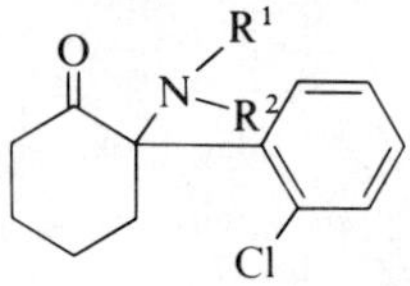

(6) $R^1 = Me, R^2 = H$
(7) $R^1 = R^2 = H$

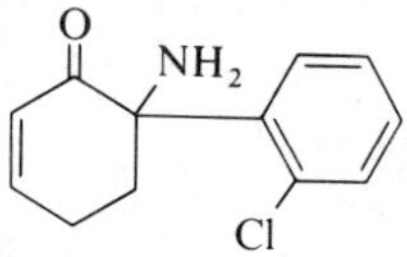

(8)

representing the drugs, metabolites, and internal standard used for quantification. Recovery was almost quantitative and the response for the drug and two metabolites was linear over the range 0.8—40 μg ml^{-1}. The results obtained using the nitrogen-specific detector were in good agreement with those using g.c.–m.s.

Disopyramide (9), an antiarrythmic drug, has been extracted from sera and determined either by g.c. using a nitrogen detector, liquid chromatography, or g.c.–m.s.[26] When an analogue standard was used all three techniques gave comparable results for the analysis of a series of samples; the coefficient of variation was 5—10% and the detection limit 0.25 mg l^{-1}. Only where interference from other drugs occurred was g.c.–m.s. recommended.

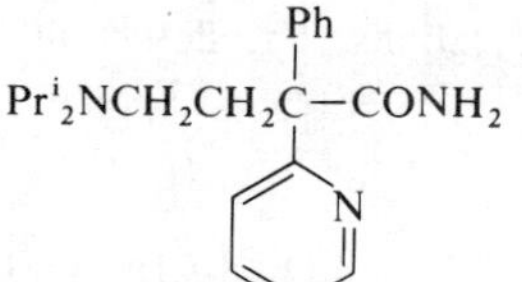

$Pr^i_2NCH_2CH_2\overset{Ph}{\underset{|}{C}}-CONH_2$

(9)

Internal Standards Yielding a Different Ion.—Carbamazepine and carbamazepine-10,11-epoxide have been simultaneously determined by g.c.–m.s. in rat plasma and brain.[27] Cytenamide, 5H-dibenzo[a,d]cycloheptene-5-carboxamide, was used as the internal standard. The linearity of the method ranged from 100 ng to 50 μg cm^{-3} or g^{-1} wet weight of tissue.

A new g.c.-m.s. chemical ionization method for the determination of the neuroleptic haloperidol in human plasma, using methane as the reactant gas, is reported.[28] The internal standard was triflupiridol and the [MH]$^+$ ions of the

---

[25] J. N. Davisson, *J. Chromatogr.*, 1978, **146**, 344.
[26] J. Vasiliades, C. Owens, and F. Ragusa, *Clin. Chem.*, 1979, **25**, 1900.
[27] C. Pantarotto, V. Crunelli, J. Lanzoni, A. Frigerio, and A. Quattrone, *Anal. Biochem.*, 1979, **93**, 115.
[28] C. L. Hornbeck, J. C. Griffiths, R. J. Neborsky, and M. A. Faulkner, *Biomed. Mass Spectrom.*, 1979, **6**, 427.

drug and standard were recorded. The response was linear over the range 2.5—15 ng haloperidol per cm^{-3} of plasma. The mean ± standard deviation for analyses of 10 ng ml^{-1} in plasma was 9.9 ± 0.75. A three-fold increase in sensitivity was achieved when a methane–ammonia reagent gas mixture was used, and this was recommended for the routine analysis of haloperidol.

A CI g.c.-m.s. method for the simultaneous determination of the analgesics hydromorphone (10), hydrocodone (11), and their 6α- (12a), 6α- (13a), 6β- (12b), and 6β-hydroxy (13b) metabolites in urine of subjects given an oral dose of 15 mg hydromorphone is described.[29] An analogue standard, cyclazocine, was added to the urine prior to acid hydrolysis, extraction, and silylation. The $[M + H]^+$ ion of the silylated drug and metabolites and the $[M + 29]^+$ of the internal standard were recorded. The assay was linear over the range 0.8—8 μg cm^{-3} for each compound and was sensitive down to 0.01 μg cm^{-3} of the compound. The 6β metabolites were distinguished from their 6α epimers by the more abundant $[M - 98]^+$ ion.

The anti-inflammatory drug ketoprofen (14) has been determined, in extracts of serum and human milk as its methyl ester, by two methods using (benzoyl-4-phenyl)-2-butyric acid as the internal standard.[30,31] The sensitivity varied from 215 μg to 100 ng cm^{-3}.

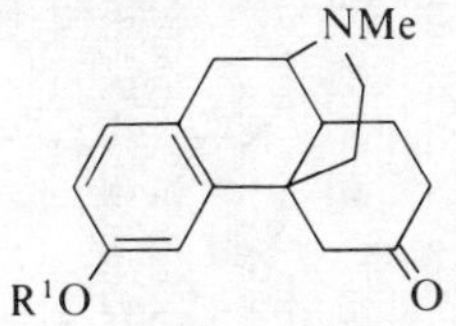

(10) R^1 = H
(11) R^1 = Me

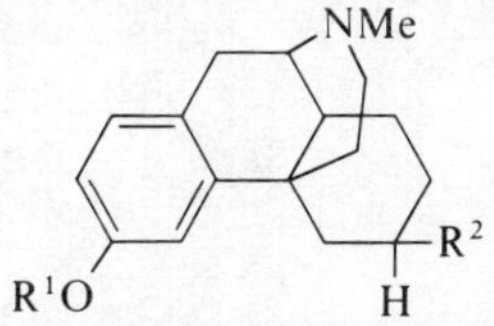

(12) a; R^1 = H, R^2 = α-OH
     b; R^1 = H, R^2 = β-OH
(13) a; R^1 = Me, R^2 = α-OH
     b; R^1 = Me, R^2 = β-OH

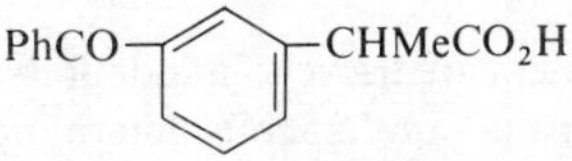

PhCO—⟨⟩—CHMeCO$_2$H

(14)

A quantitative assay for urinary propranolol and six of its thirteen known unconjugated metabolites has been developed.[32] The method is also applicable to the determination of hydrolysis products of the conjugates. The metabolites were separated from urine by XAD-2 resin column chromatography and converted into their methyl esters and trifluoroacetyl derivatives. The metatolyloxy analogue of propranolol and phenoxyacetic acid were used as internal standards for non-acidic and acidic metabolites. The metabolites measured in the urine of a dog given a 4 mg oral dose of propranolol were 4-hydroxypropranolol, α-naphthoxylactic acid, α-naphthol, and propranolol glycol.

[29] E. J. Cone and W. D. Darwin, *Biomed. Mass Spectrom.*, 1978, **5**, 291.
[30] J. De Graeve, C. Frankinet, and J. E. Gielen, *Biomed. Mass Spectrom.*, 1979, **6**, 249.
[31] D. Heusse and L. Raynaud, *Ann. Pharm. Fr.*, 1978, **36**, 631.
[32] V. T. Vu and F. P. Abramson, *Biomed. Mass Spectrom.*, 1978, **5**, 686.

Prednisone and prednisolone have been determined in serum of humans given an oral dose of 50 mg prednisone.[33] Cloprednol, the internal standard, was added to the serum and the steroids were purified by extraction into ether and either solvent partition or chromatography on magnesium silicate. Hydrocortisone differs by 2 mass units from prednisolone and the molecular ion clusters of the methoxyiminotrimethylsilyl derivatives of the two compounds overlapped each other. Therefore, the endogenous hydrocortisone present in the purified extract was removed by selective derivatization with Girard Reagent T and solvent partitioning of the prednisone and prednisolone. The two steroids were each converted into methyloximinotrimethylsilyl derivatives and quantified as their [MH]$^+$ ions by chemical ionization g.c.–m.s. To increase the sensitivity of the assay each compound was separately determined by single ion monitoring. The method could measure the steroids in the ng cm^{-3} range and offered the advantage over radioimmunoassay that there was no interference from 20-hydroxy steroids and the glucuronide conjugates of the two compounds.

An assay for the quantitative determination of the antidiabetic agent moroxydin hydrochloride (15) in human plasma and urine which used l-n-butylbiguanide (16) as the internal standard is reported.[34] The compounds were extracted from biological fluids and treated with trifluoroacetic anhydride to form the triazines (17a) and (17b), which were determined by g.c.–m.s.

$$R-C\begin{matrix}\nearrow NH \\ \searrow NHC\end{matrix}\begin{matrix}\nearrow NH \\ \searrow NH_2\end{matrix}$$

(15) R = N⌒O

(16) R = Bu^n

(17) a; R = N⌒O

b; R = Bu^n

Work on the measurement of tricyclic antidepressants in plasma continues and a g.c.–m.s. method for the simultaneous determination, at the ng cm^{-3} level in biological fluids, of imipramine and desipramine or clomipramine and N-desmethylclomipramine as the heptafluorobutyryl derivatives is described.[35] The chloro compounds are used as internal standards for the assay of imipramine and desipramine and *vice versa*.

Information on the pharmacokinetics of anti-cancer drugs is of value in designing dose schedules which can give the maximum therapeutic effect with minimum toxicity. Two papers described the measurement of 5-fluorouracil (18) in the plasma or serum of cancer patients being treated with the drug.[36,37] This

---

[33] S. B. Matin and B. Amos, *J. Pharm. Sci.*, 1978, **67**, 923.
[34] T. Doi, A. Okajima, Y. Hara, A. Murano, and A. Kato, *Yakugaku Zasshi*, 1979, **99**, 387.
[35] D. Alkalay, J. Volk, and S. Carlsen, *Biomed. Mass Spectrum.*, 1979, **6**, 200.
[36] D. B. Lakings and R. H. Adamson, *J. Chromatogr.*, 1978, **146**, 512.
[37] J. P. Cano, J. P. Rigault, C. Aubert, Y. Carcassonne, and J. F. Seitz, *Bull. Cancer (Paris)*, 1979, **6**, 67.

compound is a biologically active metabolite of the anti-tumour agent 1,3-bis(tetrahydro-2-furanyl)-5-fluoro-2,4-pyrimidinedione (19; FDI). A method for the determination of 5-fluorouracil (18) in the plasma of rats and patients given oral doses of FDI has been developed.[38] Thymine was added as an internal standard prior to derivatization. Recovery of 5-fluorouracil from spiked plasma over the range 5—1000 ng cm^{-3} was about 86%. The sensitivity of the assay was 1 ng cm^{-3} compared with 20 ng cm^{-3} for the bioassay.

(18)   (19)

The anti-cancer drug indicine-N-oxide (20; INO) was isolated by cation exchange chromatography from the urine of patients. Heliotrine-N-oxide was used as the internal standard. It was not possible to determine the INO directly by g.c. because of its decomposition on the column into indicine (21), which might be a possible metabolite of INO.[39] It was shown by g.c.–m.s. that INO and the internal standard reacted with NO-bis(trimethylsilyl)trifluoroacetamide (BSTFA) to give TMS-pyrroles (22), whereas indicine reacted with the formation of the TMS-pyrrolizidine (23). Therefore, the extracts from the cation exchange chromatography of urine were treated with BSTFA so that INO and the internal standard could be determined as the TMS-pyrroles.

(20)   (21)   (22)   (23)

R^1 = —O—C(=O)—C(OH)(CHMe)—CHMe with OH

R^2 = —O—C(=O)—C(OTMS)—CHMe with OTMS

Nicotine was measured by capillary g.c.–m.s. in the ether extracts of the plasma of dogs given an intravenous dose of 10 μg nicotine kg^{-1}. The internal standard was quinoline. The ion, m/z 84, from nicotine and the molecular ion, m/z 129,

[38] T. Marunaka and Y. Umeno, *J. Chromatogr.*, 1978, **157**, 321.
[39] J. V. Evans, A. Peng, and C. J. Nielsen, *Biomed. Mass Spectrom.*, 1979, **6**, 38.

from quinoline were recorded.[40] The sensitivity was better than 5 ng nicotine per cm^{-3}.

An assay for lidocaine (24) and its two major metabolites (25, 26), in which carbocaine is the internal standard, is reported.[41] The plasma or urine samples were treated with propionaldehyde and sodium cyanoborohydride to alkylate reductively the metabolites. The drug and propyl derivatives of the metabolites were extracted with ether and determined by recording the m/z 86, 100, 114, and 98 ions. The response was linear over the range 0.05—50 μg ml. The coefficient of variation was 10% for lidocaine, 5.5% for monoethylglycinexylidide (25), and 3.5% for glycinexylidide (26) at concentrations of 0.2 μg cm^{-3} of each compound in plasma.

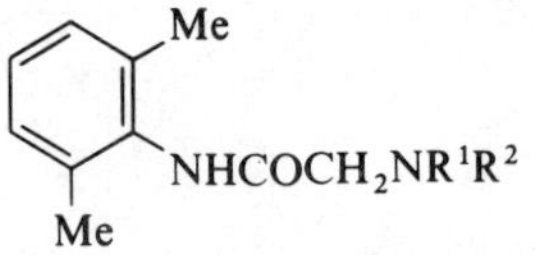

Me
⌬—NHCOCH$_2$NR^{1}R^2
Me

(24) $R^1 = R^2 = Et$
(25) $R^1 = H, R^2 = Et$
(26) $R^1 = R^2 = H$

An assay is described in which salbutamol, α'-{[(1,1-dimethylethyl)amino]methyl}-4-hydroxy-1,3-benzenedimethanol, was used as the internal standard for the measurement of terbutaline, 1-(3,5-dihydroxyphenyl)2-(t-butylamino)ethanol.[42] The drug and standard were extracted from plasma as the tetraphenylboron ion pair. The t-butyldimethylsilyl (TBDMS) ethers were formed and quantification achieved by recording the m/z 482 ion from TBDMS-terbutaline and the m/z 495 ion from TBDMS-salbutamol. The detection limit was 250 pg cm^{-3} and at 1 ng cm^{-3} the coefficient of variation was 4%.

In the measurement of benzoylecgonine, a metabolite of cocaine in urine, butylbenzoylecgonine was used as the internal standard.[43] The metabolite was extracted from urine into iodoethane–methylene dichloride as the tetrahexylammonium ion pair. Quantification was accomplished by monitoring multiple mass fragments. The coefficient of variation was 6% at 1 μg cm^{-3} concentration.

High-resolution G.C.–M.S.—When g.c.–m.s. analyses of biological fluids are carried out at resolutions of 1000 and below the sensitivity of the assay can be reduced because of interference by isobaric ions from co-eluting compounds. This interference can be overcome by using high-resolution mass spectrometry to distinguish between ions of the same nominal mass but different elemental composition.

A method using *trans*-1-(*p*-β-dimethylaminoethoxyphenyl)-1,2-diphenylprop-1-ene as an internal standard has been described for the determination

---

[40] J. Dow and K. Hall, *J. Chromatogr.*, 1978, **153**, 521.
[41] C. Hignite, C. Tschanz, J. Steiner, D. H. Huffman, and D. L. Azarnoff, *J. Chromatogr.*, 1978, **161**, 243.
[42] L. E. Martin, J. Oxford, R. J. N. Tanner, and M. J. Hetheridge, *Biomed. Mass. Spectrom.*, 1979, **6**, 460.
[43] J. E. Graas and E. Watson, *J. Anal. Toxicol.*, 1978, **2**, 80.

of the anti-oestrogen drug tamoxifen, *trans*-1-(β-dimethylaminoethoxyphenyl)-1,2-diphenylbut-1-ene,[44] which is used in the treatment of breast cancer. The drug was isolated from plasma by solvent extraction and chromatography on Lipidex 5000 and Sephadex LH 20 columns. Tamoxifen was quantified by g.c.–m.s. monitoring of the molecular ion of the internal standard, m/z 357.2093, and the molecular ion, m/z 371.2249, of tamoxifen at a resolving power of 8500. The overall recovery of tamoxifen was 78%.

Homologous Internal Standard Yielding the Same Ion.—The use of chlorpromazine as an internal standard for the determination of dibucaine in tissues removed at autopsy is reported.[45] The drug and standard were extracted from the homogenate of the tissues using the ethyl acetate–ammonium carbonate technique. A fragment ion, m/z 86, was recorded. The limit of detection was 0.8 ng per sample. A single ion monitoring technique was used to determine terbutaline in plasma and urine.[46] An internal standard which had a 1,1-dimethylpropyl group in place of the t-butyl group of terbutaline was used. The O-TMS, N-TFA derivatives were prepared and the ion m/z 355, common to the drug and its homologue, was recorded. The limit of detection was 0.3 ng cm^{-3} for a 4 ml plasma sample.

Ameneptine (27) has been determined in blood and rat brain by g.c.–m.s. Opripramol (28) which also gave an ion of m/z 192 was used as the internal standard. The method was sensitive enough to determine 50 ng cm^{-3} blood and 200 ng g^{-1} brain.[47]

HN[CH$_2$]$_6$CO$_2$Me
(27)

CH$_2$[CH$_2$]$_3$N   N[CH$_2$]$_2$OMe
(28)

Stable Isotope-labelled Analogues as Internal Standards.—The most commonly used stable isotope continues to be deuterium; only rarely were papers found in which ^{13}C, ^{18}O, or ^{15}N was used. Controversy exists as to whether stable isotope analogues exert a carrier effect but, in an assay of diphenoxylate, an antidiarrhoeal agent in plasma, a carrier effect was observed and the sensitivity of the assay was increased seven-fold by the use of a tetra-deuteriated analogue as an internal standard.[48]

Determinations of pethidine (29), an analgesic, and norpethidine (30), a metabolite with analgesic activity, by g.c. electron capture and g.c.–m.s. using [^{2}H$_4$]pethidine and [^{2}H$_5$]norpethidine as internal standards were compared.[49]

[44] S. J. Gaskell, C. P. Daniel, and R. I. Nicholson, *J. Endocrinol.*, 1978, **78**, 293.
[45] M. Kageura, K. Totoki, and T. Nagata, *Jpn. J. Legal Med.*, 1978, **32**, 188.
[46] R. A. Clare, D. S. Davies, and T. A. Baillie, *Biomed. Mass Spectrom.*, 1979, **6**, 31.
[47] C. Sbarra, P. Negrini, and R. Fanelli, *J. Chromatogr.*, 1979, **162**, 31.
[48] N. J. Haskins, G. C. Ford, S. J. W. Grigson, and K. A. Waddell, *Biomed. Mass Spectrom.*, 1978, **5**, 423.
[49] C. Linderberg, M. Berg, L. O. Boreus, P. Hartvig, K. E. Karlsson, L. Palmer, and A. M. Thornblad, *Biomed. Mass Spectrom.*, 1978, **5**, 540.

The two methods gave similar results but the g.c.–m.s. method was more rapid because the work-up procedure was simpler.

The major metabolites of pethidine are pethidinic acid (31) and norpethidinic acid (32) and their conjugates. A method based on hydrolysis of the conjugates,

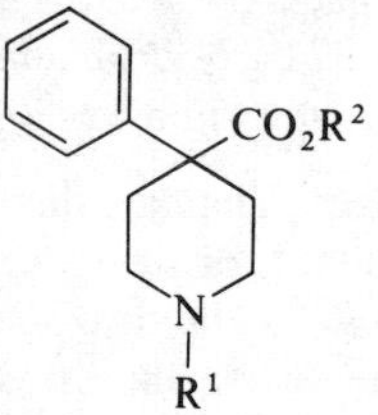

(29) $R^1 = Me, R^2 = Et$
(30) $R^1 = H, R^2 = Et$
(31) $R^1 = Me, R^2 = H$
(32) $R^1 = H, R^2 = H$

isolation of the acids by the use of XAD chromatography, and extractive alkylation of the acids using tetrapentylammonium hydroxide and pentafluorobenzyl bromide in dichloromethane is described.[50] Quantification was by g.c.–m.s. using [$^2H_4$]pethidinic acid and [$^2H_5$]norpethidinic acid as internal standards. Two papers[51,52] describing the use of a g.c.–m.s. method for quantification in plasma of both tertiary and secondary amine tricyclic antidepressants by a single ion recording procedure and using [$^2H_4$]imipramine and [$^2H_4$]desipramine as internal standards are reported. Attention was drawn to the necessity to check the purity of the labelled compound because a commercial sample of [$^2H_4$]desimipramine was reported to be only 60% pure.[51] A chemical ionization g.c.–m.s. method was used to measure separately amitriptyline, 10-hydroxyamitriptyline, nortriptyline, and 10-hydroxynortriptyline.[52] Chlorpromazine was added prior to extraction to minimize losses during analysis. The secondary amines were determined as their TFA derivatives but, to facilitate g.c. analysis, the two 10-hydroxy metabolites were subjected to a dehydration reaction. No loss of deuterium occurred during this reaction. [$^2H_6$]Amitriptyline, [^2H]10-hydroxyamitriptyline, [$^2H_3$]nortriptyline, and [^2H]-10-hydroxynortriptyline were used as internal standards. When the monodeuteriated standards were used the increase in mass from that of the drug was not sufficient to give a linear calibration and an interactive computer program was used to plot the calibration curve. The sensitivity of the assay was about 0.5 ng cm^{-3} for each of the four compounds. The results of a study carried out in patients of the pharmacokinetics of amitriptyline and the metabolites are described.

A method previously described for the g.c.–m.s. determination of clonazepam (33) and its 7-amino metabolite, using a ^{15}N-labelled analogue,[53] was found to

[50] C. Lindberg, K. E. Karlsson, and P. Hartvig, *Acta Pharm. Suec.*, 1978, **15**, 327.
[51] N. Narasimhachari, R. O. Friedel, and J. J. Saady, *Res. Commun. Psychol. Psychiatr. Behav.*, 1979, **4**, 447.
[52] W. A Garland, R. R. Muccino, B. H. Min, J. Cupano, and W. E. Fann. *Clin. Pharmacol. Ther.*, 1979, **25**, 844.
[53] B. H. Min and W. A. Garland, *J. Chromatogr.*, 1977, **139**, 121.

(33)

give poor recovery of the metabolite. This method was modified to give more consistent recoveries.[54] The NONLIN computer program was used to plot the calibration curves. The method has been used to study whether the formation of the metabolites is genetically controlled.

A g.c.-m.s. assay was used to measure cocaine and norcocaine, as its trifluoroacetyl derivative, in urine.[55] O-[^{2}H$_3$]Methylcocaine and O-[^{2}H$_3$]methylnorcocaine were used as internal standards. The assay could measure 2 ng cm^{-3} of each with about 5% precision. A report from the same laboratory describes the determination of cocaine and its metabolite benzoylecgonine using deuterium-labelled standards.[56]

A g.c.-m.s. method using deuteriated internal standards is reported for the determination of morphine and codeine in the blood of narcotic addicts.[57] A fluorimetric and a g.c.-m.s. method using a deuteriated standard were compared for the determination of morphine in blood and brain. The fluorimetric assay was less specific than the g.c.-m.s. method.[58]

Etorphine, which possesses both a phenolic hydroxy and tertiary alcohol group, was extracted from urine and silylated under conditions which gave only the TMS ether of the phenol which was determined by g.c.-m.s., using [^{2}H$_3$]etorphine as the standard.[59]

A g.c.-m.s. stable isotope-dilution assay for determining 1-butyryl-4-cinnamylpiperazine and three metabolites in urine is reported.[60] The metabolites were methylated prior to g.c.-m.s. analysis. The assay was used to study in man the pharmacokinetics and metabolism of a 5 mg oral dose of an equimolar mixture of the protio and deuterio (^{2}H$_5$) forms of the drug.

An isotope dilution assay is described for methadyl acetate, 6-(dimethylamino)-4,4-diphenyl-3-heptanol acetate, a new analgesic, in urine. A tetradeuteriated analogue was used as the internal standard. The signals from the [M − 15]$^+$ ions were recorded for quantification.[61] The analgesic 1-α-acetylmethadol (34) and its mono- and di-methylated metabolites have been determined under CI conditions by g.c.-m.s. The carrier gas was methane and the reactant gas ammonia. The trisdeuterio analogue of each compound was

---

[54] B. H. Min, W. A. Garland, K.-C. Khoo, and G. S. Torres, *Biomed. Mass Spectrom.*, 1978, **5**, 692.
[55] S. P. Jindal, T. Lutz, and P. Vestergaard, *Biomed. Mass Spectrom.*, 1978, **5**, 658.
[56] S. P. Jindal and P. Vestergaard, *J. Pharm. Sci.*, 1978, **67**, 811.
[57] D. Pearce, S. Wiersema, M. Kuo, and C. Emery, *Clin. Toxicol.*, 1979, **14**, 161.
[58] D. Reed, *Clin. Toxicol.*, 1979, **14**, 169.
[59] S. P. Jindal, T. Lutz, and P. Vestergaard, *Anal. Chem.*, 1979, **51**, 269.
[60] S. Baba, S. Morishita, and Y. Kasuya, *J. Pharmacobio.-Dyn.*, 1978, **1**, 222.
[61] S. P. Jindal, P. Vestergaard, and T. Lutz, *J. Pharm. Sci.*, 1978, **67**, 1483.

used as an internal standard. The procedure involved solvent extraction and then conversion of the respective metabolites into noracetylmethodolamide (35) and dinoracetylmethodolamide (36), by treatment with sodium hydroxide at a pH greater than thirteen. The [MH]$^+$ ions from the drug, two metabolites, and standards were recorded simultaneously. The sensitivity for plasma was 5 ng cm^{-3} with accuracy and precision of 10—15%.[62]

$$MeCO_2CHEtC(Ph)_2CHMeNMe_2 \qquad EtCHOHC(Ph)_2CHMeNRCOMe$$

(34)

(35) R = Me
(36) R = H

Carbamazepine and carbamazepine-10,11-epoxide have been determined using methane CI g.c.–m.s.[63] The ^{13}C analogues of the drug and its epoxide were used as internal standards. A non-linear least-squares program was used for preparing the standard curve and calculating the concentrations in the unknown samples.

A more rapid selected ion monitoring method for determining glutethimide and six of its metabolites has been described.[64] In this method, the three hydroxylated metabolites of glutethimide were determined as their trifluoroacetyl derivatives. The metabolism of the muscle relaxant 3-phenylpropylcarbamate (37), labelled with deuterium in the aromatic ring, has been studied

$$Ph(CH_2)_3OCONH_2$$

(37)

in rat and man.[65] In man, the metabolism of the deuterium- and ^{13}C-labelled forms of the drug were compared by measuring the percentage of dose which was excreted in the urine as either the deuterium- or ^{13}C-labelled drug and metabolites. The hydroxylated metabolites were determined as the trimethylsilyl derivatives and the acidic metabolites as their methyl esters. There were differences between rat and man in the metabolism of 3-phenylpropylcarbamate. The major metabolite in man was benzoic acid, excreted in the urine as the glycine conjugate, hippuric acid. The use of the stable isotope-labelled drug made it possible to differentiate between endogenous benzoic acid and that formed from the drug.

The determination by CI g.c.–m.s. of the anti-inflammatory agent carprofen (38) in blood has been compared using a trisdeuterio analogue and an isomeric internal standard which gave the same [MH]$^+$ ion, m/z 288, as the drug.[66] The

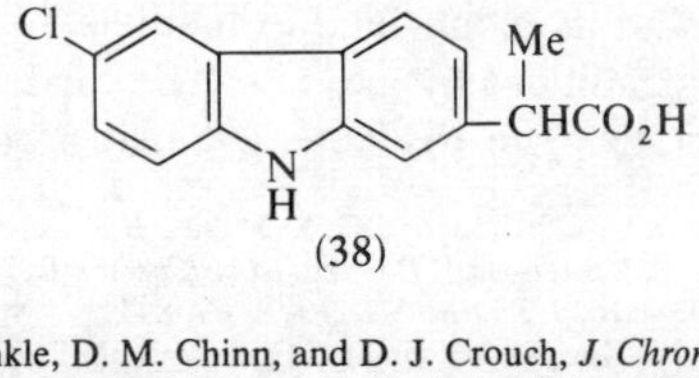

(38)

---

[62] T. A. Jennison, B. S. Finkle, D. M. Chinn, and D. J. Crouch, *J. Chromatogr. Sci.*, 1979, **17**, 64.
[63] W. F. Trager, R. H. Levy, I. H. Patel, and J. N. Neal, *Anal. Lett.*, 1978, **B11**, 119.
[64] K. A. Kennedy, J. J. Ambre, and L. J. Fischer, *Biomed. Mass Spectrom.*, 1978, **5**, 679.
[65] M. Horie and S. Baba, *Biomed. Mass Spectrom.*, 1979, **6**, 63.
[66] B. J. Hodshon, W. A. Garland, C. W. Perry, and G. J. Bader, *Biomed. Mass Spectrom.*, 1979 **6**, 325.

results indicated that better accuracy and precision was obtained using the stable isotope-labelled analogue method. A. g.c.–m.s. chemical ionization method for measuring ibuprofen as its methyl ester in serum and synovial fluid is described.[67] A quadrupole mass spectrometer was used and interference from endogenous compounds with similar nominal mass occurred. These interfering peaks and not the sensitivity of the g.c.–m.s. system caused the limit of detection to be 200 pg on column. The effect of crystal size on the bioavailability of benoxaprofen, 2-(4-chlorophenyl)-α-methyl-5-benzoxazolacetic acid, in man has been studied. The drug in capsule form and the [^{2}H$_7$]-labelled analogue of the drug in solution were given orally and the plasma and urine were analysed over 168 hours.[68] The use of the stable isotope-labelled drug halved both the time required for the bioavailability study and the number of blood and urine samples required from the subjects; 2-(3,5-dichlorophenyl)-α-methyl-5-benzoxazolacetic acid was used as the internal standard. The solvent extracts from plasma and urine were methylated and the total drug plus labelled analogue were determined by g.c. The ratio of the labelled to unlabelled drug was determined by g.c. CI mass spectrometry. The kinetics of the metabolism of the deuterium-labelled analogue were shown to be the same as those of the drug.

A radioimmunoassay for amphetamine was compared with a g.c.–m.s. method using [^{2}H$_3$]-d-amphetamine.[69] Correlation was good (R = 0.9822), the coefficient of variation being 4.9% for the g.c.–m.s. method compared with 6.9% for the radioimmunoassay. The g.c.–m.s. method was 1.9 times more sensitive than the radioimmunoassay but the latter was cheaper and faster.

A chemical ionization method using [^{2}H$_2$]warfarin as an internal standard for determining warfarin as its 4-methoxy derivative is reported.[70] The [MH]$^+$ ion, with about 20% relative ion abundance, was monitored. The assay limit was 20 ng injected.

Terbutaline has been determined in freeze-dried homogenates of tissues by a modification of an earlier ion pair–g.c. chemical ionization procedure used for plasma.[71] The deuterio analogue used as internal standard was added to the homogenate prior to analysis. The detection limit was 1.5 ng g^{-1}.

A method using $\alpha\alpha$[^{2}H$_2$]homogentisic acid is described for measuring homogentisic acid in plasma or urine from patients suffering from aspirin overdosage.[72]

An experimental drug, 16α-cyano-3β-cyclopentyloxpregn-5-en-20-one, has been assayed in dog plasma.[73] The compound was extracted by the salt–solvent pair technique, and separated from endogenous cholesterol palmitate by chromatography on Lipidex 5000. Quantification was by selected ion monitoring

---

[67] J. B. Whitlam and H. J. Vine, *J. Chromatogr.*, 1980, **181**, 463.
[68] R. L. Wolen, R. H. Carmicheal, A. S. Ridolfo, L. Thompkins, and E. A. Ziege, *Biomed. Mass Spectrom.*, 1979, **6**, 173.
[69] K. H. Powers and M. H. Ebert, *Biomed. Mass Spectrom.*, 1979, **6**, 187.
[70] P. H. Duffield, D. J. Birkett, D. N. Wade, and A. M. Duffield, *Biomed. Mass Spectrom.*, 1979, **6**, 101.
[71] J. G. Leferink, I. Wagemaker-Engels, R. A. A. Maes, and M. Van de Straeten, *J. Anal. Toxicol.*, 1978, **2**, 86.
[72] J. A. Montgomery and O. A. Mamer, *Biomed. Mass Spectrom.*, 1978, **5**, 331.
[78] C. G. Swahn, H. Beving, and G. Sedvall, *J. Chromatogr.*, 1979, **162**, 433.

of the [M − 57]⁺ base peak ion of the t-butyldimethylsilyloxime and the corresponding peak of the [^{2}H$_9$]cyclopentyloxy analogue. There was preliminary evidence that the interference from ions resulting from column bleed could be reduced by using open tubular columns.

A simple and rapid assay is reported for the determination in plasma of meclizine (39), an anti-nauseant. A pentadeuteriated aryl analogue was used as an internal standard. Quantification was by selected ion recording of the [MH]⁺ ion of the drug and standard.[74] The use of methane–ammonia instead of methane increased [MH]⁺ ion current seven-fold. The assay sensitivity was 5 ng cm^{-3}.

$$Cl-\langle\rangle-CHPhN\langle\rangle N-CH_2-\langle\rangle$$
$$Me$$
(39)

A paper on the determination of primaquine, an antimalarial, illustrates the problems which can arise in analysing biological fluids.[75] Solvent extraction of the drug from plasma and urine was unsatisfactory so the addition of the deuteriated analogue internal standard as a carrier compound to the plasma or urine was mandatory. The sample was freeze dried and extracted with ammoniacal trichloroethylene, the extract being evaporated to dryness and the residue silylated. The stability of the bis-TMS derivative was low and samples had to be analysed within four hours of the reaction. Quantification was by selected ion monitoring of the molecular ion of the drug and standard.

A selected ion method using a stable isotope analogue containing nine deuterium atoms is described for the determination of isamoxole, N-butyl-N-(4-methyloxazol-2-yl)-2-methylpropanamide, an antiallergic agent.[76] The assay was sensitive to 2 ng cm^{-3}.

A method for the determination of free and conjugated 17α-ethinyloestradiol-17β, an oral contraceptive, in plasma was reported.[77] A tritiated internal analogue standard was added and the drug and its conjugate isolated by chromatography on XAD resin. The conjugated steroid was hydrolysed with β-glucuronidase and the total free steroid was further purified by chromatography on Sephadex LH 20. The heptafluorobutyrate derivative of the steroid was formed and determined by glass capillary column g.c. and selected ion monitoring. The limit of detection was 25 pg cm^{-3}.

A method is described for the determination of baclofen, 4-amino-3-p-chlorophenylbutyric acid, a drug used for the treatment of multiple sclerosis, in serum and cerebrospinal fluid as the pentafluoropropanol derivative.[78] The presence of a chlorine atom in the molecule made it necessary to use a tetradeuteriated stable isotope analogue. Endogenous material in the extract which co-chromatographed limited the sensitivity of the assay to 5 ng cm^{-3}.

[74] H. G. Fouda, F. C. Falkner, D. C. Hobbs, and E. W. Luther, *Biomed. Mass Spectrom.* 1978, **5**, 491.
[75] J. Greaves, D. A. Price-Evans, H. M. Gilles, and J. D. Baty, *Biomed. Mass Spectrom.*, 1979, **6**, 109.
[76] D. H. Chatfield and T. J. Woodage, *Biomed. Mass Specrom.*, 1978, **5**, 466.
[77] L. Siekmann, A. Siekmann, and H. Breuer, *Fresenius Z. Anal. Chem.*, 1978, **290**, 159.
[78] C. G. Swahn, H. Beving, and G. Sedvall, *J. Chromatogr.*, 1979, **162**, 433.

Ketobemidone (40), an analgesic, has been determined in plasma using a deuteriated analogue internal standard. The sensitivity was 0.5 ng cm^{-3}.[79] 5-Fluorouracil, arising as a product of metabolism of FDI (19), has been determined in tissues from rats; [1,3-^{15}N$_2$]-5-fluorouracil was used as the internal standard. Quantification was by recording the molecular ions of the compound and iostope-labelled analogue.[80] The sensitivity was 1—5 ng g^{-1} depending upon the tissue.

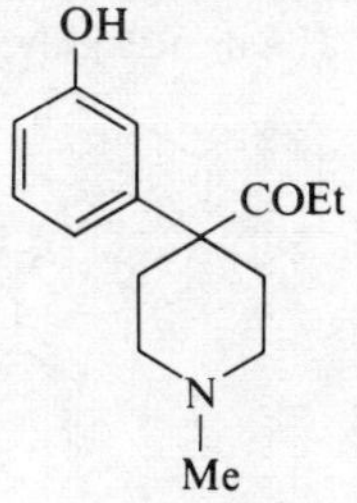

(40)

A pyrolysis method is described for determination in plasma of the quaternary ammonium compound (3-methoxycarbopropyl)trimethylammonium bromide, carpronium chloride, a parasympathomimetic agent.[81] A series of esters of the drug was studied and the isopropyl ester was chosen because it gave the best yield of isopropyl-N,N-dimethyl-γ-aminobutyrate, the pyrolysis product selected for quantification. The tri[^{2}H$_3$]methylammonium analogue of carpronium chloride was used as the internal standard. The plasma proteins were precipitated and the drug extracted as an ion pair complex, and further purified by anion exchange chromatography.

A previously described g.c.–m.s. method for determining Δ^1- and Δ^6-tetrahydrocannabinol in plasma has been improved in sensitivity and specificity by chromatographing the compounds as their trimethylsilyl derivatives.[82]

A method using g.c.–m.s. negative ion chemical ionization technique to measure clonazepam was found to be twenty times more sensitive than the corresponding positive ion technique. The sensitivity was better than 0.1 ng clonazepam per cm^{-3}. The ^{15}N-labelled internal standard used in the positive ion analysis method was replaced by a clonazepam analogue containing three ^{18}O atoms and one ^{15}N atom.[83]

When negative ions are formed by electron capture, enhanced sensitivity is reported but this is not the case when negative ions are formed by reactant negative chemical ionization. A study, in which OH$^-$ chemical ionization was compared with positive ion chemical ionization for the analysis of methadone and three metabolites and 1-α-methadol and six metabolites, showed that

---

[79] U. Bondesson and P. Hartvig, *J. Chromatogr.*, 1979, **179**, 207.
[80] T. Marunaka, Y. Umeno, and Y. Minami, *J. Chromatogr.*, 1980, **188**, 270.
[81] M. Sano, K. Ohya, and S. Shintani, *Biomed. Mass Spectrom.*, 1980, **7**, 1.
[82] A. Ohlsson, S. Agurell, J. E. Lindgren, and K. Leander, American Chemical Society Symposium Series 98, ed. J. A. Vinson, American Chemical Society, Washington, 1979, p. 73.
[83] W. A. Garland and B. H. Min, *J. Chromatogr.*, 1979, **172**, 279.

negative ionization was only marginally more sensitive.[84] A g.c.–m.s. method for the determination of the thyreostatic, methimazole, is reported. The drug was derivatized in one step by extractive alkylation using either benzyl chloride or pentafluorobenzyl bromide. Deuteriated methimazole was used as the internal standard and the precision at 5 ng cm^{-3} concentration was 6%.[85]

Selected Ion Recording Using Field Desorption.—A method for determining berberine chloride (41) in urine using the trisdeuteriomethoxy analogue as the internal standard is reported.[86] The urine was freeze-dried, the residue extracted

(41)

with ethanol, and the berberine chloride isolated by polyacrylamide gel electrophoresis and paper chromatography. The band corresponding to berberine chloride was eluted with methanol and the solution analysed by FD selected ion monitoring. Good linearity was obtained over the range 10—200 pg of the drug. The accuracy and precision of the FD and the previous g.c.–m.s. chemical ionization methods were comparable for the determination of berberine chloride.

Quantitative Liquid Chromatography–Mass Spectrometry.—On-line liquid chromatography–mass spectrometry was first reported over five years ago,[12] but it is disappointing to record that no paper was found describing the use of this technique for quantitative determination of drugs in biological fluids. Off-line h.p.l.c. and direct probe mass spectrometry have been used to determine simultaneously (R)- and (S)-warfarin in the plasma of humans given an oral dose of pseudoracemic warfarin containing an S-enantiomer labelled with ^{13}C in the 2-position of the molecule.[87] [^{2}H$_{5}$]Warfarin was used as an internal standard. The enantiomers were determined at the nanogram level with an accuracy and precision of 5%. The method was used to study, in man, the effect of treatment with phenylbutazone and seconal on the clearance of (R)- and (S)-warfarin.

The Use of Selected Metastable Peak Monitoring for Drug Analysis.—Interference in the determination of drugs by low-resolution g.c.–m.s. can occur from compounds in biological fluids which co-elute with the drug. This interference can be overcome by using high-resolution g.c.–m.s. An alternative approach, which gives improved selectivity, is that based on monitoring the metastable ions corresponding to fragmentations occurring in the first field-free region of

---

[84] A. L. C. Smit and F. H. Field. *Biomed. Mass Spectrom.*, 1978, **5**, 572.
[85] S. Floberg, K. Lanbeck, and B. Lindstrom, *J. Chromatogr.*, 1980, **182**, 63.
[86] H. Miyazaki, E. Shirai, M. Ishibashi, K. Hosoi, S. Shibata, and M. Iwanga, *Biomed. Mass Spectrom.*, 1978, **5**, 559.
[87] W. H. Howald, E. D. Bush, W. F. Trager, R. A. O'Reilly, and C. H. Motley, *Biomed. Mass Spectrom.*, 1980, **7**, 35.

a double-focusing mass spectrometer. In this procedure, the g.c.–m.s. is operated at 1000 resolving power. The ion selected for monitoring is a characteristic daughter ion formed from a parent ion in a field-free region of the mass spectrometer. Specificity is achieved by selecting both the parent ion, derived from the drug or metabolite, and a suitable daughter ion. This technique has been used to determine 20—30 pg dihydrotestosterone and testosterone in serum and tissue.[88,89]

3 Mass Spectrometry in Metabolite Identification

In many metabolic studies, radiolabelled drugs are used and this procedure has the advantage that it enables easy detection of metabolites and provides a means of following the recovery of the metabolites during their isolation from biological fluids. A variety of pH-dependent extraction, adsorption, column, thin layer, and gas chromatographic techniques, alone or in combination, has been used for the purification of metabolites for structural analysis. The quantity of pure metabolite obtained is often only a few micrograms and mass spectrometry may be the only practical physical technique which can provide information on the chemical structure of the compound from such a small sample.

Metabolites Separated by Liquid Chromatography.—An automated h.p.l.c. procedure for sample clean-up prior to g.c.–m.s. has been described.[90] Ciclobendazole (42), an anthelmintic, labelled with ^{14}C, was administered orally to animals and methanolic extracts of the faeces were purified by column and h.p.l.c. The yield of the major metabolite of the drug was 60 μg. The mass spectrum of the metabolite showed a molecular ion at m/z 275, which is 16 a.m.u. higher than that of ciclobendazole.[91] From accurate mass measurement of the molecular ion the formula was deduced to be $C_{13}H_{13}N_3O_4$. The fragmentation pattern indicated that an O-hydroxy group was present in the aromatic ring and this was confirmed by methylating the metabolite with diazomethane when a trimethyl derivative was formed.

The metabolites of ^{14}C-sulpiride (43) were isolated from rat and dog urines by column and h.p.l.c. and their structures assigned from mass spectrometry. The metabolic reactions which had occurred were O-dealkylation, pyrrolidine N-dealkylation, amide N-dealkylation, and α-C-oxidation.[92]

(42)

(43)

---

[88] S. J. Gaskell and D. S. Millington, *Biomed. Mass Spectrom.*, 1978, **5**, 557.
[89] S. J. Gaskell, R. W. Finney, and M. E. Harper, *Biomed. Mass Spectrom.*, 1979, **6**, 113.
[90] J. J. de Ridder and H. J. M. van Hal, *J. Chromatogr.*, 1978, **146**, 425.
[91] B. C. Mayo, R. R. Brodie, L. F. Chasseaud, and D. R. Hawkins, *Drug Metab. Dispos.*, 1978, **6**, 518.
[92] F. R. Sugnaux and A. Benakis, *Eur. J. Drug Metab. Pharmacokinet.*, 1978, **3**, 235.

The glucuronide conjugate of suloctidil[93] (44) was isolated by column chromatography. The silyl and silylated methylester derivatives of the metabolite were prepared and analysed by g.c.–m.s. From the fragmentation pattern of the mass spectrum it was concluded that the 2-OH group of the drug was conjugated with glucuronic acid.

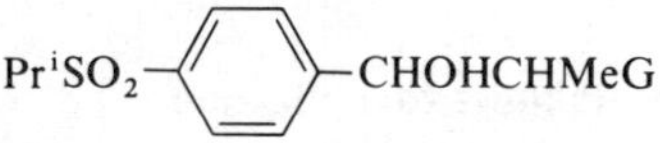

Pr^iSO_2—⟨☐⟩—CHOHCHMeG

G = Glucuronyl

(44)

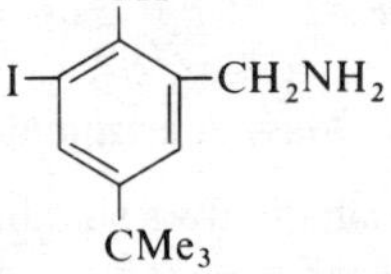

(45) [structure: iodo-tert-butyl-hydroxy-benzylamine]

The metabolism of MK 447-^{14}C (45), an antihypertensive agent, was studied. A sulphate conjugate was isolated from rat and dog urine and a glucuronide conjugate from human urine. The sulphate conjugate decomposed during direct probe analysis with loss of sulphur trioxide and formation of MK 447. In the case of the silylated human glucuronide the direct probe spectrum was complex and appeared to arise from a mixture of the N,O-di(trimethylsilyl) derivative of MK 447 and the tri-O-trimethylsilyl derivative of glucopyranurono-(6 → 1)-lactone.[94] Nuclear magnetic resonance spectroscopic evidence indicated that the human metabolite was the N-glucuronide.

The metabolism of 4'-deuteriopropranolol was investigated *in vitro* using rat liver microsomes and *in vivo* in man.[95] The 4'-hydroxypropranolol present in the microsomal incubates and the β-glucuronidase hydrolysates of urine was isolated by h.p.l.c. and the presence of the deuterium determined by direct probe m.s. and g.c.–m.s. of the trifluoroacetyl derivative. The results indicated that 4'-hydroxylation occurs both *in vitro* and *in vivo* by way of an arene oxide 'NIH shift' process.

Rats were given intravenously a mixture of [1 – ^{14}C]ellipticine (46) and either 9-deuterioellipticine or 7,9-dideuterioellipticine, and the bile was collected. Two major metabolites were isolated from the bile by the use of solvent extraction, partition chromatography, and reversed-phase h.p.l.c. methods. In the FD spectrum of metabolite A there was an [MH]$^+$ ion at m/z 343 corresponding to a sulphate of 9-hydroxyellipticine; an abundant ion at m/z 262 corresponding

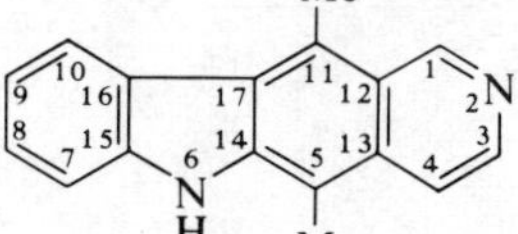

(46)

[93] M. Martens, R. Roncucci, M.-J. Simon, K. Debast, and G. Lambelin, *Eur. J. Drug Metab. Pharmacokinet.*, 1978, **3**, 223.
[94] D. J. Tocco, F. A. De Luna, A. E. W. Duncan, R. W. Walker, B. H. Arison, and W. J. A. Vandenheuvel. *Drug Metab. Dispos.*, 1979, **7**, 330.
[95] W. L. Nelson and M. L. Powell, *Drug Metab. Dispos.*, 1979, **7**, 351.

to 9-hydroxyellipticine was also present. Doublets at m/z 262 and 263 were observed in the mass spectra of both the metabolites isolated from the bile of rats given 7,9-dideuterioellipticine. These indicated hydroxylation and loss of a deuterium atom. The mass spectrum of the metabolite isolated after a dose of the mixture containing 9-deuterioellipticine showed no conservation of deuterium in either metabolite. Therefore, the site of aromatic hydroxylation was at position 9. Metabolite B gave an FD spectrum with a molecular ion at m/z 438 and $[MH]^+$ at m/z 439 corresponding to the glucuronide conjugate. At higher emission currents, the expected pyrolysis product, 9-hydroxyellipticine, was observed.[96] Hydrolysis of the respective metabolites with sulphatase or glucuronidase yielded a compound identified by mass spectral analysis as 9-hydroxyellipticine.

Three metabolites of labetalol were isolated from dog, monkey, and human urine. Two of them, (47) and (48), were easily hydrolysed by β-glucuronidase to the aglycone whilst the third was only partially hydrolysed. FD spectra confirmed that the third metabolite (49) was a glucuronide,[97] which was only hydrolysed

$$R^2O-\underset{\underset{OR^3}{|}}{\overset{\overset{CONH_2}{|}}{\bigcirc}}-CHCH_2N\overset{\overset{R^1}{|}}{C}HMeCH_2Ph$$

(47) R^1 = glucuronyl, $R^2 = R^3 = H$
(48) R^2 = glucuronyl, $R^1 = R^3 = H$
(49) R^3 = glucuronyl, $R^1 = R^2 = H$

with difficulty. The mechanism of the biosynthesis and hydrolysis of glucuronides has been studied using [^{18}O]isoborneol as a substrate and rabbit liver UDP-glucuronyltransferase and β-glucuronidase.[98] Gas chromatography–mass spectrometry was used to measure the retention of the isotope. It was found that the labelled hydroxy oxygen atom was retained during both conjugation and hydrolysis. The products of liver aldehyde oxidase catalysed oxidation of isoquinoline, 3-methylisoquinoline, phthalazine, quinazoline, quinoxaline, and cinnoline have been separated in microgram amounts using adsorption and reversed-phase h.p.l.c., and their structures were established by mass spectral and microdisc infrared analyses.[99]

Nitrofurantoin (50) has been incubated with tissue and caecal content homogenates and the major *in vitro* metabolite isolated by h.p.l.c.[100] The mass spectrum indicated that the $[MH]^+$ ion had an m/z of 209 but it was not possible from the fragmentation pattern to decide if the metabolite was the open-chain nitrile derivative resulting from the reduction of the nitro group and opening of the furan ring or the isomeric amino furan. Nuclear magnetic resonance spectroscopy confirmed that it was the open chain nitrile.

[96] A. R. Branfman, R. J. Bruni, V. H. Reinhold, D. M. Silveira, M. Chadwick, and D. W. Yesair, *Drug Metab. Dispos.*, 1978, **6**, 542.
[97] R. Hopkins, L. E. Martin, J. Oxford, and N. Scully, 'Conjugation Reactions in Drug Biotransformation', ed. A. Aitio, Elsevier, Amsterdam, 1978, p. 500.
[98] L. P. Johnson and C. Fenselau, *Drug Metab. Dispos.*, 1978, **6**, 677.
[99] C. Stubley, J. Godfrey, P. Stell, and D. W. Mathieson, *J. Chromatogr.*, 1979, **177**, 313.
[100] M. B. Aufrere, B. A. Hoener, and M. Vore, *Drug Metab. Dispos.*, 1978, **6**, No. 403.

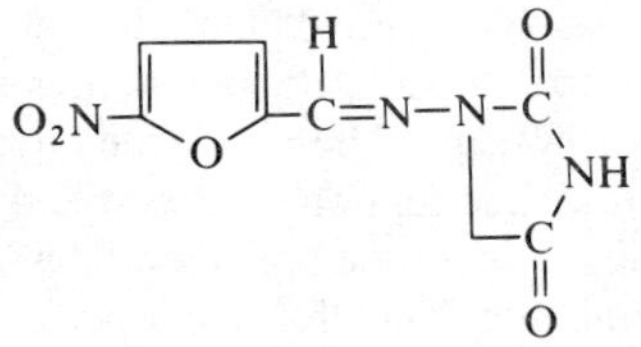

$$O_2N-\underset{O}{\underset{|}{\bigcirc}}-\overset{H}{\underset{}{C}}=N-N\underset{\underset{O}{\parallel}}{\overset{\overset{O}{\parallel}}{\underset{C}{\overset{C}{\diagup}}}}NH$$

(50)

Metabolites Isolated by Thin-layer Chromatography.—The metabolites present in the bile and β-glucuronidase hydrolysates of urine from rats given ^{14}C-labelled 4,5-dihydrodiazepam[101] have been isolated by solvent extraction, silica column, and t.l.c. Mass spectral analysis of the metabolites established that dehydrogenation at the N-4, C-5 atoms of the drug occurred with the formation of diazepam. The phenolic and N-desmethyl metabolites of diazepam and 4,5-dihydrodiazepam were also detected. Two new metabolites, one formed by exchange of a chlorine atom for a hydroxy group, and the other by the same exchange plus reduction of the lactam carbonyl group were identified from mass spectral analysis.

When incubates from rat liver microsomes and dibenzo[c,f],[1,2]diazepine were analysed by t.l.c.,[102] two of the spots present were identified from their mass spectra. These were the N-oxide, identified by direct probe analysis, and a phenol detected by g.c.–m.s. after on-column methylation. The *in vivo* metabolism of dibenzo[c,f],[1,2]diazepam was also studied in the rat. Six metabolites were isolated by t.l.c. and their structures established by mass spectrometry.[103] The metabolism of ^{14}C-iprindole (51), an antidepressant, was studied in rat,

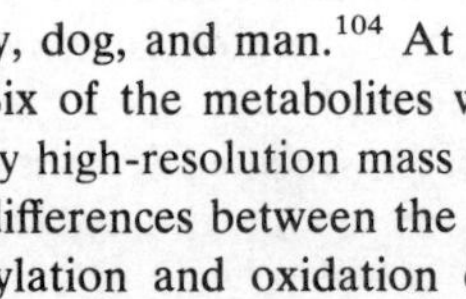

(CH$_2$)$_3$NMe$_2$

(51)

miniature pig, rhesus monkey, dog, and man.[104] At least twenty radiolabelled metabolites were detected. Six of the metabolites were isolated by t.l.c. and their structures determined by high-resolution mass spectrometry. There were quantitative and qualitative differences between the species in the metabolism of the drug, but N-demethylation and oxidation of the alicyclic ring or a combination of both were the main pathways.

Following the administration of acebutolol (52), a β-adrenergic receptor blocking drug, to rat and man two metabolites were isolated.[105] Mass spectral

---

[101] L. Otvos, Zs. Tegey, L. Vereczkey, M. Ledniczky, J. Tamas, E. Palosi, and L. Szporny, *Drug Metab. Dispos.*, 1978, **6**, 213.
[102] A. Frigerio, P. Negrini, L. Cappellini, D. Rotilio, A. de Pascale, and E. Rossi, *Eur. J. Drug Metab. Pharmacokinet.*, 1979, **4**, 35.
[103] A. Frigerio, P. Negrini, L. Cappellini, D. Rotilio, A. de Pascale, and E. Rossi, *Eur. J. Drug Metab. Pharmacokinet.* 1979, **4**, 67.
[104] S. F. Sisenwine, C. O. Tio, and H. W. Ruelius, *Xenobiotica*, 1979, **9**, 237.
[105] B. D. Andresen and F. T. Davis, *Drug Metab. Dispos.*, 1979, **7**, 360.

analysis of the metabolites showed they were the acetamide (53) and an aniline (54). It was proposed that metabolism occurred by hydrolysis of the butyramide moiety of the drug followed by reacetylation of the metabolite.

(52) R = $\underset{\text{O}}{\text{C}}$(CH$_2$)$_2$Me

(53) R = $\underset{\text{O}}{\text{C}}$Me

(54) R = H

The metabolism of a mixture of equal parts of [^{15}N$_2$]-labelled phenobarbital and ^{14}C-phenobarbital was studied in man.[106] Two metabolites were isolated from the urine. One was identified as p-hydroxyphenobarbital, the other, when examined by CI mass spectrometry, was found to give a doublet at m/z 395, 397 (MH)$^+$, which was consistent with a phenobarbital glycoside. Ions due to impurities were present in the spectrum of this metabolite in the lower mass range. A purer spectrum was obtained from the acetylated metabolite. The CI mass spectrum showed a doublet at m/z 563, 565 consistent with the tetra-acetyl derivative of phenobarbital glycoside. The metabolites of ^{14}C-praziquantel (55), an anthelmintic, were isolated by t.l.c. from human and monkey serum and urine, and rat liver homogenates. Mass spectral analysis indicated that all the major metabolites were hydroxylated products, the sites of hydroxylation being rings A, B, and D.[107] In some instances di- and tri-hydroxylation was observed.

^{14}C-Tolaxatone (56), an antidepressant, was administered orally to man. The urine and acid and β-glucuronide hydrolysates of the urine were analysed by thin-layer chromatography and two major radioactive metabolites were isolated.[108] Mass spectral analysis suggested that one was 3-(3-carboxyphenyl)-5-hydroxymethyl-2-oxazolidinone, produced by oxidation of the methyl group, and the other a phenolic derivative of tolaxatone. A series of possible phenols was synthesized for comparison and the metabolite was shown to be the 4-hydroxy compound. The metabolism of an oral dose of ^{14}C-oxaprozin (57) was studied in the beagle dog and rhesus monkey.[109] The free and conjugated metabolites were isolated from urine and bile by XAD column clean-up followed by thin-layer chromatography. Two phenolic metabolites were characterized by mass spectrometry. The conjugates of oxaprozin and of a phenolic metabolite were silylated and characterized as glucuronides from their mass spectra.

[106] B. K. Tang, W. Kalow, and A. A. Grey, *Drug Metab. Dispos.*, 1979, **7**, 315.
[107] K. U. Bühring, H. W. Diekmann, H. Müller, A. Garbe, and H. Nowak, *Eur. J. Drug Metab. Pharmacokinet.*, 1978, **3**, 179.
[108] A. Malnoë and M. Strolin Benedetti, *Xenobiotica*, 1979, **9**, 281.
[109] F. W. Janssen, S. K. Kirkman, J. A. Knowles, and H. W. Ruelius, *Drug Metab. Dispos.*, 1978, **6**, 465.

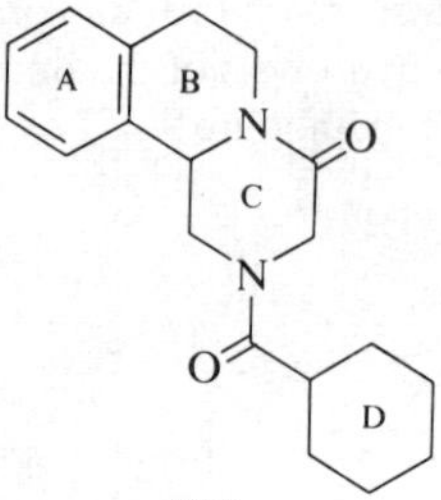

(55)

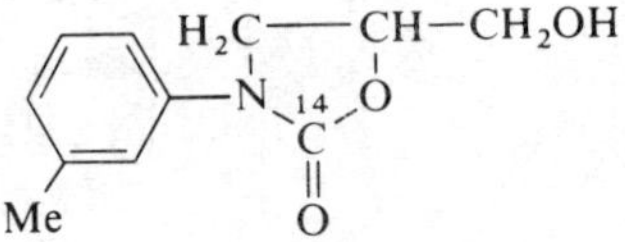

(56)

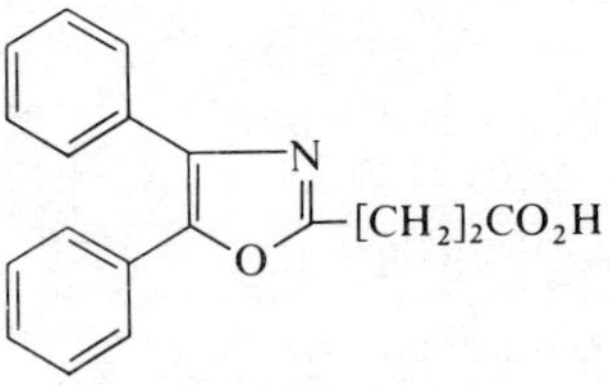

(57)

Tritium-labelled pamatolol (58), a cardioselective β-receptor antagonist, has been administered orally to mice, rats, dogs, and man and the metabolites have been isolated by t.l.c. and h.p.l.c.[110] Four urinary metabolites, (59), (60), (61), and (62), were identified by g.c.–m.s. analysis of their trifluoroacetylated and silylated derivatives. Metabolite (62) was unique to the rat.

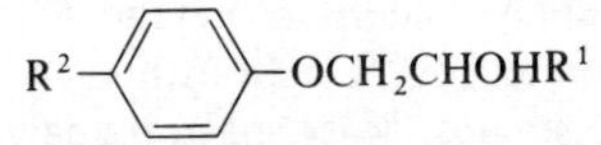

$$R^2-\text{C}_6\text{H}_4-\text{OCH}_2\text{CHOHR}^1$$

(58) $R^1 = CH_2NHPr^i$, $R^2 = (CH_2)_2NHCO_2Me$
(59) $R^1 = CO_2H$, $R^2 = (CH_2)_2NHCO_2Me$
(60) $R^1 = CH_2NHPr^i$, $R^2 = CH_2CO_2H$
(61) $R^1 = CH_2NHPr^i$, $R^2 = CHOHCH_2NHCO_2Me$
(62) $R^1 = CH_2NHPr^i$, $R^2 = CO_2H$

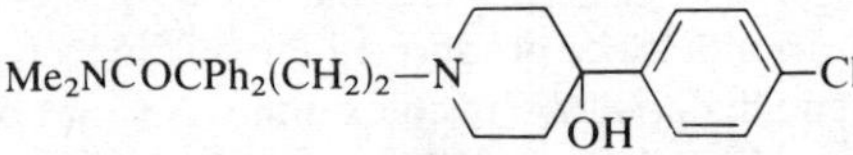

$Me_2NCOCPh_2(CH_2)_2-N\text{...}OH\text{...}C_6H_4-Cl$

(63)

An antidiarrhoeal agent, [^{14}C]loperamide (63), was administered orally to rats and six metabolites were isolated by t.l.c. from the faeces.[111] Mass spectral analysis of these metabolites showed that the main metabolic pathways were dealkylation of the dimethylamide moiety, plus monohydroxylation either in the α-phenyl ring or possibly at the α-carbon in the piperidine ring. The latter metabolites were unstable and isolated as the pyridinium metabolites. Only the desmethyl and didesmethyl metabolites were present in the urine.[111]

[110] K. J. Hoffmann, I. Skånberg, and K. O. Borg. *Eur. J. Drug Metab. Pharmacokinet.*, 1979, **4**, 163.
[111] K. Yoshida, K. Nambu, S. Arakawa, H. Miyazaki, and M. Hashimoto, *Biomed. Mass Spectrom.*, 1979, **6**, 253.

The mass spectra of the N-oxides of a range of 3-substituted pyridines and of quinoline and isoquinoline are reported. The metabolism by liver microsomes of 3-substituted pyridines and the isolation of the N-oxides and their characterization by mass spectral analysis is described.[112]

After incubation of clebopride with rabbit liver microsomes, a metabolite was isolated by t.l.c. and, from electron impact low- and high-resolution, field desorption mass spectrometry and CAT proton magnetic resonance spectroscopy, its structure was established as N-(4'-piperidyl-2'-one)-4-amino-5-chloro-2-methoxybenzamide (64).[113]

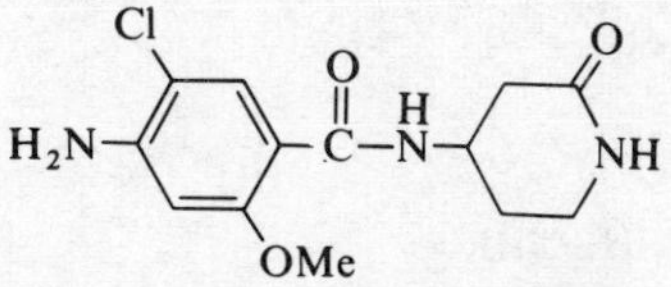

(64)

^{3}H- or ^{14}C-reproterol (65), a bronchodilator, was administered intravenously to rats and the bile collected. The conjugates were hydrolysed by glucuronidase and sulphatase and the hydrolysate chromatographed on XAD resin. The radioactive fraction was purified by thin-layer electrophoresis on cellulose and t.l.c. The mass spectrum of the metabolite suggested that it was a tetrahydroisoquinoline derivative (66) produced by interaction of reproterol with a single carbon fragment from a metabolic pool and cyclization of the product. The structure was confirmed by synthesis.[114] The metabolic fate of dl-[^{14}C]verapamil (67) has

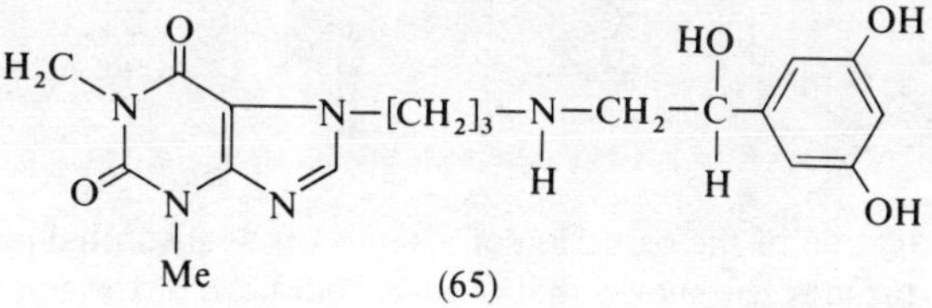

(65)

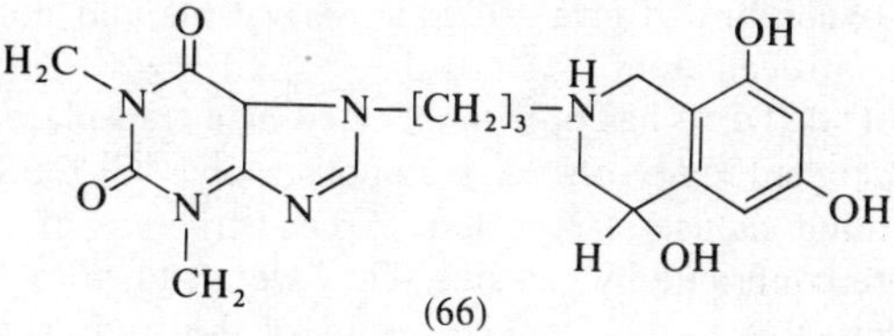

(66)

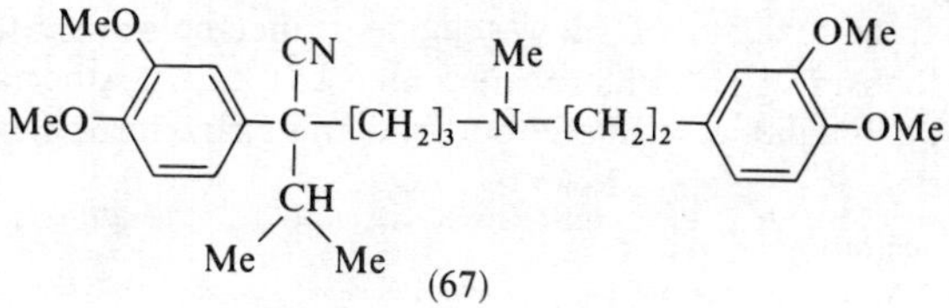

(67)

[112] D. A. Cowan, L. A. Damani, and J. W. Gorrod, *Biomed. Mass Spectrom.*, 1978, **5**, 551.
[113] G. Huizing, A. H. Beckett, and J. Sugura, *Pharm. Weekbl. (Sci. Ed.)*, 1979, **1**, 436.
[114] G. Niebch, K. H. Klingler, G. Eikelmann, and N. Kucharcyk, *Arzneim.-Forsch.*, 1978, **28**, 765.

been studied in man.[115] The metabolites were isolated and the structures elucidated by low- and high-resolution g.c.–m.s. The main metabolic step was cleavage on the C—N—C bonds by N-dealkylation preferentially at the C-atom belonging to the shorter side-chain. Verapamil and its N-dealkylated products were further metabolized by O-demethylation.

Two new metabolites of hydralazine (68) were isolated from the serum of man given oral doses of ^{14}C-labelled hydralazine.[115] Only the structure of one metabolite (69) was established by mass spectral analysis and n.m.r. spectroscopy was needed to establish the site of hydroxylation of the second (70).

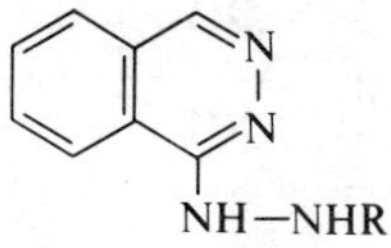

(68) R = H
(69) R = COMe

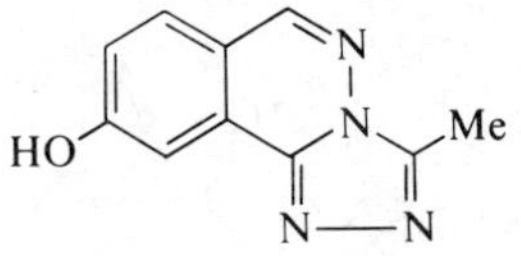

(70)

The metabolism of ^{14}C-clonidine (71) has been further studied in rat, dog, and man. Three new metabolites were isolated from the urine and from their mass spectra the following structures were proposed: 1-(2,6-dichloro-4-hydroxyphenyl)-guanidine, 2-[(2,6-dichlorophenyl)-imino]-imidazolidine-4-one, and 2-[(2,6-dichloro-4-hydroxyphenyl)-imino]-imidazolidine-4-one.[117]

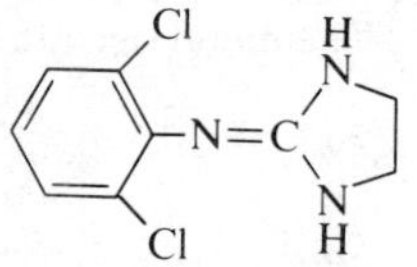

(71)

A reinvestigation of the oxidation of a series of 2-substituted phenothiazines by liver microsomes has shown that the incorporation of oxygen atoms occurs in the ring (C-oxidation) to give phenolic derivatives and not, as previously reported, at the nitrogen atom.[118]

Diethylstilbestrol (DES) has been implicated as a transplacental carcinogen in women. A detailed study on the metabolism of [^{14}C]-DES by the foetal, neonatal, and adult mouse is reported.[119] The structure of the products of metabolism were confirmed by g.c.–m.s. They were shown to be oxidized and hydroxylated derivatives of DES. The results of the study cast doubt on the proposal that the toxicity of DES was due to its metabolism via the epoxide.

The metabolism of [^{14}C]chlorpheniramine has been further studied in the dog.[120] A new metabolite isolated from urine was characterized, by direct probe

[115] M. Eichelbaum, M. Ende, G. Remberg, M. Schomerus, and H. J. Dengler, *Drug Metab. Dispos.*, 1979, **7**, 145.
[116] J. Wagner, J. W. Faigle, P. Imhof, and G. Liehr, *Arzneim.-Forsch.*, 1977, **27**, 2388.
[117] S. Darda, H. J. Forster, and H. Stahle, *Arzneim.-Forsch.*, 1978, **28**, 255.
[118] A. H. Beckett and G. E. Navas, *Xenobiotica*, 1978, **8**, 721.
[119] M. Metzler, and J. A. McLachlan, *Biochem. Pharmacol.*, 1978, **27**, 1087.
[120] J. D. Osterloh, A. Karakaya, and D. E. Carter, *Drug Metab. Dispos.*, 1980, **8**, 12.

analysis, as the alcohol (72) formed by oxidative deamination of the drug. A second new metabolite isolated from the acid hydrolysate of urine and characterized by direct probe analysis was the acid formed from oxidation of the alcohol.

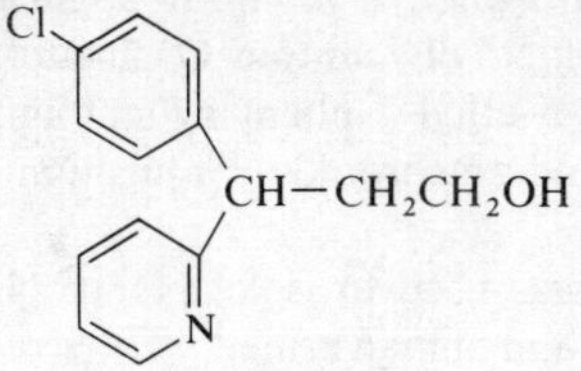

(72)

There are two reports on the metabolism of [^{14}C]antipyrine and antipyrine. In one study, two radiolabelled glucuronic acid conjugates of antipyrine were isolated from urine, and from the mass spectra of their acetyl and TMS methyl esters it was concluded that they were conjugates of 3-hydroxy and 5-hydroxy antipyrine.[121] In the other study a different conjugate was isolated. This was shown from the mass spectral analysis of the TMS derivatives to be the O-glucuronide of N-desmethylantipyrine in its enol form.[122]

Two metabolites of metyrapone (73) were isolated by t.l.c. from rat liver microsome incubates. From their chemical ionization and electron impact mass spectra it was shown that oxidation had occurred on either one or the other of the pyridine rings with the formation of two different N-oxides.[123]

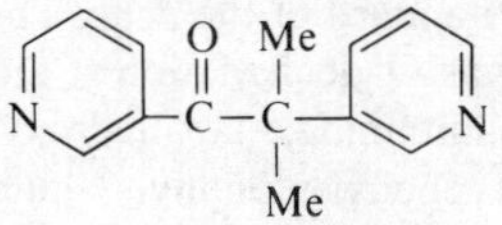

(73)

Metabolites Characterized by Gas Chromatography–Mass Spectrometry.—It is not always possible to isolate a sample of metabolite which is pure enough to characterize by direct probe mass spectrometry. However, in many cases by using g.c.–m.s. it is possible to characterize an impure metabolite or a derivative of the metabolite.

Phenytoin, an anticonvulsant, is extensively metabolized and the drug and a metabolite, 5-(p-hydroxyphenyl)-5-phenylhydantoin, were identified and estimated in rat brain by g.c.–m.s.[124]

The metabolism by man of mephenytoin has been studied in two centres.[125,126] A major metabolite in urine was identified by g.c.–m.s. as

---

[121] E. Zeitz, M. Eichelbaum, H. J. Dengler, and G. Spiteller, *Arzneim.-Forsch.*, 1978, **28**, 315.
[122] C. E. Hignite, C. Tschanz, D. H. Huffman, and D. L. Azarnoff, *Drug Metab. Dispos.*, 1978, **6**, 288.
[123] J. de Graeve, J. E. Gielen, G. F. Kahl, K. H. Tuttenberg, R. Kahl, and B. Maume, *Drug Metab. Dispos.*, 1979, **7**, 166.
[124] K. K. Midha, C. Charette, H. S. Buttar, and I. Dupuis, *J. Chromatogr.*, 1978, **157**, 416.
[125] A. Küpfer, G. M. Brilis, J. T. Watson, and T. M. Harris, *Drug Metab. Dispos.*, 1980, **8**, 1.
[126] R. K. Lynn, J. E. Bauer, W. P. Gordon, R. G. Smith, D. Griffin, R. M. Thompson, R. Jenkins, and N. Gerber, *Drug Metab. Dispos.*, 1979, **7**, 138.

p-hydroxymephenytoin.[125] Four previously unidentified glucuronide metabolites were characterized, as their methylated derivatives, by CI and EI g.c.-m.s., namely 5-ethyl-5-(hydroxyphenyl)-3-methylhydantoin-O-glucuronide, 5-hydroxyethyl-3-methyl-5-phenylhydantoin-O-glucuronide, 5-ethyl-5-(hydroxymethoxyphenyl)-3-methylhydantoin-O-glucuronide, and a metabolite tentatively identified as 5-ethyl-5-phenylhydantoin-N^3-glucuronide in which both N-demethylation and glucoronide conjugation of the hydantoin ring has occurred.[126]

Gas chromatography was used to isolate metabolites of bencyclane (74), a muscle relaxant, from rat and human urine.[127] The structures of these metabolites were established as 4-hydroxybencyclane, 4-oxobencyclane, 3-hydroxybencyclane, and monodesmethyloxobencyclane.

$$\text{(74)} \quad O-[CH_2]_3-N[CH_3]_2$$

The metabolism of an equal mixture of [$^2H_5$]-labelled and protio 3-(2′,4′,5′-triethoxybenzoyl)propionic acid, a biliary smooth muscle relaxant, administered orally to man and orally and intraveneously to rat, has been studied.[128,129] Extracts of urine from rat and man and the bile of the rat were analysed by g.c.-m.s. using the isotope cluster technique and determined quantitatively using a [$^2H_{10}$]-analogue internal standard of the drug. The primary route of biotransformation in both species was O-dealkylation at the C-4′ and C-5′ position.

Two metabolites of tolbutamide, 1-p-hydroxymethylbenzenesulphonyl-3-butylurea and 1-p-carboxyphenylsulphonyl-3-butylurea, were isolated from human plasma and identified and quantified as their methyl derivatives by g.c.-m.s. The method was sensitive enough to determine 50 pg injected on to the column.[130]

The metabolism by man of an equimolar mixture of tetradeuterio and protio potassium canrenoate (75),[131] an aldosterone antagonist, was studied in man. The metabolites were isolated from urine by XAD resin and Lipidex 5000 chromatography and then analysed by g.c.-m.s. using EI and CI and a repetitive scanning technique. Canrenoic acid and its ester glucuronide, 3β-hydroxy-3-deoxocanrenone, 3β-hydroxy-4,5α-dihydro-3-deoxocanrenone, and a 3ξ-hydroxy-4,5,6,7-tetrahydro-3-deoxocanrenone were identified.

The metabolism of befunolol (76), a β-receptor antagonist, was studied in dog and man.[132] From the g.c.-m.s. analysis of metabolites isolated from the urine, the structures of ten metabolites were established. These were produced

---

[127] K. Kimura, A. Nagata, and H. Miyawaki, *Xenobiotica*, 1979, **9**, 119.
[128] T. Kobayashi, S. Tanayama, and Y. Kanai, *Xenobiotica*, 1978, **8**, 535.
[129] T. Kobayashi, Y. Kanai, and S. Tanayama, *Xenobiotica*, 1978, **8**, 737.
[130] N. Heki, H. Hosojima, and M. Noto, *J. Jpn. Diabetic Soc.*, 1978, **21**, 835.
[131] C. W. Vose, D. R. Boreham, G. C. Ford, N. J. Haskins, and R. F. Palmer, *Drug Metab. Dispos.*, 1979, **7**, 226.
[132] M. Tohno, K. Kimura, M. Nagahara, Y. Sakai, T. Ofuji, and T. Nadai, *Yakugaku Zasshi*, 1979, **99**, 944.

by hydroxylation of the phenyl ring, reduction of the ketone group, and oxidative deamination of the dimethylamino group. Conjugation of the hydroxylated metabolites and befunolol with glucuronic acid was also observed. A species difference was observed in the metabolism of the drug. The metabolism of tritium-labelled alprenolol (77) has been studied *in vitro* using microsomes from

(75) (76) (77)

dog, rat, and guinea pig[133] and *in vivo* in the rat.[134] The metabolism of alprenolol labelled with four deuterium atoms on the propanol side-chain has been studied in dog and man using the ion doublet technique.[135] After oral administration of the drug twelve metabolites were isolated from rat bile and eight from dog and human urine. These were characterized as trifluoroacetyl, trimethylsilyl, or n-butylboronate derivatives using a g.c.–m.s. computer system. The drug was shown to be metabolized mainly by aromatic hydroxylation and alkyl oxidation but some oxidative deamination of the isopropylamine side-chain was also observed. Conjugation of alprenolol and 4-hydroxyalprenolol also occurred. To differentiate between the glucuronide of alprenolol and the two glucuronides of 4-hydroxyalprenolol the mixture of conjugates was submitted to chlordimethylsilylation when the cyclic silyl derivative was formed through interaction with the hydroxyl and amino groups of the aminopropanol side-chain of the 4-hydroxyalprenolol glucuronide. No isotopic effects were observed in the formation of the metabolites in which the aminopropanol group was unchanged. In those in which metabolism occurred at the side-chain, the deuterium-labelled drug was metabolized more slowly.

Two isomeric methoxyhydroxy metabolites of propanolol and similar metabolites with a glycol or a lactic acid side-chain have been identified by g.c.–m.s. analysis of urine from patients treated with the drug.[136]

Glucuronide metabolites of carbamazepine (5H-dibenz[*b,f*]azepine-5-carboxamide) were identified in human urine following chromatography on XAD resin, permethylation, and g.c.–m.s. using an SE-30 capillary column. Eight glucuronide metabolites, previously unidentified in man, were characterized as their permethylated derivatives.[137] These included carbamazepine *N*-glucuronide ($M^{\cdot+}$ at m/z 482), three isomers of hydroxymethoxycarbamezepine

---

[133] K. J. Hoffman, I. Skanberg, and K. O. Borg, *Xenobiotica*, 1979, **9**, 79.
[134] K. J. Hoffman, A. Arfwidsson, K. O. Borg, and I. Skånberg, *Xenobiotica*, 1979, **9**, 93.
[135] K. J. Hoffman, A. Arfwidsson, K. O. Borg, and I. Skånberg, *Biomed. Mass Spectrom.*, 1978, **5**, 634.
[136] T. Walle, E. C. Conradi, U. K. Walle, and T. E. Gaffney, *Drug Metab. Dispos.*, 1978, **6**, 481.
[137] R. K. Lynn, R. G. Smith, R. M. Thompson, M. L. Deinzer, D. Griffin, and N. Gerber, *Drug Metab. Dispos.*, 1978, **6**, 494.

O-glucuronide (M^{++} at m/z 542), and one isomer of hydroxycarbamazepine O-glucuronide (M^{++} at m/z 512). Other glucuronide metabolites, previously identified following enzymic hydrolysis, were characterized as unhydrolysed permethylated glucuronides, 10,11-dihydro-10,11-dihydroxycarbamazepine-O-glucuronide (M^{++} at m/z 544), and three isomers of monohydroxycarbamazepine-O-glucuronide (M^{++} at m/z 512).

Three major metabolites of cyclobenzaprine (78), which were present in urine of rats given the ^{14}C-labelled drug, were shown by g.c.–m.s. to be 3- and 2-hydroxycyclobenzaprine. Also identified as minor metabolites were the 10,11-epoxide of desmethyl cyclobenzaprine and a third phenolic metabolite. The phenolic metabolites were excreted as conjugates.[138]

$$HC[CH_2]_2N[CH_3]_2$$
(78)

The major human urinary metabolites of (2-cyclopentyl-6,7-dichloro-2-methyl-1-oxo-5-indanyloxy)acetic acid, a saluretic agent, have been studied in chimpanzee and man.[139] From g.c.–mass spectral analysis of the methyl ester TMS derivatives of four metabolites, it was concluded that three of the metabolites were hydroxylated at different sites in the cyclopentyl ring and the fourth was the ketone produced by oxidation of one of these alcohols.

The metabolism of ^{14}C-cannabinol has been studied in the rat.[140] Six acidic metabolites were identified in rat faeces by g.c.–m.s. and proton magnetic resonance spectroscopy. Cannabinol-7-oic acid was the major metabolite, the others in decreasing order of prominence being 1"-hydroxy-, 4"-hydroxy-, 3"-hydroxycannabinol-7-oic acid, cannabinol-3"-one-7-oic acid, and 2"-hydroxycannabinol-7-oic acid.

11-Hydroxy-Δ^9-tetrahydrocannabinol and 5'-hydroxy-Δ^9-tetrahydrocannabinol were incubated with UDP-glucuronyl transferase, immobilized on Sepharose, and two isomeric conjugates were formed from each compound.[141] Their structures were established by g.c.–m.s. of the methyl ester-trimethylsilyl and per(trimethylsilyl) derivatives using EI and CI techniques. The isomers corresponding to the phenol-linked glucuronide and the alcohol-linked glucuronide of both compounds were identified.

The following metabolites of etidocaine (79) were identified by g.c.–m.s. of the urine from a male volunteer given a 200 mg oral does of the drug: N-(2,6-dimethyl-3- and 4-hydroxyphenyl)-2-(NN-ethylpropylamino)butyramides,

---

[138] H. B. Hucker, S. C. Stauffer, A. J. Balletto, S. D. White, A. C. Zacchei, and B. H. Arison, *Drug Metab. Dispos.*, 1978, **6**, 659.
[139] A. G. Zacchei, T. I. Wishousky, B. H. Arison, and G. Hitzenberger, *Drug Metab. Dispos.*, 1978, **6**, 303.
[140] W. Yisak, M. Widman, and S. Agurell, *J. Pharm. Pharmacol.*, 1978, **30**, 554.
[141] S. Pallante, M. A. Lyle, and C. Fenselau, *Drug Metab. Dispos.*, 1978, **6**, 389.

N-(2,6-dimethyl-3- and 4-hydroxyphenyl)-2-aminobutyramides, N-(2,6-dimethyl-3- and 4-hydroxyphenyl)-2-(N-ethylamino)butyramides, and N-(2,6-dimethyl-3- and 4-hydroxyphenyl)-2-(N-propylamino)butyramides.[142]

(79)

Analysis by g.c.–m.s. of the urine from 12 persons taking amylobarbitone failed to show the presence of free N-hydroxyamylobarbitone, a previously postulated metabolite.[143] The major metabolite isolated from the bile and urine of rats given oral doses of ^{14}C-3-trifluoromethyl-α-ethylbenzhydrol was identified by direct probe and g.c.–m.s. as conjugates of 3-trifluoromethyl-4'-hydroxy-α-ethylbenzhydrol. Also present in the urine was the dihydroxy derivative.[144]

Three metabolites of ibufenac, 4-(2-carboxypropyl)phenylacetic acid, 4-(2-hydroxy-2-methylpropyl)phenylacetic acid, and 4-(1-hydroxy-2-methylpropyl)phenylacetic acid, have been identified by g.c.–m.s. in the urine of volunteers given an oral dose of the drug.[145]

Pyrolysis g.c.–m.s. was used to identify in urine the anti-inflammatory agents ibuprofen, fenprofen, naproxen, and ketoprofen.[146] The method was also used to detect metabolites of ibuprofen. A new metabolite present in the urine of a patient receiving 2-ethyl-2-methylsuccinimide[147] was identified by g.c.–m.s. as 2-carboxy-2-methylsuccinimide, formed as a result of ω1-hydroxylation of the ethyl group to form 2-(2-hydroxyethyl)-2-methylsuccinimide and oxidation of the hydroxy group of the latter to the carboxylic acid.

The metabolism of the enantiomers of 2-hydroxy-N-cyclopropylmethylmorphinan was studied in dog and man.[148] The partially purified metabolite present in urine and enzymic hydrolysis of urine were analysed directly, or as their trimethylsilyl derivatives, by g.c.–m.s. and it was established that conjugation of the isomers with glucuronic acid or sulphate occurred. The major metabolic pathway for the (−)-enantiomer was N-dealkylation and aliphatic hydroxylation, whilst the (+)-enantiomer underwent N-dealkylation and aromatic oxidation, followed by methylation of the metabolite. A study in man using the (−)-enantiomer suggested that man metabolized this isomer in a similar way to the dog.

[142] J. Vine, D. Morgan, and J. Thomas, *Xenobiotica*, 1978, **8**, 509.
[143] J. N. T. Gilbert, P. T. J. Nelmes, J. W. Powell, and S. Murray. *J. Pharm. Pharmacol.*, 1978, **30**, 595.
[144] M. Ledniczky, I. Szinai, K. Ujszaszi, S. Holly, V. Kemény, Gy. Mády, and L. Ötvös, *Arzneim.-Forsch.*, 1978, **28**, 673.
[145] J. E. Pettersen, *Biomed. Mass Spectrom.*, 1978, **5**, 488.
[146] W. J. Irwin and J. A. Slack, *Biomed. Mass Spectrom.*, 1978, **5**, 654.
[147] J. E. Pettersen, *Biomed. Mass Spectrom.*, 1978, **5**, 601.
[148] F. M. Vane, D. H. Ellis, J. Rao, M. Petrin, J. F. Blount, E. Mohacsi, A. Szuna, and J. J. Kamm, *Biomed. Mass Spectrom.*, 1978, **5**, 498.

The urine from patients treated with the antidepressant, loxapine (80), was extracted, derivatized with trifluoracetic anhydride and analysed by g.c.–m.s. Three metabolites were detected and, from their mass spectra, it was concluded that the metabolites were formed by hydroxylation of the aromatic ring, demethylation of the piperazine ring, and N-oxidation.[149]

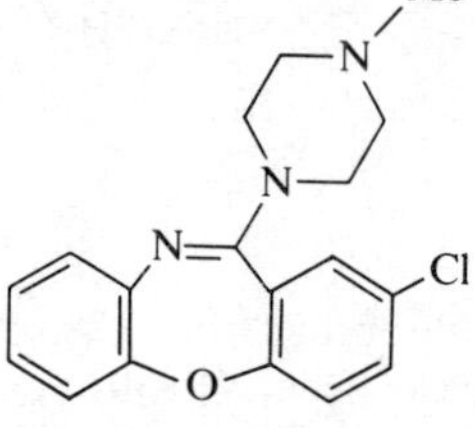

(80)

The sera and urine collected for two hours after an oral dose of sodium dipropylacetate, an anticonvulsant, to a human were anlysed by g.c.–m.s.[150] The compound was shown to be mainly metabolized by β-oxidation but some ω-oxidation also occurred.

An epoxide of the anti-inflammatory agent, alclofenac, was detected by g.c.–m.s. of the extracts of urine from mice treated with the drug.[151] The TBDMS ether derivative gave the best chromatographic separation of the chlorohydrin formed from the silylated epoxide and the silylated diol.[151]

The use of a g.c. column coated with a novel chiral polysiloxane type stationary phase, Chirasil-Val, for separating the enantiomers of several drugs and metabolites has been described.[152] The columns had high thermal stability and could be used for g.c.–m.s. The separation of the NO-pentafluoropropionyl derivatives of a number of enantiomeric sympathomimetic agents and their metabolites is described.

The metabolism of ^{14}C-isamoxole (81), an antiallergic agent, has been studied in rat and guinea pig.[153] From g.c.–m.s. analysis of urine extracts, it was established that the major routes of metabolism were deacylation, oxidative ring scission, and alkyl oxidation.

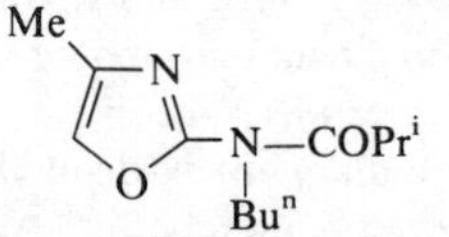

(81)

Gas chromatography–mass spectrometry was used to identify four aglycone metabolites in the butanol extract of a bile hydrolysate from a rabbit treated with an anticancer drug, doxorubicin.[154] The compounds were identified as their pertrimethylsilyl and methoxime derivatives.

[149] S. F. Cooper, R. Dugal, and M. J. Bertrand, *Xenobiotica*, 1979 **9**, 405.
[150] T. Kuhara, Y. Hirokata, S. Yamada, and I. Matsumoto, *Eur. J. Drug Metab. Pharmacokinet.*, 1978, **3**, 171.
[151] J. A. Slack and A. W. Ford-Hutchinson, *J. Pharm. Pharmacol*, 1978, **30**, 67.
[152] H. Frank, G. J. Nicholson, and E. Bayer, *J. Chromatogr.*, 1978, **146**, 197.
[153] D. H. Chatfield, J. N. Green, J. C. Kao, W. J. Ross, and T. J. Woodage, *Xenobiotica*, 1979, **9**, 585.
[154] K. K. Chan and E. Watson, *J. Pharm. Sci.*, 1978, **67**, 1748.

Chloral hydrate and [^{2}H$_6$]ethanol were administered orally to rats. When the urine was analysed by g.c.–m.s., 1-deuterio-2,2,2-trichloroethanol was detected. This was formed by a coupled redox reaction between ethanol, nicotinamide adenine dinucleotide, alcohol dehydrogenase, and chloral.[155] Two papers have appeared[156,157] which, although not dealing with drug metabolites, describe methods for preparing derivatives of sulphate esters which are suitable for g.c.–m.s.

The metabolism of the 3-[^{2}H] analogues of Δ^1-, Δ^6-, and Δ^7-tetrahydrocannabinol (THC) has been studied in mice.[158] The hydroxylated metabolites were isolated from the homogenates of the liver and purified from lipids by chromatography on Sephadex LH-20. They were converted either into their trimethylsilyl or alkane boronates and their structures determined by g.c.–m.s. Loss of the deuterium atom from the 3-position during mass spectral fragmentation was found to occur with the axial but not the equatorial isomers of some hydroxylated metabolites. The α-isomers of 5-hydroxy-Δ^6-THC, 6-hydroxy-Δ^1-THC, 1,2-dihydroxy-Δ^1-hexahydrocannabinol (HHC), and HHC-7-oic acid were identified as metabolites from the mass spectra of their derivatives. The stereochemistry of the 1,7-dihydroxy-HHC's was evaluated from mass spectra of their alkylboronates. This method was not applicable to some hydroxy metabolites such as the isomeric 1,6-dihydroxy-HHC's which appeared to fragment by ring cleavage before loss of deuterium.

[155] L. K. Wong and K. Biemann, *Biochem. Pharmacol.*, 1978, **27**, 1019.
[156] G. Paulson, M. Simpson, J. Giddings, J. Bakke, and G. Stolzenberg, *Biomed. Mass Spectrom.*, 1978, **5**, 413.
[157] S. Murray and T. A. Baillie, *Biomed. Mass Spectrom.*, 1979, **6**, 82.
[158] D. J. Harvey, *Biomed. Mass Spectrom.*, 1980, **7**, 28.

12
Organometallic, Co-ordination, and Inorganic Compounds Investigated by Mass Spectrometry

BY R. H. CRAGG

1 Introduction

This chapter reviews articles which have appeared in the literature between June 1978 and 31st May 1980. In view of the well established pattern of previous volumes the author has decided to keep to the previous format. Mass spectrometry has now become an established technique in the study of organometallic, co-ordination, and inorganic compounds although in many instances it is chiefly used as a means of determining the molecular weights of the compounds under investigation. More recently interest has increased in the applications of negative ion mass spectrometry, chemical ionization (CI), field desorption (FD), and high-temperature techniques. Throughout this chapter M^+ (or M^-) refers to the molecular ion formed by loss (or gain) of an electron.

Over the past two years a number of reviews have appeared together with a very comprehensive review on organometallic, co-ordination, and inorganic compounds covering the period June 1976 to 31st May 1978.[1] The study of trimethylsilyl (TMS) ether derivatives of compounds by mass spectrometry is well established and a review concerning the application of trialkylsilyl ether derivatives, other than TMS, for gas chromatography and mass spectrometry has been published.[2] A specialist review concerning the mass spectra of organoiron compounds[3] and also a report on the application of mass spectrometry to the determination of bond energies in transition-metal carbonyls[4] have appeared. The role of mass spectrometry in obtaining information concerning the molecular state of metal alcoholates, especially their degree of association, in the gas phase and condensed state[5] and fragmentation processes in methyl and ethyl derivatives of organophosphorus esters[6] have been reviewed. In addition, articles on negative ions,[7] mass spectrometry instrumentation for chemists and biologists,[8] and secondary ion mass spectrometry (SIMS)[9] have appeared.

---

[1] T. R. Spalding, in 'Mass Spectrometry', ed. R. A. W. Johnstone (Specialist Periodical Reports), The Chemical Society, London, 1979, Vol. 5, p. 312.
[2] C. F. Poole and A. Zlatkis, *J. Chromatogr. Sci.*, 1979, **17**, 115.
[3] J. Mueller, in 'Organic Chemistry of Iron', ed. E. A. K. von Gustoft, F. W. Grevels, and I. Fischler, Academic Press, New York, 1978, **1**, 145.
[4] G. D. Michels, *Diss. Abstr. Int. B*, 1979, **39**, 3857.
[5] N. I. Kozlova, Ya. N. Turova, and Yu. S. Nekrasov, *Koord. Khim.*, 1980, **6**, 221.
[6] C.-S. Hwang, *Diss. Abstr. Int. B*, 1979, **40**, 741.
[7] K. R. Jennings, *Philos. Trans. R. Soc. London, Ser. A*, 1979, **293**, 125.
[8] R. D. Craig, R. H. Bateman, B. N. Green, and D. S. Millington, *Philos. Trans. R. Soc. London, Ser. A*, 1979, **293**, 135.
[9] W. L. Baun, Tech. Rep. AFML-TR-Air Force Mater. Lab., 1980, TR-79-4123.

2 Main-group Organometallics

Group II.—The spectra of ketenide derivatives of mercury [*e.g.* $(AcOHgHg)_2C=C=O$] suggested that the compounds had undergone complete thermal decomposition[10] and the hard/soft acid–base theory has been successfully applied to halogen transfer observed in the spectra of diarylmercury compounds of the type $(C_6X_5)_2Hg$ (X = F, Cl, or Br).[11]

Group III.—The spectrum of tri-n-propylborane has been reported[12] and the molecular ion in the spectrum of dimesityldiphenylmethylborane has been observed.[13] In order to investigate the influence of the co-ordination number of the boron atom on fragmentation, a comparison has been made between the spectra of diphenylboron chelates [*e.g.* compounds (1) and (2)] with the analogous monophenylboron heterocycles [*e.g.* compounds (3) and (4)].[14] It was observed

that the chelates preferentially eliminate a phenyl substituent from the boron atom and it is this highly abundant fragmentation with three-co-ordinate boron which causes further fragmentation. In addition, it was found that the co-ordination number of the boron atom had no essential influence on the fragmentation process of the boron-containing ring system. All the phenylboron heterocycles studied gave highly intense molecular-ion signals in contrast to the chelates.

The spectra of the diphenylboron chelates (5)[15] and (6)[16] have been reported and in a study of chelates (7) and (8) the main fragmentation path was observed to involve loss of the alkyl group attached to boron.[17] In addition, the acetoxy

---

[10] E. T. Blues, D. Bryce-Smith, and H. Karimpour, *J. Chem. Soc., Chem. Commun.*, 1979, 1043.
[11] J. M. Miller, T. R. B. Jones, and G. B. Deacon, *Inorg. Chim. Acta*, 1979, **32**, L75.
[12] T. K. Postnikova, O. N. Druzhkov, and V. N. Glushakova, *Khim. Elementoorgan. Soedin.*, 1977, 59.
[13] N. M. D. Brown, F. Davidson, and J. W. Wilson, *J. Organomet. Chem.*, 1980, **185**, 277.
[14] E. Hohaus and W. Riepe, *Int. J. Mass Spectrom. Ion Phys.*, 1979, **31**, 113.
[15] A. S. Fletcher, W. E. Paget, K. Smith, K. Swaminathan, J. H. Beynon, R. P. Morgan, M. Bozorgzadek, and M. J. Haley, *J. Chem. Soc., Chem. Commun.*, 1979, 347.
[16] A. S. Fletcher, W. E. Paget, K. Smith, K. Swaminathan, and M. J. Haley, *J. Chem. Soc., Chem. Commun.*, 1979, 573.
[17] B. M. Zolotarev, O. S. Chizhov, A. S. Bleshinskaya, O. G. Boldyreva, V. A. Dorokhov, and B. M. Mikhailov, *Izv. Akad. Nauk SSSR, Ser. Khim.*, 1979, 226.

(7) R^1 = Pr, Bu, Me$_2$CHCH$_2$CH$_2$, or Cl
   R^2 = Pr, Bu, OH, OAc, or Cl
   R^3, R^4 = H, Me, or Et

(8) R^1 = Bu, OAc, or Cl
   R^2 = H or Me

and hydroxy derivatives were found to eliminate acetic acid and water, respectively.

Molecular ions have been observed in the mass spectra of a series of 2-allyl-1,3,2-dioxaborolanes.[18] Chemical ionization-induced fragmentations (with 2-methylpropane as reagent gas) have been reported for heterocycles (9) and (10).[19] Under the conditions of the study, the boron compound (9; $R^1 = R^2 = H$) and (10; $R^1 = R^2 = H$) did not fragment but gave only [MH]$^+$ ions. The spectra of a series of 2-p-substituted phenyl-1,3,2-dioxaborolanes (11) have been reported[20] and compared with the spectrum of the parent compound. It was found that the formation of the substituted tropylium ion [C$_7$H$_6$R]$^+$ by electron impact-induced rearrangement was influenced by the presence and nature of R.

(9)   (10)   (11) R = Cl, Br, Me, or MeO

The spectra of a series of dioxazaboracycloalkanes (12; R^1 = H, Me, or Me$_3$C; R^2 = H or Me; m, n = 1, 2) and their acyclic analogues, RNH(CH$_2$)$_n$CH$_2$OBPh$_2$ (R = H, n = 2; R = Me, n = 1), have been reported.[21] For the former (12), two primary fragmentation pathways were observed, namely cleavage of the B—Ph bond and exo- or endo-cyclic C—N cleavage. Also, transannular N → B interaction was demonstrated except for the case where R^1 = Me$_3$C. The spectra of a series of pinacol derivatives of aromatic boronic acids have been discussed[22] and the spectra of butyl boronate derivates of aldosterone,[23] inositol,[24] and benzene boronate derivatives of acyclic tetraols[25] have been reported. The molecular ion (75.4% relative abundance) has been

---

[18] R. W. Hoffmann, G. Feussner, H. J. Zeiss, and S. Schulz, *J. Organomet. Chem.*, 1980, **187**, 321.
[19] R. H. Hancock and H. Weigel, *Org. Mass Spectrom.*, 1979, **14**, 196.
[20] J. J. Kaminski and R. E. Lyle, *Org. Mass Spectrom.*, 1978, **13**, 425.
[21] I. Mazeika, A. P. Gaukhman, I. Urtane, G. Zelcans, and E. Lukevics, *Zh. Obshch. Khim.*, 1979, **49**, 2528.
[22] S. Singhawangcha, L. E. C. Hu, C. F. Poole, and A. Zlatkis, *J. High Resolut. Chromatogr., Chromatogr. Commun.*, 1978, **1**, 304.
[23] S. J. Gaskell and C. J. W. Brooks, *J. Chromatogr.*, 1978, **158**, 331.
[24] J. Wiecko and W. R. Sherman, *J. Am. Chem. Soc.*, 1979, **101**, 979.
[25] C. J. Griffiths, I. R. McKinley, and H. Weigel, *Carbohydr. Res.*, 1979, **72**, 35.

$$R^1N \begin{matrix} (CH_2)_m-CHR^2O \\ (CH_2)_n-CHR^2O \end{matrix} BPh \qquad PhB \begin{matrix} H & H \\ N-N \\ N-N \\ H & H \end{matrix} C=O$$

(12) (13)

observed in the spectrum of the carbohydrazide (13)[26] and results of a study of boron acylamidinates and acylaminopyridinates have been published.[27]

Compounds of Other Elements in Group III. The molecular ion of the ether complex of the aluminium heterocycle (14) has been observed.[28] In a study of the lower alkyl derivatives of aluminium, Et_3Al, $(Me_2CHCH_2)_3Al$, Et_2AlH, $(Me_2CHCH_2)_2AlH$, and Et_2AlCl, it was observed that the primary fragmentation path involved the loss of one substituent as a radical.[29] The probability of loss increased in the order $Cl < Et < H \sim Me_2CHCH_2$. All ions from the trialkylalanes arose from monomeric molecules but decomposition of other compounds involved dimeric species.

(14)

The results of a study of a number of 2-, 3-, and 4-substituted triarylgallanes [*e.g.* $(2\text{-}MeC_6H_4)_3Ga$] and triarylindanes have been reported.[30] The electron impact mass spectra of these compounds were found to be very similar and reflected the basic structural features of the molecules. Using carefully controlled conditions it was possible to obtain spectra in which the metal-containing ions were the major species with little contamination from hydrolysis or pyrolysis products. The results provided evidence that, for these compounds, dimers were significant gaseous species. The results of another study of alkyl derivatives of gallium have been reported[31] and it has been observed that the spectra of arylhalogallanes Ph_nGaX_{3-n} (X = Cl, Br, or I; n = 1 or 2) were consistent with the existence of halogen-bridged dimers.[32] The molecular ion, with Tl^+ as the base peak, was observed in the spectrum of C_5D_5Tl[33] and, using FD techniques, thallium has been detected at levels of 10^{-11} g in biological materials.[34]

[26] J. Bielawski and K. Niedenzu, *Inorg. Chem.*, 1980, **19**, 1090.
[27] B. M. Zolotarev, V. A. Dorokhov, O. Chizhov, L. I. Lavrinovich, and B. M. Mikhatlov, *Izv. Akad. Nauk SSSR, Ser. Khim.*, 1979, 80.
[28] H. Hoberg and F. Azwar, *J. Organomet. Chem.*, 1979, **164**, C13.
[29] V. N. Bochkarev, A. I. Belokan, N. N. Korneev, Yu, L. Lelyukhina, and I. M. Khrapova, *Zh. Obshch. Khim.*, 1979, **49**, 2294.
[30] S. B. Miller, B. L. Telus, H. J. Smith, B. Munson, and T. B. Brill, *J. Organomet. Chem.*, 1979, **170**, 9.
[31] T. K. Postnikova, O. N. Druzhkov, V. I. Bregadze, G. I. Makin, and V. A. Dodonov, *Khim. Elementoorgan. Soedin.*, 1978, 67.
[32] S. B. Miller and T. B. Brill, *J. Organomet. Chem.*, 1979, **166**, 293.
[33] G. K. Anderson, R. J. Cross, and I. G. Phillips, *J. Chem. Soc., Chem. Commun.*, 1978, 709.
[34] C. Achenbach, R. Ziskoven, F. Koehler, U. Bahr, and H. R. Schulten, *Angew. Chem.*, 1979, **91**, 944.

Group IV.—The spectra of a series of silyl and germyl acetylenes and halogeno-acetylenes ($H_3SiCCCl$, $H_3SiCCBr$, $H_3GeCCCl$, $H_3SiCCSiH_3$, and $H_3GeCCGeH_3$) were reported and the molecular ion was observed for each compound.[35] A major breakdown path for the halogenoacetylenes involved the loss of halogen followed by M—C bond cleavage. An alternative path, supported by a metastable ion, was the extrusion of C_2 leaving $[M_xH_n]^+$. The fragmentation patterns of bis(trimethylsilyl)- and bis(trimethylgermyl)-carbodi-imide have been published.[36] For the former compound, the elimination of neutral fragments containing Si=N and/or Si=C bonds were observed and confirmed by the presence of metastable ions. Although the fragmentation pathways are similar for the germanium analogue no evidence of a fragment containing a Ge=C bond was observed. In the spectra of compounds of the type o-$C_6H_4(C{\equiv}CMMe_3)_2$ (M = Si, Ge, or Sn) the $[Me_3M]^+$ ion was the base peak;[37] the intensities of the $[M]^+$ ions were 1.5, 5.0, and 18%, respectively. Ion cyclotron resonance (ICR) spectroscopy has been used to obtain a threshold for the gas-phase enthalpy of deprotonation of Me_3Si^+ and has enabled a heat of formation for $Me_2Si{=}CH_2$ (20.5 ± 4 kcal mol^{-1}) to be calculated.[38] The polysilanes $(Me_2Si)_7$, $(Me_2Si)_8$, and $(Me_2Si)_9$ have been identified.[39] The formation of the main even-electron fragment ions in the spectra of $Me_mCl_nSi(CH{=}CH_2)_{4-m-n}$ (m, n = 0—3) were found to arise from the loss of one of the substituents at silicon followed by the ejection of C_2H_2.[40] The odd-electron fragment ions became more abundant as the number of vinyl groups increased and it was suggested that they originated from cyclic species, especially silacyclopentenes. The results of a study of the negative ion spectra of a series of alkenylsilanes suggested that the silicon d-orbitals were involved in stabilizing the fragmentary ions.[41] In a study of $RSiMe_2SiMe_2CH{=}CH_2$ (R = H, Me, vinyl, allyl, or Ph), three main fragmentation pathways were identified, namely (i) elimination of a free radical without Si—Si bond cleavage, (ii) simple Si—Si bond cleavage, and (iii) Si—Si bond cleavage accompanied by rearrangement (intramolecular disproportionation).[42] The spectra of a series of alkenyl, alkadienyl, and cycloalkenyl silanes have been reported.[43] In the acyclic compounds the occurrence of an intramolecular hydrosilylation process was proposed. In addition, the position of the double bond in alkadienylsilanes was identified. The ICR spectra of a series of systems $Me_4Si–R^1OR^2$ (R^1OR^2 = linear or cyclic dialkyl ether) have been reported.[44] It was observed that nucleophilic

---

[35] S. Cradock, E. A. V. Ebsworth, and A. R. Green, *J. Chem. Soc., Dalton Trans.*, 1978, 759.
[36] J. E. Drake, B. M. Glavincevski, H. E. Henderson, and C. Wong, *Can. J. Chem.*, 1979, **57**, 1162.
[37] R. Nast and H. Gronhi, *J. Organomet. Chem.*, 1980, **186**, 207.
[38] W. J. Pietro, S. K. Pollack, and W. J. Hehre, *J. Am. Chem. Soc.*, 1979, **101**, 7126.
[39] K. Matsumura, L. F. Brough, and R. West, *J. Chem. Soc., Chem. Commun.*, 1978, 1092.
[40] A. N. Polivanov, T. F. Slyusarenko, B. I. Zhun, V. N. Bockharev, and B. J. Sheludyakov, *Zh. Obsch. Khim.*, 1979, **49**, 1311.
[41] V. P. Yuriev, A. A. Panasenko, V. I. Khvostenko, M. L. Khalilov, Ya. B. Yasman, M. M. Timoshenko, Yu, V. Cnizhov, B. G. Zykov, I. I. Furlei, and S. R. Rafikov, *J. Organomet. Chem.*, 1979, **166**, 169.
[42] A. N. Polivanov, A. A. Bernadskii, V. I. Zhun, and V. N. Bockarev, *Zh. Obshch. Khim.*, 1978, **48**, 2703.
[43] V. N. Bochkarev, E. G. Galkin, I. M. Salimgareva, and E. M. Vyrypaev, *Izv. Akad. Nauk SSSR, Ser. Khim.*, 1979, 217.
[44] V. C. Trenerry, J. H. Bowie, and I. A. Blair, *J. Chem. Soc., Perkin Trans. 2*, 1979, 1640.

attack of R^1OR^2 on Me_3Si^+ gave $Me_3Si\overset{+}{O}R^1R^2$, which generally decomposed by elimination of a neutral fragment; for example, when R > Me elimination of RH was observed. Elimination of Me_3SiF was observed from the complex containing $MeO(CH_2)_2F$. The allyldimethylsilyl cation was obtained from diallyldimethylsilane by the ICR technique.[45] A study was made of the reaction of this cation with alcohols and ethers. With the former, adducts were obtained which decomposed by loss of C_3H_6 to give the $[Me_2SiOR]^+$ ion. The silylcyclopentadienes CpSiR (R = H_3, ClH_2, Cl_2H, or Cl_3) have been characterized.[46] Molecular ions were observed for Cp_4Si, Cp_4Ge, $(MeC_5H_4)_4Si$, and $(MeC_5H_4)_4Ge$ and it was suggested that the low intensity of the molecular ion peak for the silicon compounds indicated a low Si—C bond energy.[47] In a study of alkyl-substituted silacyclobutanes (15) it was shown that fragmentation occurred by successive loss of two C_2H_4 molecules, the first from ring cleavage and the second from the R^3 group.[48] However, when R^1 = Me and R^2, R^3 = Me, Cl, the fragmentation occurred via loss of a $C_3H_5{}^{\cdot}$ radical and a C_2H_4 molecule. The EI fragmentation spectra of a series of 9-sila-9,10-dihydro-3-azaanthracenes (16) have been reported.[49] The main fragmentation route for these compounds was the formation of the $[M - RH]^+$ ions when R is a substituent at the silicon atom. It was found also that randomization occurred between the H atoms at the C-10 position and the H atoms of the pyridine and benzene rings.

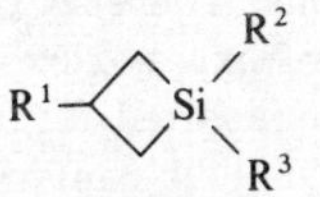

(15) 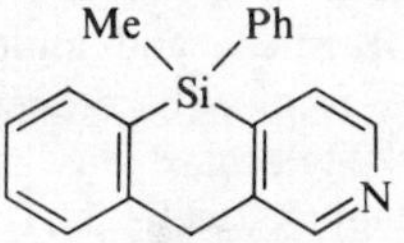 (16)

Various organosilane heterocycles (17),[50] (18),[50] (19),[51] and (20)[51] have been characterized and their respective molecular ions observed. The ready loss of Me_3SiX from compounds of the type $(Me_3Si)_3CSiPh_2X$ (where X = MeO, F, or Br) resulted in ions of moderate abundance at m/z 340 (assigned to $[Me_3Si]_2C=SiPh_2$), suggesting that these compounds may give silaolefin on

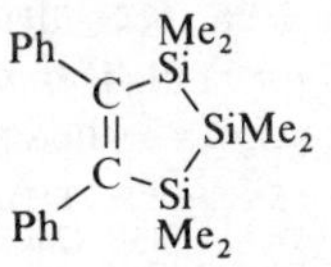

(17) 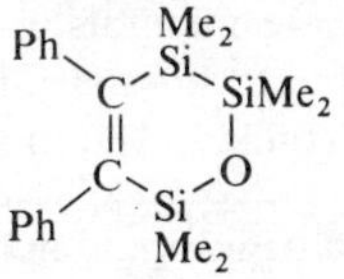 (18)

[45] I. A. Blair, V. C. Trenerry, and J. H. Bowie, *Org. Mass Spectrom.*, 1980, **15**, 15.
[46] A. Bonny, S. R. Stobart, and P. C. Angus, *J. Chem. Soc., Dalton Trans.*, 1978, 938.
[47] A. D. McMaster and S. R. Stobart, *Inorg. Chem.*, 1980, **19**, 1179.
[48] V. G. Zarkin, A. I. Mikaya, V. M. Vdovin, E. D. Babich, V. F. Traven, and B. I. Stepanov, *Zh. Obshch. Khim.*, 1979, **49**, 1307.
[49] N. S. Prostakov, N. Saxena, P. I. Zakharov, and A. V. Varlamov, *J. Organomet. Chem.*, 1980, **184**, 167.
[50] H. Watanabe, K. Higuchi, M. Kobayashi, and Y. Nagai, *J. Chem. Soc., Chem. Commum.*, 1979, 1029.
[51] C. Eaborn, D. A. R. Happer, P. B. Hitchock, S. P. Hopper, K. D. Safa, S. S. Washburne, and D. R. M. Walton, *J. Organomet. Chem.*, 1980, **186**, 309.

(19)   (20)

thermolysis.[51] The spectra of a series of β-keto silanes, p-RC$_6$H$_4$Me$_2$SiCH$_2$C(O)Me (R = H, Me, MeO, Cl, F, or CF$_3$), and their isomeric silyl enol ethers, p-RC$_6$H$_4$Me$_2$SiOC(Me)=CH$_2$, have been reported.[52] It was suggested that the compounds form a common ion structure on electron impact which decomposes by unimolecular ion decomposition. The fragmentation pathways of a series of α-silyl ketones (*e.g.* Et$_3$SiCOPh) have been compared with those of triethylsilanol and 2-phenyl-3-(triethylsilyl)oxiran.[53] For the silyl ketones, migration of phenyl groups to silicon was observed. In the spectra of ·RC$_6$H$_4$SiMe$_2$F (R = p-Me$_2$N, p-MeO, p-Cl, m-Cl, m-Br, or p-O$_2$N), the elimination of a methyl radical from the molecular ion was a major process.[54] The activation energies for this fragmentation were measured and it was suggested that the change in this parameter was due to the influence of the substituent on the charge distribution in the molecular ion. The substituent also affected the abundances of fragment and molecular ions. The spectra of a number of trimethylsilyl ethers of β-diketones and hydroxy and oxo-β-diketones have been reported.[55] Molecular ions have been reported for X(CH$_2$)$_3$SiOEtMe$_2$, p-XC$_6$H$_4$SiOEtMe$_2$ (X = OH, NH$_2$, or COOH),[56] 1-napPhSiOEtX (X = Et or p-tolyl),[57] Et$_2$Si(OEt)NEtSiH(OEt)$_2$, Ph$_2$Si(OEt)CH$_2$SiH(OEt)$_2$, and Et$_2$Si(OMe)NEtSiMe(OMe)$_2$.[58] The spectra of o-methyl and o-t-butyldimethylsilyloximes of a number of long-chain fatty acids have been reported.[59] In a study of the spectra of alkyl(silylmethyl) esters, ether cleavages induced by an anchimeric assistance of silyl functions, elimination of bivalent silicon species, and other unusual fragmentation paths (*e.g.* carbene formation) have been observed.[60] The molecular ion for Me$_3$SiN(PF$_2$)$_2$ has been observed.[61] TMS derivatives of silicate minerals and glasses have been detected by g.c.–m.s.[62] and, using the same technique, TMS derivatives of hemimorphite[63] and dioptase[64] have been studied. The molecular ions of the heterocycles (21), (22),[65]

[52] R. S. C. Tsai, G. L. Larson, and A. Oliva, *Org. Mass Spectrom.*, 1979, **14**, 364.
[53] B. Y. K. Ho, E. M. Dexheimer, L. Spialter, and L. D. Smithson, *Org. Mass Spectrom.*, 1979, **14**, 185.
[54] G. Dube and E. Gey, *Org. Mass Spectrom.*, 1979, **14**, 17.
[55] A. P. Tulloch and L. R. Hogge, *J. Chromatogr.* 1978, **157**, 291.
[56] B. N. Ghose, *J. Organomet. Chem.*, 1979, **164**, 11.
[57] C. Breliere, R. J. P. Corriu, A. De Saxce, F. Larcher, and G. Royo, *J. Organomet. Chem.*, 1979, **164**, 19.
[58] M. Elsheikh, N. R. Pearson, and L. H. Sommer, *J. Am. Chem. Soc.*, 1979, **101**, 2491.
[59] G. Phillipou and A. Poulos, *Chem. Phys. Lipids*, 1978, **22**, 51.
[60] H. Schwarz, C. Wesdemiotis, and M. T. Reetz, *J. Organomet. Chem.*, 1978, **161**, 153.
[61] W. Kruger and R. Schmutzler, *Inorg. Chem.*, 1979, **18**, 871.
[62] S. K. Sharma and T. C. Hoering, Year Book Carnegie Inst. Washington, 1977 (pub. 1978), 662.
[63] K. Kuroda and C. Kato, *J. Chem. Soc., Dalton Trans.*, 1979, 1036.
[64] H. P. Calhoun and C. R. Masson, *J. Chem. Soc., Dalton Trans.*, 1978, 1342.
[65] W. Ando and M. Ikeno, *J. Chem. Soc., Chem. Commun.*, 1979, 655.

(23),[66] and (24)[66] have been observed and for compound (24) was the base peak. Compounds of the type (25) and (26)[67] have been characterized by mass spectrometry.

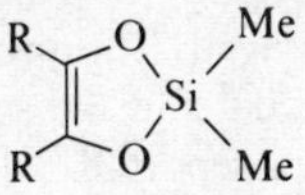

(21) R = Me, Et, Pr^n, or Pr^i

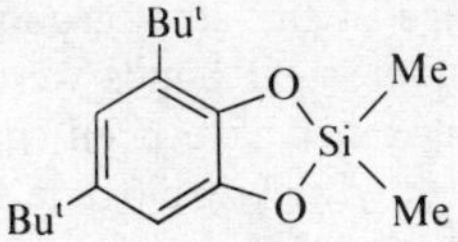

(22)

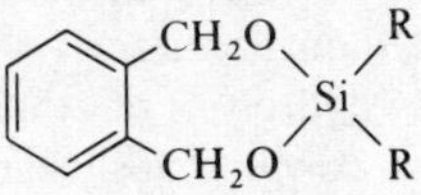

(23) R = Me or Ph

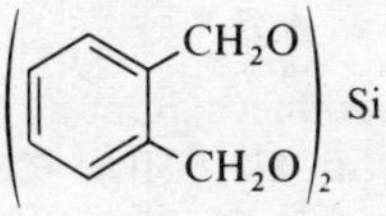

(24)

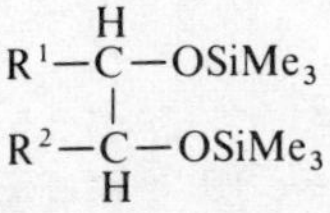

(25)

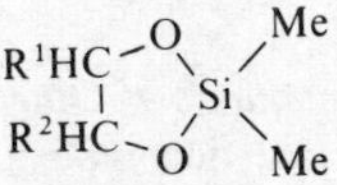

(26)

In the spectra of compounds of type (27), fragmentation was observed for both rings and the extent of N → Si transannular interaction was obtained from a study of the ratio of certain peaks in the spectra.[68] The spectra of some C-substituted silatranes have been reported.[69] The fragmentation of compounds of type (28) was observed to proceed *via* cleavage of the Si—R bond and cleavage of the silatrane moiety.[70] In the spectrum of compound (29), a transannular N → Si interaction was observed for R^3 = alkyl in contrast to (30) where there was no O → Si transannular interaction.[71] In the spectra of compound (31) and (32), fragmentation was found to proceed by loss of the substituent on silicon followed by cleavage of the rings.[72] In the case of (31; R^2 = Me),

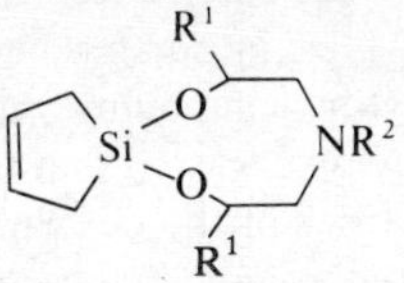

(27)

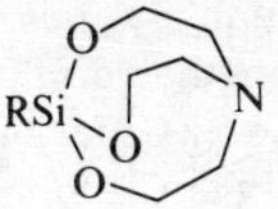

(28) R = vinyl, CH_2=CCl, C≡CH, or Ph

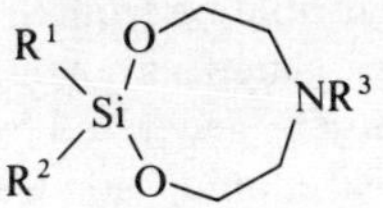

(29)

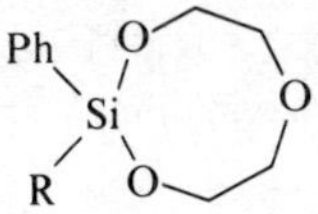

(30)

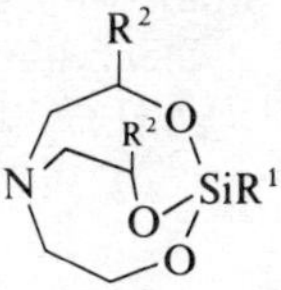

(31)

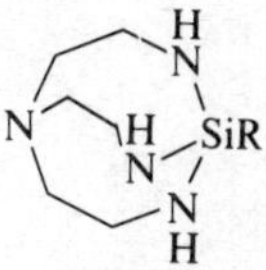

(32)

[66] L. Birkofer and O. Stuhl, *J. Organomet. Chem.*, 1979, **164**, Cl.
[67] L. Birkofer and O. Stuhl, *J. Organomet. Chem.*, 1979, **187**, 21.
[68] V. N. Bochkarev, A. A. Bernadskii, A. N. Polivanov, N. G. Komalenkova, S. A. Bashkirova, A. G. Popov, V. V. Antipova, and E. A. Chesnyshev, *Zh. Obshch. Khim.*, 1978, **48**, 2700.
[69] A. P. Gaukhman and I. Solomennikova, *Sint. Issled. Biol. Soedin., Tezisy Dokl. Konf. Molodykh. Uch., 6th*, 1978, 96.
[70] M. G. Voronkov, V. Yu, Vitkovskii, and V. P. Baryshok, *Izv. Akad. Nauk SSSR, Ser. Khim.*, 1979, 626.
[71] I. Mezeika, A. P. Gaukhman, I. Urtane, G. Zelcans, and E. Lukevics, *Zh. Obshch. Khim.*, 1979, **49**, 1327.
[72] I. Mazeika, A. P. Gaukhman, I. Jankovska, G. Zelcans, I. I. Solomennikova, and E. Lukevics, *Zh. Obshch. Khim.*, 1978, **48**, 2722.

it was suggested that ring enlargement took place by inclusion of the methyl groups, prior to fragmentation.

It has been observed that fragmentation of $(RSiO_{1.5})_n$ (R = Me or Et; $n = 8$ or 10) involved only cleavage of the alkyl groups from the Si—O skeleton[73] although where R = vinyl the Si—O bonds were also cleaved. The products of hydrolysis of ethylcyclosiloxanes have been characterized[74] and the spectra of ethylbicyclosiloxane and ethyltricyclosiloxane have been reported.[75] It was observed that fragmentation of bicyclic ions was accompanied by the formation of more stable polycyclic ions. The hydrolysis products of ethyldichlorosilane have been characterized.[76]

Compounds of Other Elements in Group IV. A quasi-radical mechanism for the thermal decomposition of R_4Ge (R = Me, Et, Pr, Bu, or $Me_2CHCH_2CH_2$), based on product analysis, has been proposed.[77] The bond-breaking sequence in the thermal decomposition was found to be analogous to that caused by EI. The results of a detailed investigation of a series of germatrans (33; R = Cl, CH_2=CH, HC≡C, hexylethynyl, PhC≡C, or Ph) have been reported.[78]

(33)

It has been observed that the spectrum of bis(2-thio-5-nitropyridine)-S-di-n-butylstannane was greatly influenced by strong Sn-ligand bonding reflecting a contribution from Sn—N interaction.[79] A benzenonium $[M + 57]^+$ ion has been postulated as an intermediate in the reaction of benzyltin compounds with Me_3C^+, generated from isobutane by CI techniques.[80] It was suggested that the benzenonium ion is stabilized by the C—Sn σ-bond electrons of the CH_2SnR_3 group and undergoes fragmentation to give p-t-butyltoluene and hence the p-t-butylbenzyl cation as well as the corresponding R_3Sn cation. In addition, it was observed that the loss of a proton from the benzenoxonium ion resulted in the formation of a p-t-butylbenzyltin compound. The spectra of $Cl_3SnCH_2CH_2COOMe$, $Cl_2Sn(CH_2CH_2COOMe)_2$, $Cl_2Sn(CH_2CH_2CONH_2)_2$,[81] and some tin derivatives of N-acylhydroxylamines (*e.g.*

---

[73] M. G. Vorenkov, V. I. Lavrentev, A. N. Kanev, V. G. Kostrovskii, and S. A. Prokhorova, *Dokl. Akad. Nauk SSSR*, 1979, **249**, 106.
[74] V. I. Lavrentev and V. G. Kostrovskii, *Zh. Obshch. Khim.*, 1979, **49**, 2013.
[75] V. I. Lavrentev, V. M. Kovrigin, and V. G. Kostrovskii, *Izv. Sib. Otd. Akad. Nauk SSSR, Ser. Khim. Nauk*, 1979, 126.
[76] V. I. Lavrentev and V. G. Kostrovskii, *Izv. Sib. Otd. Akad. Nauk SSSR, Ser. Khim. Nauk*, 1979, 134.
[77] T. A. Sladkova, O. P. Berezhanskaya, B. M. Zolotarev, and G. A. Razuvaev, *Izv. Akad. Nauk SSSR, Ser. Khim.*, 1978, 1316.
[78] M. G. Voronkov, R. G. Mirskov, A. L. Kuznetsov, and V. Yu. Vitkovskii, *Izv. Akad. Nauk SSSR, Ser. Khim.*, 1979, 1846.
[79] G. Domazetis, B. D. James, M. F. Mackay, and R. J. Magee, *J. Inorg. Nucl. Chem.*, 1979, **41**, 1555.
[80] R. H. Fish, R. L. Holmstead, M. Gielen, and B. De Poorter, *J. Org. Chem.*, 1978, **43**, 4969.
[81] P. G. Harrison, T. J. King, and M. A. Healy, *J. Organomet. Chem.*, 1979, **182**, 17.

Me$_2$Sn[ON(Ph)COPh]$_2$)[82] have been reported, and concentrations of Ph$_3$SnCl in soils at levels of 0.01 p.p.m. have been detected.[83] The heterocycles (34) and (35) have been identified[84] and the spectra of a series of trifluoromethyltin derivatives (CF$_3$)$_R$SnMe$_m$Br$_n$ (R = 1 or 2; m = 0, 1, 2, or 3; n = 0, 1, 2, or 3) have been reported.[85]

$$\text{SnPh}_2 \qquad \text{Ph}_2\text{Sn} \qquad \text{SnPh}_2$$

(34) \qquad\qquad (35)

Group V.—The spectra of Group VA ligands of the type o-C$_6$H$_4$E^{1}Me$_2$E^{2}Me$_2$ (E^1 = E^2 = P, As, N, or Sb) have been reported.[86] The spectra of Ph$_2$PMe, Me$_3$PO, Et$_3$PO, Bu$_3$PO, Ph$_2$P(O)Me, Ph$_3$PO, Ph$_2$P(S)Me, Ph$_3$P$^+$MeBr$^-$, Ph$_2$PMeI$^-$, and Ph$_4$P$^+$Cl$^-$ have been studied and an isotopic analysis was made of the vacuum pyrolysis products of the quaternary phosphonium salts.[87] The spectra of diphenylphosphino- and diarylphosphoro-hydrazides have been investigated.[88] The molecular ion of the former fragmented to give the corresponding phenol ion as the base peak. A series of diphenyl-di-p-tolyl- and di-t-butyl-methyleneamido derivatives of phosphorus (PCl$_2$L, PL$_3$, Ph$_2$PL, PL$_3$O, PClL$_2$O, and PCl$_2$LO; L = N = CR$_2$; R = Ph, p-tolyl, or Me$_3$C) have been characterized.[89] The spectra of dialkyl- and diaryl-phosphinocyclo-octatetraenes have been reported[90] and the full 70 eV spectra of a series of cyclopentadienylfluorophosphines and their borane complexes investigated.[91] In the spectrum[92] of the zwitterion F$_5$PCH$_2$PF(NMe$_2$)$_2$ a peak at m/z 259 was assigned to C$_5$H$_{14}$N$_2$P$_2$F$_5$$^+$. The spectra of 28 alkylphosphonates have been recorded.[93] It was observed that Et, Me$_2$CH, and Me$_3$C groups attached to phosphorus fragmented to the corresponding alkenes. In addition, diethyl-alkenephosphonates underwent double H rearrangements from an ethoxy group resulting in the formation of dihydroxyphosphonium ions. A characteristic fragmentation involving the loss of water was found for di- and tri-hydroxyphosphonium ions. A comparison has been made between the EI and CI spectra of MeP(O)(OP$_2$')$_2$, EtP(O)(OEt)$_2$, MeP(O)(OMe)$_2$, and MeFP(O)OPr'.[94] Methane was found to be a useful reagent gas for obtaining [MH]$^+$ ions and ethylene and isobutane gave protonated molecular ions as base peaks for all

---

[82] P. G. Harrison and J. A. Richards, *J. Organomet. Chem.*, 1980, **185**, 9.
[83] R. Kuivahara, T. Suzuki, and H. Meguno, *Agric. Biol. Chem.*, 1980, **44**, 669.
[84] A. G. Davies, M.-W. Tse, J. D. Kennedy, N. McFarlane, G. S. Pyne, M. F. C. Ladd, and D. C. Povey, *J. Chem. Soc., Chem. Commun.*, 1978, 791.
[85] J. L. Krause and J. A. Morrison, *Inorg. Chem.*, 1980, **19**, 604.
[86] W. Levason, K. G. Smith, C. A. McAuliffe, P. F. McCullough, R. D. Sedgwick, and S. G. Murray, *J. Chem. Soc., Dalton Trans.*, 1979, 1718.
[87] V. I. Mosichev and B. B. Alipov, *Khim. Tekhal. Izot. Mechenykh Soedin.*, 1977, 19.
[88] M. El-Deck, *J. Chem. Eng. Data*, 1980, **25**, 171.
[89] B. Hall, J. Keable, R. Snaith, and K. Wade, *J. Chem. Soc., Dalton Trans.*, 1978, 986.
[90] A. W. Spiegl and R. D. Fischer, *Chem. Ber.*, 1979, **112**, 116.
[91] R. T. Paine, R. W. Light, and D. E. Maier, *Inorg. Chem.*, 1979, **18**, 368.
[92] A. H. Cowley and R. C.-Y. Lee, *Inorg. Chem.*, 1979, **18**, 60.
[93] W. R. Griffiths and J. C. Tebby, *Phosphorus Sulfur*, 1978, **5**, 101.
[94] S. Sass and T. L. Fisher, *Org. Mass Spectrom.*, 1979, **14**, 257.

compounds including those for which the [MH]$^+$ ion was in low abundance in the methane spectra. Ethylene was found to be superior to isobutane insofar as it acted both as a carrier and a reagent gas and gave better sensitivity. The EI and FI spectra of 3-oxo-2-butylphosphate and its mono and dimethyl esters show components which arose from thermal reactions taking place both during sample storage and in the instrument.[95] Prominent molecular ion peaks and high P—N bond stability were observed for a series of phosphorus-containing carbamates of the type $R^1R^2P(O)NPhCOR^3$ (R^1, R^2 = alkyl, alkoxy; R^3 = alkoxy or alkylthio).[96,97] The FI spectra of $RP(O)(OEt)_2$ (R = Me or Et) were found to be more informative than the EI spectra.[98] The FI spectra gave evidence suggesting that the main reaction was due to intermolecular condensation followed by cleavage of the H atom attached to the carbon atom adjacent to phosphorus. The spectra of a series of dialkylphosphinic azides[99] and *ortho*-substituted oxides of t-arylphosphines, arylphosphinic, and phosphonic acids[100] have been recorded. A detailed study has been reported on the products of the interaction of amino-acids and amines with dimethylthiophosphinic chloride.[101] The fragmentation pathways of the compounds were elucidated by precise mass measurements and decoupled metastable transition determinations. Dimethylthiophosphinic esters of hydroxy-acids such as salicyclic acid have been characterized[102] and the effects of EI on R_2PSPSR_2 (R = Me, Et, Pr, Bu, $CH_2CH=CH_2$, or Ph) and MePSRPSRMe (R = Et, Pt, Bu, Ph, or CH_2Ph) have been investigated.[103] Fragmentations involved initial cleavage of a P—C bond and loss of R˙ followed by subsequent elimination of RPS resulting in the formation of $[R_2PS]^+$. In phenyl-substituted compounds, $[Ph_3P]$ and $[Ph_2PMe]$ ions were very abundant and were formed by migration of Ph in the molecular ion. The rearrangement of the molecular ion to a 1,3,2,4-dithiadiphosphetan ion prior to fragmentation has been suggested. Surprisingly, loss of sulphur was not a common process in contrast to the behaviour of R_3PS compounds. The spectra of the cyclic compounds (36; X = O or S), (37), and (38; X = O or S) have been reported[104] and it was observed that the intensity of the molecular

[95] S. Meyerson, E. S. Kuhn, F. Ramirez, J. F. Marecek, and H. Okazaki, *J. Am. Chem. Soc.*, 1980, **102**, 2398.
[96] V. A. Kolesova and Yu. A. Strepikheev, *Zh. Obshch. Khim.*, 1979, **49**, 2213.
[97] V. A. Kolesova, Yu. A. Strepikheev, and V. A. Valovoi, *Tr. Mosk. Khim. Tekhnol. Inst. im. D. I. Mendeleeva*, 1977, **94**, 38.
[98] V. B. Labintsev, Yu. K. Gusev, N. N. Grishin, and A. A. Petrov, *Zh. Org. Khim.*, 1978, **14**, 1371.
[99] H. Schroeder and J. Mueller, *Z. Anorg. Allg. Chem.*, 1979, **451**, 158.
[100] R. V. Poponova, E. N. Dolzhnikova, E. N. Tsvetkov, and G. S. Petrova, *Zh. Obshch. Khim.*, 1978, **48**, 1956.
[101] K. Jacob, C. Falkner, and W. Vogt, *J. Chromatogr.* 1978, **167**, 67.
[102] K. Jacob, W. Vogt, G. Schwentfeger, E. Maier, A. Ohnesorge, and N. Knedel, *Mass Spectrom. Comb. Tech. Med., Clin. Chem. Clin. Biochem., Symp.*, 1977, 246.
[103] H. Keck and W. Kuchen, *Org. Mass Spectrom.*, 1979, **14**, 149.
[104] A. E. Lyuts, V. V. Zamkova, A. P. Logunov, Z. A. Abramova, B. M. Butin, and Yu. G. Bosyakov, *Izv. Akad. Nauk SSSR, Ser. Khim.*, 1979, 20.

ion peak was greater for the sulphides than for the oxides perhaps due to two types of molecular ion being formed, one in which the charge is localized on the carbonyl group and the other with the charge round the phosphorus atom. The presence of a sulphur atom was observed to reduce the effect of the carbonyl group on fragmentation. The spectra of the macrocyclic esters (39; R = Me or Ph) and (40) have been reported[105] and the tetrachlorophosphonitrilic trimers (41; R^1, R^2 = Me; R^1 = Me, R^2 = Et; R^1 = Pr, R^2 = Me; R^1 = Bu, R^2 = Me; R^1 = Bu^i, R^2 = Me) have been characterized.[106] The spectra of compounds (42; R = H or Me)[107] have been reported and a series of compounds of the type (43; R = Me, Et, Pr, Me_2CH, Ph, p-Me-C_6H_4O, p-NO_2-C_6H_4O, 3,4-MeCl-C_6H_3O) has been investigated.[108] Ions at m/z 229 and m/z 211, assigned to the species $C_{12}H_{10}N_2OP^+$ and $C_{12}H_8N_2P^+$, found in the spectra of all the compounds studied, except in the methyl and ethyl derivatives, suggested that the dibenzodiazophosphine cyclic system has considerable stability. The compounds $(C_6H_{11})_2P(=O)-CH_2CH_2CH_2Cl$, $PH_2P(=O)-CH_2CH_2CH_2P(=O)$-PhOH, $PHPhCH_2CH_2CH_2NH_2$,[109] and $Ph_2PF_2[C(O)Ph]$[110] have been characterized and in the case of the last compound the base peak was observed at m/z 225 $[M - PhCO]^+$. The spectra of the aromatics (44; R = Me_3C, Ph, or C_6H_5) have been reported (see ref. 174).

The spectra of arsanthrene (45) and some 5,10-dihydroarsanthrenes have been reported[111] and compounds of the type R_3AsX_2 (R = Me_2CH or $PhCH_2$; X = halogen) have been characterized.[112]

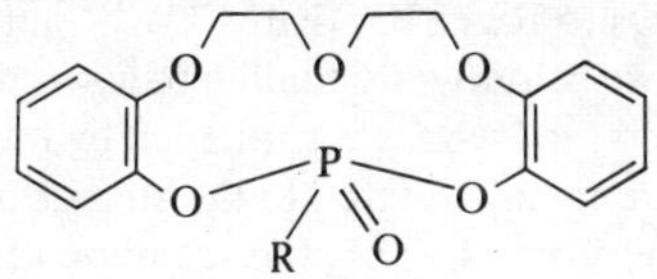

(39) 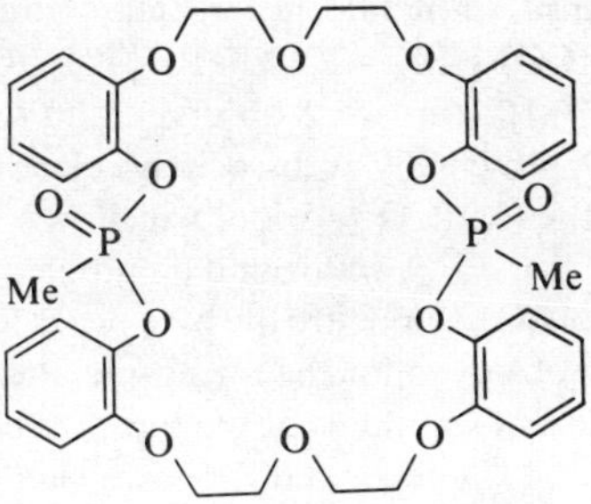   (40)

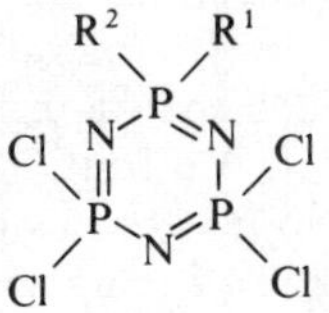

(41) 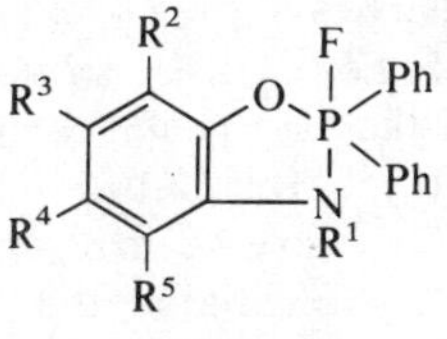   (42) 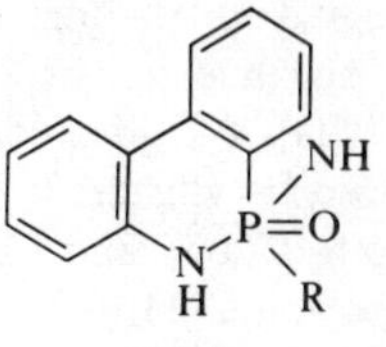   (43)

[105] K. B. Yatsimirskii, E. N. Korol, V. G. Golovatyi, T. N. Kudrya, and G. G. Talanova, *Dokl. Akad. Nauk SSSR*, 1979, 244, 1359.
[106] P. J. Harris and H. R. Allcock, *J. Chem. Soc., Chem. Commun.*, 1978, 714.
[107] H. B. Stegmann, H. V. Dumm, and K. Scheffler, *Phosphorus Sulfur*, 1978, 5, 159.
[108] M. S. R. Naidu and C. D. Reddy, *Bull. Chem. Soc. Jpn.*, 1978, 51, 2156.
[109] R. Uriarte, T. J. Mazanec, K. D. Tau, and D. W. Meek, *Inorg. Chem.*, 1980, 19, 79.
[110] S. Neumann, D. Schomburg, G. Richtarsky, and R. Schmutzler, *J. Chem. Soc., Chem. Commun.*, 1978, 946.
[111] H. Vermeer, R. J. M. Weustink, and F. Bickelhaupt, *Tetrahedron*, 1979, 35, 155.
[112] L. Verdonck and G. P. Van der Kelen, *Spectrochim Acta, Part A*, 1979, 35, 861.

(44)   (45)

The spectra of the organoantimony compounds $Bu_2SbSbBu_2$, $Pr'_2SbSbPr'_2$, and $Pr'SbBr_2$ have been examined.[113]

Compounds of Other Elements in Group V. The origin and composition of several characteristic ions formed on EI fragmentation of $PhTeCH_2CH(OMe)_2$ have been determined by high-resolution analysis[114] and compounds $PhTeCH_2CH(OR)_2$ (R = Me, CD_3, or Et) fragmented to give the rearranged ions $[PhTeOR]^+$.

The spectra of $Xe(CF_3^-)_2$ showed peaks assignable to Xe^+, Xe^{2+}, CF_3^+, and $C_2F_5^+$.[115]

3 Transition-metal Organometallics

Transition-metal Cluster Compounds.—The negative ion spectra for the clusters $M_3(CO)_{12}$ (M_3 = Fe_3, $FeRu_2$, Ru_3, Ru_2Os, $RuOs_2$, or Os_3) have been reported.[116] No $[M]^-$ ions were observed and the trinuclear fragments were due to loss of CO, principally by dissociative electron-capture processes. A decreasing intensity of the $[M_3(CO)_{11}]^-$ ion, which was observed as the base peak in the spectrum of $Os_3(CO)_{12}$, was found on progressing from Os to Ru to Fe. $[Ru_3(CO)_{10}]^-$ and $[Fe_3(CO)_9]^-$ ions are the base peaks for the last two compounds. Similar behaviour was observed for the mixed-metal carbonyls and the trend was explained by the different stabilities of the ions, which in turn are mainly related to the strengths of the metal–metal bonds. The negative ion spectra of the compounds $M_3(CO)_{11}PEt_2Ph$ (M = Ru or Os) indicated that neither the metal–CO bond strengths nor the ionization energies of the metal played an important role. In the negative ion spectra of $Fe_3(CO)_9X_2$ (X = S, Se, or Te), negative $[Fe_3(CO)_7X_2]^-$ ions composed the base peak. The negative ion spectra of the carbonyl clusters $Co_4(CO)_{12}$, $Rh_4(CO)_{12}$, $Ir_4(CO)_{12}$, $Co_3Rh(CO)_{12}$, $Co_2Rh_2(CO)_{12}$, and $CoRh_3(CO)_{12}$ have been reported.[117] No $[M]^-$ ions were found and a $[M_4(CO)_{10}]^-$ ion was found to be the base peak in all compounds with the exception of $CoRh_3(CO)_{12}$. However, tetrametal fragments were formed by the loss of carbonyl groups. Similarly, $[M]^-$ ions were not observed in the spectra of $Os_3(CO)_{12}X_2$ or $Os_3(CO)_{10}X_2$ (X = Br or I) and only Os_3 fragments due to the loss of carbonyl groups were identified.[118]

[113] H. J. Breunig and W. Kanig, *J. Organomet. Chem.*, 1980, **186**, C5.
[114] F. F. Knapp, *Org. Mass Spectrom.*, 1979, **14**, 341.
[115] L. J. Turbini, R. E. Aikman, and R. J. Lagow, *J. Am. Chem. Soc.*, 1979, **101**, 5833.
[116] R. P. Ferrari, G. A. Vaglio, and M. Valle, *J. Chem. Soc., Dalton Trans.*, 1978, 1164.
[117] P. Michelin-Lausarot, G. A. Vaglio, and M. Valle, *Inorg. Chim. Acta*, 1979, **35**, 227.
[118] G. A. Vaglio, *J. Organomet. Chem.*, 1979, **169**, 83.

In the spectrum of $Os_3(CO)_{10}I_2$, $[M - CO]^-$ was the base peak but prominent molecular ions were observed in the spectra of $Os(CO)_{12-n}CNR_n$ (n = 1 or 2; R = Me, C_6H_4OMe, Bu^n, Bu^t and n = 3 or 4; R = Bu^n) for which a stepwise loss of carbonyl and isonitrile groups, down to the Os_3^+ core, was observed.[119] Some idea of the stability of the osmium triangle was gleaned by the observation of the doubly charged species $[Os_3(CO)_9(CNBu^t)_3]^{2+}$. The spectra of $HRu_3(CO)_9HC\equiv NBu^t$,[120] $Os_5(CO)_{15}P(OMe)_3$,[121] and $(C_{10}H_{12})Ru_3(CO)_8$[122] have been reported. The EI and FD spectra of the compounds $Pt_3(CO)_3(PPh_3)_4$, $Pt_4(CO)_5(PPh_3)_4$, and $Pt_5(CO)_6(PEt_3)_4$ have been investigated[123] and, although the compounds were thermally labile, FD gave molecular and $[M - nCO]^+$ ions.

Metal Carbonyl and Related Compounds.—The spectrum of iron pentacarbonyl has been reinvestigated and a new fragment $[FeC_2]^+$ ion reported;[124] the appearance energies for minor fragments were determined. The spectra, ionization energies, fragmentation energies, ionic and neutral dissociation energies, and heats of formation of Group VIB (Cr, Mo, and W) hexacarbonyls and their pentacarbonylthiocarbonyls have been measured and compared.[125] Substitution of CS for CO resulted in a lowering of the ionization energies by 0.1—0.4 eV and a reduction of about 0.1 eV in the average metal–CO bond energies in both the molecule and molecular ion. Also, both the metal–CS and metal–CO bonds increase in strength in going from Cr to Mo to W. The spectra of $M(CO)_6$ and $M(CO)_5CS$ compounds of Group VIB metals and the $M_2(CO)_{10}$ compounds of Group VIIB metals have been reported.[126] It was suggested that in the molecular ions the metal and mixed-metal decacarbonyls could be considered as $(CO)_5M^+$—$M(CO)_5$ compounds in which there are five strong and five weak metal–CO bonds. For $Mn_2(CO)_{10}$ and $Re_2(CO)_{10}$ the M^+—M dissociation energies are 3.0 ± 0.1 and 4.0 ± 0.3 eV, respectively. The charge-exchange spectrum of nickel tetracarbonyl, using a typical petrol engine exhaust gas as the reactant gas, has been reported.[127] As a result of a study of the spectra of cis-$LM(CO)_4(^{13}CO)$ (L = piperidine, M = Cr or W) it was concluded that the initial loss of CO proceeded with complete scrambling of the label between axial and equatorial sites.[128] The $[M]^+$ ion was observed to be of low abundance for $Co(CO)_4Ge_2H_5$ but mass spectral peaks for ions of the type $[Co(CO)_nGe_2H_n]^+$ were intense[129] and, in the spectrum of $[Fe(CO)_4(GeMe_2)]_2$, the major part of the ion current was carried by $[(FeGeMe_2)_2(CO)_n]^+$ (n = 0—7) and $[Fe_2C_yH_nGe_2]^+$ (y = 0—3) fragments.[130]

---

[119] M. J. Mays and P. D. Gavens, *J. Chem. Soc., Dalton Trans.*, 1980, 911.
[120] M. I. Bruce and R. C. Wallis, *J. Organomet. Chem.*, 1979, **164**, C6.
[121] J. M. Fernandez, B. F. G. Johnson, J. Lewis, and P. R. Raitby, *J. Chem. Soc., Chem. Commun.*, 1978, 1015.
[122] O. Gambino, E. Sappa, A. M. L. Manotti, and A. Tiripicchio, *Inorg. Chim. Acta*, 1979, **36**, 189.
[123] T. Wurminghausen, H. J. Reinecke, and P. Braunstein, *Org. Mass Spectrom.*, 1980, **15**, 38.
[124] B. R. Conard and R. Sridhar, *Can. J. Chem.*, 1978, **56**, 2607.
[125] G. D. Michels, G. D. Flesch, and H. J. Svec, *Inorg. Chem.*, 1980, **19**, 479.
[126] G. D. Michels, Report 1979, 13-T-837.
[127] J. E. Campana and T. H. Risby, *Anal. Chem.*, 1980, **52**, 468.
[128] R. Davis and D. J. Darensbourg, *J. Organomet. Chem.*, 1978, **161**, C11.
[129] F. S. Wong and K. M. Mackay, *J. Chem. Soc., Dalton Trans.*, 1978, 1752.
[130] A. Bonny and K. M. Mackay, *J. Chem. Soc., Dalton Trans.*, 1978, 722.

The spectra have been reported for cis-Fe(CO)$_4$(HgCH$_2$SiMe$_3$)$_2$, Co(CO)$_4$(HgCH$_2$SiMe$_3$), W(CO)$_3$Cp(HgCH$_2$SiMe$_3$),[131] and (CO)$_4$Fe[=C(NH$_2$)$_2$],[132] and Fe$_2$(CO)$_6$[μ-P(CF$_3$)H] has been characterized by high-resolution mass spectrometry.[133] The spectra of a number of carbonyl complexes such as HM$_3$(μ_2-COMe)(CO)$_{10}$ (M = Fe, Ru, or Os),[134] [MeN(PF$_2$)P]Mo$_2$Cp$_2$(CO)$_4$,[135] and Co$_2$(CO)EtC≡CC(O)Me[136] have been recorded as a means to their identification. The complexes (46),[137] (47),[137] (48),[138] and (49)[138] have been characterized.

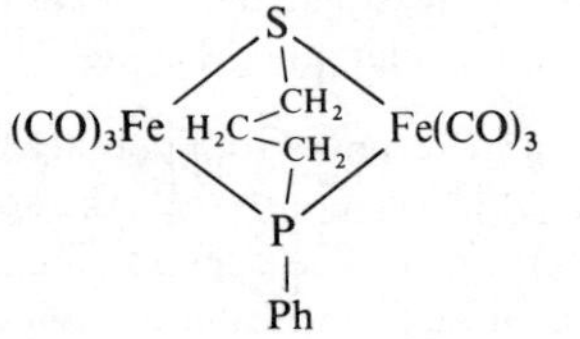

(46)

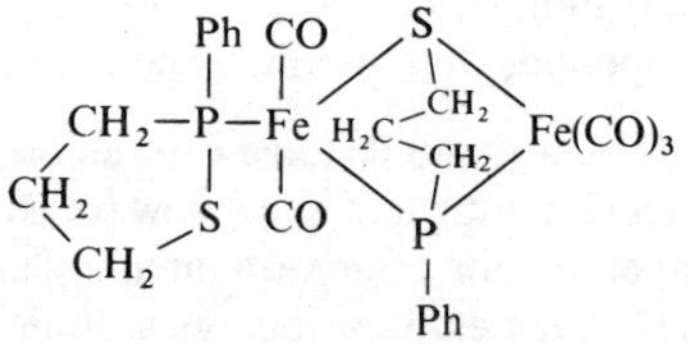

(47)

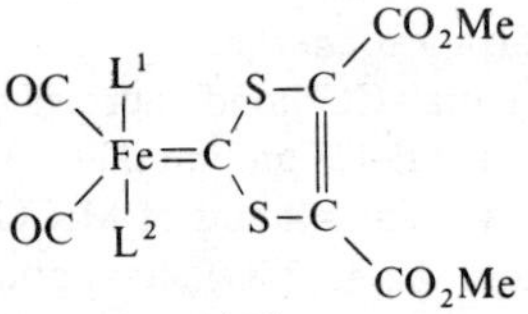

(48)

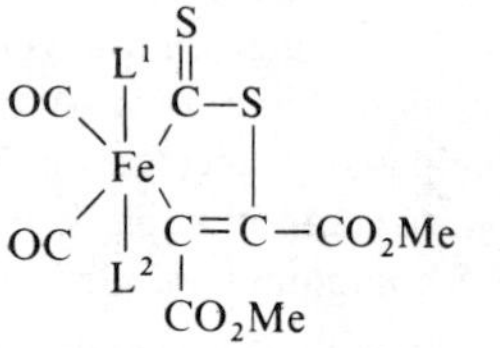

(49)

L^1 = P(OMe)$_3$, PPh$_3$, P(OMe)$_3$, PMe$_2$PL, or PMe$_3$
L^2 = P(OMe)$_3$, PPh$_3$, PMe$_3$, PMe$_2$PL, or PMe$_3$

Complexes of Two-, Three-, and Four-electron Donors.—Peaks above m/z 196 in the spectrum of C$_2$H$_4$Fe(CO)$_4$ suggested that the complex had partially decomposed into C$_2$H$_4$ and Fe$_3$(CO)$_{12}$.[139] In contrast, the molecular ion for tetramethylallene iron tetracarbonyl was observed. The spectra of compounds (50), (51), and (52) indicated that CO was ejected more easily from σ-bonded iron atoms than from π-bonded iron atoms.[140] In the organosilyl ironcarbonyl complex (53), the molecular ion was observed as well as fragmentation involving the successive loss of six carbonyl groups and two iron atoms.[141] Similarly, for complex (54) the molecular ion ejected nCO atoms (n = 1, 2, or 3).[142] The fragmentation paths of some α-π-bonded dinuclear alkenecarbonyl derivatives

---

[131] F. Glockling, V. B. Mahale, and J. J. Sweeney, *J. Chem. Soc., Dalton Trans.*, 1979, 767.
[132] W. Petz, *J. Organomet. Chem.*, 1979, **172**, 415.
[133] W. Clegg and S. Morton, *Inorg. Chem.*, 1979, **18**, 1189.
[134] J. B. Keister, *J. Chem. Soc., Chem. Commun.*, 1979, 241.
[135] R. W. Light, R. T. Paine, and D. E. Maier, *Inorg. Chem.*, 1979, **18**, 2345.
[136] S. Aime, L. Milove, and D. Osella, *J. Chem. Soc., Chem. Commun.*, 1979, 704.
[137] F. Mathey, M. B. Comarmond, and D. Moras, *J. Chem. Soc., Chem. Commun.*, 1979, 417.
[138] H. Le Bozec, A. Gorgues, and P. Dixneuf, *J. Chem. Soc., Chem. Commun.*, 1978, 537.
[139] W. E. Hill, C. H. Ward, T. R. Webb, and S. D. Worley, *Inorg. Chem.*, 1979, **18**, 2029.
[140] E. Sappa, M. L. Nanni-Marchino, and V. Raverdino, *Ann. Chim. (Rome)*, 1978, **68**, 349.
[141] R. J. P. Corriu and J. J. E. Moreau, *J. Chem. Soc., Chem. Commun.*, 1980, 278.
[142] A. D. Cian, R. Weiss, J. P. Haudegond, Y. Chauvin, and D. Commereuc, *J. Organomet. Chem.*, 1980, **187**, 73.

of iron, of the types $Fe_2(CO)_6[(CO)_2C_2Et_2]$ and $Fe_2(CO)_6[C_2OHEt]_2$, have been reported.[143] The spectra of η^1- and η^3-allylrhenium tetracarbonyl have been reported[144] and in the spectrum of (hydroxymethyl)trimethylenemethanetricarbonyl iron the molecular ion decomposed by loss of OH and CO.[145] Electron attachment by secondary electron capture in the gas phase for a series of cyclic and acyclic conjugated η^4-dienetricarbonyl iron compounds resulted in the formation of $[M]^-$ which fragmented by loss of CO and diene groups.[146] η-Cyclobutadienetricarbonyliron underwent dissociative electron capture to give the $[M-CO]^-$ ion in its negative ion spectrum and decomposed only by elimination of CO ligands. The $[M]^+$ ions in both the *endo*- and *exo*-iron tricarbonyl complexes (55) were found to yield a base peak arising from loss of CO and 2CO groups, respectively.[147] The $[M]^+$ ion and also the major fragment ions have been identified for the cobalt complexes (56) and (57).[148]

[143] E. Sappa and V. Raverdino, *Ann. Chim. (Rome)*, 1979, **69**, 349.
[144] B. J. Brisdon, D. A. Edwards, and J. W. White, *J. Organomet. Chem.*, 1979, **175**, 113.
[145] P. A. Dobosh, C. P. Tillya, E. S. Magyar, and G. Scholes, *Inorg. Chem.*, 1980, **19**, 228.
[146] M. R. Blake, J. L. Garnett, I. K. Gregor, and S. B. Wild, *J. Chem. Soc., Chem. Commun.*, 179, 496.
[147] E. Meier, O. Cherpillod, T. Boschi, R. Rovlet, P. Vogel, C. Mahaim, A. A. Pinkerton, D. Schwarzenbach, and G. Chapus, *J. Organomet. Chem.*, 1980, **186**, 247.
[148] M. D. Rausch, G. F. Westover, E. Mintz, G. M. Reisner, I. Bernal, A. Clearfield, and J. M. Troup, *Inorg. Chem.*, 1979, **18**, 2605.

Cyclopentadiene Complexes.—The fragmentation patterns of a series of alkylferrocenes (*e.g.* 1,1'-dimethylferrocene) have been reported[149] and for symmetrically substituted methylferrocenes $(Me_nC_5H_{5-n})_2Fe$ (n = 1—5) have been discussed.[150] Low- and high-resolution spectra have been used to investigate mixtures of alkyl-substituted dicyclopentadienylcarbonyl complexes of iron formed from the interaction of $Fe_3(CO)_{12}$ and petrol from thermal cracking.[151] The spectra of the thiocarbonyl-bridged dimers $[(\eta^5-C_5H_5)Fe(CO)CS]_2$, $[(\eta^5-C_5H_5)_2Fe_2CS(CO)_3]$, $[(\eta-C_5H_5)Mn(CS)NO]_2$, and $[(\eta^5-C_5H_4Me)Mn(CS)NO]_2$ have been reported and the appearance energies of most of the metal-containing fragment ions, especially those with intact ligands, have been determined.[152] Fragment ions resulting from the loss of CO or NO were found to be more abundant than those from the loss of CS. The main fragmentation processes of the ferrocinyl and cymantrenyl analogues of chalcones $R^1COCH=CHR^2$ [R^1, R^2 = $C_5H_4FeC_5H_5$ or $C_5H_4Mn(CO)_3$] were due to a cleavage of the metal–ligand bond.[153] In the spectra of a series of ferrocenyl-, cymantrenyl-, and benchrotrienyl-carbinols, the elimination of H_2O suggested a 1,2-elimination mechanism involving hydroxyl transfer to the metal atom.[154] It was proposed that the $[M-16]^+$ ions arose from thermal decomposition of the compounds in the inlet system. Field desorption has been found to be of considerable value for the determination of molecular weights of neutral transition-metal complexes and the cation molecular weight for salt complexes such as $Fe(\eta^5-C_7H_9)$-$P(OMe)_3]^+NF_4^-$.[155] High-resolution spectra and fragmentation pathways have been reported for $RC_5H_4FeC_5H_5$ [R = $C(CF_3)C(CN)_2$, $CH=C(CN)_2$, $C(Cl)=C(CN)_2$, or $C(CN)C(CN)_2$], $(CO)_3MnC_5H_4C(CN)C(CN)_2$, and $Ph_3PCuC_5H_4C(CN)C(CN)_2$[156] and the spectra of compounds (58), (59), and (60)[157] reported. The 19-electron sandwich complexes of the type $C_5H_5FeC_6Me_6$ have been identified.[158] Abundant molecular ions have been observed for

(58)   (59)   (60)

---

[149] B. V. Zhuk, G. A. Domrachev, N. M. Semenov, E. I. Mysov, R. B. Materikova, and N. S. Kochetkova, *J. Organomet. Chem.*, 1980, **184**, 231.

[150] E. I. Mysov, I. R. Lyatifov, R. B. Materikvoa, and N. S. Kochetkova, *J. Organomet. Chem.*, 1979, **164**, 301.

[151] V. F. Sizoi, D. V. Zagorevskii, Yu. S. Nekrasvo, V. D. Tyurin, and V. S. Lupandin, *Neftekhimya*, 1979, **19**, 876.

[152] A. Efraty, D. Liebman, M. H. A. Huang, C. A. Weston, and R. J. Angelici, *Inorg. Chem.*, 1978, **17**, 2831.

[153] A. N. Nesmeyanov, D. V. Zagorevskii, Yu. S. Nekrasov, V. F. Sizoi, V. M. Postnov, A. M. Baran, and E. I. Klimova, *J. Organomet. Chem.*, 1979, **169**, 77.

[154] Yu. S. Nekrasov, V. F. Sizoi, D. V. Zagorevskii, and N. I. Vasyukova, *Org. Mass Spectrom.*, 1979, **14**, 22.

[155] C. N. McEwen and S. D. Ittel, *Org. Mass Spectrom.*, 1980, **15**, 35.

[156] M. B. Freeman and L. G. Sneddon, *Inorg. Chem.*, 1980, **19**, 1125.

[157] J. E. Geurchais, F. Le Floch-Perennou, F. Y. Petillon, A. N. Keith, L. Muir, K. W. Muir, and D. W. A. Sharp. *J. Chem. Soc., Chem. Commun*, 1979, 40.

[158] D. Astruc, J. R. Hamon, G. Althoff, E. Roman, P. Batail, P. Michaud, J. P. Mariot, F. Varret, and D. Cozak, *J. Am. Chem. Soc.*, 1979, **101**, 5445.

titanium complexes of the type $(\eta^5\text{-MeC}_5H_4)_2TiL$ (L = 2,4-dithiouracil, 2-thiouracil, 2,4-dithiopyrimidine, or 2,6-dihydroxypyrimidine),[159] and bis(η^5-C_5H_5)(2-hydroxypyridinato)Ti and uracilatobis[bis(η^5-C_5H_4Me)Ti] were identified.[160]

The spectra of $(\eta^5\text{-}C_5HC)Cr(CO)_2NS$ and its nitrosyl analogue have been reported together with ionization and appearance energies.[161] The compounds had sufficient thermal stability to indicate that the observed fragmentation patterns resulted from electron bombardment and not thermal decomposition. The appearance energies of some fragment ions were used to obtain a relative estimate of the Cr^+—Co and Cr^+—NX ionic bond dissociation energies. The spectra of cyclopentadienyl complexes of chromium such as $Cp_2Cr_2(\mu\text{-}OBu^t)_2$,[162] $Cp_2Cr_2(OBu^t)_2(NO)_2$,[162] $[CpCr(CO)_3]$,[163] and $CpCrMn(CO)_2P(OMe)_3$ have been reported and dicyclopentadienyl complexes of V, Cr, Mn, Co, and Ni investigated.[164] The complexes $(\eta\text{-}C_5H_5)_2V(SiCl_3)_2$,[165] $(\eta\text{-}C_5H_5)_2W(H)SiCl_3$,[165] and $(\eta\text{-}C_5Me_4Et)_2VC\equiv CC_6H_2Me_3$[166] have been identified. The spectra of $(\eta\text{-}C_5H_5)_nM(CO)L^1L^2$ (M = Mn or Re; L^1 = CO, Ph$_3$P, or (PhO)$_3$P; L^2 is a vinylidene group) have been reported and dehydrogenation of the Re complexes was observed.[167] In the bimetallic analogues, the Re—Re bond was found to be stronger than the Mn—Mn bond. High-resolution spectra and assignments for the major ions in the spectra of $(\eta^5\text{-RC}_5H_4)Mn(CO)NOI$ (R = H or Me), $(\eta^5\text{-MeC}_5H_4)_2Mn_2(NO)_3I$, and $(\eta^5\text{-RC}_5H_4)MnNOPh_3PI$ (R = H or Me) have been reported.[168] $(\eta^5\text{-}C_5H_5)_2Mn_2(CO)_4CCH_2$,[169] $(\eta^5\text{-}C_5H_5)COM[\mu(\eta^5:\eta^1)C_5H_4]Mn(CO)_4$ (M = Mo or W),[170] and complexes (61)

(61) R = H or Me

$$L = \text{Ph-C=C(Ph)-P-C=C-Ph}$$

(62)

[159] D. R. Corkin, L. C. Francesconi, D. N. Hendrickson, and G. D. Stucky, *J. Chem. Soc., Chem. Commun.*, 1979, 248.
[160] D. F. Fieselmann, D. N. Hendrickson, and G. D. Stucky, *Inorg. Chem.*, 1978, **17**, 1841.
[161] T. J. Greenhough, B. W. S. Kolthammer, P. Legzdins, and J. Trotter, *Inorg. Chem.*, 1979, **18**, 3548.
[162] M. H. Chisholm, F. A. Cotton, M. W. Extine, and D. C. Rideout, *Inorg. Chem.*, 1979, **18**, 120.
[163] L.-Y. Goh, M. J. D'Aniello, S. Slater, E. L. Muetterties, I. Tavanaiepour, M. I. Chang, M. F. Fredrich, and V. W. Day, *Inorg. Chem.*, 1979, **18**, 192.
[164] Yu A. Andrianov and O. N. Druzhkov, *Khim. Elementoorgan. Soedin.*, 1977, 78.
[165] A. M. Cardoso, R. J. H. Clark, and S. Moorhouse, *J. Organomet. Chem.*, 1980, **186**, 241.
[166] F. H. Kohler, W. Prossdorf, U. Schubert, and D. Neugebauer, *Angew. Chem., Int. Ed. Engl.* 1978, **17**, 850.
[167] V. F. Sizoi, Yu. S. Nekrasov, Yu. N. Sukharev, N. E. Kolobova, O. M. Khitrova, N. S. Obezyuk, and A. B. Antenova, *J. Organomet. Chem.*, 1978, **162**, 171.
[168] B. W. S. Kolthammer and P. Legzdins, *Inorg. Chem.*, 1978, **18**, 889.
[169] K. Folting, J. C. Huffman, L. N. Lewis, and K. G. Caulton, *Inorg. Chem.*, 1979, **18**, 3483.
[170] R. J. Hoxmeier, J. R. Blickensderfer, and H. D. Kaesz, *Inorg. Chem.*, 1979, **18**, 3453.
[171] F. Nief, C. Charrier, F. Mathey, and M. Simauy, *J. Organomet. Chem.*, 1980, **187**, 277.

and (62)[171] have been identified and μ-(2-3-η:4-5-η)-2,4-hexadiyne-bis[dicarbonyl(η-methyl-cyclopentadienyl)manganese(I)] has been characterized.[172]

A series of organometallic complexes of cobalt, formed from the reaction of cobalt atoms with cyclopentadiene and other alkynes such as diphenylacetylene and hexafluorobutyne, have been characterized by precise mass determination.[173] Fragmentation pathways and rearrangement processes have been proposed for the metalloles (63).[174] Of special interest is the migration of ring F to the central atom during fragmentation. The cobalt complexes 1-(η-C_5H_5)CoB_5H_9 and 2-(η-C_5H_5)CoB_9H_{13} have been characterized.[175] The cobalt and iron divinylborane complexes (64), (65), and (66) have been studied and their fragmentation paths reported.[176] The tungsten complex

(63) M = Co, Rh, Ir, or Tl
R = CO or η-C_5H_5

(64) (65) (66)

(η^5-C_5H_5)WNO$_2$(O)SO$_2$C$_6$H$_4$Me has been identified.[177] Methylcyclopentadienyl and cyclopentadienyl nickel carbonyl dimers have been characterized.[178] The results of a study of tris(isopropylcyclopentadienyl) π-complexes of lanthanum, praseodymium, and neodymium showed that the isopropyl group caused major complications in molecular fragmentations and plausible pathways were proposed.[179] In the spectrum of [(η^5-C_5H_5)$_2$Rh]$_2$, the ion m/z 233, assigned to [(η^5-C_5H_5)$_2$Rh]$^+$, was observed.[180]

Complexes Containing C_6 and Higher Rings.—The effects of substituents on the fragmentation of benchrotrenyls (67; R = H, F, Cl, I, Me, MeO, MeOCO, Et, Me$_2$N, NH$_2$, Ph, Me$_3$C, or CH$_2$Ph) and (68; R = H or Me) have been investigated.[181] For (67), log[Cr]$^+$/[RC$_6$H$_5$Cr]$^+$ was found to be linearly related to the number of degrees of freedom of [RC$_6$H$_5$Cr]$^+$. Decarbonylation of the molecular ions of (67) were found to be unaffected by the nature of R. The results of an investigation of the first field-free region metastable fragmentation of (η^6-C_6H_5Me)Cr(CO)$_3$, using the linked-scan technique, have shown that the [M]$^+$ ion fragmented entirely by single and multiple CO loss and the [C_7H_8Cr(CO)$_2$] ion fragmented directly to the [C_7H_8Cr]‡ ion.[182] Studies of the

[172] G. Cash and R. C. Pettersen, *J. Chem. Soc., Dalton Trans.*, 1979, 1630.
[173] M. B. Freeman, L. W. Hall, and L. G. Sneddon, *Inorg. Chem.*, 1980, **19**, 1132.
[174] T. R. B. Jones, J. M. Miller, S. A. Gardner, and M. D. Rausch, *Can. J. Chem.*, 1979, **57**, 335.
[175] R. Wilczynski and L. G. Sneddon, *Inorg. Chem.*, 1979, **18**, 864.
[176] G. E. Herberich, E. A. Mintz, and H. Müller, *J. Organomet. Chem.*, 1980, **187**, 17.
[177] P. Legzdins and D. T. Martin, *Inorg. Chem.*, 1979, **18**, 1250.
[178] L. R. Byers and L. F. Dahl, *Inorg. Chem.*, 1980, **19**, 680.
[179] P. E. Gaivoronskii, E. M. Gavrishchuk, N. P. Chernyaev, and Yu. B. Zverev, *Zh. Neorg. Khim.*, 1978, **23**, 3139.
[180] N. E. Murr, J. E. Sheats, W. E. Geiger, and J. D. L. Holloway, *Inorg. Chem.*, 1979, **18**, 1443.
[181] Yu. S. Nekrasov and N. I. Vasyukova, *Org. Mass Spectrom.*, 1979, **14**, 422.
[182] R. Davis, M. L. Webb, D. S. Millington, and V. Parr, *Org. Mass Spectrom.*, 1979, **14**, 289.

[(C$_7$H$_8$)Cr] ion, generated from η^6-C$_6$H$_5$MeCr(CO)$_3$ and η^6-cycloheptatrienetricarbonylchromium, have been reported.[183] Measurements of appearance energies and fragmentation behaviour demonstrated that the ions from the two precursors did not give a common structure. Similar results were obtained for the [(C$_8$H$_{10}$)Cr]$^+$ ion generated from (η^6-C$_6$H$_5$Et)Cr(CO)$_3$ and from (η^6-7-exo-methylcycloheptatriene)tricarbonylchromium.

The results of a ^{2}H-labelling study suggested that the ion generated from (η^6-C$_6$H$_5$Me)Cr(CO)$_3$ is [MeC$_6$H$_4$CrH]$^+$ which decomposes to [CrH]$^+$. A peak at m/z [M − CO]$^+$ was observed in the spectrum of (69).[184]

(67)   (68)   (69)

Bisarene complexes of vanadium VL^{1}L^2 (L^1, L^2 = C$_6$H$_5$Me, C$_6$H$_4$Et$_2$) fragmented via dehydrogenation and cleavage of the methyl and ethyl groups.[185] Examination of the spectra of fac-LMo(^{13}CO)(CO)$_3$ (L = norbornadiene or bicyclo[6.1.0]nona-2,4,6-triene) showed that loss of CO from the former complex occurred with complete scrambling in contrast to the latter complex for which almost exclusive loss of the labelled group was observed.[186] The fragmentation of the latter compound was suggested to involve the formation of a new metal-olefin bond on loss of CO. Contrary to expectations, η^4- and η^6-cycloheptatriene derivatives of zero-valent Fe, Mo, and W tricarbonyls gave abundant [M]$^-$ ions in their negative ion spectra.[187] The complexes Hf(η^6-C$_6$H$_5$Me)$_2$PMe$_3$ and Hf(η^6-C$_6$H$_6$)$_2$PMe$_3$ have been identified.[188] Arene cyclo-octatetraene complexes of zero-valent Fe, Ru, and Os (arene = C$_6$H$_6$, C$_6$H$_3$Me$_3$, or C$_6$Me$_6$) have been characterized[189] and the high-resolution spectra for Co$_2$(CO)$_6$C$_8$H$_{12}$ and Co$_2$(CO)$_5$(C$_8$H$_{12}$)$_2$ reported.[190]

The spectra of some bis(triphenylphosphine)chloroplatinum(II) acetylides have been compared with those of (Ph$_3$P)$_2$PtCl$_2$ and R$_3$PPtR$_2^1$ (R = alkyl or aryl).[191] In most cases, the [M]$^+$ ions were present and the Pt—C and Pt—Cl bonds had considerably less stability than the Pt—P bonds. In addition, evidence for polymeric ions was obtained. In a study of some bis(triphenylphosphine)-platinum(II) bisacetylides, differences were observed between cis and trans complexes [e.g. (Ph$_3$P)$_2$Pt(C≡CPh)$_2$] and the Pt—P bond was found to have

---

[183] R. Davis, I. A. Ojo, and M. L. Webb, Org. Mass Spectrom., 1978, **13**, 547.
[184] E. L. Murr and M. Riveccie, J. Chem. Soc., Chem. Commun., 1978, 552.
[185] I. L. Agafonov, N. V. Shushunov, E. M. Gavrishchuk, P. E. Gaivoronskii, and E. M. Karataev, Izv. Akad. Nauk SSSR, Ser. Khim., 1979, 2134.
[186] R. Davis, M. L. Webb, and D. J. Darensbourg, J. Chem. Soc., Chem. Commun., 1979, 409.
[187] M. R. Blake, J. L. Garnett, I. K. Gregor, and S. B. Wild, J. Organomet. Chem., 1979, **178**, C37.
[188] F. G. N. Cloke and M. L. H. Green, J. Chem. Soc., Chem. Commun., 1979, 127.
[189] M. A. Bennett, T. W. Matheson, G. B. Robertson, A. K. Smith, and P. A. Tucker, Inorg. Chem., 1980, **19**, 1014.
[190] M. A. Bennett and P. B. Donaldson, Inorg. Chem., 1978, **17**, 1995.
[191] A. Furlani, P. Carusi, M. V. Russo and A. Santi, Ann. Chim. (Rome), 1979, **69**, 127.

greater stability than the Pt—C bond.[192] In the spectrum of compound (70), the molecular ion at m/z 564 proved that the molecule contained only one platinum atom[193] and complexes of the type (71) have been characterized.[194] The complex $Me_3Ta(CH_2SiMe_3)_2$ has been characterized[195] and $MeReO_3$,[196] $MeZr[N(SiMe_3)_2]_3$, and $MeHf[N(SiMe_3)_2]_3$ have been identified.[197]

```
        Me
H₂C—N—Me                  MeO
 |   |                         ╲    ╱═NOH
H₂C—Pt—Cl                       ╲  ╱
     ↑                           ╱╲
    PPh₃                  MeO   ╱   PdCl
                                    PMe₂Ph
    (70)                        (71)
```

4 Co-ordination and Metal–Organic Compounds

Compounds with Metal–Oxygen or Metal–Sulphur Bonds.—β-*Diketones and Related Complexes.* The spectra of a series of fluorinated copper(I) diketonates L_2Cu (L = fluorinated 1,3-diketone) fragmented by a stepwise ejection of alkyl/perfluoroalkyl substituents.[198] Replacement of an alkyl group by perfluoroalkyl resulted in more extensive fragmentation of the molecular ions. All complexes showed a large peak due to L^+ which could have arisen from a valence change of the copper atom in the complex, *i.e.* $[L_2Cu]^+ \rightarrow L^+ + LCu$. Experiments have been described for the determination of calcium in blood plasma and urine, the calcium being detected in a volatile ^{43}Ca acetylacetonate complex.[199] In the negative ion spectra of the complex $M(CF_3COCHCOCF_3)$ (M = Sc, Ti, V, Cr, Mn, Fe, Co, Al, Ga, or In) the $[M]^+$ and ligand ions were the most abundant.[200] Mechanisms have been proposed for the formation of all major fragment ions, many resulting from F transfer processes. The relative instabilities of $[M]^-$ ions exhibited an approximately linear dependence on the increasing 3*d*-electron population of the metal ions from Ti^{III} through to Co^{III}. EI and negative ion spectra have been reported for a series of tris and bis chelates, of the type M^1L_3 and M^2L_2 (L is an enolate ion of 2,2,6,6-TM-3,5-heptanedione; M^1 = Sc, Cr, Mn, Fe, Co, Al, Ga, or In; M^2 = Co, Ni, Cu, or Zn).[201] The spectra have been obtained at both normal and elevated pressures for the chelates of 2,2,6,6-TM-heptanedione with Cr^{III}, Fe^{III}, Co^{II}, Ni^{II}, Cu^{II}, Zn^{II}, and Al^{III}.[202] At higher pressures, it was suggested that the observed

---

[192] A. Furlani, P. Carusi, M. V. Russo, and A. Santi, *Ann. Chim. (Rome)*, 1979, **69**, 101.
[193] I. M. Al-Najjar, M. Green, S. J. S. Kerrison, and P. J. Sadler, *J. Chem. Soc., Chem. Commun.*, 1979, 311.
[194] B. L. Shaw and I. Shepherd, *J. Chem. Soc., Dalton Trans.*, 1979, 1634.
[195] C. S. Scampucci and J. G. Riess, *J. Organomet. Chem.*, 1980, **187**, 331.
[196] I. R. Beattie and P. J. Jones, *Inorg. Chem.*, 1979, **18**, 2318.
[197] R. H. Andersen, *Inorg. Chem.*, 1979, **18**, 1724.
[198] K. C. Joshi, V. N. Pathak, S. Bhargava, K. G. Das, and I. S. Mulla, *J. Indian Chem. Soc.*, 1978, **55**, 759.
[199] C. Belon, R. Huguet, and M. Audran, *Trav. Soc. Pharm. Montpellier*, 1979, **39**, 11.
[200] D. R. Dakternieks, I. W. Frazer, J. L. Garnett, and I. K. Gregor, *Org. Mass Spectrom.*, 1979, **14**, 330.
[201] J. L. Garnett, I. K. Gregor, and M. Giulhaus, *Org. Mass Spectrom.*, 1978, **13**, 591.
[202] S. M. Schildcrout, *Inorg. Chem.*, 1980, **19**, 244.

polynuclear ions were due to the interaction of primary fragment ions with the neutral mononuclear complex. The results of electron attachment reactions and the negative ion spectra have been reported for a series of nickel(II) β-diketonate complexes of the type Ni(R^1COCHCOR^2)$_2$ (where R^1 is a perfluoroalkyl group and R^2 an alkyl or aryl group).[203] It was found that the degree of fragmentation was dependent upon R^1 and R^2, and [M]$^-$ together with ligand ions provided the major contribution to the total ion currents for each complex.

A study of the effects of oxygen and sulphur donor atoms on electron attachment reactions for a series of bischelates of Ni^{II} with β-diketones and their mono and dithio analogues indicated that high yields of [M]$^-$ could be achieved by low-energy EA and that a greater ease of reduction occurred as the number of sulphur atoms bonded to Ni increased.[204] EuL$_3 \cdot$3H$_2$O (HL = trifluoroacetylacetone), EL$_3$Q (Q = 1,10-Phenanthroline or 2,2'-bipyridine), Eu$_3$L^{1}Q (HL^1 = divaloyl-TFA), and EuL$_3^2$Q (HL^2 = trifluoroacetone) have been characterized[205] and a series of fluorinated europium β-diketonates (e.g. Eu[1-(2-thienyl)-trifluorobutane-1,3-dione]$_3 \cdot$3H$_2$O) have been investigated.[206] The spectra of a series of Ni^{II}, Cu^{II}, Zn^{II}, and Co^{II} complexes of difluoromonothio-β-diketones, RCSCH$_2$COCHF$_2$ (R = 2-thienyl, Ph, or p-tolyl), have been recorded and the spectra of the Ni complex compared with the corresponding β-diketone complex.[207] The spectra of the mixed chelate complex PrL^1(L^2)$_2$ [HL^1 = 1,1,1-trifluoro-4-(2-thienyl)-butan-2-one-thione, HL^2 = 1,1,1-trifluoro-4-(2-thienyl)-2,4-butanedione] have been discussed.[208] The spectrum of Mg-2,2',6,6'-tetramethylheptadione has been reported.[209] Results of electron attachment reactions in the gas phase and the negative ion spectra for a series of bis(1,1,1,5,5,5-hexafluoropentane-2,4-dionato)metal(II) complexes, ML$_2$ (M = Mg, Mn, Co, Ni, Cu, or Zn), have been reported.[210] [ML$_2$] or [ML]$^-$ and [L]$^-$ ions were observed and the labile ligand C—F and C—CF$_3$ bonds enabled rearrangement processes, involving atom transfers, to be significant in the decomposition of molecular and fragment ions. The volatile uranium complex UO$_2$[(CF$_3$CO)$_2$CH$_2$]$_2 \cdot$THF has been characterized[211] and the spectra of tris(1,1,1-trifluoropentane-2,4-dionate) complexes of Al, Cr, Ru, and Rh[212] as well as a series of volatile metal chelate complexes of Ca, Mg, Fe, and Cd have been reported.[213,214] Di-2-pyridylketone complexes of Zn^{II}, Cd^{II}, and Hg^{II} halides have been identified.[215] Reaction schemes have been proposed for the

---

[203] D. R. Dakternieks, I. W. Frazer, J. L. Garnett, and I. K. Gregor, *Org. Mass Spectrom.*, 1979, **14**, 676.
[204] J. L. Garnett, I. K. Gregor, M. Giulhous, and D. R. Dakternieks, *Inorg. Chem. Acta*, 1980, **44**, L121.
[205] V. S. Khomenko, T. A. Rasshinina, and V. P. Suboch, *Vestsi Akad. Navuk BSSR, Ser. Khim. Navuk*, 1979, 63.
[206] V. S. Khomenko, T. A. Rasshinina, and V. P. Suboch, *Vestsi Akad. Navuk BSSR, Ser. Khim. Navuk*, 1979, 36.
[207] M. Das, *Transition Met. Chem.*, 1980, **5**, 17.
[208] N. Dowling M. Cefola, and J. G. White, *J. Inorg. Nucl. Chem.*, 1979, **41**, 106.
[209] R. Schwartz and C. C. Giesecke, *Clin. Chem. Nutr. Sci., Correll*, 1979, **97**, 1.
[210] D. R. Dakternieks, I. W. Frazer, J. L. Garnett, and I. K. Gregor, *Aust. J. Chem.*, 1979, **32**, 2405.
[211] G. M. Kramer, M. B. Dinis, R. B. Hall, A. Kaldor, A. J. Jacobon, and J. C. Scanlon, *Inorg. Chem.*, 1980, **19**, 1340.
[212] J. E. Campana, T. H. Risby, and P. C. Jurs, *Anal. Chim. Acta*, 1979, **112**, 321.
[213] D. L. Hachey, J. Johnston, and P. D. Klein, Report 1978, CONF-780501-7.
[214] D. L. Hackey, J. Johnston, and P. D. Klein, *Stable Isot. Proc. Int. Conf. 3rd.* 1978, 157.
[215] J. D. Ortego, S. Upalawanna, and S. Amanollahi, *J. Inorg. Nucl. Chem.*, 1979, **41**, 593.

formation of ligand and fragment ions formed in the electron capture and negative ion spectra of ML_2 and ML_3 complexes (M = Sc^{III}, Cr^{III}, Mn^{III}, Fe^{III}, Co^{III}, Al^{III}, Ga^{III}, Tn^{III}, Co^{III}, Ni^{II}, Cu^{II}, or Zn^{II}; L = dipivaloylmethane).[216]

Carboxylates and Related Compounds. The spectra of a series of carboxylic acid salts RCO_2M [R = Me, CF_3, Me_3C, or $CF_3(CF_2)_2$; M = Li, Na, K, Rb, or Cs] exhibited cluster ions $[M_n(O_2C)_{n-1}]^+$ (n = 2—8).[217] The observation of $[Re_2(piv)_4Cl_2]^+$, $[Re_2(piv)_4Cl]^+$, and $[Re_2(piv)_3Cl_3]^+$ ions in the spectrum of $Re_2(Me_3CCO_2)_3Cl_3$ suggested that the complex undergoes ligand redistribution on heating.[218] The spectra of $Be_4O(CO_2R)_6$ (R = H, Me, or Et) and $BeO(CO_2R^1)_5OR^2$ (R^1 = Me, R^2 = H, Me, Et, or Pr^n; R^1 = R^2 = Et) show major fragmentations involving the elimination of $(R^1CO)_2O$, $R^1CO_2R^2$, $Be(CO_2R^1)_2$, or $Be(CO_2R^1)$ or from the molecular ions.[219] It was suggested that the fragmentation patterns could be rationalized by stereochemical considerations and were practically independent of the organic substituent. The spectrum of $Zn_4O(MeCO_2)_6$ has been reported.[220] Ni^{II} and Zn^{II} bis(*NN*-dialkyldithiocarbamates) gave $[M]^+$ ions for all compounds and this ion decreased in abundance with increasing size of the alkyl substituent.[221] The dithiocarbamate complexes $Os(Et_2dtc)_3$ and $Os_2N(Et_2dtc)_5$ have been identified.[222] In the spectrum of $Pd(MeCS_2)_2$, the major species in the gas phase was the monomer with the $PdMeCS_2^+$ ion forming the base peak.[223] However, ions of appreciable abundance containing two palladium ions were detected. EI and FD spectra have been reported for a series of zinc dihydrocarbyldithiophosphates and their fragmentation patterns compared with their possible modes of decomposition.[224]

Alkoxides and Related Compounds. The isopropoxides of K, Rb, and Cs have been examined[225] and the compounds $BaTi_4(OEt)_{18}$ and $BaTi(OEt)_6$ identified.[226] The fragmentation of compound (72; R^1, R^2 = H, Me; m, n = 1, 2) proceeded *via* ring cleavage and loss of an alkoxy radical followed by further elimination of ethylene and acetylene fragments.[227] The spectra of $Al_3Cl_5(OPr^1)_4$ and of $AlCl(OPr^1)_2$ have been reported.[228,229] The spectra of pentafluorodimethylsilyl derivatives of alcohols, phenols, carboxylic acids, and

---

[216] J. L. Garnett, I. K. Gregor, and M. Giulhaus, *Org. Mass Spectrom.*, 1978, **13**, 591.
[217] E. V. White, *Org. Mass Spectrom.*, 1978, **13**, 495.
[218] F. A. Cotton, L. D. Gage, and C. E. Rice, *Inorg. Chem.*, 1979, **18**, 1138.
[219] Y. S. Nekrasov, S. Yu. Sil'vestrova, A. I. Grigor'ev, L. N. Reshetova, and V. A. Sipachev, *Org. Mass Spectrom.*, 1978, **13**, 491.
[220] V. A. Sipachev, L. N. Reshetova, Y. S. Nekrasvo, and S. Y. Sil'vestrova, *Org. Mass Spectrom.*, 1980, **15**, 192.
[221] J. Krupcik, P. A. Leclercq, J. Garaj, and J. Masaryk, *J. Chromatogr.*, 1979, **171**, 285.
[222] K. W. Given, S. H. Wheeler, B. S. Jick, L. J. Mahen, and L. H. Pignolet, *Inorg. Chem.*, 1979, **18**, 1261.
[223] O. Piovesana, C. Bellitto, A. Flamini, and P. F. Zanozzi, *Inorg. Chem.*, 1979, **18**, 2259.
[224] V. Raverdino, G. Natoli, and E. Sappa, *Chim. Ind. (Milan)* 1979, **61**, 643.
[225] T. Greiser and E. Weiss, *Chem. Ber.*, 1979, **112**, 844.
[226] E. P. Turevskaya, N. Ya. Turova, and A. V. Novoselova, *Dokl. Akad. Nauk SSSR*, 1978, **242**, 883.
[227] I. Mazeika, A. P. Gaukhman, I. Jankovska, I. I. Solomennikova, G. Zelcans, and E. Lukevics, *Zh. Obshch. Khim.*, 1978, **48**, 2728.
[228] A. I. Yanovskii, A. V. Kozunov, N. Ya. Turova, N. G. Furmanova, and Yu. T. Struchkov, *Dokl. Akad. Nauk SSSR*, 1979, **244**, 119.
[229] V. A. Kozunov, N. I. Kozlova, N. Ya. Turova, and Yu. S. Nekrasov, *Zh. Neorg. Khim.*, 1979, **24**, 1526.

amines were simple and contained a few abundant ions at high mass, characteristic of the parent compounds.[230] Principal fragment ion clusters of $Ge(OEt)_2 \cdot 2H_2O$ have been reported[231] and the major feature in the spectrum of compound (73) was the persistence of nitrogen in the fragment ions even after the loss of two OCH_2 units,[232] strongly suggesting a GeN bond in the parent compound.

A study of the bicyclic compounds (74) and (75) (R = H, Me, Et, or CH_2OH) was made in an attempt to establish the effect of the bridgehead phosphorus atom on the losses of neutral species from the molecule.[233] The expected loss

of HCHO was observed for the former compound in contrast to loss of C_2H_4 for the latter. The differences in abundance between the $[M - HS]^+$ ions from diastereoisomeric 2-dimethylamino-2-thiono-4-methyl-1,3,2-dioxaphosphoranes have been rationalized in terms of fragmentation mechanisms and ion structures.[234] The rationalizations were supported by ^{2}H-labelling experiments and metastable chracteristics. The compounds (76),[235] (77),[235] (78),[235] (79),[236] and (80)[236] have been characterized. The molecular ion, accompanied by the loss of one alkoxide group, was observed for a series of alkoxides of titanium.[237] A study of compound (81; $R^1 = R^2 = R^3 = H$ or Me; R^1 = Me, R^2 = H or Me, R^3 = H) suggested successive elimination of $OCHR^1$, $OCHR^2$, and $OCHR^3$ moieties,[238] with the ease of fragmentation of the molecular ion increasing with the number of methyl groups. Complexes of Os^{VI} with catechol and sub-catechols have been characterized.[239] In the cases of alkyl-substituted compounds, fragments due to loss of a methyl group were observed.

---

[230] A. J. Francis, D. E. Morgan, and C. F. Poole, *Org. Mass Spectrom.*, 1978, **13**, 671.
[231] L. D. Silverman and M. Zeldin, *Inorg. Chem.*, 1980, **19**, 270.
[232] L. D. Silverman and M. Zeldin, *Inorg. Chem.*, 1980, **19**, 272.
[233] K. J. Voorhees, F. D. Hileman, and D. L. Smith, *Org. Mass Spectrom.*, 1979, **14**, 459.
[234] W. J. Stec., B. Zielinska, and B. van der Graaf. *Org. Mass Spectrom.*, 1980, **15**, 105.
[235] R. D. Kroshefsky, R. Weiss, and J. G. Verkade, *Inorg. Chem.*, 1979, **18**, 469.
[236] Yu. Ya. Efremov, R. Z. Musin, E. J. Mukmenev, and N. A. Makarova, *Khim. Geterotsikl. Soedin.*, 1979, 177.
[237] M. Bochmann, G. Wilkinson, G. B. Young, M. B. Hursthouse, and K. M. A. Malik, *J. Chem. Soc., Dalton Trans.*, 1980, 901.
[238] M. G. Voronkov, I. S. Emel'yanov, V. Yu. Vitkovskii, and A. Lapsina, *Zh. Obshch. Khim.*, 1978, **48**, 2490.
[239] A. J. Nielson and W. P. Griffith, *J. Chem. Soc., Dalton Trans.*, 1978, 1501.

```
    O—CH₂                   H₂C—O
   /   |                        \
X—P   HC—O               HC—O    P—X
   \   |    \             |    \/
    O—CH    P            HC—O   \              OCHR¹CH₂
    ‖    \  |             |     \           /
    O=CH    |            HC—O    P      OV—OCHR²CH₂—N
        |   |             |     /           \
        H₂C—O            H₂C—O               OCHR³CH₂
(79) X = Cl, MeO, or EtO  (80) X = Cl, MeO, or EtO      (81)
```

Compounds Containing Metal–Nitrogen Bonds.—FD spectra have been found to provide a novel and easy method for the characterization of salts of monocationic complexes.[240] In the majority of compounds studied, the base peaks corresponded to the molecular ion for the cation and no fragmentation was observed. However, in the case of *trans*-[Co(en)₂]Cl and some related complexes, one amine ligand was lost from the base peak and further loss of a halide ligand was observed. Chelates of 1,10-phenanthroline and cobalt chloride and its derivatives have been obtained by ion bombardment of a mixture of the organic compound and a transition-metal salt.[241] SIMS has been found to be a valuable technique for the study of metal chelates in normally inaccessible oxidation states in the absence of a solvent. The macrocyclic Schiff-base complexes [82; M = Ca, Sr, or Ba; X = NCS or ClO₄; R = (CH₂)₂ or O—C₆H₄][242] and Schiff-base complexes between ZnI and β-ketoimines[243] have been characterized. Fragmentations in the spectrum of bis[*N*-(3-hydroxy-1-propyl)-2-hydroxy-1-naphthaldiminato]germanium(IV) have been confirmed by the appropriate metastable peaks.[244]

(82)

The application of mass spectrometry for the estimation of transition metals (Ni, Co, Cu) by measurement of the isotopic peaks in the spectra of the chelate complexes {chelate = 2,3-bis[(4,5-dihydro-3-methyl-5-oxo-1-phenyl-1*H*-pyrazol-4-yl)azo]quinoxaline} has been reported.[245] A study has been made on the application of mass spectrometry to the thermal deoxygenation of cobalt–amino-acid–imidazole complexes.[246] The gaseous negative ion reactions of four- and five-co-ordinate Co^{II}, Ni^{II}, and Cu^{II} Schiff-base complexes with gas mixtures

[240] D. E. Games, J. L. Gower, and L. A. P. Kane-Maguire, *J. Chem. Soc., Chem. Commun.*, 1978, 757.
[241] R. J. Day, S. E. Unger, and R. G. Cooks, *J. Am. Chem. Soc.*, 1979, **101**, 499.
[242] D. H. Cook and D. E. Fenton, *J. Chem. Soc., Dalton Trans.*, 1979, 810.
[243] F. A. Bottino, E. Libertini, O. Puglisi, and A. Recca, *J. Inorg. Nucl. Chem.*, 1979, **41**, 1725.
[244] R. V. Singh and J. P. Tandon, *Curr. Sci.*, 1979, **48**, 802.
[245] R. S. Kuzanyan, R. V. Poponova, M. S. Chupakhin, and M. N. Stopnikova, *Otkrytrya, Izobret. Prom. Obraztsy, Tovarnye Znaki*, 1980, 97.
[246] V. A. Pokrovskii, Yu. L. Bratushko, K. B. Yatsimirskii, and E. N. Korol, *Teor. Eksp. Khim.*, 1978, **14**, 754.

containing CH_4 (~90%) and O_2, NO, CO, and CF_3 (~10%) gave abundant [M]⁻ ions.[247] The negative ion spectra of multidentate Schiff-base complexes with Co^{II}, Ni^{II}, and Cu^{II} have been obtained by CI using methane, [²H₄]methane, and isobutane reagent gases; [M]⁻ ions were observed for all complexes studied.[248] The negative ion spectra of Schiff-base complexes of Cu^{II}, Ni^{II}, and Co^{II}, using fluorine-containing reagent gases (CF_4, SO_2F_2, and SF_6), have been studied.[249] The Cu^{II} complexes were unreactive with any of the reagent gases but addition of F_2^- and F^- was observed for the Ni^{II} complexes when SO_2F_2 was the reagent gas. With the Co^{II} complexes, reactions of the [M]⁻ ion with CF_4 resulted in the formation of the [M + 69]⁻ ion as the major secondary ion. The negative ion spectra, under CI conditions, for a series of Cu^{II}, Ni^{II}, and Co^{II} Schiff-base complexes (83; Q = NH or O; Z = ethylene or phenylene) have been determined.[250] The reagent gases methane, [²H₄]methane, and isobutane were used. For all complexes, the [M]⁻ ions were observed with all three reagent gases and it was suggested that ion–molecule reactions between the [M]⁻ ion and the neutral gas were the major processes for the formation of secondary ions. Negative ion spectra have been reported for the chelate (84; X^1 = NH or O, X^2 = CH_2CH_2; X^1 = O, X^2 = o-C_6H_4; M = Cu, Ni, or Co) by CI with CF_4, SO_2F_2, and SF_6 reagent gases.[251] The formation of [M] was dependent upon the reagent gas. In the complex (85; R^1 = Me, R^2 = Ph, o-MeC_6H_4, or p-MeC_6H_4; R^1 = Ph, R^2 = Ph or o-MeC_6H_4), the loss of CO with migration of methyl to the nucleus of the chelate ring was observed.[252]

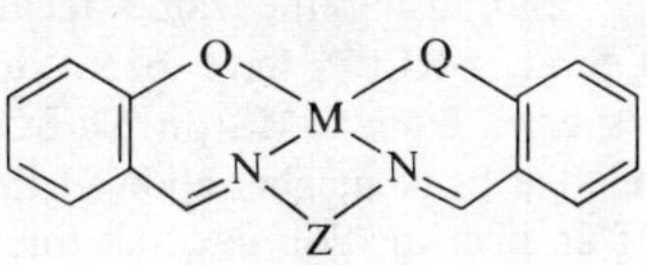

(83)

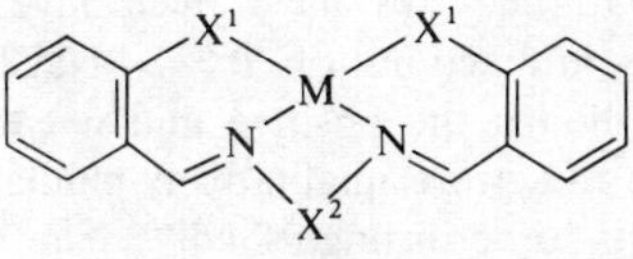

(84)

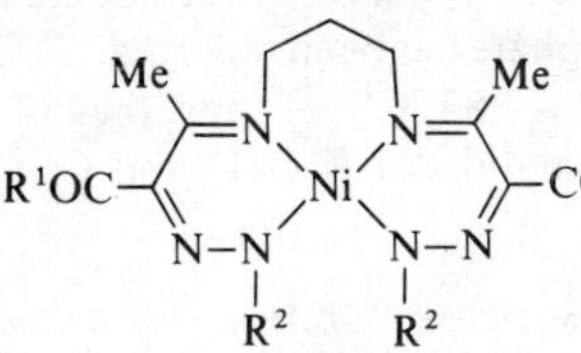

(85)

5 Inorganic Compounds

Group I.—The mechanism of ion formation in FD of alkali halides has been commented on.[253] The formation of cluster ions was attributed to the effect of a field-and-temperature-dependent charging of salt layers by alkali ions.

---

[247] E. Baumgartner and J. G. Dillard, *Inorg. Chim. Acta*, 1979, **32**, 11.
[248] J. G. Dillard and E. Baumgartner, *Inorg. Chim. Acta*, 1979, **35**, 119.
[249] J. G. Dillard and E. Baumgartner, *Inorg. Chim. Acta*, 1979, **35**, 279.
[250] E. Baumgartner, T. C. Rhyne, and J. G. Dillard, *J. Organomet. Chem.*, 1979, **171**, 387.
[251] E. Baumgartner and J. G. Dillard, *Org. Mass Spectrom.*, 1979, **14**, 360.
[252] K. P. Balakrishnan and V. Krishnan, *J. Inorg. Nucl. Chem.*, 1979, **41**, 37.
[253] F. W. Roellgen and K. H. Ott, *Int. J. Mass Spectrom. Ion Phys.*, 1980, **32**, 363.

Differences in the FD of alkali halides and Group II metal halides (*e.g.* $CaCl_2$) were reported. The vaporization of melts of several alkali metal halides, alkaline earth halides, and their mixtures has shown that the vapour species in alkali halides are composed of dimers, trimers, and tetramers in progressively decreasing amounts.[254] However, only small amounts of dimers were found in alkaline earth bromide vapours. The appearance energy of ions formed during ionization of $LiNO_2$ and $NaNO_3$ has been studied and information concerning the ionization of KNO_2, $RbNO_2$, and $CsNO_2$ has been obtained.[255] Thermodynamic characteristics of these molecules and ions have been tabulated and also the ionization energies of some monoxides and oxides of alkali metals. Between 550 and 710 K, alkali metal nitrates and nitrites gave, in addition to the metal ions, $M_2NO_3^+$, MO^+, MNO_2^+, MNO^+, M_2^+X, and $M_2NO_3^+$ for the nitrates and $M_2NO_2^+$, MO^+, M_2O^+, M_2^+, MNO^+, M_2NO^+, and MNO_2^+ for the nitrites.[256] Heterogeneous halogen exchange reactions between alkali halides and a number of mono- and di-halogenated alkanes have been probed by FD mass spectrometry.[257] The results of halogen exchange between LiI and alkanes (PrCl + Bu^{t}Br) were reported. The isotopic composition of Li in LiCl has been determined[258] and the major vapour species over solid $LiCrO_2$ have been identified.[259] The thermodynamic activity of $LiReO_4$ in the $LiReO_4$–$CsReO_4$ system at 833 K has been determined[260] and the compounds $LiReO_4$, $CsReO_4$, $(LiReO_4)_2$, $(CsReO_4)_2$, and $LiCs(ReO_4)_2$ have been shown to exist in the gas phase. The equilibria of gases coming from Knudsen cells containing LiCl, $ScCl_3$, or LiCl–$ScCl_3$ mixtures have been investigated[261] and, using the same technique, the interactions of $ScCl_3$---NaCl[262] and $ScCl_3$---KCl[263] have been studied. Although the positive and negative ion spectra from SIMS on Na_2SO_4 and Na_2SO_3 were qualitatively similar, the relative peak heights allowed the two salts to be distinguished.[264] The results of an investigation into the formation of $[Na_3SO_4]^+$ in SIMS of solid Na_2SO_3 by ^{18}O-labelling suggested that $[Na_3SO_4]^+$ was formed in the gas phase by statistical recombination.[265] The spectrum of the vapour of potassium sulphate has been reported.[266] The spectrum of Na_2CrO_4 has been measured at 670—785 °C[267,268] and that of K_2CrO_4 in the vapour phase.[268] The heats of formation of K_2OH^+ and Cs_2OH^+ ions in the vapour

---

[254] H. H. Emons, W. Horlbeck, and D. Pohl, *Ext. Abstr. Meet. Int. Soc. Electrochem.*, 1979, **30**, 193.
[255] E. N. Verkhoturov and O. T. Nikitin, *Vestn. Mosk. Univ., Ser. 2, Khim.*, 1979, **20**, 423.
[256] N. V. Bagarat'yan, E. N. Verkhoturov, M. K. Il'in, A. V. Makarov, and O. T. Nikitin, *Vses. Konf. Kalorim., [Rasshir. Tezisy Dokl.] 7th*, 1977. **2**, 313.
[257] F. W. Rollgen and K. H. Ott, *J. Chem. Soc., Chem. Commun.*, 1978, 612.
[258] V. E. Baraboshkin and V. A. Dubinin, *Zavod. Lab.*, 1979, **45**, 347.
[259] T. Ohinichi, H. Takeshita, S. Nasu, T. Sasayama, A. Malda, M. Miyake, and T. Sano, *J. Nucl. Mater.*, 1979, **82**, 214.
[260] W. Lukas, C. Chatillon, and M. Allibert, *J. Less. Common Met.*, 1979, **66**, 211.
[261] H. Schaefer and K. Wagner, *Z. Anorg. Allg. Chem.*, 1979, **450**, 88.
[262] K. Wagner and H. Schaefer, *Z. Anorg. Allg. Chem.*, 1979, **450**, 107.
[263] K. Wagner and H. Schaefer, *Z. Anorg. Allg. Chem.*, 1979, **450**, 115.
[264] J. Marien and E. DePauw, *Bull. Soc. Chim. Belg.*, 1979, **88**, 115.
[265] E. DePauw and J. Marien, *Bull. Soc. Chim. Belg.*, 1980, **89**, 83.
[266] A. G. Efimova and L. N. Goroknov, *Teplofiz. Vys. Temp.*, 1978, **16**, 1195.
[267] C. Hirayama and C. Y. Lin, *NBS Spec. Publ. (U.S.)* 1979, 561 and 1539.
[268] C. A. Stearns, F. J. Kohl, R. A. Miller, and G. C. Fryburg, *NASA Tech. Memo*, 1979, NASA-TM-79210, E-095.

above KOH and CsOH have been calculated.[269] The saturated vapour over the RbF–GaF$_3$ system, containing 25—100% GaF$_3$, was investigated at 810—1112 K and the partial pressures of RbF, GaF$_3$, RbGaF$_4$, and (RbGaF$_4$)$_2$ were determined.[270] In a study of the ion–molecule equilibria in the vapour above CsI the ions Cs$^+$, Cs$_2$I$^+$, Cs$_3$I$_2^+$, I$^-$, CsI$_2^-$, and CsI$_3^-$ were detected.[271] In a similar study of NaF, no negative Na$_n$F$_{n+1}$ ions were observed. The distances over which single atoms in a lattice can combine to form dimers during sputtering have been studied for KCl and CsCl with He$^+$ and Ar$^+$. The limiting distance for formation of Cs$_2^+$ was found to be ~200 and ~400 Å for He$^+$ and Ar$^+$, respectively.[272] The properties of CsAu have been studied and a bond energy of 460 ± 30 kJ mol^{-1} determined.[273]

The dissociation energies of the gaseous hydrides CuH, AgH, and AuH have been determined[274] and the equilibria involving gaseous species above a Cu–Sn alloy have been determined by high-temperature techniques.[275] The bond dissociation energies of Cu$_2$ and CuSn were calculated and the gaseous molecules Cu$_2$, CuSn, Cu$_2$Sn, and CuSn$_2$ identified. The vapour phases in equilibrium with the condensed mixture Cu$_x$Ag$_{1-x}$Cl and AgCl$_x$Br$_{1-x}$ ($0 \le x \le 1$) have been investigated and the trimeric species Cu$_3$Cl$_3$, Cu$_2$AgCl$_3$, CuAg$_2$Cl$_3$, Ag$_3$Cl$_2$Br, Ag$_3$ClBr$_2$, and Ag$_3$Br$_3$ identified; fragmentations Cu$_{3-y}$Ag$_y$Cl$_3^+$ → Cu$_{2-y}$Ag$_y$Cl$^+$ + CuCl$_2$ ($y = 0, 1, 2$, or 3) were observed.[276]

A value of about 3 eV has been estimated for the Au—O bond.[277] Reactions involving Au, V, Au$_2$, and AuV have been investigated and the bond dissociation energy of AuV has been calculated.[278] SIMS has been used to detect the migration of Au from the electrode into the oxide layer of Au–Y$_2$O$_3$–Y functions during electroforming processes.[279] The gaseous equilibria involving the molecules AuSi, AuSi$_2$, and Au$_2$Si have been studied by means of the Knudsen effusion technique combined with mass spectral analysis of the vapours.[280]

Group II.—The formation of gas complexes MgSc$_x$Cl$_y$ from MgCl$_2$ and ScCl$_3$ have been studied and heats of formation of MgScCl$_5$, Mg$_2$ScCl$_7$, and MgSc$_2$Cl$_3$ calculated.[281] Using negative ion mass spectrometry ^{26}Mg and ^{26}Al have been separated from mg samples with an ^{26}Al–^{27}Al ratio sensitivity of better than 10^{-14}.[282] The activities of CaF and CaO in CaF$_2$–CaO–Al$_2$O$_3$ slags at 1700 K have been measured[283] and a mass spectral analysis has been made of emitted

---

[269] L. S. Kudin, A. V. Gusarov, and L. N. Gorokhov, *Vses. Konf. Kalorim.*, [*Rasshir. Tezisy Dokl.*], 7th, 1977, **2**, 319.
[270] M. V. Korobov and L. N. Sidorov, *Vestn. Mosk. Univ., Ser. 2, Khim.*, 1979, **20**, 185.
[271] I. V. Sidorov, A. V. Gusarov, and L. N. Gorokhov, *Int. J. Mass Spectrom. Ion Phys.*, 1979, **31**, 367.
[272] F. Honda, Y. Fuknda, and J. W. Rabalais, *J. Chem. Phys.*, 1979, **70**, 4834.
[273] B. Busse and K. Weil, *Angew. Chem.*, 1979, **91**, 664.
[274] A. Kant and K. A. Moon, *High Temp. Sci.*, 1979, **11**, 55.
[275] J. E. Kingcade, D. C. Dufner, S. K. Gupta, and K. A. Gingerich, *High Temp. Sci.*, 1978, **10**, 213.
[276] O. Bernauer, B. Busse, and K. G. Weil, *Ber. Bunsenges. Phys. Chem.*, 1979, **83**, 603.
[277] A. Hecq, M. Vandy, and M. Hecq, *J. Chem. Phys.*, 1980, **72**, 2876.
[278] S. K. Gupta, M. Pelino, and K. A. Ginerich, *J. Chem. Phys.*, 1979, **70**, 2044.
[279] A. Noya, S. Kuriki, G. Matsumoto, and M. Hirano, *Thin Solid Films*, 1979, **59**, 143.
[280] K. A. Gingerich, R. Haque, and J. E. Kingcade, *Thermochim. Acta*, 1979, **30**, 61.
[281] K. Wagner and H. Schaefer, *Z. Anorg. Allg. Chem.*, 1979, **452**, 83.
[282] L. R. Kilius, R. P. Benkens, K. H. Chang, H. W. Lee, A. E. Litherland, D. Elmore, R. Ferrago, and H. E. Gove, *Nature (London)*, 1979, **282**, 488.
[283] M. Allibert and C. Chatillon, *Can. Metall. Q.*, 1979, **18**, 349.

positive ions from CaO during reactive recombination of H atoms on its surface.[284] In the spectrum of CaO containing Bi (2.72%), Ca^+, Bi_6^+, CaO_2^+, and Ca_2O^+ ions were observed. High-temperature mass spectrometry has been used to study the equilibrium behaviour of a series of IIB/VIA compounds of the type MX (M = Zn, Cd, or Hg; X = O, S, Se, or Te).[285] Evidence for the stable gaseous molecules ZnO, CdO, HgO, CdS, HgS, ZnSe, and CdSe was presented. A study has been made of the emissions which resulted from collisions of 5—200 eV electrons with gaseous HgX_2 and MeHgG (X = Cl, Br, or I).[286]

Group III.—*Compounds Containing B—O Bonds.* The thermal dissociation of gaseous boron oxides has been studied and the heat of formation of BO_2 determined.[287] The emission of negative ions from Ta and W filaments, coated with a number of boron compounds, has been studied and BO_2^- ion was formed at high temperatures from $Na_2B_4O_7$, $B + O_3$, $B + B_2O_3$, $B + Al_2O_3$, and $B + La_2O_3$.[288] A study of the spectra of the vapours of the 1:2:1 molar mixture B_2O_3–B_4C–NaF indicated that B^+, BF^+, BF_2^+, and BF_3^+ ions were present at temperatures above 450 °C.[289]

Thermodynamic properties of melts of GeO_2–B_2O_3,[290] Na_2O–B_2O_3,[291] Na_2O–B_2O_3–GeO_2,[292,293] and B_2O_3–$NaBO_2$[294] have been studied. SIMS of B_2O_3 and H_3BO_3 indicated that they had similar structural units with weaker B—O bonds in H_3BO_3.[295]

Boron–Nitrogen Compounds. High-resolution spectra have been reported for $[Me_3Si]_2NB(NH_2)NR_2$ (R = Et or Pr)[296] and $(Bu^tMe_2)SiN(R)B(NMe_2)_2$ (R = H or Me),[297] and cyclic systems of the type $(Bu^tMe_2SiNMeB—NH)_3$ have been characterized.[297] Ion cyclotron resonance trapped-electron spectra have been reported for borazine and trimethylborazine.[298]

Boron Hydrides and Related Compounds. The spectra and ion current *vs.* pressure have been determined for B_{2-10} boranes and argon and a method has been developed for the continuous gas-phase quantitative analysis of borane mixtures;[299] cothermolysis of B_4H_{10} with B_2H_6, B_5H_9, B_5H_{11}, B_6H_{10}, and B_6H_{12} at

---

[284] V. P. Grankin and V. V. Styrov, *Pis'ma Zh. Tekh. Fiz.*, 1979, **5**, 736.
[285] M. Grade and W. Hirschwald, *Z. Anorg. Allg. Chem.*, 1980, **460**, 106.
[286] J. Allison and R. N. Zare, *Chem. Phys.*, 1978, **35**, 263.
[287] N. V. Bagarat'yan, O. T. Nikitin, and L. N. Gorokhov, *Vestn. Mosk. Univ., Ser. 2, Khim.*, 1980, **21**, 139.
[288] P. B. Paic, M. M. Ceric, and K. F. Zubov, *Glas. Hem. Drus., Beograd*, 1979, **44**, 195.
[289] V. A. Ashuiko, V. P. Kryukov, and I. A. Rat'Kovskii, *Khim. Khim. Tekhnol. (Minsk)*, 1979, **14**, 155.
[290] M. M. Shul'ts, V. L. Stolyarova, and G. A. Semenov, *Fiz. Khim. Stekla*, 1978, **4**, 653.
[291] M. M. Shul'ts, V. L. Stolyarova, and G. A. Semenov, *Dokl. Akad. Nauk SSSR*, 1979, **246**, 154.
[292] M. M. Shul'ts, G. A. Semenov, and V. L. Stolyarova, *Izv. Akad. Nauk SSSR, Neorg. Mater.*, 1979, **15**, 1002.
[293] V. L. Stolyarova and G. A. Semenov, *Fiz. Khim. Stekla*, 1979, **5**, 127.
[294] M. M. Shul'ts, V. L. Stolyarova, and G. A. Semenov, *Fiz. Khim. Stekla*, 1979, **5**, 42.
[295] D. J. Joyner and D. M. Hercules, *J. Chem. Phys.*, 1980 **72**, 1095.
[296] D. M. Graham, J. R. Bowser, C. G. Moreland, R. H. Neilson, and R. L. Wells, *Inorg. Chem.*, 1978, **17**, 2028.
[297] J. R. Bowser, R. H. Neilson, and R. L. Wells, *Inorg. Chem.*, 1978, **17**, 1882.
[298] C. E. Doiron, M. E. MacBeath, and T. B. McMahon, *Chem. Phys. Lett.*, 1978, **59**, 90.
[299] T. C. Gibb, N. N. Greenwood, T. R. Spalding, and D. Taylorson, *J. Chem. Soc., Dalton Trans.*, 1979, 1392.

75 °C has been investigated.[300] The variation, with time, of the concentrations of the major products was compared with information from thermolysis of each borane in an attempt to establish the role of B_4H_{10} and possible reactive species obtained from B_4H_{10} in the cothermolysis. The hydrides 1,2-$(B_5H_8)_2$,[301] 2,2'-$(B_5H_8)_2$,[301] and $H_2\bar{B}(CH_2)_2\overset{+}{N}Me_2$[302] have been examined by high-resolution mass spectrometry. A cut-off peak at m/z 111 for the $[M]^+$ ion was observed in the spectrum of $Me_3N \cdot B_4H_8$.[303] An analysis of the intensity ratio of the high mass cut-off region in the spectrum of $B_4H_8 \cdot 2PMe_3$ was made in order to identify the composition of ions arising from the successive loss of a pair of hydrogen atoms from the parent.[304] No molecular ion was observed in the spectrum of Me_2GaBH_4.[305] Gas-phase ion chemistry of $U(BH_4)_4$ has been studied by ion cyclotron resonance spectroscopy,[306] giving an electron impact ionization energy of 9.0 ± 0.5 eV and appearance energies of several positive fragment ions.

The borane compounds $(\eta^5\text{-}C_5H_5)FeB_5H_{10}$, $(\eta^5\text{-}C_5H_5)FeB_8H_8$, $2\text{-}(\eta^5\text{-}C_5H_5)FeB_{10}H_{15}$,[307] $B_9H_9S_2$, B_9H_9SSe, and B_7H_9SSe[308] have been characterized and the spectra of 1,2-$B_{10}H_{10}AsSb$, 1,2-$B_{10}H_{10}Sb_2$, $(\eta^5\text{-}C_5H_5)CoB_9H_9AsSb$, and $(\eta^5\text{-}C_5H_5)CoB_9H_9Sb_2$ have been reported.[309] The spectra of $PhEMe_3$ and o-, m-, and p-$HCB_{10}H_{10}CEMe_3$ (E = C, Si, Ge, or Sn) have been compared; the results indicate the differences in bonding in the compounds.[310] Spectra for the halogenated *closo*-carbaboranes 2-Cl-1,6-$C_2B_4H_5$ and 2,4-Cl_2-1,6-$C_2B_4H_4$ have been reported; the $[M]^+$ ion was observed in each case.[311] *nido*-Decacarbaboranes $B_{10}H_{11}C_6H_{11}SEt_2$ and $B_{10}H_{11}C_6H_9SMe_2$[312] have been studied and high-resolution spectra of 2-$C_6H_5CH_2B_5H_8$ and 2-$CH_2CHCH_2B_5H_5$[313] obtained. A series of *nido*-metallo carbaboranes containing four skeletal carbon atoms [*e.g.* $(\eta^5\text{-}C_5H_5)CoMe_4C_4B_7H_7]$[314] have been reported. A series of cobalt carbaborane clusters $C_{13}B_7H_{26}Co\text{—}C_{18}B_8H_{18}Co$ has been characterized.[315] In a study of the spectra of (86; $X^1 = X^2 = NMe_2$ or F; $X^1 = F$, $X^2 = NMe_2$), a facile cleavage of

$$Ph_2PC\text{—}C\overset{X^1}{\underset{X^2}{\diagdown}}$$
$$B_{10}H_{10}$$

(86)

[300] T. C. Gibb, N. N. Greenwood, T. R. Spalding, and D. Taylorson, *J. Chem. Soc., Dalton Trans.*, 1979, 1398.
[301] D. F. Gaines, M. W. Jorgenson, and M. A. Kulzick, *J. Chem. Soc., Chem. Commun.*, 1979, 380.
[302] G. F. Warnock and N. E. Miller, *Inorg. Chem.*, 1979, **18**, 3620.
[303] A. R. Dodds and G. Kodama, *Inorg. Chem.*, 1979, **18**, 1465.
[304] G. Kodama and M. Kameda, *Inorg. Chem.*, 1979, **18**, 3032.
[305] A. J. Downs and P. D. P. Thomas, *J. Chem. Soc., Dalton Trans.*, 1978, 809.
[306] P. B. Armentrout and J. L. Beauchamp, *Inorg. Chem.*, 1979, **18**, 1349.
[307] R. Weiss and R. N. Grimes, *Inorg. Chem.*, 1979, **18**, 3291.
[308] G. D. Friesen, A. Barriola, P. Daluga, P. Ragatz, J. C. Hoffman, and L. J. Todd, *Inorg. Chem.*, 1980, **19**, 458.
[309] J. L. Little, *Inorg. Chem.*, 1979, **18**, 1598.
[310] A. V. Belokon, T. V. Klimova, Yu. S. Nekrasov, and I. Stanko, *Dokl. Akad. Nauk SSSR*, 1979, **245**, 363.
[311] C. Takimoto, G. Siwapinyoyos, K. Fuller, A. P. Fung, Li Liauw, W. Jarvis, G. Millhauser, and T. Onak, *Inorg. Chem.*, 1980, **19**, 107.
[312] E. I. Tolpin, E. Mizusawa, D. S. Becher, and J. Venzel, *Inorg. Chem.*, 1980, **19**, 1182.
[313] D. F. Gaines and M. W. Jorgenson, *Inorg. Chem.*, 1980, **19**, 1398.
[314] W. M. Maxell and R. N. Grimes, *Inorg. Chem.*, 1979, **18**, 2174.
[315] J. S. Plotkin and L. G. Sneddon, *Inorg. Chem.*, 1979, **18**, 2165.

the P—C bond was observed.[316] The fingerprint spectrum of $B_{11}Cl_{11}$ has been determined[317] and the results of a direct measurement, by SIMS, of self diffusion of boron in an iron–nickel–boron glass ($Fe_{40}Ni_{40}B_{20}$) have been reported.[318]

Other Compounds. The activation energy for release of hydrogen from an AlH_3 pellet by u.v. radiation has been calculated.[319] AlO^+, AlO_2^+, Al_2O^+, and $Al_2O_2^+$ ions are formed from an Al_2O_3–graphite (1:1) electrode.[320] Halide transfer, dehydrohalogenation, and oxidation reactions have been observed in ion-molecule reactions of $[Al]^+$ ion with a number of alkyl halides using an ion beam source.[321] Reaction of $[Al]^+$ with CH_2ClCH_2Cl gave $AlCl_2^+$ and C_2H_4 – a process in which the metal was formally oxidized from the +1 to the +3 state. The molecular structures of some donor–acceptor complexes of Al and Ga chlorides and bromides with NH_3 have been investigated and fragmentation pathways proposed.[322] A study of a Ge–$GaCl_3$ mixture indicated that the gaseous complex $GeCl_2 \cdot GaCl_3$ was formed.[323] The vapour pressure and thermodynamics of TlCl and Tl_2Cl_2 have been investigated.[324]

Group IV.—The ionic products $Cs^+ECl_4^-$, ECl_3^-, and Cl^- were obtained when an atomic beam of caesium, with kinetic energy of 1—150 eV, intersected a thermal energy beam of ECl_4 (E = Si, Ge, Sn, or Ti); the electron affinities of ECl_4 and ECl_3 (E = Ti or Sn) were calculated.[325] The heat and free energy of formation of silicon oxynitride (Si_2N_2O) have been determined.[326] The heats of formation of $HSiCl_3$, H_2SiCl_2, H_3SiCl, $HSiF_3$, H_2SiF_2, and H_3SiF have been determined from high-temperature thermochemical reactions of $SiH_{4(g)}$ with H_2, $HSiCl_3$ with H_2, and $SiF_{4(g)}$ with H_2.[327] Ion abundances for the species involved in the reactions were measured. The reaction between $SiBr_{4(g)}$ and H_2, between 900 and 1143 K, has been studied.[328] Mass spectral studies of the exchange between $ClSiH_3$ and N-methyldisilazane and trisilazane indicated a very low energy of activation and suggest a mechanism involving a quaternary intermediate.[329] The spectra of $(H_3SiNH)_2SiH_2$ and $(H_3Si)_2NSiH_2NHSiH_3$ contain highest mass peaks at m/z 124 and m/z 152, respectively, which were assigned to the ion species $^{28}Si_3N_2H_{10}^+$ and $^{28}Si_4N_2H_{12}^+$.[330] The products of the reaction of solid Ge or Sn bombarded with H atoms have been identified.[331] The germanium

---

[316] W. E. Hill and L. M. Silva-Trivino, *Inorg. Chem.*, 1979, **18**, 361.
[317] S. B. Awad, D. W. Prest, and A. G. Massey, *J. Inorg. Nucl. Chem.*, 1978, **40**, 395.
[318] R. W. Cahn, J. E. Evetts, J. Patterson, R. F. Somekh, and C. K. Jackson, *J. Mater. Sci.*, 1980, **15**, 702.
[319] Yu. I. Mikhailov, Yu. G. Galitsyn, V. I. Poshevnev, and V. V. Boldyrev, *Kinet. Katal.*, 1979, **20**, 330.
[320] I. Cornides and T. Gal, *High Temp. Sci.*, 1978, **10**, 171.
[321] R. V. Hodges, P. B. Armentrout, and J. L. Beauchamp, *Int. J. Mass Spectrom. Ion Phys.*, 1979, **29**, 375.
[322] M. Hargittai, J. Tamas, M. Bihari, and I. Hargittai, *Acta Chim. Acad. Sci. Hung.*, 1979, **99**, 127.
[323] H. Schaefer, *Z. Anorg. Allg. Chem.*, 1980, **461**, 29.
[324] D. L. Drapcho and G. M. Rosenblatt, *J. Chem. Thermodyn.*, 1979, **11**, 335.
[325] B. P. Mathur, E. W. Rothe, and G. P. Reck, *Int. J. Mass Spectrom. Ion Phys.*, 1979, **31**, 77.
[326] T. C. Ehbert, T. P. Dean, M. Billy, and J. C. Labbe, *J. Am. Ceram. Soc.*, 1980, **63**, 235.
[327] M. Farber, *Gov. Rep. Announce. Index (U.S.)*, 1979, **79**, 82.
[328] M. Farber and R. D. Srivastava, *Chem. Phys. Lett.*, 1979, **60**, 216.
[329] M. L. Thompson, *Inorg. Chem.*, 1979, **18**, 2939.
[330] A. D. Norman and W. L. Jolly, *Inorg. Chem.*, 1979, **18**, 1594.
[331] Yu, A. Shishkin, V. V. Marusin, and V. V. Yagzhev, *Izv. Sib. Otd. Akad. Nauk SSSR, Ser. Khim. Nauk*, 1979, 66.

compounds H_3GeECF_3 and $H_2Ge(ECF_3)_2$ (E = Se or S) have been characterized and in the case of the former compound $[M]^+$ was more abundant for E = Se, whereas the reverse was the case for the latter.[332] The vapour, consisting of $Pb_2Re_2O_9$, over a lead(II) oxide–rhenium(VII) oxide system has been investigated.[333,334] The results of a study of gaseous lead nitrate have been reported.[335]

Group V.—The spectra of $F_2PN(SiH_3)_2$, $(F_2P)_2NSiH_3$,[336] and $HN(PF_2)PF_4$[337] have been reported and fragmentation pathways proposed for compound (87).[338] Using FD techniques, levels of the cyclophosphamide (88) in samples of urine, serum, and cerebrospinal fluid, after being taken orally, have been determined.[339] The spectrum of the phosphadiazine (89) has been reported.[340]

A series of mixed compounds [*e.g.* $F_3P_3N_3Me_6$, $F_5P_3N_3H_3Me_3$, and $F_4P_4N_2(NHBu)_4$], products of partial aminolysis of $(F_2PN)_{3 or 4}$ with $MeNH_2$, Me_2NH, or $BuNH_2$, have been characterized.[341] The spectra of $N_3P_3Cl_6$ and $Cl_3P=NPOCl_2$ have been reported[342] and the ring structures (90)[343] and (91)[344]

[332] R. D. H. Smith and S. R. Stobart, *Inorg. Chem.*, 1979, **18**, 538.
[333] G. A. Semenov, E. N. Nikolaev, and K. V. Orchinnikov, *Vestb. Leningr. Univ., Fiz., Khim.*, 1978, 85.
[334] Yu. A. Gel'man, Yu, N. Lyubitov, V. I. Mikhailov, E. N. Nikolaev, and G. A. Semenov, *Zh. Fiz. Khim.*, 1978, **52**, 2674.
[335] T. P. Radus, H. R. Udseth, and L. Friedman, *J. Phys. Chem.*, 1979, **83**, 2969.
[336] E. A. V. Ebsworth, D. W. H. Rankin, and J. G. Wright, *J. Chem. Soc., Dalton Trans.*, 1979, 1065.
[337] D. W. H. Rankin and J. G. Wright, *J. Chem. Soc., Dalton Trans.*, 1979, 1070.
[338] K. Utvary, M. Kubjacek, and K. Varmuza, *Z. Anorg. Allg. Chem.*, 1979, **458**, 281.
[339] H. R. Schulten, *GC-MS News*, 1979, **7**, 74.
[340] L. D. Garaeva, O. M. Nesterova, I. S. Levi, B.-S. Kikot, and M. N. Preobrazhenskaya, *Khim. Geterotsikl. Soedin.*, 1979, 1415.
[341] T. L. Evans and H. R. Allcock, *Inorg. Chem.*, 1979, **18**, 2342.
[342] R. D. Jaegar and D. R. Taylor, *J. Chem. Soc., Dalton Trans.*, 1980, 851.
[343] C. C. Chang, R. C. Haltiwanger, M. L. Thompson, H. J. Chen, and A. D. Norman, *Inorg. Chem.*, 1979, **18**, 1899.
[344] M. L. Thompson, R. C. Haltiwanger, and A. D. Norman, *J. Chem. Soc., Chem. Commun.*, 1979, 647.

characterized. Mass spectral studies on the polymer poly[bis(trifluoroethoxy)phosphazene] indicated that gaseous products were formed from removal of a side-group on pyrolysis.[345] A mass spectral study on bis(2-ethylhexyl)phosphate has been compared with that from γ-radiolysis.[346] A $[M]^+$ ion of low abundance was observed and also double rearrangement ions $(RO)P(OH)_3^+$ (R = 2-ethylhexyl) and $P(OH)_4^+$; the appearance energies of the principal ions were determined.

The spectra of pertrimethylsilyl derivatives of 3'- and 5'-monophosphoric acids of guanosine, uridine, cytidine, and adenosine have been reported and the ion at m/z 501, observed in all cases, was found to be more abundant for the 3'-isomer than the 5'-isomer.[347] The collisional activation spectra of phenylalkylphosphonium ions and chlorine-containing quaternary phosphonium ions have been discussed.[348]

Saturated and unsaturated vapours of arsenic iodides have been studied and the standard heats of dissociation, sublimation, and formation calculated[349] and the appearance energies of a number of positive ions formed by EI upon AsF_3, $AsCl_3$, $AsBr_3$, and AsF_3 have been determined. It was observed that the translational energies of the fragment ions from AsF_3 gave a linear dependence on electron energy.[350] The spectra of a series of triarylthioantimonates $Sb(SAr)_3$ have been studied and general fragmentation pathways, which included loss of SC_6H_4X-p (X = H, Br, NH_2, NO_2, Me, or Bu^t) and S, proposed.[351] Vapour species Se_2, SbSe, and complex molecules of composition Sb_2Se_n ($n = 2—4$), Sb_3Se_n ($n = 1—4$), and Sb_4Se_n ($n = 3$ or 4) have been identified in the spectrum of Sb_2Se_3 at 765—885 K.[352]

Group VI.—The products of decomposition of SF_6 have been investigated[353] and the products from interaction of $SiCl_4$ and NH_3 indicated that vapour-phase nucleation and condensation occurred in this system at high temperatures.[354] The pyrolysis products of S_4N_4 have been investigated.[355] At about 200 °C, the major product was S_2N_2; S_3N_3 and S_4N_2 were also identified. On increasing the temperature to above 300 °C, S_7NH and N_2 were observed as the major products. The $[M]^+$ ion was observed in the spectrum of $S_5N_7SiMe_3$ (Me_3Si was the base peak)[356] and a series of aminosulphurtrifluorides R_2NSF_3 have been characterized.[357] The spectra of a series of organoselenium compounds

---

[345] H. Hiraoka, W. Y. Lee, L. W. Welsh, and R. W. Allen, *Macromolecules*, 1979, **12**, 753.
[346] S. Tachimori, *J. Radioanal. Chem.*, 1979, **49**, 179.
[347] H. Budzikiewicz and G. Feistner, *Biomed. Mass Spectrom.*, 1978, 512.
[348] R. Weber, F. Borchers, R. Levsen, and F. W. Roellgen, *Z. Naturforsch., Teil A*, 1978, **32**, 540.
[349] A. S. Alikhanyan, A. V. Steblevskii, V. B. Lazarev, V. T. Kalinnekov, Ya. Kh. Grinberg, E. G. Zhukov, L. M. Agamirova, and V. I. Gorgorakai, *Izv. Akad. Nauk SSSR, Neorg. Mater.*, 1980, **16**, 73.
[350] R. E. Pabst, M. C. Sharpe, J. L. Margrave, and J. L. Franklin, *J. Mass Spectrom. Ion Phys.*, 1980, **33**, 187.
[351] R. A. Howie, D. W. Grant, and J. L. Wardell, *Inorg. Chim. Acta*, 1978, **30**, 233.
[352] C. L. Sullivan, J. E. Prusaczyk, R. A. Miller, and K. D. Carlson, *High Temp. Sci.*, 1979, **11**, 95.
[353] G. Bruno, P. Capezzuto, and F. Cramarossa, *J. Fluorine Chem.*, 1979, **14**, 115.
[354] S.-S. Lin, *J. Electrochem. Soc.*, 1978, **125**, 1877.
[355] R. D. Smith, *J. Chem. Soc., Dalton Trans.*, 1979, 478.
[356] W. S. Sheldrick, M. N. S. Lao, and H. W. Roesky, *Inorg. Chem.*, 1980, **19**, 538.
[357] C. Braun, H. E. Sasse, and M. L. Ziegler, *Z. Anorg. Allg. Chem.*, 1979, **450**, 139.

$(RCHClCH_2)_2SeCl_2$ (R = hydrocarbon)[358] and tellurium compounds $F_nTe(OTeF_5)_m$ (n = 0, 1, 2, or 4; m = 2, 4, 5, or 6)[359] have been reported.

Group VII.—The bond energies of F_2Cl-F and $FCl-F$ and heat of formation of ClF_2 have been calculated as well as a lower limit of the electron affinity of ClF.[360]

Group VIII.—The heat of formation of KrF_2 has been obtained (15.5 kcal mol^{-1}) indicating an average $Kr-F$ bond energy of 0.46 eV.[361] ArH^+ and Ar_2H^+ ions have been detected in a study of a mixture of argon and hexamethyldisiloxane.[362]

Miscellaneous Compounds.—A study has been made on a chemical transport reaction of VO_2 using $TeCl_4$ as a transport reagent,[363] and the products of decomposition of a mixture of NH_4VO_3 and $NH_4VO_3-TiO_2$ have been investigated.[364]

The spectrum of chromium(V) trichloride oxide has been reported[365] and PtTi identified in the vapour of the Pt–Ti–C system at high temperatures.[366] SIMS has been used to study unsupported FeRu alloy catalysts.[367] The ligand exchange reactivity of $Fe(PF_3)_5$ has been observed to parallel the exchange behaviour with the same ligands under conditions of CI mass spectrometry in which the ligand was used as the reagent gas.[368] $Fe(PF_2NMe_2)_5$ has been characterized[369] and cis-bis(dicyclophosphine)dihalogenonickel(II) complexes have been characterized.[370] The complexes $PdAl_2Cl_8$, $PdAlCl_5$, and $Pd_2Al_2Cl_{10}$ have been identified as products of the interaction of Al_2Cl_6 and solid $PdCl_2$.[371] Spectra of K_2PtF_6 have been recorded at temperatures between 980 and 1120 K.[372] The molecules $RhUC_2$ and $RhThC_2$ have been observed in the gas phase above a Th–U–Rh–C system at high temperature.[373] From their heats of atomization and that of Rh–E–C–C (E = U or Th) possible structures for the molecules were suggested. The spectrum of $Rh(NO_3)_2NO(PPh_3)_2$ has been reported[374] and a study of the evaporation of solid solution components in a system of zirconium and praseodymium oxides has been made.[375]

---

[358] L. Futekov, S. Stoyanov, and H. Specker, Z. Anal. Chem., 1979, **295**, 7.
[359] D. Lentz, H. Pritzkow, and K. Seppett, Inorg. Chem., 1978, **17**, 1926.
[360] A. V. Dubin, L. N. Gorokhov, and A. V. Baluev, Izv. Akad. Nauk SSSR, Ser. Khim., 1979, 2408.
[361] J. Berkowiz and J. H. Holloway, J. Chem. Soc., Faraday Trans. 2, 1978, **74**, 2077.
[362] M. Schmidt, R. Seefeldt, and G. Pohle, J. Phys. Colloq. (Orsay, Fr.) 1979, **(C7)**, 461.
[363] Y. Bando, M. Kyoto, T. Takada, and S. Muranaka, J. Cryst. Growth, 1978, **45**, 20.
[364] L. Dziembaj and R. Dziembaj, J. Therm. Anal., 1979, **17**, 57.
[365] W. Levason, J. S. Ogden, and A. J. Rest, J. Chem. Soc., Dalton Trans., 1980, 419.
[366] S. K. Gupta, M. Pelino, and K. A. Gingerich, J. Phys. Chem., 1979, **83**, 2335.
[367] G. L. Ott, W. N. Delgass, N. Winograd, and W. E. Baitinger, J. Catal., 1979, **56**, 174.
[368] R. C. Dougherty, M. A. Krevalis, and R. J. Clark, J. Am. Chem. Soc., 1979, **101**, 2642.
[369] R. B. King and M. Change, Inorg. Chem., 1979, **18**, 364.
[370] R. A. Palmer, H. F. Giles, and D. R. Whitcomb, J. Chem. Soc., Dalton Trans., 1978, 1671.
[371] U. Floerke and H. Schaefer, Z. Anorg. Allg. Chem., 1979, **459**, 140.
[372] M. V. Korobov, E. B. Badtiev, E. F. Voronin, A. A. Bondarenko, and L. N. Sidorov, Vestn. Mosk. Univ., Ser. 2, Khim., 1980, **21**, 200.
[373] S. K. Gupta and K. A. Gingerich, J. Chem. Soc., Faraday Trans. 2, 1978, **74**, 1851.
[374] P. B. Critchlow and S. D. Robinson, Inorg. Chem., 1978, **17**, 1896.
[375] A. N. Belov, G. A. Semenov, I. V. Vinokurov, and M. D. Krasil'nikov, Izv. Akad. Nauk SSSR, Neorg. Mater., 1979, **15**, 1629.

The dissociation energy of Nb_2 has been measured and its heat of formation calculated.[376] High-temperature techniques have also been used to study gaseous carbides of uranium UC_n ($n = 1$—6),[377] TaF_n ($n = 1$—5),[378] WBr_4 and U_2Br_6,[379] UO_2,[380] Lu_3S_4,[381] TcC,[382] and the dimerization equilibrium $2UF_5 \rightleftarrows U_2F_{10}$.[383] Ion cyclotron resonance techniques have been used to study the formation and reactions of negative ions derived from WF_6[384] and SIMS has been reported for $CdTe$.[385,386] RuO_4 has been identified as a hydrolysis product of $RuOF_4$, RuF_5, and RuF_4[387] and the formation of rhenium compounds (e.g. ReO, ReO_2, ReO_3, and ReO_4) in a mass spectrometer with a rhenium filament has been studied.[388]

---

[376] S. K. Gupta and K. A. Gingerich, *J. Chem. Phys.*, 1979, **70**, 5350.
[377] S. K. Gupta and K. A. Gingerich, *J. Chem. Phys.*, 1979, **71**, 3072.
[378] K. H. Lau and D. L. Hildenbrand, *J. Chem. Phys.*, 1979, **71**, 1572.
[379] A. Popovic, J. Marsel, and O. Kaposi, *J. Inorg. Nucl. Chem.*, 1979, **41**, 1289.
[380] R. J. Ackermann, E)G. Raup, and M. H. Rand, Report 1979, INKA-Conf-79-003-036, 'AEA-SM-236/59.
[381] H. F. Franzen and A. V. Hariharan, *J. Chem. Phys.*, 1979, **70**, 4907.
[382] G. H. Rinehart and R. G. Behrens, *J. Phys. Chem.*, 1979, **83**, 2052.
[383] P. D. Kleinschmidt and D. L. Hildenbrand, *J. Chem. Phys.*, 1979, **71**, 196.
[384] P. M. George and J. L. Beauchamp, *Chem. Phys.*, 1979, **36**, 345.
[385] M. Hage-Ali, R. Stuck, A. N. Saxena, and P. Siffert, *Appl. Phys.*, 1979, **19**, 25.
[386] C. Scharager, R. Stuck, P. Siffert, J. Cailleret, C. Heitz, G. Lagarde, and D. Tenorio, *Nucl. Instrum. Methods*, 1980, **168**, 367.
[387] T. Sakurai and A. Takahashi, *J. Inorg. Nucl. Chem.*, 1979, **41**, 681.
[388] M. A. Grayson, *Int. J. Mass Spectrom. Ion Phys.*, 1979, **30**, 383.

Author Index

Abbey, L. E., 180, 226
Abbott, S. J., 245
Abd El-Kader, F. H., 46
Abdel-Moety, E. M., 202
Abdullaev, U. A., 249
Abdullayeva, M. K., 123
Aberth, W., 164
Abouaf, R., 102, 104, 116
Abraham, B., 170, 197
Abramova, Z. A., 304
Abramson, F. P., 267
Abroyan, I. A., 144
Achenbach, C., 297
Achenbach, H., 247
Achiba, Y., 10
Ackermann, R. J., 328
Adam, G., 249
Adams, C. T., 148
Adams, F., 182
Adams, N. G., 87
Adams, R. P., 190, 200
Adamson, R. H., 268
Adesomu, A. D., 210
Adley, T. J., 72
Agafonov, I. L., 313
Agamirova, L. M., 326
Agmon, N., 40
Agurell, S., 277, 290
Ahlborg, U. G., 216
Ahling, B., 219
Aikman, R. E., 306
Aime, S., 308
Ajello, J. M., 18
Akaishi, K., 214
Akhren, A. A., 257
Akimori, N., 260
Akimoto, H., 167, 200
Akopyan, M. E., 45
Albaiges, J., 218
Albers, H.-K., 209
Albrecht, P., 218, 221
Albrect, A. C., 36
Albritton, D. L., 10, 85, 86, 87, 91, 96, 102, 116
Al-Derzi, A., 3
Aldrich, J. B., 181
Aleksanyan, V. T., 236
Alexander, M. H., 98
Alexander, M., 220
Alexandru, G., 40, 47
Algosin, M., 226
Ali, M. E., 249
Alikhanyan, A. S., 326
Alipov, B. B., 303

Alkalay, D., 268
Allan, M., 106, 111
Allcock, H. R., 305, 325
Allen, R. W., 326
Allibert, M., 320, 321
Allinger, N. L., 6
Allison, J., 25, 322
Allman, D. J., 163
Al-Najgar, I. M., 314
Alsberg, T., 219
Althoff, G., 310
Alton, G. D., 148
Aluyi, H. A. S., 206
Amanollahi, S., 315
Ambe, Y., 217
Ambre, J. J., 274
Ambrus, G., 260
Amme, R. C., 176
Amos, B., 268
Anan, T., 222
Anbar, M., 165
Anderegg, R. J., 212, 258, 259
Andersen, C. A., 120
Anders, M. W., 214
Andersen, C. A., 123
Andersen, H. H., 118, 148
Andersen, R. H., 314
Anderson, B. A., 224, 241
Anderson, G. K., 297
Anderson, G. S., 147
Anderson, P. J., 214
Anderson, R. J. M., 32
Anderson, S. L., 13
Anderson, T. G., 117
Anderson, W. R., jun., 158, 160, 161
Anderssen, B. A., 202
Andlauer, B., 43
Ando, W., 300
Andressen, B., 131, 132
Andresen, B. D., 282
Andrianov, Yu. A., 311
Angelici, R. J., 310
Angus, P. C., 299
Ansari, G. A. S., 208
Antenova, A. B., 311
Antipova, V. V., 301
Arai, T., 253
Arakawa, S., 284
Araki, E., 259
Ardeleau, P., 22
Arfwidsson, A., 289
Arifov, V. A., 21

Ariga, T., 205, 260, 261
Arijis, E., 179
Arikawa, T., 139
Aringer, L., 208
Arison, B. H, 280, 290
Arjmand, M., 211
Armentrout, P. B., 323, 324
Armour, D. G., 132
Armstrong, D., 215
Arn, H., 226
Arnoux, B., 252
Arpino, P. J., 171, 228. 229, 242, 245, 262
Arshadi, M. R., 235
Asao, Y., 222
Asaro, C., 26
Asbrink, L., 7
Ashendel, C. L., 190
Ashley, J. W., 193
Ashley, T. R., 225
Ashuiko, V. A., 322
Aspinal, M. L., 154, 243
Ast, T., 78, 79, 171, 172
Astruc, D., 310
Asubiojo, O. I., 238
Aszalos, A. A., 252
Atwater, B. L., 176, 195
Aubagnac, J.-L., 66
Aubert, C., 268
Auciello, O., 132
Audier, H. E., 61, 72, 252
Audran, M., 314
Aue, D. H., 13, 14, 55, 60, 237
Aue, W. A., 199
Auerbach, A., 18
Auffret, A. D., 258
Aufrere, M. B., 281
Aurelle, H., 261
Ausloos, P., 11, 14, 50, 63
Avouris, P., 32, 33, 36
Awad, S. B., 324
Axelson, M., 208
Ayukhanov, A. K., 123
Azami, T., 72
Azarnoff, D. L., 270, 287
Azman, A., 6
Azwar, F., 297

Baas, W. J., 204
Baba, S., 274
Babich, E. D., 299
Bacon, R. N. R. E., 71
Bacskay, G. B., 16

Author Index

Baczynskyj, L., 181
Bader, G. J., 274
Badtiev, E. B., 327
Baer, T., 8, 9, 10, 20, 22, 24, 29, 41, 43, 44, 46, 48, 60
Bagarat'yan, N. V., 320, 322
Baghal-Vayjooee, M. H., 11
Bahr, U., 297
Bailey, D. N., 177
Bailey, E., 198
Baille, T. A., 60, 153, 199, 206, 227, 241, 260, 271, 293
Baitinger, W. E., 327
Baker, J. E., 121
Baker, R., 224, 225
Bakke, J., 293
Bakke, T., 227
Balakrishnan, K. P., 319
Baldwin, M. A., 62, 72, 77, 155, 229, 243
Ballantine, J. A., 208
Ballard, J., 244
Balletto, A. J., 290
Ballhorn, L., 207
Baltes, W., 222
Baluev, A. V., 327
Bando, Y., 327
Baraboshkin, V. E., 320
Baran, A. M., 310
Baranowska, E., 249
Barassin, A., 23
Barbachyn, M. R., 13
Barbalas, A. P., 78
Barber, M., 4
Bardsley, J. N, 30, 86, 91
Baril, M., 122
Barker, R. A., 53
Barlic, B., 176
Barnes, W. S., 85
Barofsky, D. F., 160, 258
Barofsky, E., 258
Barriola, A., 323
Barrow, W. L., 5
Bartl, P., 203
Bartle, K. D., 221
Bartmess, J. E., 13, 14, 60, 63, 237, 239
Barton, V., 178
Baryshok, V. P., 301
Bashkirova, S. A., 301
Bass, L., 54, 55
Batail, P., 310
Batemann, R. H., 154, 171, 294
Bates, D. R., 55
Bather, W., 65
Batra, S. W. T., 225
Batrokov, S. G., 260, 261
Batta, A. K., 208
Batten, C. F., 45
Baty, J. D., 213, 276
Bauer, J. E., 287
Bauma, W. J., 5

Baumgärtel, H., 18, 19, 234
Baumgartner, E., 237, 319
Baun, W. L., 294
Baxter, R. L., 246
Bay, H. L., 151
Bayer, E., 198, 211, 216, 292
Bayly, A. R., 138, 139, 140
Beak, P., 14
Bearman, G. H., 54
Beattie, I. R., 314
Beauchamp, G. K., 224
Beauchamp, J. L., 8, 10, 13, 17, 55, 63, 112, 113, 114, 162, 239, 323, 324, 328
Becher, D. S., 323
Beck, B., 223
Beck, S. M., 36
Beckett, A. H., 263, 285, 286
Beckey, H. D., 158, 163, 164, 233
Beckner, C. F., 211
Bedford, R. G., 178
Beenakker, J. J., 94
Begue, R.-J., 209
Behrens, R. G., 328
Beimer, R. G., 178
Beinzer, M., 220
Bellas, T. E., 225
Bellitto, C., 316
Belokon, A. I., 297
Belokon, A. V., 323
Belon, C., 314
Belov, A. N., 327
Benakis, A., 279
Benes, E., 147
Ben-Galim, E., 167
Ben'iaminovich, M. B., 119
Benkens, R. P., 321
Benn, M., 226
Bennemann, P., 222
Bennett, M. A., 313
Benninghoven, A., 118, 121, 122, 123, 139, 168, 232, 243
Benoit, F. M., 220
Benson, P. A., 164
Benson, S. W., 11
Bentley, T. W., 60
Benz, R. C., 106, 108
Berezhanskaya, O. P., 302
Berg, H., 235
Berg, J. O., 34
Berg, M., 271
Bergel'son, L. D., 260, 261
Bergstrom, G., 224
Berkley, R. D., 217
Berkowitz, D. S., 4
Berkowitz, J., 9, 10, 27, 31, 45, 327
Berman, E., 252
Bernadskii, A. A., 235, 298, 301
Bernal, I., 309
Bernard, E. M., 215

Bernauer, O., 321
Bernhard, H. O., 246
Bernheim, M., 122, 123, 138
Bernstein, R. B., 34, 35, 162
Berry, R. S., 36
Bertelsen, O., 202
Bertrand, M. J., 292
Beske, H. E., 133, 183
Bespalov, V. Ya., 5
Besset, A., 248
Besson, E., 248
Besson, F., 252
Best, D., 202
Best, F. W., 198
Beswick, J. A., 37, 102
Betowski, L. D., 14
Betty, K. R., 185, 200
Beuhler, R. J., 155, 254
Bevenue, A., 220
Beving, H., 276
Beyerman, H. C., 211
Beynon, J. H., 47, 60, 66, 77, 78, 79, 80, 81, 101, 162, 167, 169, 171, 172, 186, 240, 244, 295
Bhargava, S., 314
Bhattacharjee, R. C., 56
Bianchi, G., 223
Bickelhaupt, F., 305
Bidder, T. G., 215
Bielawski, J., 297
Biemann, K., 177, 212, 241, 257, 258, 259, 293
Bier, D. M., 167
Bierbaum, V. M., 238
Bieri, G., 11, 111
Bierl-Leonhardt, B. A., 225, 226
Bierman, D. J., 134
Bihari, M., 324
Bill, J. C., 180
Biller, J. E., 177
Billy, M., 324
Bilton, J. N., 172
Binkley, J. S., 2
Biodi, M. A., 86
Birkett, D. J., 275
Birkofer, L., 301
Biro, Z., 212
Bishop, S. H., 207, 256
Biswas, K. M., 249
Bjorseth, A., 217
Blackley, C. R., 245
Blair, I. A., 207, 298, 299
Blair, L. K., 239
Blais, N. C., 10, 20
Blaisdell, B. E., 214
Blaise, G., 118, 122, 135, 138, 139
Blake, M. R., 236, 309, 313
Blakley, C. R., 159, 230
Blandin, A., 137
Blaser, W. W., 219
Blattner, R. J., 123

Author Index

Blau, K., 227
Blazey, G. C., 148
Blekh, L. M., 235
Bleshinskaya, A. S., 295
Blickensderfer, J. R., 311
Blickensderfer, R. P., 16
Blomquist, G. J., 224
Bloor, J. E., 234
Blotney, G., 228, 257
Blount, J. F., 291
Blues, E. T., 295
Blum, M. S., 225, 226
Blum, P., 147
Blum, W., 80
Blurock, E. S., 5
Boand, G., 165
Bochkarev, V. N., 235, 236, 297, 298
Bochmann, M., 317
Bockhoff, F. M., 9, 61, 78, 244
Boerboom, A. J. H., 172, 180, 182
Böse, N., 29
Boesl, U., 32, 34, 161
Boettger, H. G., 182, 215
Boggs, J. E., 2
Bohme, P. K., 22, 81, 239
Boisson, D., 211
Boldyrev, V. V., 324
Boldyreva, O. G., 295
Bolton, P. D., 181, 186, 201
Bomse, D. S., 113, 114, 162
Bondarenko, A. A., 327
Bondesson, U., 277
Bondybey, V. E., 57, 110, 111
Boniforti, L., 219
Bonny, A., 299, 307
Boon, J. J., 208
Borchers, F., 48, 65, 256, 326
Borden, W. T., 5
Borders, D. B., 241
Boreham, D. R., 210, 251, 288
Boreus, L. O., 271
Borg, K. O., 284, 289
Borg, R., 210
Borges del Castillo, J., 157
Borkman, R. F, 7, 20
Borowiak, D., 207
Borowski, E., 253
Borstnik, B., 6
Borthwick, J. H., 209
Borwitzky, H., 199, 219
Bos, R., 249
Boschi, T., 309
Bose, A. K., 237, 245, 250
Bosshardt, H., 246
Bosshardt, H. P., 220
Bosyakov, Yu. G., 304
Both, F. K., 155

Bottcher, C., 57
Botter, R., 44, 70
Bottino, F. A., 318
Botz, F. K., 242
Boucher, L. J., 255
Bouillant, M. L., 248
Bouma, W. J., 5, 62
Bounds, D. G., 4
Bounds, P. J., 6
Bourne, G. R., 205
Boutagy, J., 213, 250
Bowen, D. V., 187
Bowen, R. D., 50, 52, 59, 63, 66, 68, 78
Bowers, M. T., 13, 14, 53, 54, 55, 58, 60, 237
Bowers, W. D., 227
Bowie, J. H., 77, 233, 234, 235, 239, 240, 298, 299
Bowie, L. J., 177
Bowser, D. V., 206
Bowser, J. R., 322
Boyd, G. S., 210
Boyd, R. K., 172, 248, 249
Bozorgzadeh, M. H, 79, 171, 172, 186, 244, 295
Bozynov, B. V., 261
Brachmann, K., 184
Bradford, D. C., 245
Bradley, C. V., 258
Bradley, R., 136
Braem, D., 72
Braenden, O. J., 246
Brand, H. R., 178
Brand, K., 148
Brandalise, B., 184
Brandenberger, H., 237, 262
Branfman, A. R., 281
Brash, A. R., 206, 260
Bratushko, Yu. L., 318
Brauman, J. I., 14, 16, 17, 19, 55, 71, 107, 114, 115, 234, 238
Brauman, J. J., 52
Braun, C., 326
Braun, W. A., 216
Braun-Elwert, G., 148
Braunstein, P., 307
Bravo, R. V. F., 246
Breckenridge, W. C., 204
Bregadze, V. I., 297
Brehm, B., 44
Breimer, M. E., 259, 261
Breliere, C., 300
Brenner, M. L., 220
Brent, D. A., 155, 242
Brenton, A. G., 47, 66, 79, 169, 171, 172, 186, 244
Breton, J., 29
Brettell, T. A., 264
Breuer, H., 276
Breunig, H. J., 306
Briggs, C. D., 218
Brilis, G. M., 287

Brill, T. B., 297
Brion, C. E., 6, 25, 26
Brisdon, B. J., 309
Brockman, A., 184
Brodie, R. R., 279
Brodskii, E. S., 260
Broer, W. T., 72
Bromund, R. H., 217
Brooks, C. J. W., 157, 204, 207, 209, 241, 275, 296
Brooks, J. N., 176
Broom, A. P., 253
Brophy, J. H., 34
Brophy, J. J., 224
Broquist, H. P., 247
Brough, L. F., 298
Brousseau, K., 243
Brown, C. L., 234
Brown, F. J., 251
Brown, G. K., 214
Brown, H. L., 165
Brown, K. F., 123
Brown, L., 250
Brown, L. D., 2
Brown, N. M. D., 295
Brown, R. D., 110
Brown, S. A., 248
Browne, L. E., 226
Browner, R. F., 218
Browning, R., 43
Bruce, M. I., 307
Bruckmuller, R., 147
Bruins, A. P., 156, 166, 186, 242, 249
Brumbaugh, D. V., 37
Brumley, W. C., 216
Bruna, P. J., 4
Bruni, R. J., 281
Brunnee, C., 184
Bruno, G., 326
Bryant, P. J., 163
Bryant, W. F, 201
Bryce-Smith, D., 295
Buchi, R., 187
Buchs, A., 72
Buck, R. P., 163
Buckley, J. A., 166, 178
Budde, W. L., 176, 216
Budzikiewicz, H., 48, 233, 241, 253, 254, 257, 326
Büchler, U., 62
Bühring, K. U., 213, 283
Buenker, R. J., 4, 6
Buker, I., 198, 209
Buko, A. M., 244
Bukow, H. H., 101
Bulk, K. T., 213
Bulota, K., 177
Burger, G. V., 224
Burger, U., 5, 72
Burgers, P. C., 10, 51, 62, 67, 72, 78
Burke, D. G., 264
Burke, R. R., 23

Author Index

Burlingame, A. L., 60, 65, 153, 171, 179, 199, 201, 214, 241, 242, 253
Burnier, R. C., 107
Burns, P., 171
Burroughs, P., 178
Burrow, P. D., 234
Burrows, M. D., 46
Bursey, J. T., 217
Bursey, M. M., 25, 60, 75, 77, 163, 220, 237, 240
Busch, K. L., 220, 237
Buser, H. R., 219, 220, 226
Bush, E. D., 278
Bushell, M. J., 255
Busse, B., 321
Butcher, A. R., 72
Butin, B. M., 304
Butscher, W., 4
Buttar, H. S., 287
Buttenshaw, J., 115
Buttery, R. G., 222, 223
Buttrill, S. E., 164, 165, 166
Byers, L. R., 312
Bylec, E., 253
Byrne, K. J., 224

Cahn, R. W., 324
Cailleret, J., 328
Cairo, P. R., 215
Calder, A. G., 202
Colhoun, H. P., 300
Calimli, A., 221
Calvin, M., 51, 62, 209
Calvino, R., 234
Cameresi, G. G., 179
Campana, J. E., 179, 201, 307, 315
Campbell, I. M., 173
Campbell, M. M., 157
Campion, P., 66
Camoni, I., 219
Cano, J. P., 268
Cant, C. S. T., 41
Capezzuto, P., 326
Capellen, J. M., 159
Cappellini, L., 282
Caprace, G., 10
Caprioli, R. M., 60, 192, 211, 244, 265
Cappuccino, N. F., 237, 245
Carcassonne, Y., 268
Cardoso, A. M., 311
Carle, J. S., 226
Carlsen, S., 268
Carlson, K. D., 326
Carmichael, R. H., 275
Carmody, J. J., 47, 79, 172
Carnevale, J., 249
Carnovale, F., 10, 20
Carpenter, A. P., 221
Carre, M., 101, 103
Carrick, A., 175, 183

Carrington, A., 25, 102, 103, 115
Carroll, D. I., 155, 229, 243
Carroll, P. R., 226
Carson, L. J., 220
Carter, A. E., 35
Carter, D. E., 286
Carter, G., 118, 132
Carter, J. A., 169
Carter, J. G., 234
Carter, P., 208
Casey, R., 259
Cash, G., 312
Casky, G. T., 148
Castaing, R., 138, 139
Castleman, A. W., jun., 19, 21
Catalán, J., 15
Cathcart, R. E., 176
Caulton, K. G., 311
Cavill, G. W. K., 224, 225
Cederbaum, L. S., 6
Cefola, M., 315
Cerchar, B. P., 217
Ceric, M. M., 322
Ceyer, S. T., 13, 20
Chadwick, M., 281
Chait, E. M., 177, 178, 179
Chalmers, R. A., 214
Chambaz, E. M., 155
Chan, B. G., 223, 226
Chan, C. L., 234
Chan, I. Y., 32, 36
Chan, K. K., 292
Chan, K. W., 217
Chang, C. C., 325
Chang, D., 258
Chang, K. H., 321
Chang, M. I., 311
Change, M., 327
Chapman, C. J., 259
Chapman, D. J., 255
Chapman, J. R., 176, 179, 183
Chapman, K. R., 148
Chapman, O. L., 225, 226
Chapus, G., 309
Charette, C., 287
Charrier, C., 311
Chashchin, V. L., 257
Chasseaud, L. F., 279
Chatfield, D. H., 276, 292
Chatillon, C., 320, 321
Chau, M., 55
Chauvin, Y., 308
Chebroux, P., 155
Chelgren, J. E., 123
Chen, E. C. M., 207, 256
Chen, H. J., 325
Cheng, C.-C., 206
Cheng, F. W., 208
Chen, Hu, L.-E., 227
Cherlnyshev, E. A., 235

Chernikov, V. A., 247
Chernoff, D. A., 57
Chernoff, N., 216
Chernyaev, N. P., 312
Cherpillod, O., 309
Chesnavich, W. J., 54, 55, 58
Chesnyshev, E. A., 301
Chheda, G. B., 215
Chiabrando, C., 219
Chien, R. L., 36
Child, P., 209
Childs, R., 7, 20
Chinn, D. M., 274
Chisholm, M. H, 311
Chizov, O. S., 60, 153, 199, 241, 255, 295, 297
Cho, A. K., 213
Chopin, J., 248
Choudhury, G., 216
Choy, Y.-M., 206
Chriki, M., 72
Christ, V., 203
Christensen, R. G., 171
Christie, J. R., 46, 49, 68, 76, 77
Christie, V., 122
Christie, W. W., 202
Christodoulides, A. A., 234
Christopherson, C., 226
Christophorou, L. G., 233, 234
Chupakhin, M. S., 318
Chupka, W. A., 29
Chutjian, A., 18
Cian, A. D., 308
Ciccioli, P., 201
Claeys, M., 213
Clair, R. L., 54, 239
Clare, R. A., 271
Clark, D. A., 260
Clark, R. J., 327
Clark, R. J. H., 311
Clausen, K., 234
Claussen, C., 126
Clay, K. L., 211
Clearfield, A., 309
Clegg, W., 308
Clement, R. E., 217, 219
Clementi, E., 5
Clerc, J. T., 187
Clevenstine, E. C., 247
Cloke, F. G. N., 313
Clutton, D. W., 222
Cnizov, Yu. V., 298
Cobb, M., 7, 20
Cochrane, W. P., 220
Coen, S., 227
Coffelt, J. A., 225
Coffin, R. D., 215
Coggiola, M. J., 114
Cohen, A. I., 265
Colbourne, D., 10
Colby, B. N., 184, 227

Author Index

Cole, B. E., 27
Cole, E. R., 249
Cole, N. W., 46, 77
Cole, W. J., 255
Coleman, E. C., 222
Coles, J. N., 129
Colli, G., 219
Colligon, J. S., 118
Collin, J. E., 10
Colson, S. D., 34
Colton, R. J., 168
Comarmond, M. B., 308
Comer, J., 30
Comisarow, M. B., 173, 181
Commereuc, D., 308
Compson, K. R., 179
Compton, R. N., 17, 18, 32, 35, 234
Conard, B. R., 307
Cone, E. J., 267
Conradi, E. C., 289
Conzemius, R. J., 159
Cook, A. M, 220
Cook, D. H., 318
Cook, J. C., 199, 200
Cook, J. C., jun., 250, 252
Cook, K. D., 21
Cooke, M., 220, 221
Cooks, R. G., 47, 50, 60, 73, 75, 77, 79, 107, 168, 171, 172, 233, 236, 238, 243, 244, 318
Cooper, C. B., 139
Cooper, C. D., 17, 234
Cooper, C. V., 216
Cooper, S. F., 292
Cooper, W. J., 221
Coplan, M. A., 22
Coppens, P., 10
Corriu, R. J. P., 300
Corderman, R. R., 117
Corey, R. C., 260
Corkin, D. R., 311
Cornides, I., 324
Corongiu, G., 5
Correia, M. A., 242
Corriu, R. J. P., 308
Corval, M., 66
Cory, D. A., 255
Cory, H. T., 213, 253
Cosby, P. C., 24, 102, 103, 104, 114, 116
Cossart, D., 111
Cossart-Magos, C., 111
Costa, B., 179
Cotter, R. J., 155, 178, 243
Cotton, F. A., 311, 316
Cotton, R. G. H., 215
Coufal, H., 8
Coulson, D. R., 40
Coulston, F., 221
Court, W. A., 203
Coutts, R. T., 216
Covington, M. A., 161

Cowan, P. A., 263, 285
Cowley, A. H., 303
Cox, J. D., 12
Cozak, D., 310
Cradock, S., 298
Craig, R. D., 154, 294
Crain, P. F., 213, 215, 244, 254
Cramarossa, F., 326
Cramers, C. A. M. G., 214
Cranberg, L., 152
Crance, M., 32
Crawford, O. H., 16
Crawford, R. W., 178
Cremaschi, P., 31
Crespi, H. L., 255
Crewe, R. M., 224
Critchlow, P. B., 327
Crofts, J. G., 110
Crombie, L., 246
Crombie, W. M. L., 246
Cross, R. J., 18, 297
Crouch, D. J., 274
Crow, F. W., 166, 169, 200
Crump, D. R., 224
Crunelli, V., 266
Cullis, P. G., 165
Cummins, J. T., 215
Cupano, J., 272
Cutler, R. I., 148

DaGragnano, V., 183
Dahl, L. F., 312
Daiker, K. C., 15
Dakternieks, D. R., 236, 314, 315
Dale, F., 103
Dalgarno, A., 30
Daluga, P., 323
Damani, L. A., 285
Damen, H., 185, 189
Dan, P., 258
Danby, C. J., 41, 43
Dandeneau, R., 197
D'Angona, J. L., 173, 180
Daniel, C. P., 271
Daniel, F. B., 218
D'Aniello, M. J., 311
Daniewski, M. M., 199
Danks, D. D., 215
Danks, D. M., 214
Dannacher, J., 43, 106
Darcy, M. G., 164
D'Arcy, P. H., 180
Darda, S., 286
Darensbourg, D. J., 307, 313
Darvill, A., 255
Darwin, W. D., 267
Das, B. C., 252, 258
Das, K. G., 314
Das, M., 315
Das, R., 4
Dasgupta, A., 254

Das Neves, H. J. C., 211
Datwyler, P., 246
Daubach, R. O., 181
Daughton, C. G., 220
Dauphin, B., 213
Daves, G. D., jun., 158, 160, 161, 242, 258
Davidson, E. M., 205
Davidson, E. R., 5
Davidson, F., 295
Davidson, W. R., 14, 55, 166
Davies, A. G., 303
Davies, D. S., 271
Davies, M. M., 202
Davies, N. W., 224, 225
Davis, D. W., 15
Davis, F. T., 282
Davis, R., 155, 307, 312, 313
Davisson, J. N., 266
Dawkins, B. G., 228, 229, 245
Dawson, J. H. J., 238, 239
Day, R. A., 257
Day, R. J., 47, 50, 73, 75, 168, 243, 318
Day, V. W., 311
Dayal, B., 208
Dayriger, H. E., 176
De, B. R., 4
Deacon, G. B., 295
Dean, T. P., 324
de Angelis, L., 219
Debast, K., 280
de Boer, J., 94
De Ceuninck, F., 182
Decorpo, J. J., 162, 168
Deelder, R. S. , 198
DeFrees, D. J., 5, 11
De Grave, J., 226, 267, 287
DeHaseth, J. A., 188
de Heer, F. J., 45
Dehmelt, H., 107
Dehmer, J. L., 19, 21, 27
Dehmer, P. M., 21, 29
Deinzer, M. L., 237, 289
de Jong, E. G., 191, 256
DeJongh, D. C., 241, 252
DeKock, R. L., 13
deKoning, B. R., 126
Delange, C. A., 10
De Lange, F., 208
Del Bene, J. E., 15
Delcambe, L., 252
DeLeeuw, D. M., 10
De Leon, I. R., 221
Delgado-Barrio, G., 37
Delgass, W. N., 327
Deline, V. R., 123
Dell, A., 159
Delseth, C., 208, 250
De Luna, F. A., 280
Delwiche, J., 10

de Meyer, C., 80
Demitt, T. F., 207
Demole, E., 222
Dencker, W. D., 177
Dendramis, N., 214
Dengler, H. J., 286, 287
Denisovich, L. I., 236
Denne, D. R., 154, 179
Denney, D. W., 217
Dennis, E., 139
Dennis, W. H., 221
Denson, K. B., 184, 201
De Paritosh, K., 264
de Pascale, A., 282
DePauw, E., 320
De Poorter, B., 302
DePristo, A. E., 38
DePuy, C. H., 238
Derai, R., 22, 112
de Ridder, J. J., 279
Derrick, P. J., 46, 49, 60, 65, 68, 74, 76, 77, 153, 164, 165, 199, 241
De Saxce, A., 300
Desgres, J., 211
Desiderio, D., 249
Desiderio, D. M., 211, 257, 260
Desouter-Lecomte, M., 38, 39
Detwiler, K. N., 164
Deutsch, J., 72
Devant, G., 229, 242
Devoret, M., 110
Devos, R. H., 218
Dewar, M. J. S., 2
DeWit, A. G. J., 146
Dewitt, D., 214
Dexheimer, E. M., 300
De Zeeuw, R. A., 212
Dhaenens, L., 73
Dias, J. R., 71, 251
Diaz, A. V., 177
Dibeler, V. H., 43
di Domenico, A., 219
Diekmann, H. W., 283
Dielmann, G., 155, 198, 242
Diercksen, G. H. F., 6
Dietz, L. A., 169
Dietz, T. G., 35
Dijkstra, A., 189
Dijkstra, G., 76, 256
Dill, D., 19
Dill, J. D., 14, 52, 78
Dillard, J. G., 237, 319
Dillon, A. F., 168
Dinis, M. B., 315
Dinizo, S. E., 242
Dispert, H., 17, 234
Ditchfield, R., 2
Dixneuf, P., 308
Dixon, D. A., 15, 37
Dixon, T. A., 117
Djerassi, C., 208, 250, 251
Dlodzian, G., 136

Dobosh, P. A., 309
Dobretsov, E. N., 119
Dodds, A. R., 323
Dodonov, V. A., 297
Dodson, M. H., 184
Doérfler, D. L., 173
Doi, T., 268
Doiron, C. E., 322
Dolzhnikova, E. N., 304
Domazetis, G., 302
Domcke, W., 6
Domrachev, G. A., 310
Donaldson, P. B., 313
Donchi, K. F., 49, 68
Donohue, D. L., 169
Donovan, J. M., 207
Doolittle, R. E., 225
Doonan, S., 258
Dorado, M., 5
Dorman, D. E., 252
Dorokhov, V. A., 295, 297
Dotan, I., 87, 91
Doucas, G., 123, 148
Dougherty, R. C., 241, 255, 327
Douglas, R. A., 148
Douse, J. M. F., 222
Dow, J., 270
Dowling, N., 315
Downey, R., 178
Downs, A. J., 323
Draffan, G. H., 262
Drake, J. E., 298
Drapcho, D. L., 324
Drawert, F., 222, 223
Draxl, K., 7
Dreifuss, P. A., 255
Dressler, M., 213
Dromey, R. G., 188
Dropkin, D., 219
Drowart, J., 10
Drucker, D. B., 206
Druetta, M., 101
Druzhkov, O. N., 295, 297, 311
Dube, G., 300
Dubin, A. C., 327
Dubinin, V. A., 320
Dubs, P., 222
Duchamp, D. J., 181
Duffield, A. M., 211, 226, 275
Duffield, P. H., 226, 275
Dufner, D. C., 321
Dugal, R., 292
Dugan, R. J., 181
Duhan, H. A., 215
Dumail, M., 148
Dumas, M., 209
Dumm, H. V., 305
Dunbar, R. C., 48, 52, 57, 60, 100, 106, 108, 109, 110, 113, 114, 117
Duncan, A. E. W., 280
Duncan, M. A., 35

Duncan, W. R. H., 203
Dunlavely, S. J., 10
Du Plessis, D., 197
Dupuis, I., 287
Dupuy, H. P., 222, 223
Durand, B., 213
Durand, J., 155
Durbeck, H. W., 198, 209
Durden, D. A., 172, 227
Durup, J., 100, 102, 115
Duthoit, C., 80
Dutta, S. P., 215
Dutton, G. G. S., 206
Dwyer, L. A., 224
Dye, J., 158
Dyke, J. M., 10
Dyson, C. B., 184
Dzhemilev, U. M., 234
Dzidic, I., 155, 229, 243
Dziembaj, R., 327
Dzioba, S., 132

Eaborn, C., 299
Eades, R. A., 15
Eadon, G., 71
Eagles, J., 202, 210, 259
Eastes, W., 56
Eastham, D. A., 152
Eberly, J. H., 33
Ebert, M. H., 275
Ebsworth, E. A. V., 298, 325
Eckers, C., 155, 245
Ederer, D. L., 27
Edmonds, C. G., 204, 209
Edwards, D. A., 309
Efimova, A. G., 320
Efraty, A., 310
Efremov, Yu. Ya., 317
Egan, W., 206, 257
Egge, H., 259
Egger, B., 222
Egli, H. P., 179
Eglinton, G., 218, 245, 255
Egsgaard, H., 202
Eguchi, S., 248
Ehbert, T. C., 324
Eiceman, G. A., 219
Eichelbaum, M., 286, 287
Eichelberger, J. W., 176, 216
Eichelberger, T. S., 35
Eigendorf, G. K., 156
Eikelmann, G., 285
Eisele, F. L., 92
Eisner, T., 224
Eklund, G., 217, 218
Ekong, D. E. U., 210
Eland, J. H. D., 9, 10, 31, 41, 43, 44, 110
El-Deck, M., 303
Elford, M. T., 85, 90
Ellenberger, M. R., 15
Elli, G., 219
Elliott, W. H., 241

Author Index

Ellis, B. E., 248
Ellis, D. H., 291
Ellis, H. W., 85, 92, 98
Ellison, G. B., 17, 117
Elmore, D., 321
El-Sayed, M. A., 32, 34
Elsheikh, M., 300
Emel'yanov, I. S., 317
Emery, C., 273
Emmert, G. A., 150
Emons, H. H., 320
Ende, M., 189, 251, 286
Endoh, M., 44
Engel, L. L., 208
Engelking, P. C., 17, 107, 117
Enggist, P., 222
English, J. H., 111
Enke, C. G., 77, 172, 244
Entsch, B., 254
Enzell, C. R., 241
Era, V., 209
Erdahl, W. L., 231, 245, 259
Erikson, M. D., 219
Ermler, W. C., 21
Esherick, P., 32
Eshuis, W., 180
Etters, R. D., 21
Eusinsky, W., 147
Evans, C. A., jun., 118, 121, 122, 123, 243
Evans, D. A., 224
Evans, F. J., 247
Evans, J. V., 269
Evans, M., 215, 222
Evans, N., 173
Evans, S., 186, 230
Evans, S. L., 210
Evans, T. L., 325
Evetts, J. E., 324
Evlasheva, T. I., 10
Ewing, G. E., 38
Extine, M. W., 311
Eyem, J., 165
Eyler, J. R., 105, 181

Faigle, J. W., 286
Fales, H. M., 210, 226
Falick, A. M., 65
Falkner, C., 304
Falkner, F. C., 204, 276
Fal'ko, V. S., 234, 235
Falter, H., 258
Fanelli, R., 219, 271
Fann, W. E., 272
Fano, U., 31
Farber, M., 324
Farmer, P. B., 198
Farmery, B. W., 138
Farneth, W. E., 15
Farrington, J. W., 221
Faulkner, M. A., 266
Fausett, D. W., 194
Favrebonvin, J., 248

Favre-Bouvin, J., 248
Fayad, N. K., 10
Fedin, E. I., 236
Fedor, D. M., 79, 172
Fedorova, I. G., 144
Feeney, J., 159
Fehsenfeld, F. C., 87, 91
Feistner, G., 253, 326
Felton, R. H., 5
Fenical, W., 250
Fenistein, S., 22, 24, 112
Fennessey, P. V., 257
Fenselau, C., 206, 243, 281, 290
Fenton, D. E., 318
Fenwick, G. R., 202, 210, 259
Ferguson, E. E., 86
Fernandez, J. M., 307
Ferrago, R., 321
Ferrand, R., 217
Ferrari, R. P., 234, 306
Ferreira, L. F. A., 22, 44
Ferrito, V., 210
Feser, K., 199
Feussner, G., 296
Field, F. H., 21, 166, 169, 187, 236, 251, 257, 278
Fieselmann, D. F., 311
Finney, R. W., 279
Finnie, J., 257
Finkle, B. S., 274
Fisanick, G. J., 35
Fischer, R. D., 303
Fish, R. H., 226, 302
Fischer, C. L., 14, 78
Fischer, G. S., 222
Fischer, L. J., 274
Fischer, N. H., 249
Fisher, T. L., 303
Fitch, W. L., 214
Fitzgerald, J. F., 215
Flad, J., 36
Flamini, A., 316
Flammang, R., 75, 78, 80, 81
Flammang-Barbieux, M., 80
Flamme, J. P., 46
Flanigan, E., 155
Flannery, M. R., 24
Flath, R. A., 222, 226
Fletcher, B. S., 225
Fleetwood, J. A., 209
Flesch, G. D., 307
Fletcher, A. S., 295
Flippin, L. A., 238
Flippen-Anderson, J. L., 250
Floberg, S., 278
Floerke, U., 327
Florêncio, H., 72, 76
Fluckiger, R., 179
Fluit, J. M., 146
Flynn, R. G. A., 238
Fogarasi, G., 2
Fogel, Y. A., 118
Folkers, K., 258

Folting, K., 311
Foltz, R. L., 171, 230
Ford, G. C., 251, 271, 288
Ford, G. P., 2
Ford-Hutchinson, A. W., 205, 292
Fore, S. P., 222
Förner, W., 6
Forrey, R. R., 226
Forst, W., 56
Forster, H. J., 286
Forsth, L. R., 90
Forys, M., 22
Fouda, H. G., 276
Fraley, D. F., 163
Franchetti, V., 107
Francesconi, L. C., 311
Francis, A. J., 317
Frank, H., 198, 211, 216, 292
Franke, P., 249
Francke, W., 226
Frankinet, C., 267
Franklin, J. L., 10, 19, 326
Franklin, R. T, 226
Franklin, W. E., 173
Franzen, H. F., 328
Franzen, J., 165
Fraser, B. A., 206, 257
Fravel, H. G., 219
Frazer, I. W., 236, 314, 315
Frazier, J. R., 234
Fredrich, M. F., 311
Freidel, R. O., 272
Freeman, C. G., 14
Freeman, M. B., 310, 312
Frees, L. C., 12
Freiser, B. S., 81, 107, 109, 110, 112, 113
French, J. B., 166, 178
Freudenthal, J., 181, 191, 200
Freuler, J., 226
Frey, R., 9, 44
Frey, W. F., 17, 234
Frick, W., 158, 258
Fridh, C., 7
Friedman, L., 155, 254, 255, 325
Friesen, G. D., 323
Friesen, M. D., 166, 219
Frigerio, A., 266, 282
Fritzche, H., 235
Fromme, I., 207
Fromson, J. M., 216
Frost, D. C., 10
Fryburg, G. C., 320
Fu, E., 255
Fu, E. W., 48, 106
Fujii, T., 182, 200, 217
Fujikawa, T., 26
Fujimaki, M., 223
Fujiwara, H., 237, 245, 250
Fuknda, Y., 321
Fukui, K., 14
Fukunaga, K., 215

Author Index

Fukushima, K., 253
Fukutome, H., 44
Fukuyama, H., 203
Fuller, K., 323
Fullmer, P. G., 220
Fung, A. P., 323
Funke, P. T., 265
Furlani, A., 313, 314
Furlei, I. I., 298
Furmanova, N. G., 316
Furrer, A., 222
Furst, D. E., 264
Futekov, L., 327
Futrell, J. H., 17, 56, 107, 194, 230
Fuwa, K., 200

Gäumann, T., 49, 65, 72
Gaffney, T. E., 289
Gage, L. D., 316
Gagosian, R. B., 221
Gaillard, M. L., 101, 103
Gaily, T. D., 18
Gaines, D. F., 323
Gaivoronskii, P. E., 312
Gal, T., 324
Galatsis, P., 248
Galitsyn, Yu. G., 324
Galkin, E. G., 298
Galli, G., 219
Galloy, C., 4, 38
Galyashin, V. N., 260
Gambino, O., 307
Games, D. E., 155, 199, 229, 241, 242, 245, 247, 249, 259, 318
Ganguly, A. K., 237, 245
Ganjei, J. D., 133
Gara, A., 220
Garaeva, L. D., 325
Garaj, J., 316
Garbe, A., 283
Gardner, S. A., 312
Garland, W. A., 237, 272, 273, 274, 277
Garman, A. J., 258
Garnett, J. L., 236, 309, 313, 314, 315, 316
Garretson, A. L., 252
Garrett, R. F., 122, 131
Garrett, W. R., 16
Garrison, B. J., 127
Garton, G. A., 202
Gardella, J. A., jun., 168
Gasco, A., 234
Gaskell, S. J., 201, 204, 207, 241, 251, 271, 275, 279, 296
Gaskin, P., 210
Gates, S. C., 190, 213, 214
Gatland, I. R., 85, 92, 98
Gattabeni, F., 219
Gaudin, D., 254

Gaukhman, A. P., 296, 301, 316
Gauyacq, D., 111
Gavens, P. D., 307
Gavrishchuk, E. M., 312, 313
Gehlhoff, M., 215
Geiger, W. E., 312
Geisler, C., 214
Geiss, F., 191
Gelbart, W. M., 111
Gel'man, Yu. A., 326
Gelpi, E., 204
Geltman, S., 32
Gender, P. L., 215
Gentry, W. R., 54
Geoghegan, R., jun., 252
George, P. M., 17, 328
Gerard, M., 22, 112
Gerard-Ain, M., 112
Gerber, G. E., 212, 259
Gerber, N., 287, 289
Gero, S. D., 252
Gerwer, A., 26
Gerwig, G., 207
Ghaderi, S., 165, 173, 181
Ghisalberti, E. L., 262
Ghose, B. N., 300
Ghosh, A. C., 212, 258
Gibb, T. C., 322, 323
Gibbs, P. A., 223
Giddings, J., 227, 293
Gielen, J. E., 267, 287
Gielen, M., 302
Giesecke, C. C., 315
Giessmann, U., 163, 256
Giffin, C. E., 182
Gilbert, J. N. T., 291
Gilbert, H. T., 262
Giles, H. F., 327
Gilbert, J. D., 262
Gilbert, J. R., 60
Gilmore, F. R., 11
Gillen, K. T., 18
Gilles, H. M., 276
Gingerich, K. S., 321, 327, 328
Giordani, A. B., 77, 244
Gioumousis, G., 85
Girard, J.-P., 63, 71
Giraud, A., 233
Giulhaus, M., 314, 315, 316
Given, K. W., 316
Glass-Maujean, M., 29
Gläumann, T., 165
Glavincevski, B. M., 298
Glick, M. R., 215
Glish, G. L., 75, 79, 171
Glockling, F., 308
Glushakova, V. N., 295
Green, A. R., 298
Gnanasunderam, C., 226
Gobby, P. L., 167
Gochman, N., 177

Goddard, J. D., 39
Godfrey, J., 281
Godfrey, P. D., 110
Goh, L.-Y., 311
Golab, T., 220
Gold, J. W. M., 215
Golden, D. M., 14
Goldstein, W., 148
Golik, J., 253
Gollnow, B. I., 254
Golovatyi, V. G., 305
Golovin, A. V., 45
Gomez, L. D., 202
Gomez-Sanchez, C. E., 209
Gooden, R., 52
Gooding, K. M., 224
Goodman, L., 34
Goodman, M., 224
Good-Zamin, C. J., 126
Gordoetshii, I. G., 229
Gordon, M. S., 5
Gordon, W. P., 287
Gorgorakai, V. I., 326
Gorgues, A., 308
Gorni, A., 219
Gorokhov, L. N., 10, 320, 321, 322, 327
Gorrod, J. W., 285
Gorth, R. G., 230
Goss, S. P., 104
Gotchev, B., 29
Goto, G., 260
Goudard, M., 248
Gough, D. W., 92
Gough, T. A., 183, 216
Gough, T. E., 38
Gove, H. E., 321
Govers, T. R., 22, 44, 45, 112
Gower, J. L., 155, 199, 245, 318
Graas, J. E., 270
Grabowski, J. J., 238
Grade, M., 322
Graham, D. M., 322
Graham, R. A., 190
Gramberg, L. G., 181
Granat, M., 190, 200
Grankin, V. P., 322
Grant, D. W., 326
Grassi, P., 219
Gray, C. P., 212, 259
Grayson, M. A., 328
Greaves, J., 276
Green, B. N., 154, 171, 294
Green, C. R., 198
Green, J. N., 292
Green, M., 314
Green, M. L. H., 313
Greene, L. J., 155
Greenhough, T. J., 311
Greenway, T. J. I., 148
Greenwood, N. N., 322, 323
Gregor, I. K., 236, 309, 313, 314, 315, 316

Author Index

Greiser, T., 316
Gress, M. E., 20
Greving, J. E., 212
Grey, A. A., 283
Grey, E., 300
Griffin, D., 220, 237, 287, 289
Griffin, G. W., 218
Griffith, G. L., 152
Griffith, W. P., 317
Griffiths, A. G., 192
Griffiths, C. J., 296
Griffiths, I. W., 77, 101, 162
Griffiths, J. C., 266
Griffiths, W. R., 303
Grigor'ev, A. I., 236, 316
Grigson, S. J. W., 271
Grimes, R. N., 323
Grimsrud, E. P., 167
Grinberg, Ya. Kh., 326
Grinsted, M. J., 184
Grishin, N. N., 304
Grishin, V. D., 229
Grob, G., 198
Grob, K., 198
Grob, K., jun., 198
Gronhi, H., 298
Gross, H. J., 254
Gross, M. L., 49, 51, 65, 72, 75, 165, 173, 181, 219, 254, 261
Grote, A. A., 217
Grote, H., 247
Grove, M. D., 250
Gruber, V. F., 264
Gruen, D. M., 127, 131, 138, 139, 143, 144, 151
Grützmacher, H. F., 46, 52, 65, 72, 75, 79
Grula, J. W., 225
Grupe, A., 251
Grynberg, G., 32
Gschwend, P. M., 217
Guadagni, P. G., 222
Guberman, S., 30
Gudeman, C. S., 117
Gudzinowicz, B. J., 262
Gudzinowicz, M. J., 262
Gülacar, F. O., 72
Guenther, F. R., 218
Guerchais, J. E., 310
Gürtler, P., 29
Guest, J. A., 104
Guilhaus, M., 236
Guichon, G., 171, 228, 229, 245, 262
Gunstone, F. D., 259
Gupta, N. K., 254
Gupta, S. K., 321, 327, 328
Gusarov, A. V., 10, 321
Gusev, Yu. K., 304
Gustafsson, J.-A., 184, 211
Gutiev, G. G., 260
Gutshall, P. L., 163

Guyon, P. M., 22, 27, 29, 30, 44
Gygax, R., 17, 19, 107

Ha, T. K., 3
Haan, E. A., 214, 215
Habermeier, H., 250
Hachey, D. L., 315
Hackett, H. M., 221
Haddad, R. Y., 23
Haddon, W. F., 171, 186, 244, 249
Hadzi, D., 176
Haegele, K. D., 229
Haese, N. N., 6
Hage-Ali, M., 328
Hager, A., 215
Hagge, D. E., 183
Haginaka, J., 228
Hagler, A. T., 19
Hagstrum, H., 134
Hahnel, R., 214
Haider, K., 247
Hajlbrah, S. K., 245
Halbacher, H., 258
Haley, M. J., 295
Halgren, T. A., 2
Halket, J. M., 194, 200, 208, 228
Hall, B., 303
Hall, J. H., 2
Hall, K., 270
Hall, L. W., 312
Hall, R. B., 315
Halpern, B., 211, 214, 215, 264
Haltiwanger, R. C., 325
Hamill, R. L., 252
Hamilton, R. H., 211
Hamlet, P., 21
Hammarstrom, S., 205, 260
Hammer, D., 147
Hammerstinol, H., 223
Hammond, G. S., 40
Hammond, J., 214, 215
Hammond, R. C., 202
Hamnett, A. F., 203
Hamon, J. R., 310
Han, B. H., 249
Hancock, R. A., 165, 166
Hancock, R. H., 296
Handjieva, N. V., 249
Hanessian, S., 252
Hangartner, M., 218
Hanisch, R. C., 216
Hanley, P. R., 123
Hanrahan, L. R., 169
Hansen, G., 156, 243
Hanson, G. R., 42
Hanson, J. M., 258
Hansoul, J. P., 4
Hansson, G. C., 259, 261
Hanuš, V., 74
Hao, C.-T., 222

Happer, D. A. R., 299
Haque, R., 321
Hara, Y., 268
Hardy, M., 203
Hargittai, I., 324
Hargreaves, R. T., 241
Hariharan, A. V., 328
Harland, P. W., 14
Harmon, K., 171
Harper, D. B., 223
Harper, M. E., 279
Harrington, M. M., 92
Harris, D., 249
Harris, F. M., 77, 101, 162
Harris, J. A., 81
Harris, J. H., 183
Harris, M., 177
Harris, P. J., 305
Harris, T. M., 247, 287
Harrison, A. G., 10, 11, 14, 23, 56, 63
Harrison, D. E., jun., 127
Harrison, P. G., 302, 303
Hart, R. G., 139
Hartman, K. G., 12
Hartvig, P., 271, 272, 277
Harvan, D. J., 25, 75, 77, 166, 219, 240
Harvey, D. J., 208, 213, 251, 293
Harvey, T. M., 183
Harwood, J. L., 203
Hase, S., 206, 255
Hase, W. L., 38
Hashimoto, K., 216
Hashimoto, M., 284
Hashizume, A., 185
Hashizume, T., 213
Haskins, N. J., 210, 213, 251, 271, 288
Hass, J. R., 25, 75, 77, 166, 219, 220, 237, 240
Hasted, J. B., 23, 86
Hastings, C. R., 199
Hatch, F. W., 170, 197
Hattox, S. E., 185, 201
Haudegond, J. P., 308
Havemann, V., 57
Haverkamp, J., 256
Hawkins, D. R., 279
Hawthorne, S. J. A., 218
Hayashi, K., 208
Hayashi, M., 207
Hayashi, S., 248
Hayes, J. M., 169, 170, 184, 201
Hays, J. D., 109, 114
Hayward, E. J., 165
Hazelby, D., 183, 243
Hazi, A. U., 27
Healy, M. A., 302
Heath, B. A., 34, 35
Heath, R. R., 225
Heckman, R. A., 198

Hecq, A., 321
Hedfjäll, B., 169, 182, 200
Heerma, W., 72, 76, 256
Hefetz, A., 225
Heffler, H., 259
Heftmann, E., 249
Hehre, W. J., 2, 5, 11, 298
Heilbronner, E., 6
Heimbach, H., 78
Heimerl, J., 86
Heimermann, W. H., 202, 259
Heinemeier, J., 148
Heinen, G., 184
Heinen, H. J., 160, 163, 164
Heininger, J., 215
Heinrich, K. F. J., 122
Heinzer, I., 221
Heitbaum, J., 163
Heitz, C., 328
Heki, N., 288
Helbig, H. R., 185
Helland, P., 227
Heller, E. J., 37
Heller, S. R., 176, 193
Helm, H., 102, 103
Hemberger, P. H., 47, 66, 79, 172
Hendel, J. G., 203
Henderson, D. E., 221, 298
Hendricks, L. J., 52
Hendrickson, D. N., 311
Hendriks, H., 249
Hendry, L. B., 226
Henion, J. D., 171, 245
Henneberg, D., 185, 189, 197
Hennequin, J. F., 139
Henrichs, S. M., 221
Henrichs, U., 197
Henry, F. T., 228
Herberich, G. E., 312
Herbst, E., 55, 220
Hercules, D. M., 168, 322
Heresh, F., 247, 264
Herlihy, W. C., 212, 259
Herman, J. E., 181
Herrman, K., 222
Herrmann, A., 10, 32
Hermann, Z., 57
Herron, J. T., 7
Herschbach, D. R., 18, 37
Herschberger, R. L, 148
Hertz, H. S., 171, 218
Herzig, U., 74, 171
Herzog, L. F., 178
Herzog, R. F. K., 120
Hess, J., 230
Hesse, M., 241, 246
Hetheridge, M. J., 270
Heubers, W., 169
Heusse, D., 267
Hicks, P. J., 30
Hidy, B., 171, 230
Higgins, I. J., 202
Hignite, C., 241, 270, 287

Higuchi, K., 299
Higuchi, R., 242, 251
Hilderbrand, D. L., 328
Hileman, F. D., 317
Hilker, D. R., 261
Hill, H. H., 199
Hill, W. E, 308, 324
Hillenkamp, F., 160
Hillig, H., 178
Hilmer, R. M., 193
Hilt, E., 163
Hilton, P. R., 26
Hinchliffe, A., 4
Hinchman, B., 201
Hinde, A. L., 19
Hinthorne, J. R., 120
Hippler, R., 147
Hirano, M., 321
Hirao, H., 20
Hiraoka, H., 326
Hiraoka, K., 19
Hirata, Y., 167, 183, 230, 234, 236, 243, 259
Hirayama, C., 320
Hirokata, Y., 292
Hirooka, T., 13
Hirota, A., 78
Hirschfeld, T., 170, 199
Hirschwald, W., 322
Hirst, D. M., 3, 4
Hirter, P., 155, 245, 258
Hitchcock, A. P., 25
Hitchock, P. B., 299
Hites, R. A., 216, 218
Hitzenberger, G., 290
Ho, B. Y. K., 300
Hobbs, D. C., 276
Hobbs, R. H., 21
Hoberg, H., 297
Hodges, R. V., 13, 104, 239, 324
Hodshon, B. J., 274
Hoecker, J. L., 265
Höhne, G., 78
Hoelzle, E., 224
Hoener, B. A., 381
Hoering, T. C., 300
Hötzel, Ch., 164
Hof, P. J., 170
Hofer, W., 127
Hofer, W. O., 138, 151
Hoffman, J., 207
Hoffman, J. C., 323
Hoffmann, K. J., 284, 289
Hoffmann, R. W., 296
Hofmann, M., 32
Hogge, L. R., 190, 200, 203, 300
Hohara, T., 242, 251
Hohaus, E., 295
Hohenstatt, M., 107
Hohne, B. A., 195
Holkeuoer, D. H., 178
Holland, J. K., 155, 158, 190

Holland, O. B., 209
Holland, P. M., 21
Holland, P. T., 210, 249
Holloway, J. D. L., 312
Holloway, J. H., 10, 327
Holly, S., 291
Holman, R. T., 202, 259
Holmbom, B., 29
Holme, T., 256
Holmes, J. H., 46
Holmes, J. L., 10, 11, 46, 48, 50, 61, 62, 67, 68, 78
Holmgren, J., 256
Holmstead, R. L., 220, 302
Holsboer, H. H., 169
Holzer, G., 214
Holzmann, G., 234
Hommes, H., 72, 78
Honda, F., 321
Hone, D. W., 137
Hong, Y.-M., 258
Honig, R. E., 118
Honma, K., 58
Honovich, J., 114
Hopkins, R., 281
Hopkinson, J. A. V., 185
Hopper, D. G., 4
Hopper, S. P., 299
Horani, M., 101, 103, 111
Horheinz, W., 208
Horie, M., 274
Horlbeck, W., 320
Hornbeck, C. L., 266
Horning, E. C., 155, 229, 243
Horning, M. G., 155, 229, 243
Hortig, G., 148
Horvath, G., 260
Hoshita, T., 208
Hosoi, K., 278
Hosojima, H., 288
Houle, F. A., 8, 13, 63
Houriet, R., 56, 72, 165
Howald, W. H., 278
Howard, R. W., 224
Howard, W. E., 39
Howarth, D. W., 177
Howe, I., 59, 77, 80, 81, 167, 240, 258
Howells, S., 172
Howie, R. A., 326
Howorka, F., 25, 87
Howse, P. E., 224, 225
Howton, C., 104
Hoxmeier, R. J., 311
Hoyer, A., 251
Hoyer, G. A., 189
Hoyt, D. W., 217
Hribar, J. D., 248
Hsieh, T., 20, 43
Hsiung, H. H., 243
Hsu, C. S., 164
Hu, L. E. C., 296
Huang, M. H. A., 316

Author Index

Huard, D., 10
Huber, B. A., 102, 104, 116
Huber, H., 20
Huber, J., 148
Huber, J. W., 228
Hubin, M. J., 10
Hubin-Franskin, M. J., 10
Hucker, H. B., 290
Hudgens, J. W., 162
Huestis, D. L., 103
Huffman, D. H., 270, 287
Huffman, J. C., 311
Huguet, R., 314
Huizing, G., 285
Hull, C. W., 178
Hummel, R. A., 219
Huneck, S., 249
Hunt, D. F., 77, 155, 166, 169, 200, 242, 244
Hunt, S., 215
Hunter, E. P., 21, 251
Hunter, R. L., 173
Hunter, R. O., 104
Hurley, A. C., 16
Hurst, G. S., 162
Hursthouse, M. B., 317
Husek, P., 211, 257
Husen, J., 36
Hush, N. S., 16, 26
Husinsky, W., 147
Husmann, H., 197, 199
Hussain, M. G., 259
Hutchins, R., 226
Hvistendahl, G., 50, 52
Hwang, C.-S., 186, 294
Hyder, H. R. McK., 148
Hylin, J. W., 220

Iacobucci, A. C., 248
Ichimura, T., 215
Ikenaka, T., 206, 255
Ikeno, M., 300
Il'in, M. K., 320
Illenberger, E., 18, 234
Illies, A. J., 166
Illing, H. P. A., 216
Imhof, P., 286
Imura, M., 213
Ingelman-Sundberg, M., 184, 211
Inghram, M. G., 42, 45, 49
Ingles, D. L., 257
Ingels, J., 179
Inoue, M., 139
Inoue, S., 256
Invernizzi, G., 219
Irwin, W. J., 226, 227, 262, 291
Isenhour, T. L., 176, 188, 191, 193, 195
Ishibashi, M., 207, 260, 278
Ishiguro, S., 222
Isolani, P. C., 239
Itabashi, Y., 208

Ito, K., 29
Ito, M., 35
Ittel, S. D., 165, 310
Ivashkiv, E., 265
Iwanga, M., 278
Iwase, H., 257
Izumi, Y., 164

Jackson, A. H., 242, 246, 249
Jackson, C. K., 324
Jackson, R. L., 19, 107, 234
Jacob, J., 204
Jacob, K., 304
Jacobon, A. J., 315
Jaeger, R. D., 325
Jahr, D., 222
Jakobsen, R. T., 217
James, B. D., 302
Janak, J., 213
Janousek, B. K., 16, 17, 19, 107
Jankovska, I., 316
Jankowski, K., 254
Janssen, F., 199
Janssen, F. W., 283
Janssen, G., 202
Jansson, B., 219, 220
Jarnagin, R. C., 10
Jarvis, W., 323
Jasinski, J. M., 14, 55, 71, 114, 115
Jason, A. J., 9
Jaspers, D., 168
Jayasimhulu, K., 257
Jefford, C. W., 5
Jellum, E., 214, 215
Jemma, A., 219
Jen, T.-L., 226
Jenden, D. J., 213
Jenkins, R., 287
Jennings, K. R., 66, 186, 236, 294
Jennison, T. A., 274
Jick, B. S., 316
Jindal, S. P., 273
Johansson, C. E., 220
John, J. O., 226
Johne, S., 246
Joho, M., 222
Johnsen, R., 86
Johnson, B. F. G., 307
Johnson, J. C., 160
Johnson, K. E., 37
Johnson, K. M., 41
Johnson, L. P., 206, 281
Johnson, P. M., 31, 33
Johnson, R., 24
Johnston, J., 315
Johnstone, R. A. W., 185
Jolley, M., 193
Jolly, W. L., 324
Jonathan, N. B. H., 10
Jones, A., 249
Jones, D. W., 221

Jones, G. R., 216
Jones, P. J., 314
Jones, P. L., 17, 19, 107, 108
Jones, P. R., 230
Jones, S., 252
Jones, S. R., 245
Jones, T. B., 111
Jones, T. H., 226
Jones, T. R. B., 295, 312
Jonsson, B. O., 11, 22
Jordan, K. D., 16, 20, 233, 234
Jorgensen, W. L., 47, 50, 75
Jorgenson, M. W., 323
Jortner, J., 37
Josefson, B., 217
Joselau, J.-P., 206
Joshi, K. C., 314
Josreth, A. B., 218
Jost, C., 187
Jourdain, D., 203
Jowko, A., 22
Joyes, P., 135
Joyner, D. J., 322
Judge, D. L., 30, 110
Jungen, C., 27
Jurela, Z., 122, 139
Jurenitsch, J., 247
Jurs, P. C., 179, 191, 201, 315
Juttner, F., 223

Kaberline, S. L., 179, 192
Kaesz, H. D., 311
Kageura, M., 271
Kahl, R., 287
Kakis, G., 204
Kaldor, A., 315
Kalinin, Yu. T., 167
Kalinnekov, V. T., 326
Kalita, C. C., 252
Kalman, J. R., 52
Kalman, O. F., 29
Kalow, W., 283
Kambara, H., 166, 171, 242
Kameda, M., 323
Kamerling, J. P., 207, 256
Kaminski, J. J., 296
Kaminsky, M., 118
Kamm, J. J., 291
Kamps, L. R., 220
Kanai, Y., 288
Kaneko, Y., 86
Kane-Magiure, L. P. A., 318
Kanev, A. N., 302
Kanig, W., 306
Kanomata, I., 166
Kant, A., 321
Kao, J. C., 292
Kapila, S., 199
Kaposi, O., 328
Kappus, G., 184
Karakaya, A., 286

Karasek, F. W., 178, 185, 199, 200, 217, 219, 227
Karataev, E. M., 313
Karger, B. L., 171, 230, 231
Karimpour, H., 295
Karlsson, A. L., 259
Karlsson, K. A. 257, 259, 261
Karlsson, K. E., 272
Karmen, A., 264
Karni, M., 47, 72
Karpas, Z., 19
Karpov, G. V., 167, 229
Kase, Y., 215
Kasuya, Y., 273
Kasymov, Sh. Z., 249
Katakuse, U.; 164, 173, 243, 258
Katayama, D. H., 104, 111
Kates, M., 261
Kato, A., 268
Kato, C., 300
Kato, T., 109
Katori, M., 260
Katsir, Z., 224
Katz, J. J., 255
Katz, W., 123
Kavlock, R.-J., 216
Kawamura, M., 251
Kawasaki, T., 242, 251
Kay, K. G., 35, 36, 37
Keakle, J., 303
Kebarle, P., 14, 19, 63
Keck, H., 304
Keesee, R. G., 19
Keister, J. B., 308
Keith, A. N., 310
Keith, L. H., 176, 216
Keller, G. E., 176
Kelly, R., 126, 132
Kemény Gy. Mády, V., 291
Kemper, K. W., 148
Kenne, L., 206, 256
Kennedy, J. D., 303
Kennedy, J. F., 206
Kennedy, K. A., 274
Kenny, J. E., 37
Kenny, M. D., 181
Kenyon, C. N., 214, 255
Kerrison, S. J. S., 314
Khafizov, R. S., 167
Khalilov, M. L., 298
Khallef, K. D., 221
Khitrova, O. M., 311
Khmel'nitskii, R. A., 233, 247, 260
Khomenko, V. S., 315
Khoo, K.-C., 273
Khorana, H. G., 212, 259
Khrapova, I. M., 297
Khusnutdinov, R. I., 234
Khvostenko, O. G., 236
Khvostenko, V. I., 234, 235, 236, 298
Kiehn, T. E., 215

Kihira, K., 208
Kihara, T., 89
Kikot, B.-S., 325
Kilius, L. R., 321
Kim, J. H., 249
Kim, M. S., 48, 52, 57, 77, 106, 109, 113
Kim, S. H., 167
Kim, W. S., 217
Kimble, B. J., 219
Kimura, K., 10, 288
Kindt, T. J., 258
King, G. K., 238
King, G. S., 210, 227, 251
King, R. B., 327
King, T. J., 302
Kingcade, J. E., 321
Kingdon, K. H., 119
Kinh, P.-Q., 208
Kinoshita, K., 216
Kirby, D. P., 171, 230, 231
Kirby, K., 30
Kirkman, S. K., 283
Kiser, R. W., 186
Kisic, A., 204
Kistemaker, J., 172
Kistemaker, P. G., 160, 180, 182, 242
Kitaura, K., 20
Kjaer, A., 223
Klass, G., 239
Kleier, D. A., 2
Klein, F. S., 19
Klein, P. D., 184, 315
Klein, R., 113
Klein, R. A., 204, 259
Klein, W., 220
Kleingeld, J. C., 66
Kleinschmidt, P. D., 328
Kligman, A., 224
Klimova, E. I., 310
Klimova, T. V., 323
Klingler, K. H., 285
Kloster-Jensen, E., 43, 111
Klots, C. E., 39, 41
Klun, J. A., 225, 226
Knapp, F. F., 306
Knapp, H. R., 205
Knedel, N., 304
Knevel, A. M., 244
Knight, D. W., 249
Knight, M. E., 245, 249
Knight, W. E., 204
Knoche, H. W., 261
Knoeppel, H., 191
Knol-Kalkman, A., 211
Knorr, F. J., 194
Knowles, D. R., 253
Knowles, J. A., 283
Knowles, J. R., 245
Knox, A. B., 148
Knutti, R., 229
Kobayashi, A., 224
Kobayashi, M., 299

Kobayashi, T., 288
Koch, E. E., 29
Koch, L., 184
Kocher, C. W., 219
Kochetkova, N. S., 310
Kocovsky, P., 251
Kodama, G., 323
Koehler, F., 297
Köhler, H. J., 5, 6, 15
Köllmann, K., 29
Koeniger, N., 225
Koenitzer, H., 187
Köppel, C., 72
Kogler, W., 199
Kohl, F. J., 320
Kohl, W., 247
Kohler, F. H., 311
Kokke, W. C. M. C., 250
Kolesova, V. A., 304
Koller, J., 6
Kolobova, N. E., 311
Kolthammer, B. W. S., 311
Komalenkova, N. G., 301
Komori, K., 139
Komori, T., 242, 251
Kompa, K. L., 162
Kondow, T., 25
Kondrat, R. W., 77, 236, 244
Konig, W. A., 222
Kononenko, G. P., 248
Konovalova, O. A., 249
Konowalow, D. D., 20
Koransky, W., 216
Korneev, N. N., 297
Korobov, M. V., 321, 327
Korol, E. N., 305, 318
Koronelli, T. V., 261
Korschinek, G., 148
Korte, F., 207, 220, 221
Korth, J., 264
Koski, W. S., 12
Koss, G., 216
Kostrovskii, V. G., 302
Kovac, P., 255
Kovacik, V., 255
Kovalev, I. D., 159
Kovrigin, V. M., 302
Kowalski, B. R., 176
Koyama, Y., 180
Koyano, I., 58
Kozlova, N. I., 294, 316
Kozunov, A. V., 316
Kraemer, W. P., 6
Kraft, R., 246
Kralj, B., 171, 172
Kramer, G. M., 315
Kramer, S. D., 162
Kramer, V., 171, 172
Kramer, W. G., 265
Krasil'nikov, M. D., 327
Krass, E., 258
Krause, D. A., 47, 50, 75
Krause, J. L., 303

Author Index

Kraus, A. R., 118, 127, 138, 139, 143, 144, 151
Krenmayr, P., 74, 171
Krevalis, M. A., 327
Krien, P., 229, 245
Krienen, U., 178, 201
Krishnan, V., 319
Kritskaya, I. I., 236
Krivit, W., 214
Kroes, H. H., 255
Krogh-Jespersen, K., 34
Krohn, V. E., 119, 121, 123
Kroshefsky, R. D., 317
Kruger, W., 147, 300
Kruit, P., 35
Krupcik, J., 316
Kruse, J. R., 172
Krusius, T., 257
Kruskal, J. B., 193
Krutzsch, H. C., 211, 258
Kryukov, V. P., 322
Kubec, F., 136
Kubjacek, M., 325
Kucharczyk, N., 285
Kuchen, W., 304
Kuck, D., 65
Kudin, L. S., 321
Kudo, H., 252
Kudo, Y., 258
Kudrya, T. N., 305
Kuebler, N. A., 34
Küpfer, A., 287
Küstler, A., 44
Kuhara, T., 292
Kuhn, E. S., 304
Kuhn, U., 202
Kuivahara, R., 303
Kuksis, A., 209
Kulander, K. C., 37
Kulkarni, P. S., 165, 181
Kulmacz, R. J., 204
Kulzick, M. A., 323
Kumakura, M., 46
Kumar, K., 89
Kump, R., 227
Kuo, M., 273
Kupel, R. E., 217
Kuper, H., 178
Kupka, K.-D., 160
Kuppermann, A., 30
Kupriyanov, S. E., 47
Kuramoto, T., 208
Kurata, T., 223
Kuriki, S., 321
Kuroda, H., 26
Kuroda, K., 300
Kuroki, H., 220
Kurz, E. A., 170
Kusaka, T., 260
Kusmierz, J., 228
Kuster, T., 186
Kusunoki, I., 110
Kutschera, W., 148
Kuzanyan, R. S., 318

Kuznetsov, A. L., 302
Kuznetsov, M. A., 5
Kwiatkowski, J., 189, 194
Kyoto, M., 327
Kyrala, G. A., 103
Kyriakidis, N., 172, 256

Labbe, J. C., 324
Labintsev, V. B., 304
Labows, J., 224
Lacave, C., 261
Lacey, M. J., 172, 185
Lacmann, K., 17, 234
Ladd, M. F. C., 303
LaFleur, A. L., 177
Lafont, R., 203
Lagarde, G., 328
Lagergren, C. R., 170
Lagow, R. J., 306
Lai, S.-T. F., 243
Lakings, D. B., 268
Lamb, W. E., jun., 29, 46, 103
Lambelin, G., 280
Lambropoulous, P., 33
Lamparski, L. L., 219
Lanbeck, R., 278
Langhoff, P. W., 26
Langmuir, I., 119
Langsdorf, A., 152
Langworthy, T. A., 204
Lanzoni, J., 266
Lao, C., 138
Lao, M. N. S., 326
Lapeyre, M., 206
Lapsina, A., 317
Laramée, J. A., 47, 75, 79, 172
Larcher, F., 300
Larin, N. V., 159
Larsen, E., 202
Larson, G., 259
Larson, G. L., 300
Larsson, F. C. V., 235
Larsson, K., 216
Larzilliere, M., 103
Laseter, J. L., 183, 218, 221
Latner, A. L., 209
Lau, K. H., 328
Lau, P.-Y., 259
Lau, Y. K., 14
Laub, R. J., 210
Lauer, G., 6
Laurov, V. P., 144
Lavanchy, A., 72, 251
Lavis, A., 208
Lavrentev, V. I., 302
Lavrinovich, L. I., 297
Law, J. R., 218
Lawesson, S.-O., 234, 235
Lawrence, R. H. J., 246
Lawson, A. M., 214
Lawson, G., 166
Lax, P. M., 227

Lazarev, V. B., 326
Leach, S., 110, 111
Leander, K., 277
Le Bozec, H., 308
Leclercq, P. A., 206, 220, 316
Ledford, E. B., jun., 165, 173, 181
Ledniczky, M., 282, 291
Lee, C., 221
Lee, C. R., 201
Lee, H. W., 321
Lee, J., 14
Lee, K. Y., 214
Lee, L. C., 30, 104, 110
Lee, M. G., 199, 245, 247
Lee, R. C.-Y., 303
Lee, T. D., 160
Lee, W. Y., 326
Lee, Y. S., 21
Lee, Y. T., 13, 20
Leeuw, J. W., 208
Leferink, J. G., 275
Leffler, J., 259, 261
Le Floch-Perennou, F., 310
Legendre, M. G., 222, 223
Legzdins, P., 311, 312
Lehman, T. A., 60
Lehmann, H., 235, 256
Lehmann, J.-P., 167
Lehmann, K. K., 34
Lehmann, W. D., 159, 182, 257, 262
Lehrer, M., 264
Lehrle, R. S., 168
Leibich, H. M., 213
Lelyukhina, Yu. L., 297
Lentz, D., 327
Leplyanin, G. V., 234
Le Roux, M., 224
Lester, G. R., 60
Leta, D. P., 133
Letham, D. S., 254
Leung, H.-W., 11, 63
Leutwyler, S., 10, 32
Lev, N. B., 114
Levason, W., 303, 327
Levi, B. A., 5
Levi, I. S., 325
Levin, V. A., 264
Levine, R. D., 57
Levins, R. J., 218
Levinson, A. R., 226
Levinson, B. Z., 226
Levsen, K., 45, 47, 48, 51, 59, 65, 66, 78, 256
Levsen, R., 326
Levy, D. H., 37
Levy, L. A., 75
Levy, R. H., 274
Lewin, P. K., 209
Lewis, E., 155, 242, 245
Lewis, I. A. S., 199, 245
Lewis, J., 307
Lewis, L. N., 311

Lewis, R. K., 123
Lewis, S., 65, 214
Lewy, A. J., 237
Leyden, J., 224
Leyman, W., 198
Li, Y. H., 23
Liang, Y., 209
Liao, J. C., 218
Liardon, R., 202, 211, 257
Lias, S. G., 7, 11, 12, 14, 63
Liauw, Li, 323
Libby, L. M., 225
Libertini, E., 318
Lichtenstein, C., 172
Lichtin, D. A., 35, 162
Lichty, D. G., 228
Liebl, H., 118, 120, 138
Liebman, D., 310
Liehr, G., 286
Liehr, J. G., 213, 253, 265
Lifshitz, C., 43, 45, 48, 78
Light, J. C., 96
Light, R. W., 303, 308
Lightman, A. J., 35
Ligon, W. V., jun., 163, 169, 170, 173, 243
Limburg, K., 254
Lin, C. Y., 320
Lin, S. L., 85, 86, 91, 93, 94, 95, 98
Lin, S.-S., 326
Lin, Y. Y., 166, 236, 245, 251
Lincoln, K. A., 161
Lindahl, R., 220
Lindberg, B., 206
Lindberg, C., 271, 272
Linden, H. B., 163
Linder, F., 29
Linderberg, N., 256
Lindgren, J. E., 277
Lindholm, E., 7
Lindinger, W., 25, 91
Lindsay, D. M., 21
Lindstrom, B., 278
Lineberger, W. C., 17, 19, 107, 108, 117
Ling, L. C., 222, 223
Linn, S. H., 20
Linscheid, M., 253
Lipscomb, W. N., 2
Liptak, A., 250
Lisboa, B. P., 208, 209
Lischka, H., 5, 6, 15
Liston, S. K., 43
Litherland, A. E., 321
Little, J.L., 323
Liu, N. W. K., 178
Liverman, M. G., 35, 36
Livett, M. K., 20
Lloyd, H. A., 209, 210
Locht, R., 30, 46
Loew, G. H., 4
Lofqvist, J., 224
Logunov, A. P., 304

Long, F. A., 43
Longevialle, P., 63, 70, 71
Longhurst, C., 224, 225
Lopez, V., 216
Lorents, D. C., 18
Lorquet, J. C., 4, 38, 39
Lory, E. R., 159
Los, J., 196
Lossung, F. P., 8, 11, 46, 62
Loudon, A. G., 74, 258
Lounasmaa, M., 247
Louter, G. J., 172
Lovins, R. E., 257
Lowry, S. R., 188, 193
Loy, M. M. T., 32, 36
Lubman, D. M., 32
Lucas, S. V., 217
Luciani, L., 219
Ludwig, B., 221
Luederwald, I., 258
Luijten, W. C. M. M., 81
Lukas, W., 320
Lukashenko, I. M., 247, 268
Luken, W. L., 16
Lukevics, E., 296, 301, 316
Lum, R. M., 177
Lun, S.-W., 229
Lundblad, A., 215
Lundgren, L., 224, 241
Lundquist, T. R., 139
Lundsten, J., 215
Lunel, J., 252
Lupandin, V. S., 310
Lupotto, E., 223
Luther, E. W., 276
Lutz, T., 273
Lyashko, A. N., 255
Lyatifov, I. R., 310
Lyle, M. A., 290
Lyle, R. E., 296
Lynn, R. K., 287, 289
Lyon, P. A., 254
Lyubitov, Yu. N., 325
Lyuts, A. E., 304

Mabry, T. J., 241
McAdams, M. J., 159, 230, 231, 245
McAdoo, D. J., 52
McAuliffe, C. A., 303
MacBeath, M. E., 322
McCaman, M. W., 184
McChesney, J. D., 225
McCloskey, J. A., 212, 213, 215, 244, 253, 254
McClusky, G. A., 77, 236, 244
McConnell, M. L., 214
McCorkle, D. L., 234
McCormack, A., 176
McCourt, F. R., 94
McCowan, J. C., 248
McCrery, D. A., 109
McCulloh, K. E., 9, 10

McCullough, P. F., 303
McCusker, M. V., 104
McDaniel, C. A., 224
McDaniel, E. W., 85, 92, 98
McDavid, L., 177
Macdonald, C. G., 185, 172
MacDonald, R. J., 122, 126, 130, 131, 135, 138, 139, 140
McDowell, C. A., 10
McDowell, P. G., 224
McDowell, R. A., 183
McEwan, M. J., 14
McEwen, C. N., 165, 237, 310
McFadden, W. H., 199, 228, 230, 245
McFarland, M., 86, 91
McFarlane, N., 303
Macfarlane, R. D., 160, 243
McFarlane, S. H., 180
McGill, J. R., 176
McGilvery, D. C., 77, 102, 103, 104, 162, 172, 181, 243, 244
McGilvery, E., 77
McGinnis, P. F., 178
McHugh, J. A., 118
McIlrath, W. O., 223
McIntyre, L. C., 29, 46
McIver, R. T., 11, 13, 14, 60, 63, 237
McIver, R. J., jun., 173
Mackay, G. I., 22, 81, 239
Mackay, K. M., 307
Mackay, M. F., 302
McKee, R. E., 211
McKeen, L. W., 45, 194
Mck. Halket, J., 209
McKibben, J. L., 152
Mackie, K. L., 206
Mackie, P. R., 221
McKinley, I. R., 296
McKinnon, S., 249
McKoy, B. V., 26
McLachlan, J. A., 286
McLafferty, F. W., 9, 14, 25, 47, 52, 60, 61, 73, 77, 78, 155, 159, 172, 176, 187, 195, 201, 228, 229, 243, 244, 245
McLaughlin, J. R., 225
McLean, A. D., 4, 10
MacLeod, J. K., 5, 62, 254
McLoughlin, R. G., 8, 9, 10, 61, 62, 63
McMahon, B., 220
McMahon, T. B., 54, 239, 322
McMaster, A. D., 299
McMaster, B. N., 185
McMillan, J., 210
McMillan, M. R., 22
McNeal, C. J., 243
McNeil, M., 255

Author Index

McPeters, H. L., 19
McPherron, R., 171, 179, 201
Madani, C., 155
Madden, J. K., 206
Madsen, J. O., 223
Madyastha, K. M., 241
Maeda, K., 180
Maes, R. A. A., 275
Magnusson, S., 258
Magyar, E. S., 309
Mahaim, C., 309
Mahajan, V. K., 257
Mahale, V. B., 308
Mahan, B. H., 13, 20, 110
Mahen, L. J., 316
Mahle, N. H., 193
Maier, D. E., 303, 308
Maier, E., 304
Maier, H. J., 148
Maier, J. P., 43, 100, 106, 110, 111
Maier, R., 148
Mainzer, N., 66
Maitland, G. C., 92
Makarov, A. V., 320
Makarova, N. A., 317
Makin, G. I., 297
Maksimov, G. A., 159
Malda, A., 320
Malik, K. M. A., 317
Malinowski, E. R., 195
Malinski, E., 228
Malley, M. F., 265
Malnoë, A., 283
Malterud, K. E., 202
Mamer, O. A., 275
Mandelbaum, A., 47, 66, 71, 73
Mandli, H., 186
Mandova, N., 250
Manmade, A., 212, 258
Manotti, A. M. L., 307
Mansell, P. I., 41
Mansfield, W. K., 152
Mansvelt, F. J. W., 212
Manura, J. J., 264
Manvelyan, R. V., 47
Maquestiau, A., 75, 78, 80, 81
Marai, L., 204, 209
Marcelle, G. B., 249
March, J. F., 259
Marchington, A. F., 5
Marecek, J. F., 304
Mareda, J., 5
Marekov, N. L., 249
Marfat, A., 260
Margrave, J. L., 10, 326
Marian, C. M., 4
Marien, J., 320
Mariot, J. P., 310
Markey, S. P., 237
Marlowe, S., 180
Marmet, P., 10

Marriott, T. D., 164
Mars, J. R., 163
Marsel, J., 171, 172, 328
Marshall, J. C., 191
Marshall, P. J., 208
Martens, M., 280
Marthaler, O., 43, 110, 111
Martin, D. T., 312
Martin, D. W., 85
Martin, L. E., 270, 281
Martin, P. J., 122, 130, 131, 138, 151
Marunaka, T., 269, 277
Marusin, V. V., 324
Marynick, D. S., 15
Marx, R., 22, 24, 110, 112
Masaryk, J., 316
Mason, E. A., 85, 86, 89, 91, 92, 93, 94, 95, 98
Massey, A. G., 324
Massey, G. A., 160
Massey, I. J., 250
Masson, C. R., 300
Masuda, Y., 220
Matcha, R. L., 21
Materikova, R. B., 310
Mather, R. E., 165, 166
Matheson, T. W., 313
Mathey, F., 308, 311
Mathieson, D. W., 281
Mathur, B. P., 17, 35, 324
Matin, S. B., 268, 275
Matsuda, H., 164, 173, 243, 258
Matsuda, K., 5, 224
Matsumoto, G., 321
Matsumoto, H., 215
Matsumoto, I., 292
Matsumoto, K., 183, 234, 236, 259
Matsumura, G., 256
Matsumura, K., 298
Matsuo, T., 164, 173, 243, 258
Matsushima, Y., 206, 255
Matteo, G., 72
Matthews, D. E., 167, 169, 170, 184
Matveentsev, V. D., 257
Mauchamp, B., 203
Mauclaire, G. H., 22, 24, 102, 112
Maume, B., 287
Mavrodiev, V. K., 236
Maxwell, D. C., 176
Maxwell, W. M., 323
May, R. J., 170
May, W. E., 218
Mayar, M. S. B., 252
Mayee, R. J., 302
Mayer, H., 207
Maynard, J. B., 227
Mayo, B. C., 279
Mayr, H., 6
Mays, M. J., 307

Mazanec, T. J., 305
Mazeika, I., 296, 316
Mazur, Y., 252
Mazza, M., 217
Mead, P. J., 49, 68
Medved, M., 171, 172
Meek, D. W., 305
Megill, L. R., 86
Meguno, H., 303
Meier, E., 309
Meier, P. F., 21
Meier, R. W., 201
Meier, S., 160, 198
Meili, J., 179, 201, 214
Meinschein, W. G., 201
Meinwald, J., 224
Meisel, W. S., 193
Meiselman, S., 171
Meisels, G. G., 20, 43, 45, 166
Meissner, G., 257
Meister, W., 203
Melera, A., 171, 245
Mellon, F. A., 174
Mental, J. E., 30
Menu, A., 78
Meot-Ner, M., 21, 56, 257
Mercea, V., 22
Merli, F., 219, 227
Merritt, C., jun., 190
Mes, G. F., 165
Messens, E., 213
Metzger, J., 227, 247
Metzler, M., 286
Meuzelaar, H. L. C., 160, 180, 182, 242, 247
Meyer, H. J., 235
Meyer, H.-W., 222
Meyer, W., 16
Meyerson, S., 304
Meyrant, P., 80
Mezeika, I., 301
Miasek, P. G., 56
Micha, D., 56
Michaelis, W., 221
Michaud, P., 310
Michel, G., 252
Michelin-Lausarot, P., 306
Michels, G. D., 294, 307
Michels, H. H., 21
Mico, B. A., 242
Middleton, R., 148
Midha, K. K., 287
Migahed, M. D., 46, 233
Mihalov, V., 255
Mikaya, A. I., 299
Mikhailov, B. M., 295
Mikhailov, V. I., 325
Mikhailov, Yu. I., 324
Mikhatlov, B. M., 297
Mikkers, F. E. P., 214
Milberg, R. M., 199, 200
Milewich, L., 209
Millard, B. J., 263

Miller, A. R., 40, 75
Miller, C. D., 172
Miller, J. C., 35
Miller, J. M., 295, 312
Miller, R. A., 320, 326
Miller, R. E., 38
Miller, R. K., 207
Miller, S. B., 297
Miller, T., 220, 237
Miller, T. A., 57, 110, 111
Miller, T. M., 104
Miller, W. E., 323
Miller, W. H., 39
Milleur, M. B., 21
Millhauser, G., 323
Milliet, A., 61, 72
Millington, D. S., 154, 171, 201, 207, 251, 279, 294, 312
Milloy, H. B., 90
Milne, G. H., 213
Milne, G. W. A., 176, 193
Milner, R. G., 23
Milove, L., 308
Milverton, D. R. J., 25, 103
Milz, S., 228
Min, B. H., 237, 272, 273, 277
Minami, Y., 277
Minato, T., 21
Minderman, G., 247
Minnikin, D. E., 202, 259
Mintz, D. M., 30, 41
Mintz, E., 309
Mintz, E. A., 312
Mintzer, D., 90
Mioskowski, C., 260
Mirskov, R. G., 302
Miskiewicz, M. A., 223
Mispreuve, H., 80, 81
Mitachek, P., 201
Mitsui, Y., 166
Miyagawa, S., 139
Miyahara, K., 251
Miyakake, T., 260
Miyata, T., 215
Miyatake, T., 205, 261
Miyawaki, H., 288
Miyazaki, H., 207, 260, 278, 284
Mizugaki, M., 203
Mizusawa, E., 323
Mó, O., 5, 15
Möhlmann, G. R., 30
Mohacsi, E., 291
Mohraz, M., 110, 111
Molenaar-Langeveld, T. A., 66, 72
Molton, P. M., 207
Momigny, J., 30, 46
Momoi, T., 214
Monahan, J. E., 31
Monchick, L., 95
Montag, R. A., 13

Montgomery, F. E., 155, 243
Montgomery, J. A., 275
Monts, D. L., 36
Moon, K. A., 321
Moor, J., 256
Moore, A., 172
Moore, S. C. R., 5
Moorhouse, S., 311
Mootoo, B. S., 249
Mooyman, R., 10
Morabito, J. M., 123
Moran, T. F., 7, 20, 22, 24, 226
Moras, D., 308
Moreau, J. J. E., 308
Morel, A. F., 246
Moreland, C. G., 322
Morgan, A. E., 118, 122
Morgan, D., 291
Morgan, D. E., 317
Morgan, R. P., 47, 66, 74, 78, 165, 169, 171, 172, 186, 295
Morgan, T. F., 180
Morgans, D., 226
Morgenthaler, L. N., 105, 181
Mori, A., 215
Moriniere, M., 209
Morishita, S., 273
Morita, I., 205, 260
Morokuma, K., 14
Morris, A., 10
Morris, H. R., 159, 257, 258, 260
Morris, R. J., 208
Morrison, E. R., 202
Morrison, G. H., 122, 133, 139, 168, 179
Morrison, J. A., 303
Morrison, J. D., 8, 9, 61, 77, 102, 103, 104, 162, 172, 181, 244
Morrison, W. F., 98
Mortarini, V., 234
Morton, S., 308
Morton, T. H., 52
Morton, T. L., 69
Mosbach, E. H., 208
Moseley, J. T., 100, 102, 103, 104, 116
Mosichev, V. I., 303
Moss, S. R., 205
Motley, C. H., 278
Muccino, R. R., 272
Muckel, H. J., 223
Mudgett, M., 187
Müller, H., 283, 312
Mueller, E., 294, 304
Mueller, M., 148
Mueller, M. D., 253
Mueller, W. F., 207
Muenow, D. W., 178
Muetterties, E. L., 311
Muir, K. W., 310

Muir, L., 310
Mujake, M., 320
Mukhtar, E. S., 77, 101, 162
Mukmenev, E. J., 317
Mulla, I. S., 314
Muller, E., 178, 201
Muller, M. D., 197, 223
Muller, W., 221
Mulloney, T. J., 19, 108
Mumma, R. O., 211
Mungheer, R., 73
Munroe, M., 230
Munson, B., 156, 243, 297
Munthe, E., 215
Murai, A., 257
Murakami, J., 35
Muranaka, S., 327
Murano, A., 268
Murao, K., 215
Murata, T., 206, 259, 261
Murcray, D. G., 176
Murday, J. S., 168
Murota, S., 205
Murota, S. I., 260
Murphy, R. C., 185, 201, 205, 211, 260
Murr, E. L., 313
Murr, N. E., 312
Murray, K. E., 257
Murray, P. T., 10, 24, 29
Murray, S., 227, 291, 293
Murray, S. G., 303
Murrell, J. N., 3
Murry, B. A., 209
Muschik, G. M., 209, 251
Musin, R. Z., 317
Myher, J. J., 204, 209
Myklebust, R. L., 122
Mysov, E. I., 310

Naaman, R., 32
Nadai, T., 288
Nagahara, M., 288
Nagai, Y., 299
Nagata, A., 288
Nagata, T., 271
Nageli, P., 179
Nagy-Felsobuki, E., 10
Naidu, M. S. R., 305
Nakagawa, T., 228
Nakanishi, K., 253
Nakata, H., 248
Nakatsuji, H., 5
Nakayama, M., 248
Nakayama, N., 156, 252
Nambu, K., 284
Nanni-Marchino, M. L., 308
Narang, S. A., 243
Narasimhachari, N., 272
Narushima, K., 139
Nast, E., 298
Nassin, B., 71, 251
Nasu, S., 320
Natalis, P., 10

Author Index

Natoli, G., 316
Navas, G. E., 286
Navinsek, B., 144
Nazarenko, V. A., 16
Nazarova, T. V., 247
Neal, J. N., 274
Neborsky, R. J., 266
Negrini, P., 271, 282
Neilson, P. V., 55
Neilson, R. H., 322
Nekrasov, Yu. S., 236, 294, 310, 311, 312, 316, 323
Nelmes, P. T. J., 291
Nelson, D., 249, 250
Nelson, D. R., 224
Nelson, S. D., 258
Nelson, W. L., 280
Nenner, I., 22, 44
Nesmeyanov, A. N., 236, 310
Nesterova, O. M., 325
Nestrick, T. J., 219
Neugebauer, D., 311
Neuhauser, W., 107
Neukom, H. P., 198
Neumann, G. M., 165
Neumann, S., 305
Neuschäfer, D., 25
Neusser, H. J., 32, 34, 161
Newbury, D. E., 122, 133
Newns, D. M., 136
Newton, D. L., 219
Ng, C. Y., 13, 20
Ng Ying Kin, N. M. K., 215
Nibbering, N. M. M., 47, 49, 51, 65, 66, 72, 75, 78, 165, 233, 238, 239, 257
Nicholls, T. M., 214
Nicholson, G. J., 198, 216, 292
Nicholson, R. I., 271
Nickless, G., 220, 221
Nicolaides, N., 245
Niebch, G., 285
Niedenzu, K., 297
Niederwieser, A., 212
Nief, F., 311
Nielsen, C. J., 269
Nielsen, P., 209
Nielsen, A. J., 317
Nieman, G. C., 34
Niemann, C. J., 204
Nieuwpoort, W. C., 72
Nihel, K., 212
Nikitin, O. T., 320, 322
Nikolaev, E. N., 21, 325
Nilsson, B., 207
Nilsson, C.-A., 220, 237
Ninkov, Z., 110
Nishi, I., 178
Nishimura, H., 209
Nishimura, S., 254
Nishimura, T., 45
Nishimura, Y., 24, 44
Nivois, C., 209

Niwa, Y., 45
Noest, A. J., 238
Nogradi, M., 248
Nohynek, G., 221
Nomoto, K., 10
Nomura, H., 258
Norayer, I., 191
Norden, N. E., 215
Nordholm, S., 26
Norman, A. D., 324, 325
Norman, G. O., 255
Norris, D. D., 182
Norstrom, A., 220, 237
Nosaka, Y., 109
Noto, M., 288
Nourtier, A., 118, 137
Novak, F. A., 39
Novick, S. E., 17, 19, 107, 108
Novoselova, A. V., 316
Novotny, M., 214, 227
Nowak, H., 283
Nowak, P. J. C. M., 169
Nowlin, J. G., 155, 243
Noya, A., 321
Nozoye, H., 45
Numrich, R. W., 37
Nunomura, N., 222

Oates, J. A., 205
Obezyuk, N. S., 311
Obt, K. H., 319
Occolowitz, J. L., 220, 252, 253
Ockerman, P. A., 215
Odham, G., 241
Odom, R. W., 165
Oehlschlager, A. C., 208
Oelz, O., 205
Oertel, H., 19
Oertli, C. U., 229
Oesterhelt, G., 203, 208
Ofuji, T., 288
Ogden, J. S., 327
Ogilvie, K. K., 253
Ohashi, M., 156, 252
Ohinichi, T., 320
Ohloff, G., 222
Ohlsson, A., 277
Ohnesorge, A., 304
Ohno, K., 5
Ohta, T., 26
Ohumura, Y., 230
Ohya, K., 277
Ojo, I. A., 313
Okada, K., 257
Okajima, A., 268
Okano, J., 139
Okano, Y., 215
Okazaki, H., 304
O'Keefe, A., 110
Okogon, J. I., 210
Okouchi, Y. Y., 217
Okuda, M., 167, 200

Okuda, T., 207
Okutani, T., 122, 123
Okuyama, F., 163
Olah, G. A., 48, 109
Olcay, A., 221
Oldenberg, E. J., 259
Oldenettel, J., 104
Olinger, R., 158
Oliva, A., 300
Olivier, J. L., 30
Ollis, W. D., 252
Olson, J. R., 176
Olson, K. W., 177
Onak, T., 323
O'Neil, S. V., 27, 33
Onkenhout, W., 81
Ono, Y., 20
Onyiruka, O. S., 246
Ordelman, J. E., 184
Oref, I., 36
O'Reilly, R. A., 278
Orel, A. E., 26, 27
Orlowski, T. E., 113
Ormerod, R. C., 43
Orr, J. C., 208
Orr, K. H., 320
Ortego, J. D., 315
Orth, R. G., 110, 117
Ortiz de Montellano, P., 242
Ory, R. L., 223
Osamura, Y., 21
Osborne, A. D., 46, 50
Osella, D., 308
Oshima, M., 261
Osterdahl, B.-G., 206
Osterloh, J. D., 286
Ott, A. Y., 255
Ott, G. L., 327
Ott, K. H., 165
Otten, A., 212
Ottinger, C., 25, 43, 110
Ott-Kuhn, U., 211, 257
Ötvös, L., 282, 291
Ourisson, G., 221
Ovchinnikov, K. V., 325
Owen, T., 177
Owens, C., 266
Oxford, J., 270, 281
Ozenne, J.-B., 102, 115

Pabst, R. E., 10, 326
Pacakova, V., 220
Pacansky, J., 8
Pace-Asciak, C. R., 205
Padieu, P., 209, 211
Paget, W. E., 295
Pagano, F. P., 78
Pahle, J., 215
Pai, R. Y., 85, 234
Paic, P. B., 322
Pailer, M., 208
Paine, R. T., 303, 308
Pak, C. S., 250
Pakdel, H., 221

Pallante, S., 290
Palmer, L., 271
Palmer, R. A., 327
Palmer, R. F., 210, 251, 288
Palosi, E., 282
Pan, R. P., 21
Panasenko, A. A., 298
Pandey, R. C., 252
Pandolfi, R., 34
Pang, F., 2
Pantarotto, C., 266
Papamichalis, P., 180, 226
Parfitt, W., 178
Parkanyi, C., 227
Parker, C. E., 25, 77, 166, 219, 240
Parker, D. H., 32, 33, 34
Parker, J. E., 23
Parmentier, G., 202
Parr, A. C., 9, 10, 27, 62
Parr, V., 312
Parrish, M. E., 170, 197
Parsley, K. R., 198, 259
Pascard, C., 252
Pascher, I., 257, 259
Pascuel, C., 185
Patel, I. H., 274
Patel, J. R., 218
Pathak, V. N., 314
Patil, G., 252
Patterson, D. G., 251
Patterson, J., 324
Patterson, R. L. S., 222
Paulson, G., 227, 293
Paulson, J. F., 103, 104
Payen, P., 217
Payne, M. G., 162
Paysen, R. A., 234
Pearce, D., 273
Pearl, P. L., 12
Pearson, N. R., 300
Peatman, W. B., 29
Pedersen, B. S., 234
Pedersen, L. G., 25, 77, 240
Pedley, J. B., 12
Peel, J. B., 10, 20
Pegues, R. F., 78
Pelino, M., 321, 327
Pellin, M. J., 131
Pellizzari, E. D., 217, 219
Penca, M., 176
Peng, A., 269
Peppard, T. L., 222
Perera, C., 263
Pérez, P., 15
Perlberger, J. C., 5
Pernot, C., 102
Perone, S. P., 192, 201
Perov, A. A., 47
Perovic, B., 139
Perreira, N. B., 164
Perret, C., 61, 72
Perry, C. W., 274
Perry, D. L., 249

Pettersen, J. E., 291
Pesyna, G. M., 176
Peters, H. L., 107
Petersen, E. M., 193
Petersen, T. E., 258
Peterson, J. R., 104, 114
Peterson, K. I., 21
Petillon, F. Y., 310
Petit, M., 233
Petrin, M., 291
Petrov, A. A., 304
Petrova, G. S., 304
Pettersen, R. C., 312
Pettit, B. R., 210, 227, 251
Petz, W., 308
Peyerimhoff, S. D., 4, 6
Peypoux, F., 252
Pfleger, K., 249
Pfleiderer, W., 253
Pham, D., 115
Phillipou, G., 203, 207, 300
Phillips, E., 30, 110
Phillips, I. G., 297
Phillips, J. S., 9
Phillips, P. J., 205
Pickering, L. K., 265
Piel, A., 191
Pierce, A. M., 208
Pierce, H. D., 208
Pierce, R. C., 254
Pietro, W. J., 298
Pignolet, L. H., 316
Pike, A. W., 251
Pilcher, G., 12
Pimlott, W., 259, 261
Pincus, P. A., 160
Pinkerton, A. A., 309
Piovesana, O., 316
Piper, L. G., 10
Piper, P. J., 260
Pirone, A. J., 159
Pisana, J. J., 211
Pitzer, K. S., 21
Plimmer, J. R., 225, 226
Plall, H. M., 221
Plotkin, J. S., 323
Poelman, J., 191
Pohl, D., 320
Pohle, G., 327
Pokhodenko, V. D., 16
Pokrovskii, V. A., 318
Polanyi, J. C., 38
Politzer, I. R., 218
Politzer, P., 15
Polivanov, A. N., 235, 298, 301
Pollak, E., 57
Pollack, S. K., 5, 298
Pomernacki, C., 178
Ponomarev, G. V., 255
Poole, C. F., 214, 227, 228, 294, 296, 317
Poon, C., 178
Poon, C. C., 166

Pope, W. M., 92
Pople, J. A., 2
Poponova, R. V., 304, 318
Popov, A. G., 301
Popov, S. S., 249
Popovic, A., 328
Poppinger, D., 19
Popravko, S. A., 248
Portenschleg, R., 118
Porter, C. J., 78
Poshevnev, V. I., 324
Posthumus, M. A., 160, 242
Postnikova, T. K., 295, 297
Postnov, V. M., 310
Potapov, V. K., 10, 47
Pottie, R. F., 138
Poulos, A., 300
Povey, A., 220
Povey, D. C., 303
Powell, J. W., 291
Powell, M. L., 280
Powell, R. A., 122
Powers, K. H., 275
Powers, P., 180, 185
Powis, I., 41
Pozharov, S. L., 21
Prager, M. J., 223
Pramanik, B. N., 237
Pramanik, P. N., 237, 245, 250
Pratt, G. E., 203
Preobrazhenskaya, M. N., 325
Prescott, S. R., 166
Prest, D. W., 324
Prest, H. F., 20
Prestwich, G. D., 224
Pretch, E., 5
Preti, G., 213, 224
Preuss, H., 36
Price, P. C., 166
Price-Evans, D. A., 276
Prichard, B. A., jun., 152
Priddle, J. D., 257
Primiani, M., 225
Pritchard, H. O., 37
Pritzkow, H., 327
Privett, O. S., 231, 245, 259
Proctor, C. J., 172
Profous, C. Z., 112
Prokhorova, S. A., 302
Prome, J. C., 257, 261
Prossdorf, W., 311
Prostakov, N. S., 299
Prusacyzk, J. E., 326
Przybylski, M., 258
Pugh, M. E., 199, 245
Puglisi, O., 318
Pujar, B. G., 250
Pulay, P., 2
Pullin, C. J., 214
Purnell, J. H., 210
Puzo, G., 253, 257, 261
Pyne, G. S., 303
Pyrek, J. S., 249

Author Index

Quattrone, A., 266
Quilliam, M. A., 253
Quirijns, J. K., 218
Quirke, J. M. E., 255
Qureshi, N., 259

Raba, M., 254
Rabalais, J. W., 321
Rabbih, M. A., 72
Rabinovitch, B. S., 36
Rache, H., 184
Radermacher, L., 183
Radford, T., 241, 248
Radom, L., 5, 19, 62
Radovick, S., 15
Radus, T. P., 325
Radziejewska-Lebrecht, J., 207
Raff, L. M., 58
Rafikov, S. R., 234
Rafter, J. J., 184, 211
Ragaraj, G., 205
Ragatz, P., 323
Ragusa, F., 266
Rahn, W., 222
Raisys, V. A., 263
Raitby, P. R., 307
Ramachandra, R., 71
Ramaekers, J. J. M., 198
Ramakrishna Rao, T. V., 10
Ramana, D. V., 72
Ramaswamy, R., 38
Ramendik, G. L., 167
Ramirez, F., 304
Rampacek, J., 177
Rand, M. H., 328
Randall, L. G., 170, 199
Rankin, D. W. H., 325
Rao, A. R., 255
Rao, J., 291
Rao, K. R. N., 155, 245
Rapoport, H., 215
Rapp, U., 155, 198, 242
Rappe, C., 220
Rashkes, Ya. V., 249
Rasmussen, G. T., 176, 191, 193, 195
Rasshinina, T. A., 315
Rat'Kovskii, I. A., 322
Raup, E. G., 328
Rausch, M. D., 309, 312
Rauvala, H., 257
Rava, R. P., 34
Raverdino, V., 308, 309, 316
Ravi, B. N., 250
Rawlings, G. D., 217
Raynaud, L., 252, 267
Rayner, E. T., 222
Razinger, M., 176
Razuvaev, G. A., 302
Rebière, J., 122
Recca, A., 318
Rechsteiner, C. E., 163
Reck, G. P., 35, 324

Redant, G., 200
Reddy, C. D., 305
Reddy, R., 10
Reed, D., 273
Reed, K. J., 17, 107
Reed, R. I., 156, 158
Reents, W. D., 81, 110
Rees, M. W., 259
Reetz, M. T., 300
Regner, F., 224
Reid, I., 138
Reid, N. M., 166, 178
Reid, W. K., 158
Reilly, J. P., 162
Reinecke, H. J., 307
Reiner, E., 180, 226
Reinhardt, W. P., 27
Reinhart, P. W., 17, 27
Reinhold, V. H., 257, 281
Reinhold, V. N., 208
Reis, F. De. A. M., 246
Reisner, G. M., 309
Reitnauer, P. G., 235
Remaley, S., 224
Remberg, G., 286
Renaud, R. L., 208
Renberg, L., 220
Rerai, R., 24
Rescigno, T. N., 26, 27
Reshetova, L. N., 236, 316
Rest, A. J., 327
Rettner, C. T., 34
Reuss, J., 20
Reynaert, J. C., 10
Rhodes, G., 214
Rhodin, T., 136
Rhyage, R., 262
Rhyne, T. C., 319
Rice, C. E., 316
Rice, S. A., 39, 57
Richard, G. J., 46
Richards, H. T., 148
Richards, J. A., 192, 303
Richards, W. G., 5
Richtarsky, G., 305
Richter, J., 165
Richter, W. J., 80, 233
Rickard, G. J., 76, 77
Rideout, D. C., 311
Ridge, D. P., 53
Ridolfo, A. S., 275
Riedl, P., 208
Riepe, W., 178, 189, 295
Riess, J. G., 314
Rigaud, M., 155
Rigault, J. P., 268
Rigby, L. J., 187
Rijkeboer, G., 184
Rijpstra, W. I. C., 208
Riley, S. J., 18
Rimmer, J., 255
Rimpler, H., 228
Rinaldi, G., 201
Rinehart, G. H., 328

Rinehart, K. L., jun., 252
Ringo, G. R., 123
Ringoir, S., 214
Ringsdorf, H., 258
Risby, T. H., 166, 179, 201, 307, 315
Ritter, G. L., 193
Ritter, H. P., 178
Riveccie, M., 313
Riveros, J. M., 239
Roach, J. A. G., 220
Robb, J. C., 168
Robbiani, R., 19, 186
Roberts, D. J., 220, 221
Roberts, G. C. K., 159
Roberts, P. G., 25, 102, 103, 115
Robertson, D. H., 190
Robertson, G. B., 313
Robertson, P. L., 225
Robertson, S. M., 206
Robin, M. B., 34, 35
Robinson, S. D., 327
Robson, R. E., 89, 90, 94
Roche, A. L., 44
Rodionov, A. N., 10
Röllgen, F. W., 158, 160, 163, 165, 233, 242, 256, 319, 326
Roesky, H. W., 326
Rofikov, S. R., 298
Rogers, D. E., 164, 165
Rohmer, M., 250
Rohwedder, W. K., 250
Roman, E., 310
Romkowski, M., 184
Roncucci, R., 280
Roos, G., 138
Rose, M. E., 227
Rose, T. L., 104
Rosello, J., 204
Rosenberg, A., 264
Rosenblatt, G. M., 324
Rosenfeld, R. N., 14, 114, 115
Rosenkrantz, M. E., 20
Rosenstock, H. M., 7, 9, 60, 62
Rosmus, P., 16
Ross, J., 96
Ross, U., 56
Ross, W. J., 292
Rossi, E., 282
Rossi, J.-C., 63, 71
Rossiter, M., 199, 245, 249
Rosynov, B. V., 260
Rothe, E. W., 17, 35, 324
Rothkopf, H. W., 234
Rotilio, D., 282
Rotter, H., 192, 193
Rovlet, R., 309
Roy, N., 255
Roy, R. L., 170
Roy, T. A., 166, 236, 251

Royo, G., 300
Rozett, R. W., 193
Rozynov, B. V., 248
Rubinstein, I., 221
Rudat, M. A., 122, 139, 168, 179, 237
Rudenauer, F. G., 187
Rudinsky, J. A., 225
Rüdenauer, F. G., 118, 121
Ruelius, H. W., 282, 283
Ruff, G. A., 103
Rusbüldt, D., 131
Rushnek, D. R., 177
Russell, D. H., 49, 51, 65, 75
Russell, N. J., 203
Russell Helbig, H., 201
Russo, M. V., 313, 314
Russo, S., 223
Rutherford, L., 179
Ruveda, E. A., 246
Ruzo, L. O., 220
Ryan, S. R., 29, 46
Rybalko, K. S., 249
Rye, R. T. B., 68, 72, 78
Ryhage, R., 169, 182, 200, 237
Ryker, L. C., 225
Rylance, J., 12

Saady, J. J., 272
Sadana, K. L., 253
Sadler, P. J., 314
Sadovskaya, V. L., 260, 261
Saeed, T., 200
Safa, K. D., 299
Saferstein, R., 264
Saha, S. K., 253
Saile, V., 29
St. John, G. A., 164
Saito, K., 203
Saito, S., 207
Sakai, A., 208
Sakai, Y., 288
Sakamoto, T., 203
Sakuma, H., 222
Sakuno, A., 257
Sakurai, T., 328
Salen, G., 208
Salimgareva, I. M., 298
Saluja, P. P. S., 19, 63
Sambe, H., 5
Samfield, M., 217
Samhoun, M. N., 260
Sammons, M. C., 163
Samuelsson, B., 205, 260
Samuelsson, B. E., 257, 259, 261
Sander, R., 138
Sandra, P., 197, 198, 200, 217
Sannigrahi, A. B., 4
Sano, M., 258, 277
Sano, T., 320
Santi, A., 313, 314
Sappa, E., 307, 308, 309, 316

Sargeson, A. M., 211
Sariaslani, F. S., 202
Sarre, P. G., 102
Sarre, P. J., 25
Sarre, P. R., 103
Sasaki, K., 178
Sasaki, M., 222
Sasaki, S., 258
Sasayama, T., 320
Sass, S., 303
Sasse, H. E., 326
Sastry, S. D., 213, 241
Sathyamurthy, N., 38
Satouchi, K., 203
Saunders, G. A., 249
Saunders, M., 18
Savage, A. V., 206
Sawahata, T., 209
Saxena, A. N., 328
Saxena, N., 299
Saxon, R. P., 102, 104
Sbarra, C., 271
Scampucci, C. S., 314
Scanlon, J. C., 315
Schaaf, A., 203
Schaefer, H., 320, 321, 324, 327
Schaefer, H. F., tert., 39
Schaefle, J., 221
Schafer, R. W., 180, 226
Schalcher, M., 179
Schaldach, B., 46, 52, 75, 172
Schamp, H. W., 89
Scharager, C., 328
Scharfner, K. H., 147
Scharmann, A., 147
Schauenburg, G., 191
Scheffler, K., 305
Scheibye, S., 234
Schell, J., 213
Schels, H., 249
Schelten, J., 120
Schenk, H., 19
Scheppele, S. E., 164
Scheuer, P. J., 250
Scheulen, B., 209
Scheunemann, H. V., 18
Scheutwinkel-Reich, M., 203
Scheuvert, I., 220
Scheibel, H. M., 242, 255
Schildcrout, S. M., 314
Schiller, Ch., 160
Schirmer, J., 6
Schlag, E. W., 32, 34, 39, 161
Schlereth, F. H., 178
Schleyer, P. von R., 6
Schlunegger, U. P., 167, 258
Schmeltekopf, A. L., 10, 86
Schmelzer, A., 6, 43
Schmid, E. R., 264
Schmid, M., 222
Schmid, P. P., 197, 223
Schmidt, J., 235, 246, 249
Schmidt, M., 327

Schmitt, J. A., 223
Schmitt, R. J., 238
Schmutzler, R., 300, 305
Schnautz, N., 197
Schneider, F., 57
Schnoes, H. K., 259
Schoeller, D. A., 169, 184
Schoen, A. E., 244
Scholes, G., 309
Schomburg, D., 305
Schomburg, G., 197, 199, 219
Schomerus, M., 286
Schootbrugge, G. A. V. D., 146
Schoots, A. C., 206, 214
Schram, B. L., 255
Schram, K. H., 212, 253
Schreier, P., 222
Schroeder, H., 304
Schroeer, J. M., 134, 136
Schroepfer, G. T., 204
Schroer, J. A., 209, 251
Schubert, R., 72
Schubert, U., 311
Schuetzle, D., 183
Schuler, A., 203
Schuller, W. H., 222
Schulte, H., 9
Schulte, K. W., 6
Schulte-Elte, K. H., 222
Schulten, H.-R., 159, 162, 182, 242, 255, 257, 262, 297, 325
Schulz, S., 296
Schumacher, E., 10, 32
Schuyl, P. J. W., 211
Schwartz, H., 50, 51, 62, 65, 66, 72, 77, 78, 80, 81, 165, 233, 300
Schwartz, H. L., 230
Schwartz, R., 315
Schwarzenbach, D., 309
Schwarzenbach, R. P., 217
Schweich, D., 187
Schweig, A., 6
Schweinler, H. C., 234
Schwentfeger, G., 304
Scoles, G., 38
Scott, G. C., 230
Scott, J. A., 14, 63, 213, 237
Scott, M. F., 213
Scott, R. P. W., 230
Scott, T. W., 36
Scully, N., 281
Seaver, M., 162
Sedgwick, R. D., 4, 303
Sedvall, G., 276
Seefeldt, R., 327
Seibl, J., 186, 253
Seiders, B. A. B., 16
Seifert, K., 246
Seifert, W. E., 192, 201, 211
Seiler, E. U., 184
Seitz, J. F., 268

Author Index

Self, R., 202, 216, 259
Seligmann, O., 248
Selim, E. T. M., 72
Selva, A., 249
Semenov, N. M., 310, 322, 325, 327
Semrau, G., 163
Sen Sharma, D. K., 66
Seppelt, K., 327
Sepulchre, A. M., 252
Sergeev, Yu. L., 45
Sethi, S. K., 166
Seymour, F. R., 207, 256
Sexton, J. J., 259
Shabanowitz, J., 77, 155, 242, 244
Shaddock, V. M., 171
Shadoff, L. A., 219
Shafer, K. H., 217
Shaldach, B., 79
Shamsiyev, U. B., 123
Shannon, J. S., 249
Shapiro, J. A., 254
Shapiro, M., 48
Shapiro, M.-J., 52, 238
Sharma, S. K., 300
Sharp, D. W. A., 310
Sharpe, M. C., 10, 326
Sharvit, J., 71
Shaw, B. L., 314
Shaw, D. H., 207
Shaw, G. J., 255
Sheats, J. E., 312
Shefer, S., 208
Shehata, M. T., 126
Sheldrick, W. S., 326
Sheludyakov, B. J., 298
Shepherd, I., 314
Sherman, W. R., 228, 255, 296
Sherrod, R. E., 234
Sheunemann, H.-U., 234
Shever, P. J., 208
Sheves, M., 252
Shibata, S., 278
Shida, T., 109
Shih, S. K., 6
Shiley, R. H., 110
Shimizu, R., 122, 123
Shimizu, Y., 208
Shimojima, N., 222
Shimonishi, Y., 164, 258
Shintani, S., 277
Shirai, E., 278
Shirley, D. A., 15
Shishkun, Yu. A., 324
Shizuka, H., 72
Shold, D. M., 11, 50, 63
Shono, K., 123
Short, M. N., 259
Showmaker, D. W., 215
Shukla, V. K. S., 202
Shuler, K. E., 96
Shul'ts, M. M., 322

Shushan, B., 172
Shushunov, N. V., 313
Sicherer, C. A. X. C. F., 256
Sichtermann, W. K., 123, 168, 232, 243
Sidorov, L. N., 321, 327
Sidorova, I. V., 10
Sieck, L. W., 167
Siefert, J. H., 6
Siegel, J., 19
Siegel, M. W., 166
Siegelman, H. W., 254, 255
Siekmann, A., 276
Siffert, P., 328
Sigmund, P., 118, 126
Sigsby, M. L., 73
Silberhorn, D., 258
Silva-Trivino, L. M., 324
Silveira, D. M., 281
Silverman, L. D., 317
Sil'vestrova, S. Yu., 316
Simauy, M., 311
Simm, I. G., 43
Simmonds, D. J., 246
Simon, M.-J., 280
Simon, W., 5, 197, 223, 253
Simonis, J., 110
Simons, D. S., 121
Simons, J., 16, 107
Simpson, M., 227, 293
Singh, J., 220
Singh, M., 248
Singh, R. V., 318
Singhawangcha, S., 227, 296
Sipachev, V. A., 316
Sisenwine, S. F., 282
Siwapinyoyos, G., 323
Sizoi, V. F., 210, 310, 311
Sjoblad, S., 215
Sjovall, J., 208
Skånberg, I., 284, 289
Skinner, R., 220
Skullerud, H. R., 90, 91
Skurat, V. E., 167, 229
Slack, J. A., 226, 227, 262, 291, 292
Sladkova, T. A., 302
Slater, S., 311
Sloane, C. S., 38
Slodzian, G., 121, 122, 123, 135, 138, 139
Smale, T. C., 256
Smalley, R. E., 35, 36
Smisko, M. J., 190
Smit, A. L. C., 169, 236, 278
Smith, A., 202, 203
Smith, A. B., 224
Smith, A. G., 204
Smith, A. K., 313
Smith, C., 252
Smith, D., 9, 20, 43, 77, 87, 172, 217, 244
Smith, D. E., 217
Smith, D. H., 122, 176, 214

Smith, D. L., 102, 104, 244, 253, 317
Smith, E. B., 92
Smith, G. A., 159
Smith, G. P., 104
Smith, H. J., 297
Smith, H. P., jun., 120
Smith, H. V., jun., 148
Smith, J. N., 146
Smith, J. N., jun., 151
Smith, K., 295
Smith, K. G., 303
Smith, K. M., 255
Smith, L. L., 166, 208, 236, 245, 251
Smith, M. J. H., 205
Smith, P., 177
Smith, P. A., 209
Smith, R. D., 326
Smith, R. D. H., 325
Smith, R. G., 287, 289
Smith, R. M., 213, 224
Smith, S. O., 221
Smithson, L. D., 300
Smolarek, J., 34
Snaith, R., 303
Snape, C. E., 221
Sneddon, L. G., 310, 312, 323
Snowdon, K. J., 122, 126, 131, 132, 139
So, S. P., 15
Sochtig, I., 222
Sofranko, S., 201
Sogn, J. A., 187
Sokolova, O. S., 234
Solch, J. G., 172
Solomennikova, I., 301, 316
Soltmann, B., 155, 158
Soman, S. M., 206
Somekh, R. F., 324
Sommer, L. H., 300
Sonnet, P. E., 225
Sottrup-Jensen, L., 258
Sourisseau, C., 112
Spalding, T. R., 322, 323
Sparks, A. N., 225, 226
Speck, D. D., 187
Specker, H., 327
Spencer, G. F., 250
Spencer, R. B., 165, 181
Spezeski, J. J., 29, 103
Sphon, J., 255
Sphon, J. A., 216, 220
Spialter, L., 300
Spiegl, A. W., 303
Spindt, C. A., 164
Spiteller, G., 189, 214, 247, 251, 287
Spiteller, M., 189, 214, 251
Splitter, J. S., 51, 62
Spolaore, P., 148
Sponholtz, M., 223
Sponsel, V. M., 210
Spyckerelle, C., 221

Author Index

Squires, D. B., 126
Squires, L., 24
Sridhar, R., 307
Srivastava, R. D., 324
Sroubek, Z., 136
Stadelmann, J. P., 43, 111
Stadler, E., 226
Stahl, D., 15, 49, 63, 65, 77, 236
Stahle, H., 286
Stalmeier, P. F. M., 172
Stan, H.-J., 170, 197, 203
Standisavljevic, M., 170
Stanko, I., 323
Stanley, G., 227
Stannard, P. R., 34, 111
Stapleton, B. J., 52, 63, 68, 235
Staudenmaier, G., 138, 139
Stauffer, S. C., 290
Stearns, C. A., 320
Steblevskii, A. V., 326
Stec, W. J., 317
Steel, G., 165
Steele, K. W., 184
Steffen, D. A., 176
Steffens, G. L., 250
Stegmann, H. B., 305
Stehl, R. H., 219
Steiger, W., 118, 121, 187
Stein, J. D., 123
Steinbach, K., 216
Steiner, B. W., 7
Steiner, J., 270
Steiner, W., 212
Steinfelder, K., 234
Stell, P., 281
Stenberg, U., 219
Stenhagen, G., 224, 241
Stepanov, B. I., 299
Stephenson, M. D., 206
Stepikheev, Yu. A., 304
Sternowsky, H. J., 215
Steve, L., 214
Stevens, W. J., 20
Stevenson, D. P., 85
Stillwell, R. N., 155, 229, 243
Stobart, S. R., 299, 325
Stockbauer, R., 9, 10, 27, 29, 41, 42, 45, 49, 62
Stockdale, J. A. D., 17, 18
Störi, H., 25
Stoffels, J. J., 170
Stokke, O., 214, 215, 227
Stoll, H., 36
Stoll, R., 160, 242
Stolyarova, V. L., 322
Stolze, R., 48
Stolzenberg, G., 227, 293
Stopnikova, M. N., 318
Storms, H. A., 122, 123
Storp, S., 123
Storstet, P., 215
Stoubek, Z., 147

Stoyanov, S., 327
Strand, M. P., 36
Strathdee, S., 43
Straub, K., 171
Straub, K. M., 242, 253
Strauss, P. A., 177
Strausz, O. P., 221
Strel'chenok, O. A., 257
Strelisky, J., 248
Strolin Benedetti, M., 283
Stroud, C., 58
Struble, D. L., 226
Struchkov, Yu. T., 316
Stuart, R. V., 147
Stubley, C., 281
Stuck, R., 328
Stucky, G. D., 311
Stuhl, O., 301
Sturm, R. E., 176
Styrov, V. V., 322
Su, E. C. F., 53
Su, T., 14, 53, 54, 58
Subba Rao, S. C., 206
Subden, R. E., 208
Suboch, V. P., 255, 257, 315
Suchkov, A. I., 159
Suchy, M., 203
Sudo, M., 214
Sufet, I. H., 215
Sugai, S., 178
Sugawara, F., 224
Sugawara, S., 222
Sugimura, Y., 227
Sugiura, T., 46
Sugiyama, T., 213
Sugnaux, F. R., 279
Sugura, J., 285
Sukharev, Yu. N., 311
Sullivan, C. L., 326
Sullivan, S. A., 239
Sultanov, A. Sh., 234, 236
Summerhays, K. D., 5
Summers, D. M., 210
Summons, R., 264
Summons, R. E., 254
Sundaram, N., 72
Sundstrom, G., 219, 220
Sunner, J., 22
Suzuki, I. H., 180
Suzuki, M., 205, 260
Suzuki, T., 303
Suzuki, Y., 214
Svendsen, A. B., 208
Svensson, S., 207, 215
Swahn, C. G., 276
Swain, S., 33
Swaminathan, K., 295
Swanson, A. R., 214
Sweeley, C. C., 155, 158, 190, 213, 214
Sweeney, J. J., 308
Sweeny, J. C., 248
Swingler, D. L., 168
Swofford, H. S., 166

Sysoev, A. A., 167
Szabo, I., 22
Szafranek, J., 228, 257
Szendrei, K., 246
Szinai, I., 291
Szporny, L., 282
Szulejko, J. E., 77, 167, 240
Szuna, A., 291

Taagepera, M., 5
Tabché-Fouhaillé, A., 22, 44, 102
Tabenchi, T., 183
Tachimori, S., 326
Tadjeddine, M., 116
Taft, R. W., 5, 13
Taggi, F., 219
Tai, H.-H., 205
Taira, T., 167
Tajima, S., 72
Takada, T., 5, 327
Takagi, H., 167, 200
Takagi, T., 208
Takahama, K., 215
Takahashi, A., 328
Takahashi, G., 216
Takahashi, J. M., 222
Takahashi, S., 206
Takayama, T., 259
Takeshita, H., 320
Takeuchi, T., 167, 230, 234, 236, 259
Takeuchi, Y., 257
Takhistove, V. V., 234
Takimoto, C., 323
Talanova, G. G., 305
Talrose, V. L., 167, 229
Tamas, J., 282, 324
Tan, K. H., 6, 25, 26
Tan, Y. L., 218
Tanaka, K., 44, 58
Tanaka, O., 251
Tanayama, S., 288
Tandon, J. P., 318
Tang, B. K., 283
Tang, S.-Y., 248
Taniguchi, Y., 212, 253
Tanioka, K.-I., 214
Tanner, J. T., 220
Tanner, R. J. N., 270
Tanner, S. B., 208
Tanner, S. D., 22, 81, 239
Tantsyrev, G. D., 21, 229
Tarasov, V. A., 249
Tarchini, C., 250
Tatsuta, L., 178
Tattje, D. H. E., 249
Tau, K. D., 305
Tavanaiepour, I., 311
Tayler, C., 154
Taylor, A., 179, 258
Taylor, D. R., 325
Taylor, F., 202
Taylor, G. W., 260

Author Index

Taylor, J. A., 45
Taylor, J. W., 21, 31, 45, 193, 194
Taylor, K. T., 166, 179, 243
Taylor, M. L., 172
Taylor, O. R., 225
Taylor, P. L., 205
Taylor, P. R., 16
Taylorson, D., 322, 323
Tchir, M. F., 227
Tebby, J. C., 303
Tecon, P., 49, 65
Tedder, J. M., 22
Teece, R. G., 206
Tegyey, Zs., 282
Teichman, J., 220
Telegina, N. P., 235
Telin, B., 209
Teller, G., 213
Telliard, W. A., 216
Tellinghuisen, J., 44, 91
Telus, B. L., 297
Teng, H. H., 106
Ten Noever de Brauw, M. C., 160, 199, 218, 242
Tennyson, J., 3
Tenorio, D., 328
Terent'ev, P. B., 233, 249
Terlouw, J. K., 10, 51, 62, 67, 68, 72, 78
Thackston, M. G., 92, 98
Thenot, J.-P., 155, 243
Thiecke, R. J., 198
Thiel, W., 2
Thogersen, H. C., 258
Thomas, C. B., 72
Thomas, D. W., 168
Thomas, G. E., 126
Thomas, J., 291
Thomas, P. D. P., 323
Thomas, R., 23, 250
Thomas, S., 143
Thomas, T. F., 103, 104
Thommen, F., 111
Thompkins, L., 275
Thompson, A. C., 204
Thompson, D. L., 38
Thompson, J. C., 197
Thompson, M., 170
Thompson, M. L., 325
Thompson, M. W., 127, 138
Thompson, R. M., 215, 255, 287, 289
Thompson, S., 218
Thomsen, M. W., 15
Thomson, B. A., 166
Thorn, R., 152
Thornblad, A. M., 271
Thorpe, T. M., 228
Thum, F., 127
Thunberg, T., 216
Thunemann, K. H., 6
Thyagarajan, B. S., 166
Tichý, M., 63, 71

Tiedemann, P. W., 13, 20
Tiernan, T. O., 172
Tillya, C. P., 309
Timoshenko, M. M., 298
Tint, G. S., 208
Tio, C. O., 282
Tippins, J. R., 260
Tiripicchio, A., 307
Tissie, G., 261
Tjoa, S. S., 257
Tocco, D. J., 280
Todd, J. F. J., 165, 166, 200
Todd, L. J., 227, 323
Todd, P. J., 77, 78, 243
Toennies, J. P., 56
Tohno, M., 288
Tolelea, L., 208, 250
Tolk, N. H., 126
Tolliver, D. E., 103
Tolpun, E. I., 323
Tolstikov, G. A., 234, 236
Tomassian, A., 177
Tomer, K. B., 219
Tomizawa, G., 178
Torgerson, D. F., 160
Torline, P., 197
Tornabene, T. G., 204
Torres, G. S., 273
Toschek, P., 107
Totoki, K., 271
Town, W. G., 191
Townsend, L. B., 213
Townsend, R. E., 216
Traeger, J. C., 8, 9, 10, 49, 61, 62, 63, 68
Trager, W. F., 274, 278
Trautmann, K. H., 203
Traven, V. F., 299
Tremblay, P.-A., 261
Trenerry, V. C., 298, 299
Trivedi, M., 201
Trott, G. W., 181, 186, 187, 201
Trott, W. M., 10, 20
Trotter, J., 311
Trotter, W. J., 220
Troup, J. M., 309
Truscott, R. J. W., 214, 215, 264
Tryhorn, S. E., 202
Tsai, B. P., 20, 45, 46
Tsai, J. C. C., 123
Tsai, R. S. C., 300
Tschanz, C., 270, 287
Tse, M.-W., 303
Tsong, I. S. T., 130, 131, 132
Tsuchiya, M., 167
Tsuchiya, T., 45, 72
Tsuge, S., 167, 227, 230
Tsugita, T., 223
Tsui, F.-P., 206, 257
Tsuji, K., 24, 44
Tsuji, M., 24, 44
Tsuji, S., 131

Tsukayama, M., 248
Tsvetkov, E. N., 304
Tucker, P. A., 313
Tümmler, R., 234, 235
Tugrul, T., 221
Tuithof, H. H., 172, 182
Tulloch, A. P., 203, 300
Tumlinson, J. H., 225
Tung, C. M., 234
Turbini, L. J., 306
Tureček, F., 74, 251
Turevskaya, E. P., 316
Turner, C. M., 152
Turner, D. W., 110
Turova, N. Ya., 294, 316
Tuttenberg, K. H., 287
Twarowski, A. J., 36
Twichell, J. E., 227
Tykesson, P., 148
Tyurin, V. D., 310

Uden, P. C., 221
Udseth, H. R., 325
Ueda, K., 243
Uhlenbeck, G. E., 94
Ujszaszi, K., 291
Ulubelen, A., 241
Umeno, Y., 269, 277
Une, M., 208
Unger, S. E., 168, 243, 244, 256, 318
Uno, T., 228
Unrau, A. M., 208
Unsworth, J. F., 255
Upalawanna, S., 315
Uriarte, R., 305
Urtane, I., 296, 301
Useth, H. R., 230
Utvary, K., 325

Vaccani, S., 110
Vaglio, G. A., 234, 236, 306
Vajda, J. H., 14, 63
Valance, A., 3
Valle, M., 306
Vallerand, P., 122
Vallicenti, A. J., 259
Valovoi, V. A., 304
van Bekkum, J., 191
van Brederode, J., 248
Van De Graaf, B., 211
Van Den Berg, J. H. M., 198
Van Den Hende, J. H., 189
Vanden-Heuvel, W. J. A., 264, 280
Van der Gaever, F., 73
van der Graaf, B., 317
van der Greef, J., 47, 49, 65, 72, 75, 165, 233, 257
Van der Hart, W. J., 106
Vanderhoff, J. A., 104
Van der Kelen, G. P., 305
van der Leeuw, Ph. E., 25
Van Der Paauw, C. G., 218

Author Index

Van der Sande, C. C., 73
van de Runstraat, C. A., 45
Van der Weg, W. F., 134
van der Wiel, M. J., 25
Van de Straeten, M., 275
van Dishoek, E. F., 106
Vandy, M., 321
Vane, F. M., 291
Vangaever, F., 197, 217
van Hal, H. J. M., 279
Van Haverbeke, Y., 78, 80, 81
Van Hoye, E., 182
Van Lumig, A., 20
Van Marlen, G., 189
Van Montagu, M., 213
van Nigtevecht, G., 248
Van Raaphorst, J. G., 184
Van Thuijl, J., 81
Van't Klooster, H. A., 189, 191
van Velzen, P. N. Tn., 106
Varenne, P., 61, 62, 247, 252, 258
Varlamov, A. V., 299
Varmuza, K., 190, 192, 193, 325
Varret, F., 310
Vasiliades, J., 266
Vasjuta, Yu. V., 167
Vastola, F. J., 159
Vasyukova, N. I., 310, 312
Vaz Pirez, M., 38
Vdovin, V. M., 299
Veith, H. J., 225
Veith, R. C., 263
Veje, E., 131, 132
Veksler, V. I., 119
Velghe, M., 103
Venkataraghavan, R., 176, 187, 195, 201, 229
Venzel, J., 323
Verboom, G. M. L., 20
Verdonck, L., 305
Vereczkey, L., 282
Verejaus, D., 179
Vergori, L., 219
Verkade, J. G., 13, 317
Verkhoturov, E. N., 320
Vermeer, H., 305
Vermeulen, N. P. E., 66
Vernin, G., 227
Vernon, C. A., 258
Vernura, D., 243
Versino, B., 191
Verzele, M., 197, 198, 217
Vestal, J. L., 105
Vestal, M. L., 102, 159, 230, 231, 245
Vestergaard, P., 273
Vetter, M., 202
Vetter, W., 203, 241
Vettori, V., 249
Vick, K. W., 225

Vidaud, P. H., 22
Viehbock, F., 147
Viehland, L. A., 85, 86, 91, 92, 93, 95, 98
Vignon, M., 206
Vijfhuizen, P. C., 51, 76
Vilallonga, E., 56
Vilatov, V. N., 167
Vil'chinskaya, V. I., 257
Vilesov, F. I., 45
Villermaux, J., 187
Vine, H. J., 275
Vine, J., 250, 291
Vinokurov, I. V., 327
Vitovskii, V. Yu., 301, 302, 317
Vliegenhart, J. F. G., 207
Voelter, W., 258
Vogel, P., 309
Vogt, H., 160
Vogt, J., 43, 62, 185
Vogt, W., 304
Voigt, D., 235, 246, 249, 256
Volk, J., 268
von Ardenne, M., 234, 235
von Niessen, W., 6
Von Rudloff, E., 190, 200
Voorhees, K. J., 317
Vore, M., 281
Vorenkov, M. G., 302
Voronin, G. F., 327
Voronkov, M. G., 301, 302, 311
Vose, C. W., 210, 251, 288
Vouros, P., 171, 208, 230, 231, 251, 257
Voyle, C. A., 222
Vrščaj, V., 171
Vu, V. T., 267
Vyrypaev, E. M., 298

Waddell, K. A., 271
Wade, D. L., 259
Wade, D. N., 275
Wade, K., 303
Waern, R., 258
Wagemaker-Engels, I., 275
Wagner, H., 248, 250
Wagner, J., 286
Wagner, K., 320, 321
Wagner, W., 45, 78
Wahlberg, I., 241
Wahrhaftig, A. L., 170, 199
Waight, E. S., 172, 256
Wakefield, C. J., 166, 179
Wakeham, W. A., 92
Walder, R., 165, 166
Waldman, M., 93
Waldmann, L., 94
Walker, J. R., 205
Walker, J. Q., 227
Walker, R. W., 264, 280
Walle, T., 289
Walle, U. K., 289

Waller, G. R., 246
Wallington, M. J., 180, 185
Wallis, R. C., 307
Walls, F. C., 171, 179, 201
Walter, P., 247
Walter, W., 202
Walters, E. A., 10, 20
Walton, D. R. M., 299
Wang, C.-L. A., 218
Wang Chang, C. S., 94
Wankenne, H. W., 46
Wannier, G. H., 88
Warburton, G., 154, 243
Ward, C. H., 308
Wardell, J. L., 326
Warmack, R. J., 17, 18
Warnock, G. F., 323
Warthen, J. D., jun., 250
Wasada, N., 185
Washburne, S. S., 299
Washida, N., 167, 200
Watabe, T., 209
Watanabe, H., 299
Watson, E., 270, 292
Watson, J. T., 287
Watson, U., 214
Watts, R. O., 90
Wauters, B., 217
Webb, K. S., 155, 183
Webb, M. L., 312, 313
Webb, T. R., 308
Weber, J. H., 194
Weber, R., 51, 66, 78, 326
Webster, J. T. A., 205
Wechsung, R., 160
Weerasinghe, N. C., 245
Weerasinghe, N. C. A., 245
Weese, G. M., 50
Wegmann, D., 187
Wehner, G. K., 147
Weibel, L. A., 248
Weidenfeld, J., 208
Weigel, H., 165, 296
Weil, K. G., 321
Weimann, B., 185, 189
Weimer, D. F., 224
Weingartshofer, A., 30
Weinkam, R. J., 180, 264
Weinken, R. J., 173
Weinmann, S. A., 245
Weise, C.-L., 213
Weiss, E., 316
Weiss, G., 165, 170
Weiss, M. J., 20, 43
Weiss, R., 308, 317, 323
Weisz, A., 66
Weiszbart, A., 264
Wellington, J. L., 224
Wells, D. E., 216
Wells, R. J., 208, 250
Wells, R. L., 322
Welsh, J. A., 104
Welsh, L. W., 326
Wen, J. H., 264

Author Index

Wendolski, J. J., 16
Wentrup, C., 78
Weringa, W. D., 72
Werner, A. S., 20, 46
Werner, H. W., 118, 121, 122
Wesdemiotis, C., 50, 51, 62, 65, 66, 78, 81, 300
West, J. B., 27
West, R., 298
Westmore, J. B., 253
Weston, C. A., 310
Westover, G. F., 309
Westwood, N. P. C., 10
Weustink, R. J. M., 305
Weyl, H., 127
Wharton, L., 37
Whealton, J. H., 86, 89
Wheeler, S. H., 316
Whitcomb, D. R., 327
White, D. F., 205
White, E., 171
White, E. V., 316
White, J. G., 315
White, J. W., 309
White, M. A., 198
White, P. Y., 240
White, R. A., 226
White, R. L., 181
White, S. D., 290
Whitehouse, D. A., 216
Whiting, D. A., 246
Whitlam, J. B., 275
Whitten, J. L., 31
Whorton, A. R., 205
Widman, M., 290
Wiebers, J. L., 79, 172, 244, 245
Wieboldt, R. C., 195
Wiecek, C., 211
Wiecko, J., 228, 255, 296
Wiersema, K. S., 273
Wightman, F., 228
Wignaendts Van Resandt, R. W., 169
Wilcox, J. B., 22, 24
Wilczynski, R., 312
Wild, S. B., 236, 309, 313
Wilden, D. G., 30
Wilhelm, J., 165
Wilhelmi, M., 184
Wilkins, A. L., 210
Wilkins, C. L., 165, 173, 179, 181, 192
Wilkinson, G., 317
Wilkinson, S. P., 214
Willett, G. D., 9, 10, 43
Williams, A. L., 249
Williams, D. H., 50, 51, 52, 59, 62, 63, 66, 68, 159, 257, 258
Williams, D. T., 220
Williams, P., 118, 123, 131
Williams, P. S., 210
Williams, R. L., 170

Williams, R. N., 217
Williams, V. P., 172
Williamson, A. D., 32
Williamson, J. E., 173
Wilson, A. T., 184
Wilson, J. M., 158, 295
Windel, B., 184
Windsor, J. G., 218
Wing, W. H., 103
Winkler, F. J., 15, 63, 236
Winmill, D. L., 170
Winn, J., 110
Winograd, N., 127, 327
Winter, M., 222
Winter, M. J., 10
Winter, N. W., 21
Wipf, H.-K., 203
Wise, D. S., 213
Wise, S. A., 218
Wishousky, T. I., 290
Witiak, D. N., 52
Wittmaack, K., 118, 139, 140
Wojinski, S. L., 72
Wolen, R. L., 275
Wolf, F. J., 264
Wolf, R. J., 38
Wolfe, L. S., 215
Wolfsberg, M., 5
Wolfschütz, R., 78, 80, 81
Wolkoff, P., 48, 50, 61
Wollenweber, E., 202
Wollnik, H., 258
Wong, C., 298
Wong, F. S., 301
Wong, J. G., 218
Wong, L. K., 293
Wong, S.-L., 206
Wood, B. J., 155
Wood, D. L., 226
Wood, G. W., 241, 259, 261
Wood, K. V., 31
Woodage, T. J., 276, 292
Woodin, R. L., 55, 113, 162
Woodruff, S. B., 38
Woodruff, H. B., 188
Woods, R. C., 6, 117
Worley, J. F., 250
Worley, S. D., 308
Wöste, L., 10, 32
Wright, D. E., 252
Wright, J. G., 325
Wright, L. A., 21
Wright, L. H., 209, 251
Wright, R., 131
Wright, R. B., 118, 131, 144
Wu, A. A., 3
Wu, A. T., 205
Wu, C. H., 10
Wu, C. Y. R., 30, 110
Wu, K. T., 44
Wuepper, J. L., 173, 227
Wulfson, N. S., 248
Wurminghausen, T., 307
Wyatt, J. R., 168

Yagzhev, V. V., 324
Yahara, S., 257
Yamabe, S., 20, 21
Yamabe, T., 14
Yamada, S., 252, 292
Yamagiwa, C., 44
Yamaizumi, K., 253
Yamaizumi, Z., 254
Yamanaka, Y., 203
Yamashita, K., 14, 207, 224, 260
Yáñez, M., 5, 15
Yang, M.-T., 206
Yang, S. K., 230
Yanovskii, A. I., 316
Yasman, Y. A. B., 298
Yasuhiko, Y., 258
Yasuhira, K., 216
Yatsimirskii, K. B., 305, 318
Yau, A. W., 37
Yelton, R. O., 172
Yencha, A. J., 44
Yesair, D. W., 281
Yeung, S. K. F., 204
Yisak, W., 290
Yoder, C. H., 13
Yoder, C. S., 13
Yokotsuka, T., 222
Yonezawa, T., 5
Yorke, D. A., 171
Yoshida, K., 284
Yoshida, Y., 167, 230
Yoshimine, M., 44
Yoshioka, M., 208
Yoshioka, Y., 20
Yost, G. S., 242
Yost, R. A., 77, 172, 244
Yotsui, Y., 258
Young, C. E., 131
Young, G. B., 317
Young, J. P., 162
Young, N. D., 190
Young, S. E., 164
Young, S. J., 220
Younginger, E. J., 167
Youngless, T. L., 163
Yousef, I. M., 209
Yu, M. L., 123
Yuan, B., 205
Yun, K. S., 95
Yuriev, V. P., 298
Yusuf, N. A., 130, 131, 132

Zacchei, A. G., 290
Zafirou, O. C., 217
Zagorevskii, D. V., 310
Zaikin, V. G., 248
Zakett, D., 238, 244
Zakharov, P. I., 249, 299
Zakheim, D. S., 33
Zamkova, V. V., 304
Zandee, L., 34, 35, 162
Zanozzi, P. F., 316
Zare, R. N., 10, 25, 32, 322

Zaretskii, Z. V. I., 251, 258
Zarkin, V. G., 299
Zaterskii, Z. V. I., 252
Zavadil, J., 136
Zdansky, K., 136
Zegarski, B. R., 111
Zeimer, J. N., 201
Zeiss, H. J., 296
Zeitz, Von. E., 287
Zelcans, G., 296, 301, 316
Zeldin, M., 317
Zeman, A., 203
Zerenner, E. H., 197
Zerilli, R. F., 228

Zhuk, B. V., 310
Zhukov, E. G., 326
Zhun, V. I., 298
Ziege, E. A., 275
Ziegler, M. L., 326
Zielinska, B., 317
Zielinski, J., 253
Ziemer, J. N., 192
Ziersel, J. F., 181
Zimmerman, A. H., 17, 19, 107, 234
Zimmermann, S., 25
Zinsmeister, H. D., 249

Ziskoven, R., 297
Zlatkis, A., 214, 227, 228, 294, 296
Zoller, P., 33
Zolotarev, B. M., 255, 295, 297, 302
Zoloty, N. B., 167
Zubov, K. F., 322
Zülicke, L., 57
Zuhrt, Ch., 57
Zupan, J., 176
Zverev, Yu. B., 312
Zykov, B. G., 298